Valence Instabilities and Related Narrow-Band Phenomena

Valence Instabilities and Related Narrow-Band Phenomena

Edited by

R. D. Parks

University of Rochester
Rochester, New York

PLENUM PRESS ☐ NEW YORK AND LONDON

Library of Congress Cataloging in Publication Data

Main entry under title:

Valence instabilities and related narrow-band phenomena.

Includes index.

1. Rare earth metal compounds – Congresses. 2. Energy-band theory of solids – Congresses. 3. Valence (theoretical chemistry) – Congresses. I. Parks, Ronald D. [DNLM: 1. Chemistry, Physical – Congresses. QD469 161v 1976]

QD172.R.2 546'.4 77-3155

ISBN 0-306-31047-3

Proceedings of an International Conference on Valence Instabilities and Related Narrow-Band Phenomena held at the University of Rochester, Rochester, New York, November 11–13, 1976

A Division of Plenum Publishing Corporation
227 West 17th Street, New York, N.Y. 10011

Printed in the United States of America

Preface

Those well-intending workers, especially theorists, who have viewed hungrily the mixed valence problem, but have not yet made the bold leap, might be comforted to learn that the Rochester conference left the virginal state of that problem essentially intact. That is not to say that the event was prosaic. Indeed, the conferees exhibited a level of effervescence appropriate to the freshness and challenge of the problem at hand.

If the meeting failed to solve major questions, it at least established several guidelines. One is that future experimental efforts, at least on a short time scale, might be spent most profitably on those substances which exhibit consistent, and hence probably intrinsic, behavior from laboratory to laboratory. A recurring message, not always subtle, to the theorists was that piecemeal approaches to the mixed valence problem, characteristic of much of the work to date, are of limited usefulness. For at the core of the problem one has a melange of boot-strapping interactions which must be sorted out and dealt with properly. Paraphrasing Phil Anderson (see Epilogue), the mixed valence problem is in the same category of problems which are failing to be done in field theory these days.

It is not at all obvious that the theoretical development of this subject will follow traditional patterns in condensed matter physics, wherein the solution of a tough but well-defined problem requires typically a decade of concentrated effort, e.g., superconductivity or the Kondo effect. What may emerge instead is a pattern much like that in particle physics, wherein the theorists will separate into an elitist group and the experimentalists will propagate their essentially empirical studies independently. Despite the progress or lack thereof on the theoretical front, the coming decade will probably see the experimental camp grow and flourish. It seems inevitable also that chemists and chemical engineers interested in catalysis (most good catalysts are mixed-valent) and developmental engineers searching for new classes of narrow-band devices will become particularly observant of our efforts.

I wish to thank my colleagues, D. Chakraborty, M. Croft, L. Kupferberg, J. Lawrence, M. Manheimer, J. Markovics, and I. Zoric for their assistance with all aspects of the conference and the members of the Organizing Committee, P. Anderson, S. Doniach, T. Penney, and C. Varma for their assistance in planning the program. On behalf of the Organizing Committee, and, indeed, all the conferees, I wish to thank the National Science Foundation, the Army Research Office and the Office of Naval Research for partially supporting the conference.

Ron Parks

Contents

INVITED PAPERS

POSTER SESSION PAPERS

INVITED PAPERS

CONFIGURATION-BASED IONIC ENERGY LEVELS

AND NARROW ENERGY BANDS+

L. L. Hirst

Institut für Theoretische Physik

Universität Frankfurt a.M., W. Germany

I. INTRODUCTION

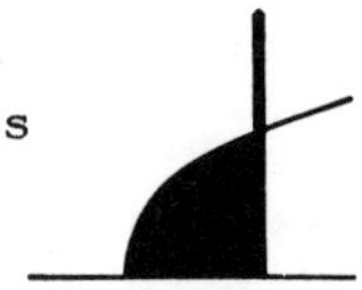

On the letterhead used by the organizers of this conference we find the ideogram reproduced at the right. It symbolizes the density of states of a metallic substance containing 4f ions in which valence fluctuations occur. The density of states contains a nearly-free-electron contribution from the conduction band and a high narrow peak due to the 4f electrons. How does the 4f density of states arise and how can it be characterized quantitatively? According to the ideas prevailing a few years ago, one would have said that it resulted from a 4f-like virtual bound state[1] which was nonmagnetic and therefore degenerate, having a total capacity of 14 electrons per ion (or perhaps 6 or 8 if spin-orbit splitting were invoked). More recently it has been recognized that, because the Coulomb interaction between 4f electrons is too large to be treated as a perturbation, a configuration-based approach is needed to obtain an adequate description of such systems[2-8]. Interconfiguration fluctuations (or valence fluctuations) are understood as arising when the energy levels are arranged such that two configurations (or valencies) compete for stability. When one goes from one configuration to the next the change in the number of 4f electrons is 1, which would seem to indicate that the integrated density of states in the 4f peak should be 1. However, a more careful analysis shows that this inference is not correct. In what follows, we shall show that a correct application of the configuration-based approach yields an integrated density of states in the 4f peak which agrees with none of the above-mentioned values. We shall also relate the concept of valence fluctuations to the density-of-states curve[8].

The 4f systems of interest may reasonably be described by a generalized Anderson model, specified in an obvious notation by

$$H = \sum_i \left(H_{ion}^{(i)} + H_{mix}^{(i)} \right) + H_{cond} \qquad (1)$$

$$H_{ion} = -Vn + Un(n-1)/2 + H_{ics} \qquad (2)$$

$$\Delta = \pi \left| V_{mix} \right|^2 \rho_{cond}(\epsilon_F) \qquad (3)$$

The eigenstates of the ionic Hamiltonian for a given 4f ion belong to definite n-electron configurations; the first two terms of eq.(2) give the mean energy of the configuration, while the remaining term specifies symbolically the intraconfigurational splittings (Hund splitting, spin-orbit splitting, etc.) which can be quantified by well-known methods. $U \sim 10$ eV represents the leading part of the Coulomb interaction between 4f electrons. The strength of the mixing interaction is characterized by the quantity Δ, which is roughly the rate at which spontaneous mixing transitions would occur if not opposed by intra-ionic energies, and is 3 orders of magnitude smaller than U in typical 4f systems.

The most familiar methods in the theory of metals are fermion-decoupling approximations, which in the present context mean treating U as a weak perturbation. Such approximations can be expected to work well only when $U \lesssim \Delta$. To make such an approximation and then set U/Δ equal to a large number,or even infinity, is a self-contradictory procedure, and one has no right to be surprised when it yields bad results. For example, Hartree-Fock solution of eq.(1) for a single 4f impurity ion, in the simplified special case where all intraconfigurational splitting is omitted, yields a discontinuous jump between different types of solution in the vicinity of a configuration crossover, i.e. exactly in the region where valence fluctuations are expected[7]. This unphysical behavior is clearly an artifact of the fermion-decoupling approximation which is not implicit in the model Hamiltonian. We therefore reject the use of such approximations in treating 4f valence-fluctuation systems.

It is indeed possible to obtain fermion-based theories from a low-density expansion even when $U \gg \Delta$ [9]. However, we emphasize that the number of 4f electrons or holes per ion must be small compared to 1, not 14, in order for such an expansion to be valid, and this excludes the cases of real interest. (The bands of highly-correlated quasi-particle excitations to be discussed below become equivalent to the results of conventional fermion-based band theory in the special case where the 4f shell is empty or full, corresponding again to the low-density limit.)

The approach which is physically appropriate for systems with $U \gg \Delta$ is the configuration-based approach, in which one takes several-electron single-ion eigenstates as unperturbed states and treats the mixing as a weak perturbation [2-8]. This approach has not yielded a complete solution for the valence-fluctuation problem, but it provides considerable illumination, as we hope to show in what follows. The configuration-based description is simple but relatively unfamiliar. In learning to use it, one must guard against the uncritical use of concepts suggested by fermion-decoupling approximations, since configuration-based and fermion-based descriptions are inequivalent except in certain very limited contexts. In the remainder of this section we shall review some elementary properties of configuration-based ionic levels in the absence of a mixing interaction.

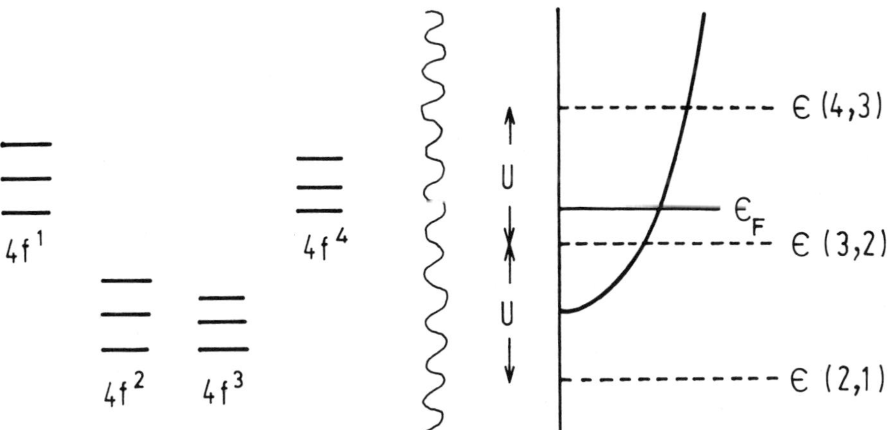

Fig. 1. Configuration-based energy levels.

The configuration-based energy levels of an ion in a metal are illustrated schematically in Fig. 1. The mean energies of the various ionic configurations, shown on the left, fall on a parabola in accordance with eq.(2). Since each ionic energy level represents the total energy of a several-electron state, a direct comparison with the conduction-electron levels shown on the right would be physically meaningless. The vertical wavy line separating the two sides of the diagram is intended as a reminder against making such comparisons. What can be compared with the conduction-electron levels is the energy difference between two ionic levels differing by one in the number of 4f electrons, $E_{n,v} - E_{n-1,v'}$ in an obvious notation. One such energy difference of particular importance, the <u>ionization stage energy</u>, is

$$\epsilon(n,n-1) = E_{n,g} - E_{n-1,g} \quad , \tag{4}$$

where g denotes the ground level within a given configuration. When the intraconfigurational splitting is neglected, these energies form a ladder with uniform spacing U. They have been indicated by dashed lines on the right of Fig. 1 .

The conduction electrons satisfy Fermi statistics, meaning that the occupation of each level may be specified independently and the mean occupation in thermal equilibrium is given by a Fermi distribution. For the configuration-based ionic levels, the rule is that one and only one state of a given ion is occupied at any time. The mean occupation in thermal equilibrium is given by a Gibbs distribution, which is normalized to 1 and depends on a chemical potential, the Fermi energy:

$$P_{n,v} = (1/Z)\, \exp\left[-(E_{n,v} - n\,\epsilon_F)/kT\right] \quad . \tag{5}$$

It seems appropriate to say that the ionic levels satisfy "Gibbs statistics". Although not one of the traditionally-recognized categories of "elementary" statistics, Gibbs statistics is not essentially more complicated than Boltzmann statistics, to which it reduces in the special case where only one configuration is thermally accessible.

The stable ionic level in the zero-temperature limit is the ground level of that configuration for which $E_{n,g} - n\,\epsilon_F$ is the smallest. Equivalently, the configuration $4f^n$ is stable if the Fermi energy is intermediate between the ionization-stage energies

$\epsilon(n,n-1)$ and $\epsilon(n+1,n)$. The ion therefore gives up an electron if the energies are shifted so that an ionization-stage energy is raised past the Fermi energy. In this sense each ionization-stage energy can be regarded as a non-degenerate effective fermion level, capable of holding one electron per ion, which is the basis of the Falicov model[5]. However, we emphasize that this model applies only in the absence of mixing, where the 4f ions act only as a population reservoir[5,6]. A realistic treatment of finite mixing, or of magnetic properties, is not possible in this model, since the effective fermion must be non-degenerate and cannot be assigned angular-momentum indices $m\sigma$ without giving the effective level the wrong capacity. In our opinion, some claims which have been advanced concerning the inclusion of finite mixing in the Falicov model [10] are far too optimistic because of this difficulty.

It should be mentioned that an alternative convention for defining the ionic energies is widely used, e.g. [11], according to which the energy of an n-electron ionic level is measured relative to the lowest level with n-1 electrons. This incremental ionic energy is given in terms of our total n-electron energies by

$$\delta E_{n,v} = E_{n,v} - E_{n-1,g} \tag{6}$$

The two conventions obviously contain the same information and will yield equivalent results if used consistently. Since the incremental energy is an energy difference corresponding to a difference of one 4f electron, it can be compared directly to the conduction-band energies. This comparison is sometimes convenient, but it can also be misleading, since the occupation rules for ionic levels remain the same as described above and are fundamentally different from those applying to fermion levels. We believe that conceptual clarity is best achieved by referring the discussion to the more fundamental total ionic energies.

II. THE QUASI-PARTICLE SPECTRUM

Valence fluctuations arise in a system at configuration crossover, meaning that the energy necessary to take an ion from one configuration to the other, by transferring an electron to the conduction band at the Fermi energy, is approximately zero. In concentrated systems this condition is often maintained over a finite range of parameters by the energy balance involved in filling the conduction band [5,6]. In the many-body ground state of a system at configuration crossover one has strong mixing of the two configurations even when $U \gg \Delta$. To obtain a good solution for this many-body ground state is a very difficult problem, which we shall not attempt to discuss. Instead, we shall show how one can characterize the spectrum of quasi-particle excitations out of the many-body ground state and use this spectrum to discuss certain properties of the system. Details which must be omitted from the present short discussion will be given elsewhere[8].

The quasi-particle spectrum is the spectrum of excitations to exact many-body eigenstates reached from the many-body ground state by adding or removing an electron. We expect this spectrum to consist of a smooth background due to the conduction electrons, upon which are superposed narrow bands of quasi-particles having predominantly 4f character. These 4f-like bands have widths roughly of order Δ and lie at energies equal to a difference of initial and final ionic energies. We emphasized earlier that the energies of a single ion cannot be treated as fermion levels or replaced by equivalent fermions. The 4f-like quasi-particles are fermions, but they are excitations of the system as a whole rather than of a single ion; and the quasi-particle picture is valid only when the spatial density of quasi-particles actually present is sufficiently small. The bands of 4f-like quasi-particles amount to highly-correlated narrow bands, where the correlations result from the large Coulomb interaction inside the 4f ions and can be taken into account by means of a configuration-based approach. The splitting pattern and the capacities of the various bands will be found to differ significantly from those of the uncorrelated narrow bands which would be yielded by conventional band theory.

We can characterize the quasi-particle spectrum by finding the integrated density of quasi-particle states within each 4f-like band. Since Δ is typically smaller than spin-orbit splittings but larger than crystal-field splittings, the resolved quasi-particle bands will usually correspond to transitions between definite LSJ manifolds of different configurations. To evaluate the band capacities we use sum-rule strengths defined in terms of single-ion matrix elements of a 4f creation-annihilation operator, such as

$$Q^{(-)}_{n \to n-1} = \sum_{v' \, m \sigma} |\langle n-1, v' \,|\, a_{m\sigma} \,|\, n, v \rangle|^2 \qquad (7)$$

or an analogous $Q^{(+)}$ involving a creator. The sum of v' extends over all states within the final LSJ manifold, if one wants to characterize a single band, or over larger manifolds if the total capacity of a group of bands is to be evaluated. The sum-rule strengths can be computed by using coefficients of fractional parentage or stretched-state techniques.

Before considering a system at configuration crossover, let us discuss the quasi-particle spectrum in a system where the 4f ions are configurationally stable with configuration n . The spectrum for such a system is indicated schematically in Fig. 2. The quasi-particle

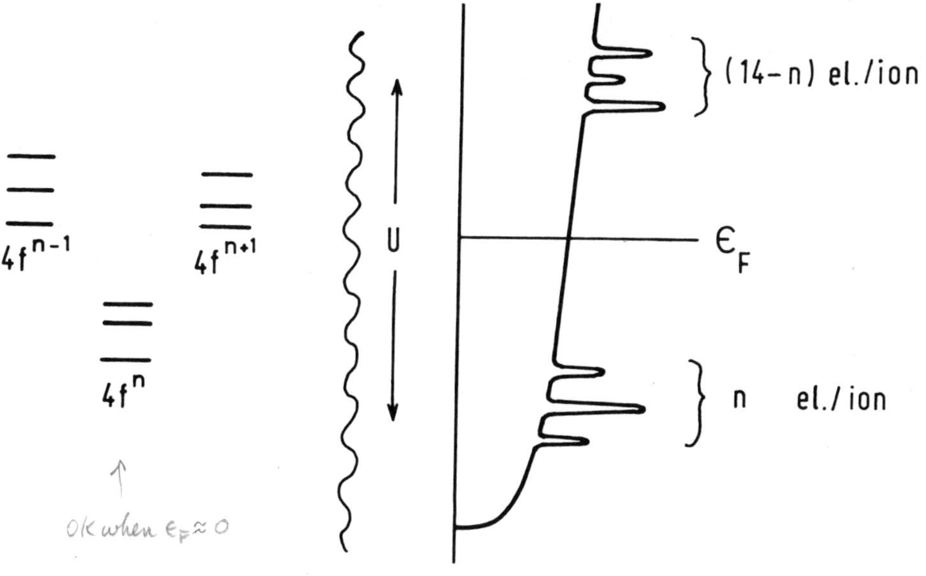

Fig. 2. Ionic energy levels and quasi-particle spectrum in a system with configurationally stable 4f ions.

bands divide into two groups: a group of bands below the Fermi energy corresponding to electron-removal excitations from the LSJ ground level of n to the various LSJ levels of n-1, and a group of bands above the Fermi energy corresponding to electron-addition excitations taking the ion from the LSJ ground level of n to the various LSJ levels of n+1. The total capacities of the bands in the two groups are found to be n and (14-n) electrons per ion respectively. According to conventional band theory one would obtain 14 narrow 4f bands, some of them degenerate, with an integrated capacity of 1 electron per ion apiece. The present highly-correlated narrow-band excitations are typically split into more than 14 bands (although we have not attempted to show this in Fig. 2), and the capacities of the individual bands are generally non-integer. A division of the bands into two groups separated by $\sim$U is obtained in conventional band theory only as a result of magnetic ordering. It should be emphasized that the division of the bands into two groups obtained in the present appoach does not depend upon magnetic ordering, and the two groups cannot be identified as containing "spin-up" and "spin-down" electrons respectively.

To apply similar considerations to a system at configuration crossover, one needs to know the probabilities of finding the two competing configurations, n and n-1, in the many-body ground state. Let z be the mean number of holes per 4f ion relative to n. Then the probability of an ion being found in the LSJ ground level of n or n-1 is (1-z) or z respectively. The population parameter z is not computed in the present theory; we assume that it can adequately be estimated from the zero-mixing theory [5,6], or determined from experimental data such as the lattice constant or XPS spectra. The form of the quasi-particle spectrum is indicated schematically in Fig. 3. When we classify the 4f-like bands

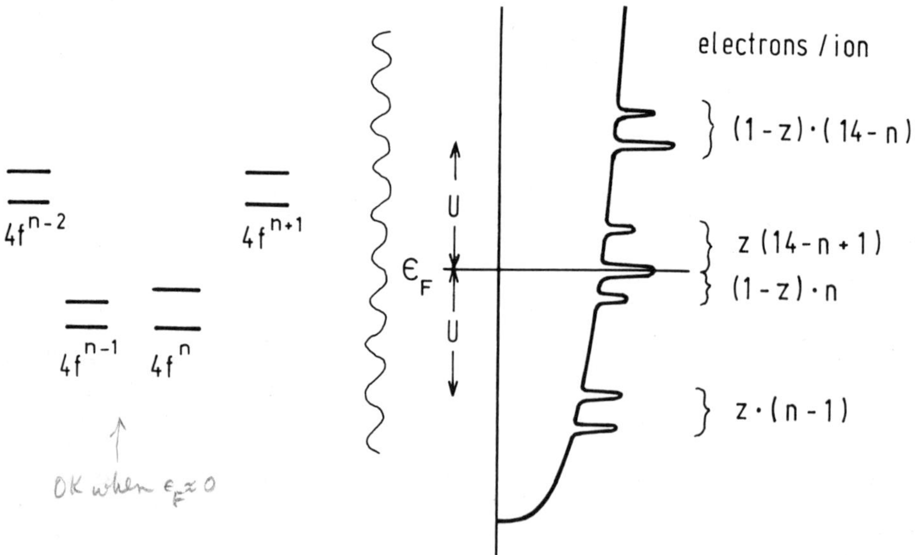

Fig. 3. Ionic energy levels and quasi-particle spectrum in a system with 4f ions at configuration crossover.

according to the interconfiguration excitation we now obtain 4 groups, namely $n-1 \rightarrow n-2$, $n \rightarrow n-1$, $n-1 \rightarrow n$, and $n \rightarrow n+1$, reading from bottom to top in Fig. 3. The integrated density of states in each group of bands can be obtained from the sum-rule strengths and has been indicated in Fig. 3. It should be noted that the quasi-particle bands are non-rigid, since these integrated densities of states depend upon the population parameter z.

The quasi-particle spectrum is particularly important for the discussion of the low-energy properties of a system at configuration crossover, properties which are associated with the concept of valence fluctuations. In this application one is interested only in the band of 4f-like quasi-particles which overlaps the Fermi energy, which will be called the principal band. The stability criteria given above imply that such a principal band exists if and only if the system is at configuration crossover. The integrated density of states in the principal band is found to be

$$\int d\epsilon \; \rho_{4f}(\epsilon) = (1-z)Q^{(-)} + zQ^{(+)} \tag{8}$$

where $Q^{(\pm)}$ are sum-rule strengths for interconfiguration excitations between the LSJ ground levels of n and n-1. The values of these sum-rule strengths for those 4f ions in which valence fluctuations are believed to have been observed are listed in Table I; also included are the whole-configuration sum-rule strengths which would apply if no intraconfigurational splitting existed. We see that the integrated density of states in the principal band is strongly affected by the intraconfigurational splitting and varies widely from one case to another, depending on the configurations involved and the value

Table I. Sum-rule strengths for principal band

	LSJ gnd levels		Whole configs	
	$Q^{(-)}$	$Q^{(+)}$	$Q^{(-)}$	$Q^{(+)}$
Ce(4+,3+)	1.000	6.000	1.	14.
Sm(3+,2+)	0.898	0.150	6.	9.
Eu(3+,2+)	0.143	1.143	7.	8.
Tm(3+,2+)	3.018	1.857	13.	2.
Yb(3+,2+)	8.000	1.000	14.	1.

of z. The integrated density of states in the principal band is not one per ion, despite the fact that the number of electrons per ion changes by one as the system moves through configuration crossover. This apparent paradox is seen to involve no real contradiction when one remembers that the quasi-particle bands are non-rigid and change as the system moves through configuration crossover.

III. INTERCONFIGURATION FLUCTUATIONS

By saying that interconfiguration (or valence) fluctuations occur in a system at configuration crossover, we mean that the system has a many-body ground state in which each ion has an appreciable probability of being found in either of the two competing configurations. ICF need not occur in all systems near configuration crossover, since a ground state in which each ion has a definite configuration might be stabilized by size effects or by Coulomb interactions between the 4f ions[15]. However, ICF do seem to occur in many such systems, as indicated by various kinds of experimental evidence. The most direct evidence for ICF as opposed to a definite configuration at each site is provided by Mössbauer isomer-shift spectra of ICF systems containing Eu ions [12]; such spectra show a single line in the ICF case, whereas a two-line spectrum would result in the case where each ion has a definite configuration.

Some caution is necessary in using the ICF concept and considering how ICF could be observed. Like the closely-related concept of "resonance" in quantum chemistry, ICF refers in the first instance to what would happen if the system were prepared in a pure-configuration initial state and then released: the probability of each ion being found in a given configuration would then indeed fluctuate in time. The many-body ground state may also be considered to contain ICF as a zero-point "motion", but it is really static. A noise output or other directly-observable fluctuation effects therefore will be seen only to the extent that the system is observed at finite temperature or frequency.

ICF concepts can conveniently be quantified in terms of the quasi-particle spectrum, whereby the <u>width</u> of the principal band is identified with the ICF <u>rate</u>. Arguments can be given which suggest that the shape of the principal band should be taken as Lorentzian and its width (full width at half maximum) should be taken as

$$W = (Q^{(-)} + Q^{(+)})\,\Delta\,. \qquad (9)$$

According to this expression and Table I, the ICF rate depends upon the amount of intraconfigurational splitting included; it is significantly smaller for ICF between LSJ ground levels as compared to the hypothetical case without intraconfigurational splitting. The density of states in the principal band being completely fixed by z, Δ , and the known sum-rule strengths, various quantities related to the quasi-particle spectrum can be calculated. For example, we obtain a low-temperature specific heat which is linear in the temperature, with a coefficient very large compared to a normal metal[8]. One also concludes that the spectral density of low-frequency noise generated by the fluctuating 4f electrons and detectable by a probe such as a nuclear spin should go to zero linearly with the temperature, in accordance with the general remarks made earlier.

The broadening effect of ICF in spectroscopy depends upon the particular type of measurement. We shall consider two examples. X-ray photoemission spectroscopy (XPS) of an ICF system yields a spectrum containing resolved lines corresponding to transitions between different configuration-based single-ion states[13]. The ICF width W, eq.(9), represents the energy resolution with which pure-configuration single-ion states are mixed into the initial many-body eigenstate of the system. The lines of an idealized XPS spectrum would therefore have a width equal to the sum of an initial-state width W and a final-state width equal to the mixing-decay rate of the highly-excited ionic final state produced in the XPS transition[14]. Both contributions are temperature-independent. In Mössbauer isomer-shift spectroscopy of an ICF system[12], contributions from the two configurations are not resolved into separate lines. The isomer shift of the gamma frequency is proportional to the instantaneous electric charge density at the Mössbauer nucleus. Fluctuations in the charge density are not driven by the measurement process; they must occur spontaneously, and will occur only to the extent that quasi-particle excitations are thermally populated. Consequently, it is found that the ICF contribution to the Mössbauer linewidth should go to zero linearly with the temperature, in accordance with the fact that the many-body ground state must be static[8].The expected ICF contributions to the linewidth in both XPS and Mössbauer spectroscopy are relatively small, and experiments up till now have sufficed only to establish upper limits on them.

References

+Project of the Sonderforschungsbereich für Festkörperspektroskopie Darmstadt-Frankfurt, supported by the Deutsche Forschungsgemeinschaft.

1. B. Coqblin and A. Blandin, Adv. Phys. 17, 281 (1968).

2. L. L. Hirst, Phys. kondens. Meterie 11, 255 (1970).

3. M. B. Maple and D. K. Wohlleben, Phys. Rev. Letters 27, 511 (1971).

4. D. K. Wohlleben and B. R. Coles, in Magnetism, Vol. V, edited by G. T. Rado and H. Suhl, Academic Press, New York, 1973.

5. R. Ramirez and L. M. Falicov, Phys. Rev. B3, 2425 (1971).

6. L. L. Hirst, J. Phys. Chem. Solids 35, 1285 (1974).

7. L. L. Hirst, AIP Conference Proceedings 24, 11 (1975).

8. L. L. Hirst, Phys. Rev. B, in press.

9. J. R. Schrieffer and D. C. Mattis, Phys. Rev. A140, 1412 (1965).

10. M. Avignon and S. K. Ghatak, Sol. State Comm. 16, 1243 (1975).

11. P. Wachter, CRC Critical Rev. Sol. State Sci., July 1972, p. 189.

12. I. Nowik, this conference.

13. M. Campagna, this conference.

14. L. L. Hirst, Phys. Rev. Letters 35, 1394 (1975).

15. Assuming all ions initially to be equivalent, we are describing a situation in which size effects or interionic Coulomb interactions break this equivalence. ICF is suppressed when the equivalence-breaking interactions are strong enough to split the principal band into parts which no longer appreciably overlap the Fermi energy. It is convenient to say that such a system is not at configuration crossover, reserving the latter term for systems in which the interconfiguration excitation energy is $\lesssim \Delta$.

COMMENTS

KASUYA: Usually people consider only the mixing of the conduction band with the f-band. I think another important contribution to the f-band width is the mixing of the 4-f state with the valence band. This mixing width can be large (e.g., 0.1 to 0.2 ev). As a result we have a 4f Bloch state which makes a so-called Anderson localization with the impurities, or potential fluctuations. Then we have another story because something different happens. For example, we have a large T-linear specific heat, but usually it is much smaller than that we expect from the 4f band width, two orders of magnitude larger than usual.

HIRST: What I tried to emphasize in this talk is what I would consider to be the essence of the valence fluctuation problem; namely, that you have on the one side band states which satisfy Fermi statistics and on the other side you have ionic states which satisfy what I call Gibbs statistics, which are not Fermi statistics; and, it doesn't really matter what the band states are. They could, for example, have a large amount of 5-d character as a lot of people assume for SmS. That doesn't matter much for the basic principle of valence fluctuations, provided you start from the Anderson-type model which I indicated here. But I think what you're saying is that one could then worry about the binding of a 5-d-like conduction electron back to the ion which admitted it; and, I agree that this can certainly make the problem much more complicated than it is. I would say this is an excitonic binding and I would accept then that Mott's general argument is applicable in this case, viz., that if you have a very small admission of electrons into the conduction band, they're going to be bound onto the admitting ion. But once you get over the hill-- get into the metallic region-- then this excitonic binding effect may be relatively unimportant.

WOHLLEBEN: (Question directed to Kasuya) Do you agree for SmB_6 or would you disagree?

KASUYA: I have studied many of the Sm compounds and found the difference in behavior between SmB_6 and SmS, with doping, with pressure, etc. The situation depends very sensitively on the circumstance. In any case, in general, I think the most important mixing effect for 4-f electrons is the valence band and the d-f mixing is most important for d-band character. Also polaronic effects are very important in determining the difference between various compounds.

WOHLLEBEN: Are there any comments with respect to the spectra which we saw?

COQBLIN: I can't see clearly what is new compared with previous theory when you either have to compute a given property or explain a given experiment, because you present the idea very generally. I don't see what is new in valence fluctuations or how we can see valence fluctuations, since you say that valence fluctuations should not be taken too literally. For example, if you take the XPS spectra, the valence fluctuation concept is not necessary to explain the two peaks-- we can explain them with a rather normal mixed valence idea without fluctuations. How can you see valence fluctuations?

HIRST: Let me say the following. Firstly, there is in fact a great difficulty in deterimining whether a given system is fluctuating, valence-- or not, because the frequency of the fluctuations fall in a very awkward spectroscopic region. What I'm doing here in this talk is assuming that fluctuations are occurring and what I'm getting out is a peak in this principal band in the density of states spectrum. This principal band obviously has a rather strong qualitative resemblence to the ideas of virtual bound states (VBS) which one had earlier but what comes out differently is that the numbers which come out are different, that the capacity of this principal band may be different by an order of magnitude from what one might have expected using the old-fashioned VBS ideas. I've tried to show how the valence fluctuation concept would come out of this spectrum. I agree that once you've got a density of quasiparticle states, whether from the old VBS theory or from the explicitly configuration-based approach, what you do with it from that point on is more or less the same. Then probably what I've said about valence fluctuations is sort of what you would have said earlier, concerning for example the specific heat of rare earth materials which are at configuration crossover. The difference is that the numbers come out quite differently and they change drastically from one material to another depending on the specific ionic configuration involved.

COEY: I'd like to ask a point about your density of states curve where you have a number of sharp peaks below as well as above the Fermi level. If you were to look at the optical transitions excited from the high density of states peak(s) below to the high density of states peak(s) above, would you expect that this should look something like the excitations between terms in individual ions or is it something quite different?

HIRST: It's just basically the same thing. From a practical point of view if one wants to discuss resonances in, e.g., optical spectra, XPS, etc., I would not recommend bothering to use this quasiparticle spectra because it contains essentially the same information you obtain directly from the configurationally-based ionic energy level picture. I just wanted to bring out the comparison to conventional band theory.

KONDO ANOMALIES DUE TO RARE EARTH IMPURITY IONS*

M. B. Maple, L. E. DeLong, W. A. Fertig,
D. C. Johnston, R. W. McCallum and R. N. Shelton
Institute for Pure and Applied Physical Sciences
University of California, San Diego
La Jolla, California 92093

I. INTRODUCTION

Rare earth (RE) ions are capable of producing "Kondo-like" anomalies in the physical properties of both concentrated and dilute RE metallic systems which are qualitatively similar. These temperature dependent anomalies scale with a characteristic temperature T_0, above which the RE metallic system behaves magnetically and below which it behaves nonmagnetically. The physical properties in which Kondo-like anomalies are found include the electrical resistivity, magnetic susceptibility, thermoelectric power, specific heat, and, in systems where appropriate, superconducting properties such as the critical temperature T_c and the jump in specific heat which occurs at T_c. The anomalies can be viewed as originating from admixture of the localized 4f electron states with the extended states of the conduction band, which can be traced to the unstable valence of certain RE ions.

In concentrated RE metallic systems, the Kondo-like anomalies are often attributed to valence or interconfiguration

*Research supported by the U. S. Energy Research and Development Administration under Contract No. ERDA E(04-3)-34 PA 227.

fluctuations (ICF), while in dilute RE metallic systems, the Kondo-like anomalies are generally ascribed to the Kondo effect which arises from the conduction electron-impurity spin exchange interaction

$$\mathcal{H}_{int} = -2\,\mathcal{J}\,\underset{\sim}{S} \cdot \underset{\sim}{s}$$

Here, $\mathcal{J}$ is the negative exchange interaction parameter, $\underset{\sim}{S}$ is the impurity spin, and $\underset{\sim}{s}$ is the conduction electron spin density at the impurity site.

Whereas anomalous behavior due to ICF in concentrated RE metallic systems has been well-documented for the RE ions Ce, Sm, Eu, Tm and Yb, anomalous behavior associated with the Kondo effect in dilute RE metallic systems has only been well-established for Ce and its "4f hole" counterpart Yb. However, Kondo-like anomalies in both the superconducting and normal state properties have recently been observed in two dilute RE metallic systems containing Sm and Eu, respectively, as the RE impurity ion. The properties of these two systems, $(\underline{La}Sm)Sn_3$ and $Eu_xMo_6X_8$, where X = S or Se, are reviewed herein.

II. THE $(\underline{La}Sm)Sn_3$ SYSTEM

The first definitive evidence for a Kondo effect due to Sm impurity ions in a metallic matrix has emerged from recent experiments on the $(\underline{La}Sm)Sn_3$ system by DeLong et al. [1] and Bakanowski et al. [2] It should be noted that prior evidence for a Kondo effect due to Sm ions has been reported by Chouteau et al.,[3] who attributed various anomalies in the physical properties of concentrated (Sm, La)S alloys to the spurious presence of small amounts of Sm^{3+} ions, the concentrations of which depended upon lattice defects and vacancies.

The first indication that the $(\underline{La}Sm)Sn_3$ system might exhibit a Kondo effect came from measurements of the initial depression of T_c, $(-dT_c/dn)_{n=0}$, of $LaSn_3$ by Sm impurities by Schmid and Umlauf[4] and Abou-Aly et al.[5] These measurements revealed that $(-dT_c/dn)_{n=0}$ is anomalously large for Sm impurities in comparison to the other rare earth impurities. This is reminiscent of the anomalously large depressions of T_c encountered in the superconducting matrices containing Ce impurities which exhibit a Kondo effect,[6] and suggests, in analogy, that the exchange interaction parameter $\mathcal{J}$ for Sm in $LaSn_3$ is large and antiferromagnetic.

Since it has been established that the behavior of the superconducting properties of a matrix-impurity system provides a powerful means of exploring the local moment problem and the Kondo effect,[7] measurements were performed to determine the detailed dependence of the superconducting transition temperature on Sm impurity concentration and the reduction of the specific heat jump at the transition due to Sm impurities in $LaSn_3$. The results are shown in the form of a plot of the reduced transition temperature T_c/T_{c_0} vs. Sm concentration n in Fig. 1(a) and a plot of reduced specific heat jump $\Delta C/\Delta C_0$ vs. T_c/T_{c_0} in Fig. 1(b). Figure 1(a) shows that the T_c/T_{c_0} vs. n curve deviates from the initial linear behavior below $T_c/T_{c_0} \simeq 0.5$ [the value of T_{c_0} for $LaSn_3$ is 6.38 K and $(-dT_c/dn)_{n=0} = 4.32$ K/at.% Sm]. Figure 1(b) illustrates the marked negative deviation of the $\Delta C/\Delta C_0$ vs. T_c/T_{c_0} curve from the behavior predicted by the BSC law of corresponding states and the Abrikosov-Gor'kov (AG) theory.[8]

The rapid and nearly linear dependence of T_c/T_{c_0} on n and the pronounced negative deviation of the $\Delta C/\Delta C_0$ vs. T_c/T_{c_0} curve from both the BCS and AG theory are very similar to the behavior exhibited by the $(\underline{La, Th})$Ce system[9, 10] near an equiatomic La, Th matrix composition. Thus a qualitative comparison of the data of $(\underline{La}Sm)Sn_3$ and $(\underline{La, Th})$Ce alloys can be made by assuming that the modification of the superconducting properties by the Kondo effect scales with T_K/T_{c_0} in the same manner in both systems. Since T_K is of the order of one to ten times T_{c_0} near the equiatomic La, Th host composition, we can therefore estimate from the $(\underline{La}Sm)Sn_3$ data that T_K for the $(\underline{La}Sm)Sn_3$ system lies in the range $T_K \sim 5$-100 K.

The incremental susceptibility $\Delta\chi = \chi\,[(\underline{La}Sm)Sn_3] - \chi[LaSn_3]$ is shown in the form of a $\Delta\chi/n$ vs. log T plot in Fig. 2. It can be seen that the Sm contribution to the susceptibility varies only weakly with temperature and is reminiscent of the behavior exhibited by the nonmagnetic concentrated RE ICF systems SmB_6 and SmS in its collapsed metallic phase[11]; the susceptibility is much smaller than the weakly temperature dependent Van Vleck susceptibility of the configuration $4f^6$, and it lacks the low temperature Curie-law divergence expected for the configuration $4f^5$. Nearly constant between room temperature and ~ 100 K, the susceptibility increases by a factor ~ 2.5 below 100 K and approaches a constant value as $T \to 0$.

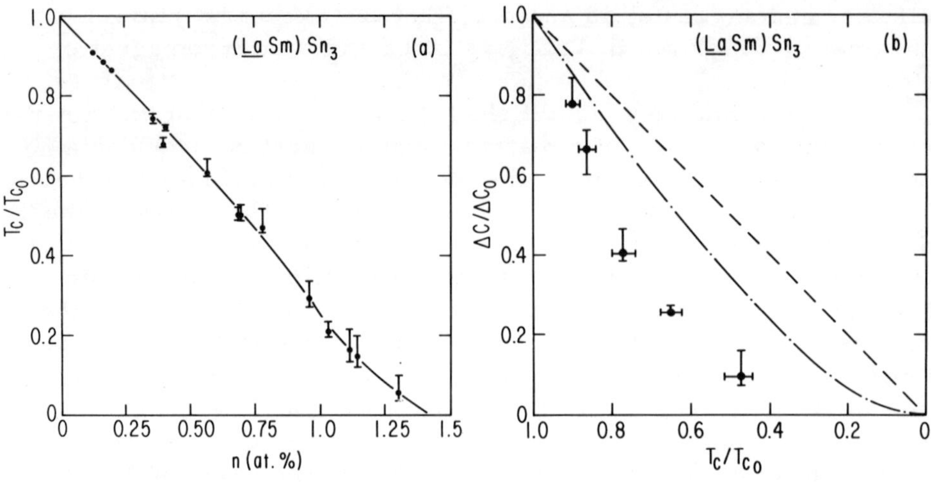

Fig. 1. (a) Reduced superconducting transition temperature T_c/T_{c_o} vs. Sm impurity concentration n for the (LaSm)Sn_3 system. The solid line is a smooth curve drawn through the data. The vertical bars denote the 10% and 90% points of inductively measured transition amplitudes. (b) Reduced specific heat jump $\Delta C/\Delta C_o$ vs. reduced transition temperature T_c/T_{c_o} for the (LaSm)Sn_3 system. The dashed line represents the BCS law of corresponding states and the dot-dashed line indicates the AG result. The vertical bars represent estimated experimental uncertainty in determining the magnitude of the specific heat jumps at T_c, and the horizontal bars represent estimates of the total width of the transitions (after ref. 1).

Measurements[1] of the specific heat in the normal state between 0.6 and 10 K on a series of (LaSm)Sn_3 alloys with concentrations between 0 and 0.8 at. % Sm reveal a large enhancement of the electronic specific heat coefficient of 1.06 ± 0.16 J/mole Sm K^2. This is equivalent to an average enhancement of the $LaSn_3$ host density of states of 230 states/eV-atom-spin direction! This is, to our knowledge, the largest enhancement of γ by a dilute impurity reported for any system to date.

A relatively high value of T_K suggested by the superconductivity data is consistent with the weak temperature dependence of the magnetic susceptibility and the value of the $n(\Delta\gamma)T$ contribution to the specific heat. Generally, the specific heat exhibits a maximum near T_K; the $n(\Delta\gamma)T$ behavior would then appear to be the low temperature tail of a Kondo specific heat

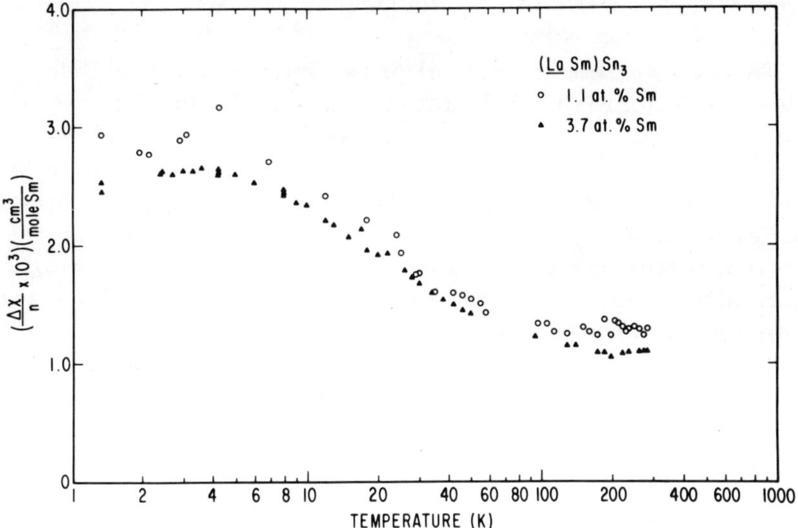

Fig. 2. Magnetic susceptibility (corrected for the $LaSn_3$ background susceptibility) of $(\underline{La}Sm)Sn_3$ alloys vs. temperature (note logarithmic scale) (after ref. 1).

anomaly with a maximum well above the 10 K upper limit of the measurements.

Bakanowski et al.[2] have reported the existence of a minimum in the normal state electrical resistivity of the $(\underline{La}Sm)Sn_3$ system. The incremental resistivity contributed by the Sm ions can be described well by a calculation based on the exchange model by Hamann[12] with an effective Sm spin of 0.27 ± 0.05 and $T_K = 110 \pm 20$ K.

The experiments of DeLong et al. and Bakanowski et al. reviewed in this section document the Kondo anomalies which occur in the superconducting and normal state physical properties of the $(\underline{La}Sm)Sn_3$ system and indicate that T_K could be relatively large, possibly of the order of 100 K. The strong resemblance between the anomalies in the normal state physical properties of the $(\underline{La}Sm)Sn_3$ system and those observed in concentrated systems such as SmS in its collapsed metallic phase clearly suggests the existence of a phenomenological relationship between the anomalies in dilute RE systems which are attributed to the Kondo effect and the anomalies in concentrated RE systems which are ascribed to ICF.

III. THE $(RE)_xMo_6S_8$ AND $(RE)_xMo_6Se_8$ SYSTEMS

The two series of Chevrel phase rare earth molybdenum sulfides and selenides, $(RE)_xMo_6S_8$ and $(RE)_xMo_6Se_8$ ($x \approx 1$), display extremely interesting superconducting and magnetic behavior. With the exception of the Ce and Eu compounds in both series, all of the $(RE)_xMo_6S_8$ [13] and $(RE)_xMo_6Se_8$ [14] compounds are superconducting, in spite of the fact that they contain nearly 7 at.% magnetic rare earth ions. The compound $La_{1.0}Mo_6Se_8$ has a relatively high critical temperature $T_c = 11.3$ K [14] and a very high zero temperature upper critical field estimated to be of the order of 460 kOe.[15] The compound $Gd_{1.2}Mo_6Se_8$ and Gd-rich pseudoternary compounds $(Gd, La)_{1.0}Mo_6Se_8$ exhibit a striking lambda-type specific heat anomaly in the superconducting state which appears to originate from the presence of the Gd magnetic moments but whose exact nature remains to be established.[16] In this section we review the normal state properties of the nonsuperconducting Ce and Eu molybdenum sulfide and selenide compounds which have been found to exhibit Kondo-like behavior.[17]

The superconducting transition temperatures of the $(RE)_xMo_6S_8$ and $(RE)_xMo_6Se_8$ compounds are plotted vs. RE in Fig. 3.[18] The dashed lines are linear interpolations between the superconducting transition temperatures of the nonmagnetic end-member La and Lu molybdenum sulfide and selenide compounds which represent a first approximation to the transition temperatures of the RE compounds with partially-filled 4f electron shells in the absence of pair breaking interactions due to the magnetic moments of the RE ions. The depressions of T_c relative to these linear interpolations are shown in Fig. 3. Here it can be seen that the systematic variation of ΔT_c with RE can be described qualitatively by the deGennes factor $(g-1)^2 J(J+1)$ (represented by the solid lines in Fig. 3) where g is the Landé g-factor and J is the total angular momentum of the Hund's rule ground state. This follows from the AG theory [8] for RE impurity ions in the limit of low impurity concentration n for which

$$\Delta T_c \propto nN(E_F)\, \mathscr{J}^2 (g-1)^2 J(J+1)$$

if n is fixed and $N(E_F)$ does not vary with RE. The value of $|\mathscr{J}|$ inferred from the ΔT_c vs. RE data for both series of compounds $(RE)_xMo_6S_8$ and $(RE)_xMo_6Se_8$ is $|\mathscr{J}| \sim 0.01$ eV.[18] The anomalously large depressions of T_c for the Ce and Eu molybdenum

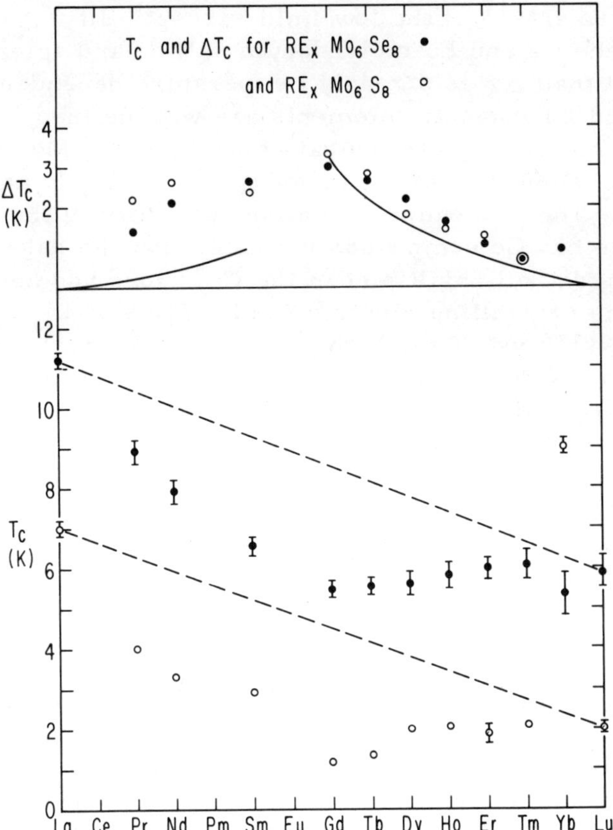

Fig. 3. Superconducting transition temperature T_c and depression of T_c (ΔT_c) vs. RE constituent for $(RE)_xMo_6Se_8$ and $(RE)_xMo_6S_8$ compounds. The dashed lines represent the interpolated T_c's of the RE molybdenum chalcogenide compounds in the absence of magnetic pair breaking interactions. The solid line gives the value of ΔT_c expected from the AG theory relative to the measured value of ΔT_c for the $Gd_xMo_6S_8$ compound (see text) (after ref. 18).

sulfide and selenide compounds (which remain normal to 50 mK) suggest that these compound should exhibit Kondo-like anomalies in their normal state properties. This has been borne out by the behavior of the magnetic susceptibility, electrical resistivity and low temperature heat capacity of these materials which are described below.

Measurements[17] of the low field susceptibility $\Delta\chi$ vs. temperature for the Ce and Eu molybdenum sulfide and selenide compounds reveal that $\Delta\chi$ is strongly temperature dependent and show that the Ce and Eu magnetic moments are well defined; magnetization vs. field measurements indicate that the $Ce_{1.0}Mo_6S_8$ and $Ce_{1.0}Mo_6Se_8$ compounds order magnetically at 2.8 K and ~ 0.4 K, respectively. The pronounced negative curvature of the $(\Delta\chi)^{-1}$ vs. T plots for the two Ce compounds indicates that the degeneracy of the $J = 5/2$ Hund's rule multiplet of the Ce^{3+} ions has been partially-lifted by the crystalline electric field. Plots of $\mu_{eff} = [3\, k_B\, \Delta\chi\, T/N]^{1/2}$ vs. T for both Ce compounds level off near room temperature to the values $\mu_{eff} = 2.34\, \mu_B$ for $Ce_{1.0}Mo_6S_8$ and $2.52\, \mu_B$ for $Ce_{1.0}Mo_6Se_8$ (the Ce^{3+} free ion value is $\mu_{eff} = 2.54\, \mu_B$). This suggests that the magnitude of the negative conduction electron-impurity spin exchange interaction is larger for $Ce_{1.0}Mo_6S_8$ than for $Ce_{1.0}Mo_6Se_8$. Plots of $(\Delta\chi)^{-1}$ vs. T for the two Eu compounds are nearly linear and both compounds have Curie-Weiss temperatures of ~ -0.6 K. These features suggest that the Kondo temperatures are relatively low (~ 0.1 K) for the latter compounds.

Figure 4 shows the electrical resistivity (normalized to its value at 294 K), $\rho(T)/\rho(294\text{ K})$, vs. temperature for the Ce and Eu molybdenum sulfide and selenide compounds and for $Lu_{1.2}Mo_6S_8$. The resistivity of the nonmagnetic reference compound $Lu_{1.2}Mo_6S_8$ decreases monotonically with decreasing temperature with some negative curvature and then drops abruptly to zero at its superconducting transition temperature of ~ 2 K. The resistivity vs. temperature curve for both the $Ce_{1.0}Mo_6S_8$ and $Ce_{1.1}Mo_6Se_8$ compounds reveals a monotonic decrease with decreasing temperature with pronounced negative curvature above a minimum which occurs at 11 K and 6-7 K, respectively, for the two compounds. The resistivities of the latter compounds increase with decreasing temperature below their respective minima; for the $Ce_{1.0}Mo_6S_8$ compound there is a sharp drop in resistivity at 2.5 K, attributed to magnetic ordering which occurs at this temperature. However, the most striking behavior of the resistivity vs. temperature is exhibited by the two Eu molybdenum sulfide and selenide compounds. As the temperature is lowered, the resistivity of the $Eu_{1.2}Mo_6S_8$ compound increases by a factor of two between 294 and 100 K; below 100 K, it increases more rapidly and then saturates to a constant value of the order of $7.2\, \rho(294\text{ K})$ as $T \to 0$. The temperature dependence of the resistivity of the $Eu_{1.2}Mo_6Se_8$ sample is equally as striking, but qualitatively different. As the

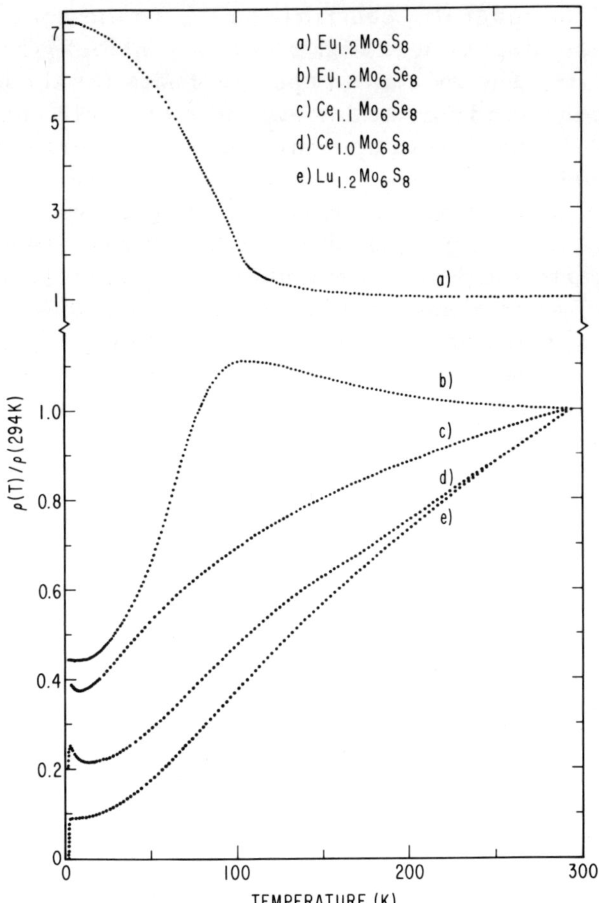

Fig. 4. Electrical resistivity ρ(T) as a function of temperature for five RE molybdenum chalcogenides (normalized to room temperature values) (after ref. 17).

temperature is lowered, the resistivity of the $Eu_{1.2}Mo_6Se_8$ compound rises about 10% between 294 and 100 K; below 100 K it decreases rapidly to a value ~ 0.44 ρ(294 K) at a temperature ~20 K, below which it again rises slightly. Although qualitatively different from each other, these features in the resistivity vs. temperature curves of the two Eu compounds are similar to those which have been observed in concentrated RE metallic systems and attributed to ICF.[11, 19]

Plots of the magnetic contribution C_m to the specific heat vs. temperature are displayed in Fig. 5 for $Ce_{1.0}Mo_6S_8$ and $Ce_{1.0}Mo_6Se_8$.[17] The measured specific heats for these compounds were corrected for the lattice and electronic contributions by subtracting the measured specific heat of the corresponding $La_{1.0}Mo_6S_8$ and $La_{1.0}Mo_6Se_8$ compounds. Figure 5 shows that above ~ 4 K, C_m rises with increasing temperature in a similar way for the two Ce compounds which probably represents the low temperature part of a Schottky anomaly due to crystal field splitting of the Ce $J = 5/2$ Hund's rule multiplet. Below ~ 4 K, C_m for the $Ce_{1.0}Mo_6S_8$ compound rises with decreasing temperature and then drops sharply at 2.5 K, the magnetic ordering temperature. In contrast, C_m for the $Ce_{1.0}Mo_6Se_8$ compound continues to increase with decreasing temperature to the ~ 1.2 K lower limit of the measurements. A measurement of C_m for $Eu_{1.2}Mo_6S_8$ [17]

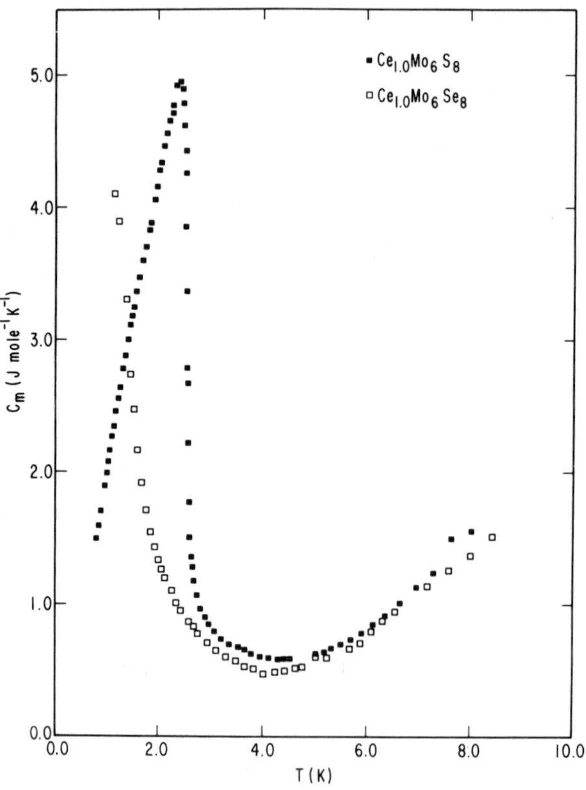

Fig. 5. Magnetic heat capacity C_m vs. temperature T for $Ce_{1.0}Mo_6S_8$ and $Ce_{1.0}Mo_6Se_8$ (after ref. 17).

yielded qualitatively the same temperature dependence as for $Ce_{1.0}Mo_6Se_8$.

The exceptionally large depressions of T_c and the striking anomalies in the resistivities of the Ce and Eu molybdenum chalcogenides are clear evidence for Kondo-like behavior in these systems. However, it is noteworthy that in contrast to the traditional Kondo systems, the RE ions in these compounds occupy specific (translationally invariant) lattice sites. This constitutes a unique feature that may have special physical consequence in spite of the relatively low RE concentrations involved.

IV. CONCLUDING REMARKS

The experiments reviewed herein show that there is a phenomenological relationship between the Kondo-like anomalies in the physical properties of concentrated and dilute RE metallic systems. The anomalies can be viewed as originating from the hybridization of the localized 4f electron states and the extended states of the conduction band as a result of the unstable valence of certain RE ions. In concentrated RE metallic systems, the Kondo-like anomalies are often attributed to ICF and have been well-documented for the RE ions Ce, Sm, Eu, Tm and Yb. However, in dilute RE metallic systems, the Kondo-like anomalies are generally ascribed to the "Kondo effect" but have only been well-established for Ce and its "4f hole" counterpart Yb. In this paper we have shown that Kondo-like behavior also occurs in the physical properties of dilute RE metallic systems for the RE ions Sm and Eu. In analogy with the concentrated RE metallic systems, we anticipate that Kondo-like anomalies will soon be found in dilute metallic systems containing Tm impurities as well.

REFERENCES

1. L. E. DeLong, R. W. McCallum, W. A. Fertig, M. B. Maple and J. G. Huber, submitted to Phys. Rev. Lett.

2. S. Bakanowski, J. E. Crow and T. Mihalisin, submitted to Phys. Rev. Lett.

3. G. Chouteau, F. Holtzberg, O. Peña, T. Penney, R. Tournier and S. von Molnar, International C. N. R. S. Colloquium on Metal-Nonmetal Transitions, Autrans, France, 6-28/7-1, 1976.

4. W. Schmid and E. Umlauf, Commun. Phys. 1, 67 (1976).

5. A. I. Abou-Aly, S. Bakanowski, N. F. Berk, J. E. Crow and T. Mihalisin, AIP Conf. Proc. No. 29, Magnetism and Magnetic Materials-1975, ed. by J. J. Becker, G. H. Lander and J. J. Rhyne, Am. Inst. of Phys. 1976, p. 358.

6. M. B. Maple, in "MAGNETISM: A Treatise on Modern Theory and Materials," ed. by H. Suhl (Academic Press, New York, 1973), Vol. V, Ch. 10.

7. M. B. Maple, Appl. Phys. 9, 179 (1976).

8. A. A. Abrikosov and L. P. Gor'kov, Zh. Eksp. Teor. Fiz. 39, 1781 (1960); Sov. Phys. JETP 12, 1243 (1961).

9. J. G. Huber, W. A. Fertig and M. B. Maple, Solid State Commun. 15, 453 (1974).

10. C. A. Luengo, J. G. Huber, M. B. Maple and M. Roth, Phys. Rev. Lett. 32, 54 (1974) and J. Low Temp. Phys. 21, 129 (1975).

11. M. B. Maple and D. Wohlleben, AIP Conf. Proc. No. 18, Magnetism and Magnetic Materials-1973, ed. by C. D. Graham, Jr. and J. J. Rhyne, Am. Inst. of Phys. 1974, p. 447; and references cited therein.

12. D. R. Hamann, Phys. Rev. 158, 570 (1970).

13. Ø. Fischer, A. Treyvaud, R. Chevrel and M. Sergent, Solid State Commun. 17, 721 (1975).

14. R. N. Shelton, R. W. McCallum and H. Adrian, Phys. Lett. 56A, 213 (1976).

15. S. Foner, E. J. McNiff, Jr., R. N. Shelton, R. W. McCallum and M. B. Maple, Phys. Lett. 57A, 345 (1976).

16. R. W. McCallum, D. C. Johnston, R. N. Shelton and M. B. Maple, to appear in the Proceedings of the Second Conference on Superconductivity in d- and f-Band Metals (edited by D. H. Douglass), held at the University of Rochester, April 30-May 1, 1976; and submitted to Phys. Rev. Lett.

17. D. C. Johnston, R. W. McCallum, W. A. Fertig, R. N. Shelton and M. B. Maple, to be published.

18. R. W. McCallum, R. N. Shelton, M. B. Maple and H. Adrian, Bull. Am. Phys. Soc. 21, 383 (1976).

19. B. C. Sales and R. Viswanathan, J. Low Temp. Phys. 23, 449 (1976).

COMMENTS

PARKS: I would like to make a comment about pinning of the narrow 4f level to the Fermi level in the case of dilute systems as opposed to concentrated systems. Since the Sm 4f wave function is not extended in space as it is for Ce, in order to get the large value of Jex implied by the experiments by Maple, et al., the narrow 4f level must lie extremely close to the Fermi level thus making the energy denominators very small. In the case of concentrated systems there exists mechanisms which tend to pin the 4f level to the Fermi level. In the dilute case it seems fortuitous that the 4f level would be so close to the Fermi level unless there is some mechanism which tends to keep it there. Such a mechanism has never been proposed for the dilute limit.

HIRST: The idea of Fermi-type quasiparticles I was talking about this morning would work as well for the concentrated as for the dilute case,except that the intensities in the dilute case will be reduced by the concentration factor. However, the question of pinning the 4f level to the Fermi level is a property of concentrated systems. If it happens in the dilute case, it is fortuitous.

VARMA: (in response to comment by Parks) In the case of an isolated impurity with a charge different from the host, the position of the local level is determined by considerations of local charge neutrality, whereas in the concentrated limit the position of the 4f level is determined by electron counting; i.e., if the renormalized density of states on the local orbitals is large that automatically pins it at the Fermi level.

(The comment by Anderson after paper by Haldane is also relevant to the points raised above.)

COQBLIN: I would like to mention that the case of LaThCe alloys, where the La (or Th) concentration is varied, is different than the case of LaCe alloys under pressure, because in the case of ternary alloys there are clearly environmental effects, as shown by magnetization (O. Pena and R. Tournier-Grenoble) or nuclear orientation (A. Benoit and J. Flouquet-Orsay) measurements.

ENERGY WIDTH OF THE MIXED VALENCE STATE FROM INELASTIC NEUTRON SCATTERING AND ELECTRON TUNNELING

D. Wohlleben

University of Cologne, 5 Cologne 41, Zuelpicher Strasse 77, West Germany

The groundstate of mixed valence compounds contains, among others, appreciable amplitudes for the groundstates of 4f shell with n and n-1 electrons (A_n and A_{n-1}) and simultaneously with A_{n-1} the amplitude $A_{k,n-1}$ of an extra conduction electron. Since these amplitudes do not refer to energy eigenstates, they must each have a separate energy spectrum. Presumably they are each spread over an energy interval of order kT_f, where T_f is the fluctuation temperature. The spectra are measurable at finite frequency and temperature. Local measurements (on a scale of order cell volume or less) can in principle measure $A_n(\omega)$ and $A_{n-1}(\omega)$, whereas transport measurements should be sensitive to $A_k(\omega)$. Diffuse neutron scattering measures locally, identifies the amplitude of the total angular momentum of the 4f shell through the 4f formfactor and is capable of measuring its energy dependence. Electron tunneling can distinguish the amplitude A_k of the usual conduction electrons (spread over eV) from $A_{k,n-1}$, since the latter is spread over only $\simeq 10^{-2}$ eV around the Fermi energy.

In Ce Pd_3 inelastic diffuse neutron scattering measures the energy spectrum of the J = 5/2 state which goes with the $4f^1$ configuration. Experiments on the IV compound Ce Pd_3 and on the isostructural reference compounds YPd_3 (no 4f electrons) and Tb Pd_3 (stable $4f^8$ shell) show (1):

a) 4f formfactor in Tb Pd_3 and Ce Pd_3, but not in Y Pd_3

b) Total magnetic crossection as expected for $4f^8$ in Tb Pd_3 but reduced by one half with respect to $4f^1$ in Ce Pd_3

c) Quasielastic Lorentzian lines with $\Gamma/2 \simeq$ 0,6 meV in Tb Pd_3 and $\Gamma/2 \simeq 20$ meV in Ce Pd_3 at 300 K

Whereas in Tb Pd_3 the quasielastic line width follows the normal $\Gamma \sim T$ (Korringa-) behaviour down to helium temperature, the Lorentzian spectrum in Ce Pd_3 has no appreciable temperature dependence down to the measuring limit of 100 K. Since $\Gamma/2 > kT$ for $T < 200$ K, the fast decay of the $4f^1$ magnetization in Ce Pd_3 cannot be thermally driven and is most likely a property of the ground state, in accordance with a recent phenomenological model for IV compounds (2), which assumes reduced $4f^1$ occupancy and temperature independent short lifetimes of the local $4f^1$ Zeeman levels due to spontaneous ICF between $4f^1$ and $4f^0$.

With respect to $A_{k,n-1}(\omega)$, the differential conductance was measured at helium temperatures on tunnel diodes of Nb, Mo or Cu against the IV compound Yb Cu_2 Si_2 (3). The conductance shows a large anomaly consisting of a temperature independent voltage asymmetry and a zero bias anomaly with logarithmic voltage and temperature dependence and with 7.5 meV half width at 1.6 K. No such anomalies were detected when Yb was replaced by Gd and Lu (stable 4f shell). The half width of the anomaly is nearly identical with the width of the Zeeman levels of the local $4f^{13}$ configuration in Yb Cu_2 Si_2 ($kT_{f13} = 7$ meV) which was extracted from the temperature dependence of the static susceptibility (2). The conductance anomaly is interpreted with a peak in the conduction state density at the Fermi level due to the extra conduction electron which appears during ICF with the $4f^{13}$ configuration.

REFERENCES

1. E. Holland-Moritz, M. Loewenhaupt, W. Schmatz, and D. Wohlleben, Verh. Dtsch. Phys. Ges. 7, 729 (1976) and to be published.
2. B. C. Sales and D. Wohlleben, Phys. Rev. Lett. 35, 1240 (1975).
3. H. P. Leppin, B. C. Sales and D. Wohlleben, Verh. Dtsch. Phys. Ges. 7, 729 (1976) and to be published.

COMMENTS

DONIACH: I would like to make a comment about different time scales. The time scale associated with magnetic fluctuations is completely different than that associated with charge fluctuations.

WOHLLEBEN: This is not the case if you assume that the magnetic moment changes every time the configuration changes.

DONIACH: This is very unlikely since, e.g., in the case of transition metal impurities like Mn in Cu the time scale for charge fluctuations is very short while that for magnetic fluctuations is very long. I think the same description would work for f-band systems.

HIRST: I would like to make the following point. In the photoemission measurement the final states are highly excited states and one is observing their decay, while in the magnetic measurements only low energy excited states are created and their decay is strongly blocked by the coulomb interaction.

ORBACH: Your neutron scattering results on $CePd_3$ look very much like what people see in Cu-Fe. That is, the Korringa rate is not proportional to temperature in the region where you're very close to the Kondo effective temperature. Why are you led to this other interpretation?

WOHLLEBEN: What I'm really trying to say is that the magnetization is 4f like, lives for a finite time of order 10^{-13}sec and is gone for a time of similar length. I'm not trying to impress on you any one single interpretation, viz., valence fluctuations, but merely offer that as a possible interpretation.

EDELSTEIN: I would like to return to the comment made earlier concerning the lifetimes associated with charge and spin fluctuations. The two lifetimes are certainly different for the transition elements; however, in the case of Ce, where there is only one potentially magnetic electron the two lifetimes may be the same. The reason for this is that when the electron hops off there is not another magnetic electron around to remember the magnetic configuration.

DONIACH: I don't agree, because when it hops off it stays around and it's still an electron and it still has a spin.

THERMODYNAMICAL BEHAVIOR AT VALENCE INSTABILITIES*

J. M. Lawrence†, M. C. Croft and R. D. Parks

Dept. of Physics and Astronomy, University of Rochester

Rochester, N. Y. 14627 U.S.A.

I. INTRODUCTION

The isomorphic valence transitions [1] in cerium and SmS are solid-state analogues of the liquid-vapor transition. A line of first-order transitions in the P-T plane terminates at a critical point $\{P_o, T_o\}$ [2]; the lattice symmetry is not broken, and density is the order parameter. The enormous cell-volume change ($\Delta V/V \sim 0.15$) is accompanied by a divergence in compressibility at the critical point [3, 4]; large changes in magnetic susceptibility χ and electrical resistivity ρ reflect the change in electronic structure underlying the phase transition. The first-order transitions can also be driven second order by intra-rare-earth alloying in such systems as $Sm_{1-x}RE_xS$ [5] and $Ce_{1-x}RE_x$[6]. In this paper we discuss primarily the system $Ce_{1-x}Th_x$. The critical point $\{x_o, T_o\}$ at P=0 in the x-T plane is located; the behavior is that of mean-field theory [7]. This presumably reflects a long-range interaction arising from the electro-elastic coupling involved in the lattice collapse. A quantitative Landau free energy functional, valid in the critical region, is exhibited[8]. We attempt to infer the existence of certain terms in the microscopic hamiltonian, corresponding to those in the free energy. We argue that: (1) in contrast to SmS the 4f spin entropy plays an important role in stabilizing integral valence in the γ-state; (2) the role of the solute is to effectively shift the 4f level E_f by means of both size-mismatch and valence-mismatch (charge-transfer); (3) the non-linear "bootstrap" interaction involves both lattice-mediated and charge-transfer interactions. The experimental properties of the intermediate valence α-state [9] are rather well described by an electron-liquid picture, involving a narrow hybridized 4f-s,d band, of width Γ; we will show that alloying apparently has only a weak effect on Γ. In the integral valence γ-state there are strong anomalies in the thermodynamic and transport properties, which vary fairly strongly with alloying [8, 10]. Presumably these have their origin in the situation of a localized magnetic 4f level E_f, degenerate with s-d band electrons and in close proximity to the fermi level $\mathcal{E}_F$. The interpretation of these anomalies is an open question; this ambiguity extends into the critical region, where the microscopic origin of the linear coupling of various quantities to the order parameter is not well understood.

II. EXPERIMENTAL ASPECTS

Thorium and cerium are completely miscible in the system $Ce_{1-x}Th_x$; for $x>0.10$ thorium dilution stabilizes the fcc phase, thus avoiding the presence of dhcp β-phase[6]. Techniques of sample preparation, measurement and analysis were standard. Details can be found elsewhere; here we would like to focus attention on certain experimental difficulties inherent in this system[11]. Our greatest problem in sample preparation is ensuring that the thorium goes homogeneously into solution. Segregation can be detected visually since the cerium rich regions blacken rapidly in air, while thorium rich regions remain shiny. The fact that X-ray diffraction patterns showed only single phase fcc up to the detection limit (5%) does not rule out finer scale inhomogeneity. The equation of state given below is in principle an invaluable tool for sample characterization - e.g. inhomogeneity causes excessive transition rounding; segregation results in wrong values of T_o. The scatter in the plots of Fig. 3 reflects this. A second problem is that the transition does not go to 100% completion. First order transitions on cooling (warming) proceed most of the way to completion but then tail out, joining the equilibrium curve at lower (higher) temperature. A small fraction $\eta_\gamma \sim 0.005$ of untransformed magnetic cerium ions is retained at 4.2K. This shows up in the low temperature susceptibility as an impurity Curie tail $\eta_\gamma C/T$ [10]. A related feature is the 1-2K "hysteresis" ΔT_c for samples with $x>x_o$[6,7,10]. In one study we found that ΔT_c roughly correlated with η_γ. We think these effects reflect incomplete transformation, arising from the enormous strain fields associated with the 15% volume collapse; and compounded by grain boundaries, inhomogeneity and thermal gradient effects. The effect on the critical exponent δ is negligible; the A and B parameters observed on warming cycles consistently differ by 5-10% from those obtained on cooling. For consistency all data given below are for cooling cycles. In pure cerium the α-γ transition triggers a degree of β-phase formation[6]. In $Ce_{1-x}Th_x$ for $0.20<x<0.27$ an anomalous increase in the sample length or resistivity occurs after partial completion of the first-order transition on cooling; it appears to indicate nucleation of the high volume β-phase (followed by subsequent decay within 0.5K), occuring right at the first-order transition. Decomposition of β-phase on leaving the transition region appears complete: No anomalies in any properties occur at 12-13K, where β-cerium goes antiferromagnetic, and no net expansion is present at room temperature. The effect of sample history, inhomogeneity and sweep rate on this process is not clear. We speculate that the softening of the bulk modulus near $\{x_o, T_o\}$ lessens the strain energy penalty for β-phase formation at γ-α domain boundaries.

III. THE PHASE TRANSITION IN $CE_{1-x}TH_x$

In Fig. 1 the resistivity ρ(T) of $Ce_{1-x}Th_x$ is shown for four concentrations in the critical region. Plots for sample volume V(T) and effective moment $\mu_{eff}=T\chi/C$ are qualitatively identical[Fig. 6]. For $x\to x_o^-$ the large first-order discontinuities and hysteresis dwindle and precursive rounding sets in; for $x>x_o$ these quantities show s-shaped curves with a maximum in slope at $T_c(x)$, which approaches a divergence as $x\to x_o^+$. This maximum has been traced out as far as x= 0.75. The role of reduced volume $\Delta V/V_o=V-V_o/V_o$ as the order parameter mandates the critical anomaly $\Delta V\sim\Delta T^{1/\delta}$ along the pathway $\Delta x= x-x_o\equiv 0$[8]. To find δ we plot $T^*(T)=\Delta V/(dV/dT)$ in Fig. 2 for $x\simeq x_o$ [7]. We obtain δ from the slope; the values above and below T_o are symmetrical about 3.0, the mean-field value. From Figs. 3 and 4 we see that the same divergence occurs for ρ and μ_{eff} - i.e. these quantities couple linearly to the order parameter near T_o. Given mean-field theory we can utilize a Landau expansion of the free energy in x,P,T space:

$$\Delta\mathfrak{F} = \sigma\left\{-a_1\frac{\Delta V}{V_o} + \frac{a_2}{2}\left(\frac{\Delta V}{V_o}\right)^2 + \frac{a_4}{4}\left(\frac{\Delta V}{V_o}\right)^4\right\} \quad . \tag{1}$$

To first approximation a_1 and a_2 are linear functions of ΔT, Δx and ΔP, changing sign at the critical point:

$$a_i = a_{iT}\Delta T + a_{iP}\left(\frac{\Delta P}{P_o}\right) + a_{ix}\left(\frac{\Delta x}{x_o}\right) \quad i=1,2 \quad a_{1T}\equiv 1 \quad . \tag{2}$$

Since a_4 must not change sign, it is approximately constant. The equation of state, obtained by minimizing $\mathfrak{F}$ with respect to ΔV is:

$$a_1 = a_2\frac{\Delta V}{V_o} + a_4\left(\frac{\Delta V}{V_o}\right)^3 \quad . \tag{3}$$

For P≡0 we have utilized the linear coupling of ρ to the order parameter; and followed the procedure of least-squares fitting the resistivity of each sample to:

$$a_4\left(\frac{\Delta\rho}{\rho_o}\right)^3 + a_{2x}\left(\frac{\Delta x}{x_o}\right)\left(\frac{\Delta\rho}{\rho_o}\right) = \Delta T + a_{1x}\left(\frac{\Delta x}{x_o}\right) \quad . \tag{4}$$

We present in Table 1 the values of the parameters, averaged over the indicated interval of fit, and the scatter. These revised values reflect adoption of a normalization per cerium atom $\rho(T) = (\rho_{total} - \rho(4.2))/(1-x)$, an improved least-square routine and additional data. In Fig. 3 we plot the data together with various predictions of Eq. 4. We utilize $a_4\equiv 118K$ in this plot. $T_c(x)$ is obtained from the midpoints of the hysteresis loops for $x<x_o$, and from the temperature of the maximum derivative for $x>x_o$. The phase boundary determines a_{1x}

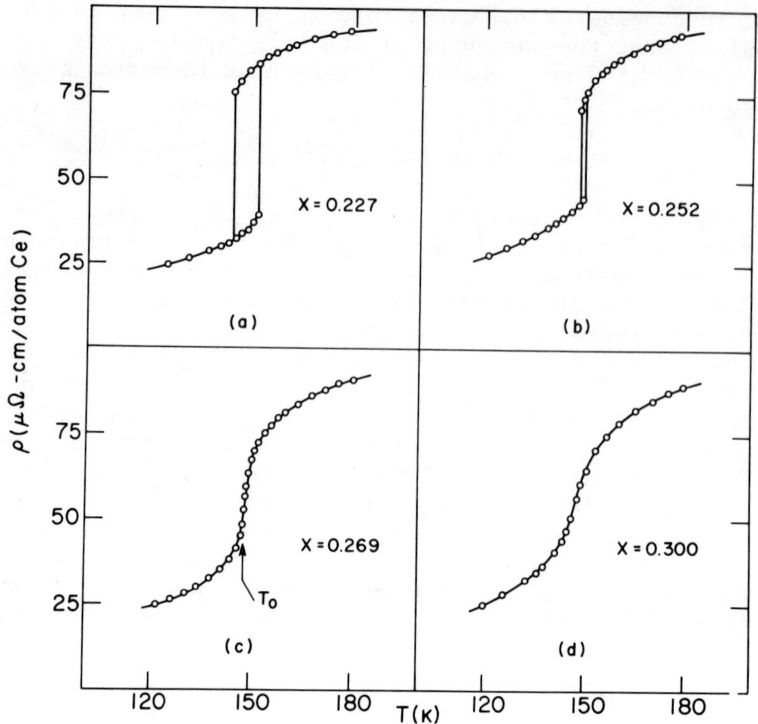

Fig. 1. Resistivity $\rho(T)$ of $Ce_{1-x}Th_x$ for four samples in the critical region.

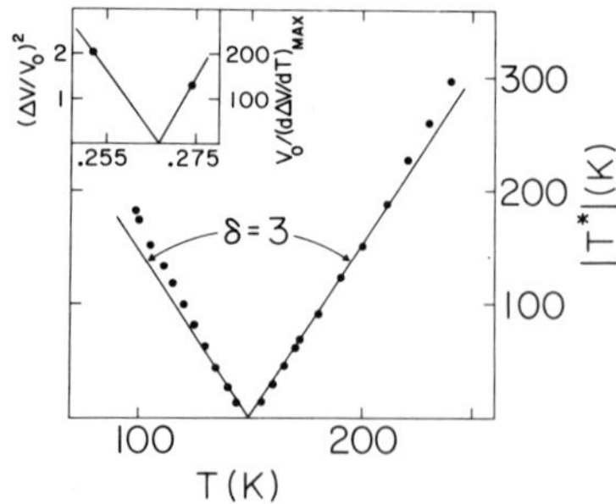

Fig. 2. $T^*(T) = \Delta L/(dL/dT)$ obtained from the sample length L of $Ce_{.726}Th_{.274}$. The slope gives the δ exponent. Inset: Squared first-order jump $(\Delta L)^2$ and inverse maximum derivative $(dL/dT)^{-1}_{max}$ vs. x.

TABLE 1

x_o	ρ_o	$T_c(x)$
0.265 ± .003	57 ± 3μΩ-cm/Ce atom 0.17 < x < 0.32	$= T_o - a_{1x}(\Delta x/x_o)$ $= 148 + 13(\Delta x/x_o) \pm 3K$ 0.1 < x < 0.5
a_4	a_{2x}	a_{2T}
$125-290(\Delta x/x_o)\pm 10K$ 0.26 < x < 0.32 60K 0.32 < x < 0.41	(118 ± 10)K 0.26 < x < 0.32	0.0 ± 0.4 0.26 < x < 0.32

through $\Delta T/\Delta x = -a_{1x}/x_o$(for $\Delta\rho=0$). For $Ce_{1-x}Th_x$, a_{1x} is small. For $a_{1x}=a_{2T}\equiv 0$ we obtain a_{2x} from the maximum slopes:

$$\rho_o/(\partial\rho/\partial T)_{max} = a_{2x}(\Delta x/x_o)^{\gamma} \tag{5}$$

with $\gamma=1$. The first order jumps then determine a_4:

$$\left(2\frac{\Delta\rho}{\rho_o}\right)^2 = 4\left(\frac{a_{2x}}{a_4}\right)\left(\frac{\Delta x}{x_o}\right)^{2\beta} \tag{6}$$

with $\beta=1/2$. These are exhibited in Fig. 3b. In Fig. 3c we plot the data directly against the predictions of Eq. 4. Note the resemblance of the $\rho(T)$ curves for x<, = , >x_o to the M(H) curves for T<, = ,>T_o in a mean-field ferromagnet. (For $a_{1x}=a_{2T}\equiv 0$ the analogy $\{\Delta V, \Delta T, \Delta x\}\rightarrow\{M, H, \Delta T\}$ is exact.) Fig. 3d shows a T* plot for a nearly critical sample; for $\Delta x\equiv 0$, $a_{2T}\equiv 0$, we expect $T^*=\delta(T-T_o)$. (The asymmetry in δ can be accounted for by a finite $a_{2T}\sim 0.3$. This term may, however, be an artifact of the 1-2K "hysteresis" observed for $x>x_o$; it has the wrong sign for an entropy term.) From Eq. 3 the form of a derivative of $v\equiv\Delta V/V_o$ with respect to a general variable y=x,T,P is:

$$\frac{\partial v}{\partial y} = \frac{a_{1y} - a_{2y}v}{a_2 + 3a_4v^2} .$$

For a perpendicular pathway of approach (e.g. $a_2\equiv 0$) v changes sign at the transition so that $\partial v/\partial y$ will change sign unless $|a_{2y}v|<|a_{1y}|$. Since stability requires $\partial v/\partial P<0$, a_{2P} must be small. Hence the compressibility diverges as v^{-2} on this pathway, i.e. $\partial v/\partial P\sim(\Delta T)^{-2/3}$[4]. Similarly since entropy S couples linearly to V and $a_{2T}\simeq 0$ the specific heat $T\partial S/\partial T(\sim T\partial V/\partial T)$ also follows $(\Delta T)^{-2/3}$ on this pathway[12].

Since $\Delta\rho$, ΔL(i.e. ΔV) and $\Delta\mu_{eff}$ all couple linearly we can obtain similar equations of state replacing V, $\mu_{eff}\rightarrow\rho$. The slopes in the insets of Fig. 2 and Fig. 4 give a_{2x} and a_4. For ΔV we obtain $a_o(T_o)=5.015$ $a_{2x}=9530$ $a_4=7(10)^5$; for μ_{eff}, $a_{2x}=150K$ $a_4=118K$ $\mu_{eff}(T_o)=$

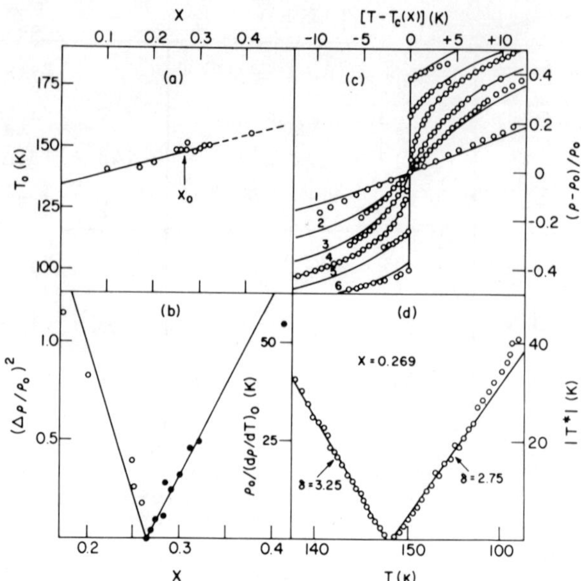

Fig. 3. a) Phase boundary $T_c(x)$ for $Ce_{1-x}Th_x$, b) Squared first-order jumps $(\Delta\rho)^2$ and inverse maximum derivatives $(d\rho/dT)^{-1}_{max}$, c) $\Delta\rho$ vs. ΔT for several values of x, d) $T^* = \Delta\rho/d\rho/dT$. ρ is the resistivity and the solid lines are the predictions of the equation of state. From Ref. 10.

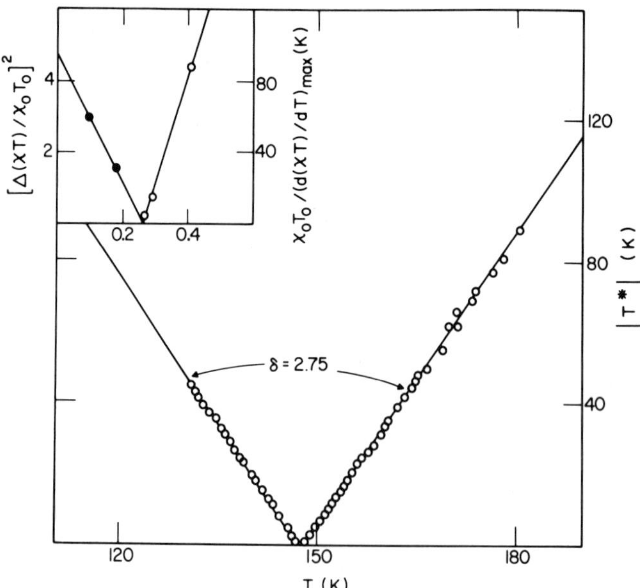

Fig. 4. $T^* = \Delta\mu_{eff}/(d\mu_{eff}/dT)$ for $Ce_{.731}Th_{.269}$. Inset: Squared first-order discontinuities and inverse maximal derivative for $\mu_{eff} = T\chi/C$. From Ref. 10.

0.24. In order to have identical a_4 and a_{2x} for the different measurements we would have to ensure that the non-critical background were subtracted correctly in each case. For resistivity we have merely subtracted a constant impurity contribution $\rho(4.2)$, neglecting phonon scattering. For effective moment no subtraction has been made. If we subtract $a_o(4.2)$ the values for ΔV become $a_{2x}=177$ $a_4=125$. Given the crudeness of the background procedure the agreement is fine.

The microscopic order parameter is presumably the valence, or the 4f occupation number η_f. (In the light of the neutron scattering results [20] this may not be true; for the sake of discussion we assume it anyway.) V, ρ and μ_{eff} apparently couple linearly to η_f near T_o. The equation of state in a given quantity (ρ, V or μ) can break down in three ways: Higher order terms in an expansion $\mathfrak{F}(\Delta\eta_f)$ can become important, e.g. causing saturation of the first order discontinuities as Δx becomes more negative. Secondly the linear approximations for the a_i can break down. Thirdly the linear coupling of the given quantity to η_f can break down, so that $\mathfrak{F}(\Delta\rho)$ is no longer equivalent $\mathfrak{F}(\Delta\eta_f)$. This shows up in the deviations in the T* plot for ρ for $T>T_o+10K$ [7] and in the T* plot for V for $T<T_o-10K$.

IV. DISCUSSION OF THE FREE ENERGY

Given a hamiltonian $\mathcal{H}$ for the system, one could statistically average and expand the resulting potential about the critical point. Since the microscopic order parameter is presumably the occupation number η_f this would give:

$$\Delta\mathfrak{F} = \sigma\left\{\frac{A}{4}\Delta\eta_f^{\ 4} + \frac{B}{2}\Delta x\Delta\eta_f^{\ 2} + (D\Delta x-\Delta T)\Delta\eta_f + E\Delta P\Delta\eta_f\right\}$$

(Implicit here is the assumption that near the critical point $\Delta\eta_f$ and ΔV are linearly coupled). There should be a correspondence (although not one-one) between terms in $\mathfrak{F}$ and terms in $\mathcal{H}$. In this spirit, we attempt to infer the microscopic origin of each term in the free energy. We consider first the term $-\sigma\Delta T\Delta\eta_f$. This has the form of an entropy term, with the order parameter conjugate to T. This would be contingent on the entropy S being linearly coupled to the order parameter, which has indeed been found to be the case[12]. We have argued [8] that this term reflects the 4f spin entropy $Nk\eta_f\ell n(2J+1)$. A priori there is no reason for assuming that lattice or configuration entropy won't contribute. However, apart from spin entropy, all other things seem equal and one would expect the phase boundaries for $Ce_{1-x}RE_x$ alloys to be comparable to those for $Sm_{1-x}RE_xS$ alloys, where $J\equiv0$ in the integral valence ground state, rather than being essentially orthogonal, as observed[Fig. 5]. Of course it is the tradeoff between the $\Delta T\Delta\eta_f$ term and the $D\Delta x\Delta\eta_f$ term that determines the slope. Although $\Delta T/\Delta x$ can be varied considerably in other cerium-based alloys [13] it is generally two orders of magnitude smaller than in the SmS systems. We should also point out that integration under the specific heat curve shows that only about half the expected

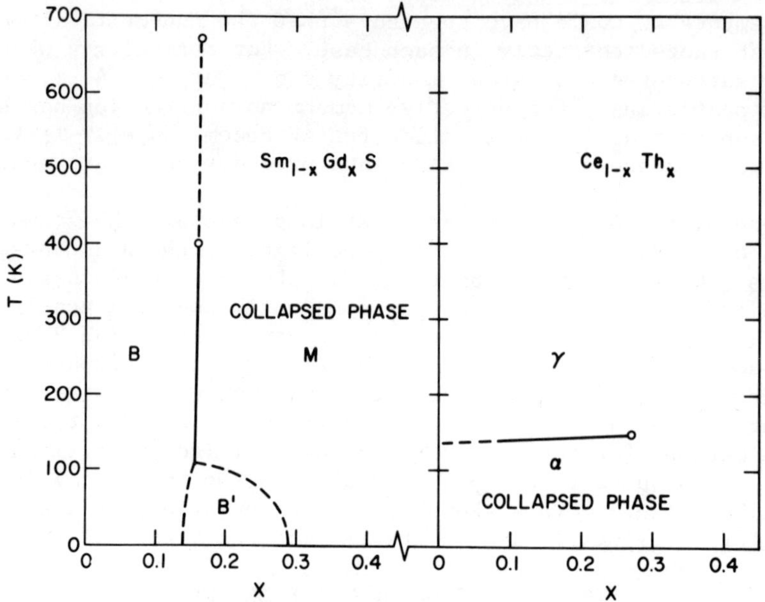

Fig. 5. Comparison of the phase boundaries of $Sm_{1-x}Gd_xS$ and $Ce_{1-x}Th_x$. From Ref. 8.

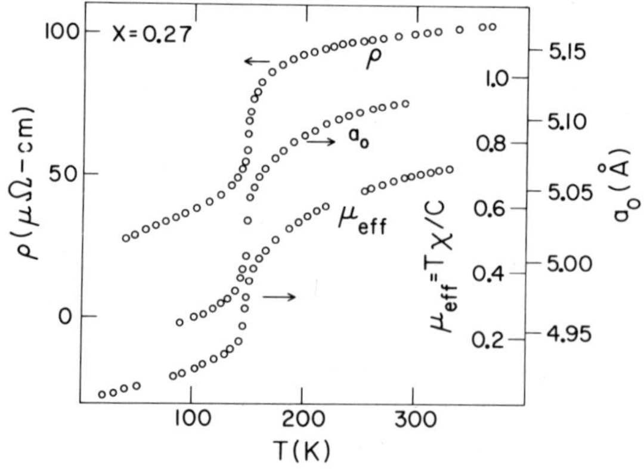

Fig. 6. Resistivity ρ, lattice constant a_o and effective moment $\mu_{eff}=T\chi/C$ for $Ce_{.73}Th_{.27}$.

$Nk\ell n6$ entropy is generated during the phase transition[12]. Of the rest some must appear as electronic entropy in the intermediate valence state where the γT linear term in the specific heat is enhanced; some is also generated at high temperatures, corresponding to a very slow saturation of μ_{eff}. This will be discussed further below. The $D\Delta x\Delta\eta_f$ term might have two origins. A size mismatch between solute and host should give rise to an effective $P_{eff}V$ term with large solutes acting like negative pressure. For solutes of identical valence $\Delta T/\Delta x$ is indeed linear in solute size[13]. However, at fixed size, the slope increases with solute valence. A term $(\mathcal{E}_F-E_f)\eta_f$ is expected in the hamiltonian. If a rigid band treatment applied, the high-valence solutes would increase the conduction electron density and hence $\mathcal{E}_F$; the integral valence state would then be further stabilized, which is inconsistent with experiment. A speculation is that solute-solvent charge transfer would further screen the 4f electrons, favoring further delocalization, stabilizing intermediate valence, as observed[13, 14]. The non-linear interaction driving the valence transition, represented by the $\Delta x\Delta\eta_f^2$ term, must involve a long force range to give rise to mean-field behavior[7]. Utilizing the Ginzburg criterion, the free energy parameters and region of fit, we have obtained [11] a lower limit of $\xi(x=0;\ T=0)\geq 7Å$. Any purely electronic mechanism would be screened out over $\sim 2Å$; hence a lattice-mediated interaction must be involved. In one treatment [15] cerium would be treated as a pseudo-binary alloy of tri- and tetra-valent atoms with 15% size mismatch; phase segregation occurs due to an infinite ranged attractive interaction between ions of the same size. This would give first-order transitions, and mean-field theory. However intermediate valence implies all atoms are identical over a time scale longer than h/Γ; the force range for a size mismatch interaction should be cut off at roughly hv_s/Γ where v_s is the velocity of sound. For $\alpha(\gamma)$ cerium taking $\Gamma/h\sim 1500K$ (150K) this would be 0.7(7)Å. It is unfortunate that extrinsic effects prevent us from measuring closer to $\{x_o,\ T_o\}$ to see whether mean-field theory begins to break down. In a second treatment [16] the lattice collapse on delocalization lowers the s,d band bottom, moving E_f closer to $\mathcal{E}_F$, promoting further delocalization. However, there are opposing tendencies - e.g. the bandwidth also increases. This treatment correctly predicts the tendency to become more continuous with increasing x or P, but requires ad hoc insertion of a phenomenological negative interaction parameter to give first order transitions. Such a term might arise, at least in part, from electronic interactions[17]. The solute valency effect mentioned above might be explained by charge-transfer amplification of the screening; and it is found [13] that x_o (which determines the scale over which the interaction changes sign) also depends independently on valence mismatch and size mismatch. This supports the existence of electronic interactions. A plausible scheme is that these create the first-order transitions, while lattice coupling helps drive the transition second order, and dominates the critical behavior.

V. THE PROPERTIES OF THE INTERMEDIATE AND INTEGRAL VALENCE STATES; THE EFFECT OF ALLOYING

The susceptibility χ, specific heat C_p and resistivity ρ have been studied in allotropically pure α-cerium, great pains being taken to avoid the presence of dhcp β-phase[e.g. 9]. χ is independent of T out to about 50K; from $\chi \sim 1/\Gamma$ one would estimate a 4f-hybrid bandwidth $\Gamma \sim 0.13$eV. C_p shows a related enhanced linear coefficient, and the resistivity is that of a decent metal with no T^2 dependence. One can also circumvent β-phase by alloying with thorium. On alloying the susceptibility remains essentially constant out to $x \simeq 0.27$. This is perhaps surprising - it is as though the hybridization is not effected much by alloying strains and charge transfer. (It would be desirable to substantiate this conclusion in $Ce_{1-x}Th_x$ of higher allotropic purity, since the observed constancy depends on subtracting out the contribution of $\sim 1/2\%$ of untransformed magnetic cerium ions [10].) The resistivity of $Ce_{1-x}Th_x$ and $(Ce_{1-x}La_x)_{.9}Th_{.1}$ is constant between 1-4K, to one part in 10^4 - no T^2 or Kondo behavior arises from the untransformed ions. Thermopower [18] and thermal expansion at low temperatures appear rather conventional. The upshot is that α-cerium is fairly well described as a pauli paramagnet, involving a narrow hybridized band[19]. The parameter Γ is consistent with the α-state neutron linewidth reported at this meeting[20].

In the integral valence state of cerium the 4f local level at E_f is degenerate with the s-d band and in close proximity to the fermi level $\mathcal{E}_F$. One expects, in this situation, strong spin fluctuations and/or valence fluctuations. γ-cerium has many similar properties to the trivalent cerium intermetallics (e.g. $CeAl_3$, $CeIn_3$) where large resistance and thermopower maxima are observed. The resistivity is large and only weakly temperature-dependent [Fig. 6] - for $x \simeq 0.27$ it can be well approximated by $\rho(T) = \rho(4.2) + 65\mu\Omega cm + 0.04T\mu\Omega\text{-cm}$ for $T > 300K$. This might be accounted for by spin-disorder scattering $\rho \sim N\eta_f J_{sf}{}^2 S(S+1)$ but the value is far too large in comparison with other rare earths, unless J_{sf} is enhanced. The thermopower [18] increases with decreasing T to a large value ($\sim 30\mu V/K$) until it is cut off by the valence transition. If the γ-state could be stabilized to T=0 the resistivity, in addition to the thermopower, might peak. The susceptibility appears to follow $\chi = C/(T + \theta(x))$ where C is approximately the free ion value, and $\theta(x)$ roughly tracks the metastable limit of the γ-state[10]. This suggests θ may represent a fundamental energy scale - e.g. a spin fluctuation energy or a bandwidth. If this latter were the case, then an important feature of the valence transition would be an order of magnitude increase in bandwidth on transformation into the α-state. The $\theta(x)$ value is, again, of the order of the γ-state neutron linewidth[20]. The specific heat [12] is enhanced over the lattice-only value of 3R; generation of entropy S appears to correlate with the slow saturation of effective moment $\mu_{eff} = T\chi/C$ which is only ~ 0.75 at 350K[Fig. 6]. It is here worth pointing out that although one is well away from the critical region at 350K, the valence transition may itself be weakly operative. High sensitivity plots L(T), $\mu_{eff}(T)$ or $\rho(T)$[Fig. 6] show weak negative curvature, even at 300K. Thus η_f may still be changing, and this would effect the interpretation of the various properties.

Further information concerning the mechanisms underlying these anomalous thermodynamic and transport properties can be obtained by examining how they alter on approaching the phase transition. We have seen that ΔV, $\Delta\rho$, $\Delta\mu_{eff}$ and ΔS all couple linearly near T_o - implying that these vary linearly with η_f in the critical region. (Unless, of course, the valence is not the microscopic order parameter.) $\Delta\rho$ couples linearly over a far wider region below T_o than above, where deviations in the T* plot occur already at T_o + 10K[7]. If the dominant mechanism were spin disorder scattering one might expect the opposite - best linear coupling above T_o where the spin is best. Perhaps J_{sf} varies rapidly with T above T_o, but saturates below. A second question concerns the linear coupling of μ_{eff}. Using phenomenology as a guide with $\chi = C\eta_f/(T + T_{VF})$ one might expect $(T + T_{VF})\chi/C$ to exibit linear coupling for some $\theta<T_{VF}<\Gamma$. We have attempted such fits, but they fail badly unless $\Delta\to 0$. Rather $\mu_{eff} = T\chi/C$ couples linearly over a wide range near T_o. The proposed mechanism for the entropy, $S\sim Nk\eta_f \ln(2J+1)$, should be called in question simultaneously. Finally it is worth noting the rapid breakdown in the T* plots for ΔV below T_o. Varma has pointed out in Ref. [1] that the lattice constant need not be linear in the valence. Our essential point here is not to promote phenomenology or particular interpretations, but rather that the underlying mechanisms determining the resistivity, thermopower, susceptibility, etc., are not transparent - neither near T_o, nor in the integral valence state, nor, really, in the cerium intermetallics.

VI. CONCLUSION

By obtaining the detailed form of the equation of state for $Ce_{1-x}Th_x$ near the critical point, we have been able to discuss the chemical physics of the phase transition. Lattice-propagated interactions, charge transfer (screening) and 4f spin entropy dominate the thermodynamic behavior. The intermediate valence state seems to be adequately described as a simple narrow-band metal; the problem being for the theorist to obtain the hybridized band in the limit of large Coulomb correlation. The properties in the integral valence state, and their behavior in the transition region, are not well understood. The inelastic neutron scattering results, presented at this meeting [20], give direct insight into the microscopic processes involved. Coupled with the bulk measurements, these results should stimulate deeper theoretical exploration of the valence instability.

REFERENCES

* Work supported by the Army Research Office and the Office of Naval Research.

† Present Address: Physics Dept. U. of California, Irvine 92717.

1. C. M. Varma, Rev. Mod. Physics 48, 219 (1976).
2. A. R. Kutsar, Fiz. Metal. Metalloved 33, 1104 (1972).
3. T. Penney, R. L. Melcher, F. Holtzberg, G. Güntherodt, AIP Conference Proceedings No. 29, 392 (1976).
4. M. C. Croft and R. D. Parks, these proceedings.
5. A. Jayaraman, P. Dernier, L. D. Longinoti, Phys. Rev. B11, 2783 (1975).
6. K. A. Gschneidner, Jr., R. O. Elliott, R. R. McDonald, J. Phys. Chem. Solids 23, I, 555; II, 1191; III, 1201 (1962).
7. J. M. Lawrence, M. C. Croft and R. D. Parks, Phys. Rev. Lett. 35, 289 (1975).
8. J. M. Lawrence, M. C. Croft, J. M. Markovics and R. D. Parks, AIP Con. Proc. No. 29, 429 (1976); R. D. Parks and J. M. Lawrence, ibid, p. 479.
9. D. C. Koskimaki and K. A. Gschneidner, Jr., Phys. Rev. B11, 4463 (1975).
10. J. M. Lawrence and R. D. Parks, Journal d'Physique, Supplement, to be published; also J. M. Lawrence and R. D. Parks, these proceedings.
11. J. M. Lawrence, Thesis, University of Rochester, 1976.
12. J. M. Markovics and R. D. Parks, these proceedings.
13. M. A. Manheimer and R. D. Parks, these proceedings.
14. B. Alascio, A. Lopez and C. F. E. Olmedo, Phys. Rev. B7, 1274 (1973).
15. P. W. Anderson and S. T. Chui, Phys. Rev. B9, 3229 (1974).
16. C. M. Varma and V. Heine, Phys. Rev. B11, 4763 (1975).
17. R. Ramirez and L. M. Falicov, Phys. Rev. B3, 2426 (1971).
18. I. Zoric and R. D. Parks, these proceedings.
19. C. M. Varma and Y. Yafet, Phys. Rev. B13, 2950 (1976).
20. R. Birgeneau and S. M. Shapiro, these proceedings.

COMMENTS

ORBACH: I don't think you're right about there being a cutoff in the force range. The interaction proceeds by exchange of a virtual phonon, and the range can be infinite. In superconductivity where $hv_f/E_g \sim \xi(T=0)$, the interaction is short-ranged but the coherence length is very long; the two should not be confused.

CAMPAGNA: It's no great surprise that the rigid band model doesn't work to describe the linear term. In XPS one can see this also. The reason is that on alloying, the bottom of the band moves down, bringing the fermi level with it. This brings E_f closer to ε_F

JEFFERSON: I don't see why configurational entropy shouldn't also be important.

LAWRENCE: An $n_f \log n_f$ term might show up in the free energy as a $-\varepsilon \Delta T(\Delta V)^2$ term. Such a non-linear albiet small term was observed, but it had the wrong sign; it is probably an extrinsic (fine scale segregation or inhomogeneity) effect.

BIRGENEAU: The answer to the lattice dilemma is given in a recent paper by Cowley, which everyone should read. For a compressional mode transition on a lattice he shows that the marginal dimensionality is zero. It is not true that the lattice isomorphic transition falls in the same universality class as the liquid gas transition, since infinitesimal translational symmetry is broken.

MAGNETIC EXCITATIONS IN MIXED VALENT MATERIALS

R. J. Birgeneau and S. M. Shapiro[†]

Department of Physics, Massachusetts Institute of Technology[*] and Brookhaven National Laboratory[†]

ABSTRACT

The spin dynamics of a number of mixed valent materials have now been extensively studied using varied experimental probes including EPR, Raman scattering, infrared spectroscopy and inelastic neutron scattering. In this paper we review briefly photon and neutron spectroscopic studies of the $^7F_0$-$^7F_1$ magnetic excitation dynamics in semiconducting SmS, SmSe and SmTe. New results for the phonon dispersion relations in SmS are presented. We also discuss recent neutron scattering experiments by Axe, Lawrence, Parks and ourselves on the γ-α transition in $Ce_{.74}Th_{.26}$. It is found that the spin fluctuation energy evolves from ~20 to >70 meV in going from the γ to the α phase.

*Work at M.I.T. supported by the National Science Foundation
†Work at Brookhaven performed under the auspices of the U.S. Energy Research and Development Agency

I INTRODUCTION

In the past two years the magnetic excitations in mixed valent materials have been probed by a variety of experimental techniques. The most direct of these is inelastic magnetic neutron scattering[1] which in the most favorable situations allows a complete study of the generalized susceptibility $\mathrm{Im}\chi(\vec{Q},\omega)$ for wave vectors throughout the Brillouin zone and for energies ranging typically from 0.1 to 100 meV. The generalized spin susceptibility, in turn provides important space and time dependent information about the microscopic electronic states of the system. In this paper we review recent neutron and optical studies of two model mixed-valent materials, SmS and $Ce_{1-x}Th_x$. Both of these have been extensively discussed in the literature and indeed, in the proceedings of this conference itself, so we shall not discuss the relevant background in any detail.

For our purposes it is sufficient to note that at room temperature SmS, which has the rock-salt structure, is a magnetic semiconductor in which the Sm^{++} ion has a well defined $4f^6$ configuration. At ~6.5 kBar it undergoes a strongly first order isostructural phase transition to a mixed valent state.[2] $Ce_{1-x}Th_x$ for $x \simeq 0.265$, on the other hand, has a second, or nearly second order, phase transition as a function of temperature from the fcc γ to the fcc α phase;[3] here $T_c(0.265) \sim 148$ K. $Ce_{.74}Th_{.26}$ is a good metal in both phases; further, as we shall show, the Ce ion also seems to exhibit mixed-valent characteristics in both phases albeit with markedly different spin fluctuation energies. Thus although these two materials represent archetypes of mixed-valence phenomena, they are in fact quite different in their details and hence offer an interesting contrast. We begin with a discussion of the magnetic exciton dynamics in the semiconducting phase of SmS.

II SmS

We consider firstly the electronic energy levels and concomitant magnetic exciton dynamics in semiconducting SmS. For an assemblage of $Sm^{2+}(4f^6)\,^7F$ ions the Hamiltonian may be written

$$H = \Delta \sum_i \vec{L}_i\cdot\vec{S}_i + \sum_i H_{CEF,i} - 2\sum_{i>j} J_{ij}\vec{S}_i\cdot\vec{S}_j, \qquad (1)$$

where Δ is the spin-orbit coupling constant, H_{CEF} is the SmS cubic-crystal-field Hamiltonian, and J_{ij} is the isotropic exchange coupling between the Sm^{2+} spins i and j. In this case, Eq. (1) is best treated by diagonalizing each term in succession. Diagonalization of the spin-orbit coupling term yields the $^7F_0$ atomic manifold lowest, $^7F_1$ at energy Δ, $^7F_2$ at energy 3Δ, etc. In the free ion[8] $\Delta = 422$ K so that for measurements below room temperature one need only consider the $^7F_0$ and $^7F_1$ J-manifolds. In SmS, because of the

O_h symmetry, the crystal-field term H_{CEF} will couple the $^7F_0$ and $^7F_1$ manifolds to the higher J-multiplets but it will not lift the $^7F_1$ triple degeneracy. The change in $E(^7F_1)-E(^7F_0)$ due to H_{CEF} in SmS is almost certainly less than 10 K.

The exchange term $J_{ij}\vec{S}_i \cdot \vec{S}_j$ serves to couple the $^7F_0 \rightarrow {}^7F_1$ ionic excitations on sites i and j. Thus, the single-ion transition becomes a propagating excitation in the crystal. Indeed, it corresponds almost exactly to the idealized singlet-triplet model which has been discussed extensively in the literature.[4] Using the value $\langle ^7F_1|S|^7F_0\rangle=2$, the $^7F_0-{}^7F_1$ singlet-triplet dispersion relation within the MF-RPA approximation is given by[5]

$$\hbar\omega(\vec{k}) = \Delta[1 - 16R(T)J(\vec{k})/\Delta]^{1/2}. \tag{2}$$

where

$$J(\vec{k}) = \Sigma_j J_{ij} \exp[i\vec{k}\cdot(\vec{r}_i-\vec{r}_j)]$$

and

$$R(T) = \frac{1 - \exp(-\Delta/k_BT)}{1 + 3\exp(-\Delta/k_BT)}$$

For a paramagnet this mode is triply degenerate. Since the $^7F_0$ state is coupled to $^7F_1$ by both L and S this mode is magnetic-dipole active. Indeed these excitations have now been studied by each of inelastic neutron scattering, Raman scattering and far-infrared absorption.[6]

We show in Fig. 1 the magnetic exciton dispersion relations as measured by Shapiro, Birgeneau and Bucher[5] using inelastic magnetic neutron scattering. The solid line through the T=77 K data represents the results of a least- squares fit of Eq. (2) to the measured energies. The spin Hamiltonian parameters so obtained are $\Delta=36.2\pm0.2$ meV, $J_1=0.043\pm0.003$ meV, $J_2=-.025\pm0.007$ meV and $J_3=-0.003\pm0.001$ meV. The dot and dot-dashed lines represent calculated dispersion relations at T=197 K and T=273 K respectively. The agreement is excellent, thus confirming that in this weak-exchange regime the RPA works very well. Measurements of the k=0 exciton energies as a function of temperature have been obtained by Nathan et al.[6] in each of SmS , SmSe, and SmTe using optical techniques. Their results are shown in Fig. 2. The solid lines represent fits of Eq. (2) at k=0 assuming the validity of the RPA theory. Nathan et al. have fixed Δ at the free ion value of 293.5 cm^{-1}=36.4 meV and they note that good fits can be obtained for a range of Δ of ±0.5 meV. The results for the exchange are given in Table I. The neutron and optical results are in quite good agreement for SmS. In all three materials the spin orbit coupling constant is identical to that of free ion Sm^{++} to within the errors. This might be considered somewhat surprising given the close proximity of the

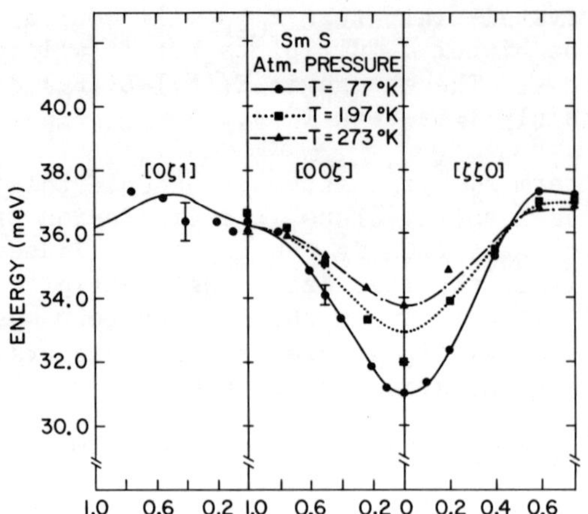

Fig. 1 - Dispersion relations of the $Sm^{2+}(4f^6)\ {}^7F_0-{}^7F_1$ singlet-triplet excitons at 77, 197 and 273 K. The lines are theoretical curves as described in the text. Figure taken from Ref. 5.

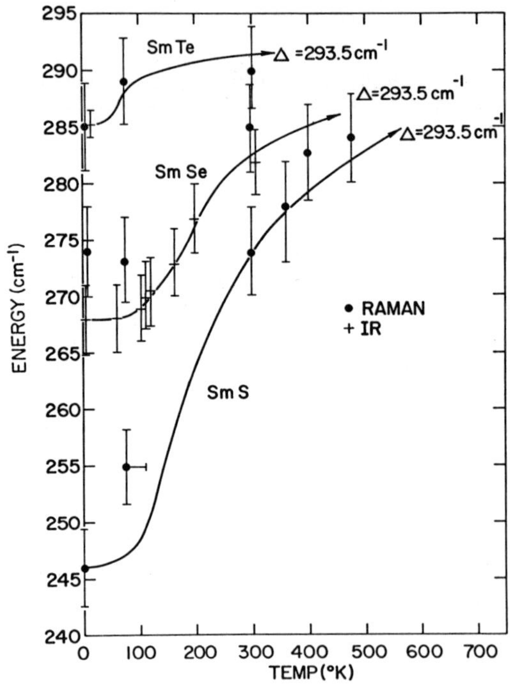

Fig. 2 - Magnetic exciton energy at k=0 as a function of temperature in SmS, SmSe, and SmTe. The solid lines are fits to Eq. (2) with $\Delta=293.5\ cm^{-1}=36.4$ meV. Figure taken from Ref. 6.

TABLE I

Exchange interactions, measured and calculated static Van Vleck susceptibilities in the Samarium Chalcogenides

Material	16 J(0) (meV)	χ calc (cm^3/mole)	χ meas (cm^3/mole)	Ratio
SmS	10.9(optical) 9.6 (neutron)	9.72	9.15	1.06
SmSe	5.7 (optical)	8.42	7.85	1.07
SmTe	2.0 (optical)	7.55	7.0	1.08

pressure-driven mixed valent phases.

From the same mean field theory one obtains for the static susceptibility at T=0

$$\chi = 8 N \mu_B^2/(\Delta - 16 J(0)) \quad (3)$$

We compare in Table I the measured Van Vleck susceptibilities[2] with those calculated from Eq. (3) using Δ, J(0) deduced from the magnetic excitons. The calculated susceptibility in all cases is about 7% higher than the measured value. Since Eq. (3) is based on free ion wave functions this may simply imply that the matrix element $\langle 1|L + 2S|0\rangle$ which equals 2 for the free ion is ~2.07 in these chalcogenides due to covalency effects.

As discussed by a number of authors the exchange interactions in these materials are quite similar to those in the europium chalcogenides[5-7] and undoubtedly arise from the same microscopic origins. It is of interest to note that Birgeneau et al. and Mehran et al.[7] have deduced the Eu^{++}-Sm^{++} exchange from Eu^{++} impurity spin resonance. An examination of Table IV of Ref. 7 shows that the Sm^{++}-Eu^{++} net exchange is ~70% of the Sm^{++}-Sm^{++} exchange in each system.

In order to complete the empirical characterization of SmS we have also carried out some limited measurements of the phonon dispersion relations at room temperature. The results are shown in Fig. 3. Unfortunately because of the intensity limitations we were unable to measure the LO branch, which according to recent theory, may play an important role in the semiconductor-metal transition. We note that the k=0 transverse optic mode is at an energy identical to that in EuS. For EuS the LO mode is around 33.0 meV, very near to the $^7F_0$-$^7F_1$ magnetic excitons already measured. These

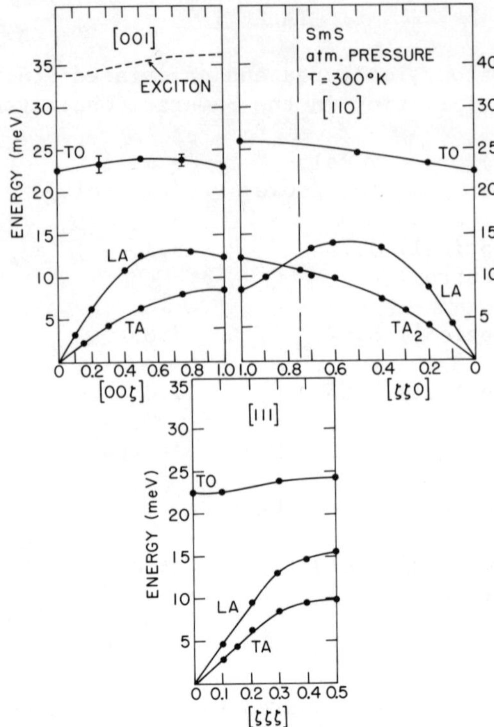

Fig. 3 - Phonon dispersion relations in SmS at atmospheric pressure at T=300 K. The ordinates are in fcc reciprocal lattice units ($2\pi/a$).

modes thence could exhibit interesting interaction effects. Unfortunately, we will have to await the availability of a rather large isotope-enriched single crystal before these measurements will be feasible.

In summary, the dynamics and general magnetic properties of semiconducting SmS, SmSe and SmTe are now quite well understood. The most remarkable result is that the behavior is quite simple; there are no obvious precursive effects of the incipient pressure-driven mixed valent transitions.

III $Ce_{1-x}Th_x$

The element cerium is perhaps the most thoroughly studied of the rare earth mixed valent systems. At room temperature as a function of pressure cerium exhibits a first order isostructural phase transition from the fcc γ phase to the more compact fcc α phase.[8] It has been assumed that in the γ phase Ce is essentially integral valent with configuration $4f^1$ and that the transition corresponds to a partial delocalization ($4f \rightarrow 5d$) of this single 4f electron. The first order transition line in cerium terminates at a critical point at P~18 kBar, T~600 K. Similar effects may be obtained by alloying cerium with the quadrivalent element thorium. As discussed most recently by Parks and Lawrence,[3] in $Ce_{1-x}Th_x$ the first order phase line terminates at a critical point defined by T=148 K, x_o=0.265. This alloy then is much easier to study experimentally since one avoids the simultaneous complications of high pressures and high temperatures. Recently, Axe, Parks, Lawrence and the authors[9] have carried out an extensive neutron scattering investigation of $Ce_{.74}Th_{.26}$ and in this section we summarize some of their salient results. In a cubic paramagnet the magnetic neutron scattering cross section may be written

$$\frac{d^2\sigma}{d\Omega d\omega} = \left(\frac{\gamma e^2}{m_e c^2}\,\frac{g}{2}\right)^2 \frac{k_f}{k_i}\,|f(\vec{Q})|^2\; 2S^{zz}(\vec{Q},\omega)$$

where

$$S^{zz}(\vec{Q},\omega) = \frac{1}{2\pi}\int dt \sum_{\vec{r}} e^{i(\vec{Q}\cdot\vec{r}-\omega t)} \langle S^z_{\vec{o}}(0)\; S^z_{\vec{r}}(t)\rangle \tag{4}$$

$$= \frac{kT}{g^2\mu_B^2}\,\chi^{zz}(\vec{Q})\,\frac{\hbar\omega/k_BT}{(1-e^{-\hbar\omega/kT})}\; F^{zz}(\vec{Q},\omega)$$

$$= \frac{\hbar}{\pi}\,\frac{1}{g^2\mu_B^2}\,(n(\omega)+1)\;\mathrm{Im}\,\chi^{zz}(\vec{Q},\omega)$$

where the symbols have their usual meaning.[1] Note that by definition $S^{zz}=S^{xx}=S^{yy}$. In the above we have assumed that the spatial part of the wave function, given in fourier transform by the form factor $f(\vec{Q})$, is factorizable from the spin density operator $S_{\vec{Q}}(0)$ (for a rare earth $\vec{S}\rightarrow\vec{J}$ the total angular momentum). We remind the reader in addition that if the spin dynamics are purely relaxational, $F(\vec{Q},t)\sim e^{-\Gamma(\vec{Q})t}$, then

$$F(\vec{Q},\omega) = \frac{1}{\pi}\,\frac{\Gamma(\vec{Q})}{(\omega^2+\Gamma^2(\vec{Q}))} \tag{5}$$

In Fig. 4 we show the lattice constant vs. temperature behavior obtained by Shapiro et al.[9] in their polycrystalline $Ce_{.74}Th_{.26}$ sample. There is a small first order jump at T=153.5 K on heating and at T~148 K on cooling but the transition is nearly continuous. A simple linear interpolation (Vegard's Law) implies that the 4f-occupancy changes from $4f^{.83}$ at 300 K to $4f^{.71}$ just above the transition to $4f^{.61}$ just below the transition to $4f^{.44}$ at 10 K. Here the largest uncertainty is in the absolute value of the valency; the relative values should be rather more accurate. The 4f-

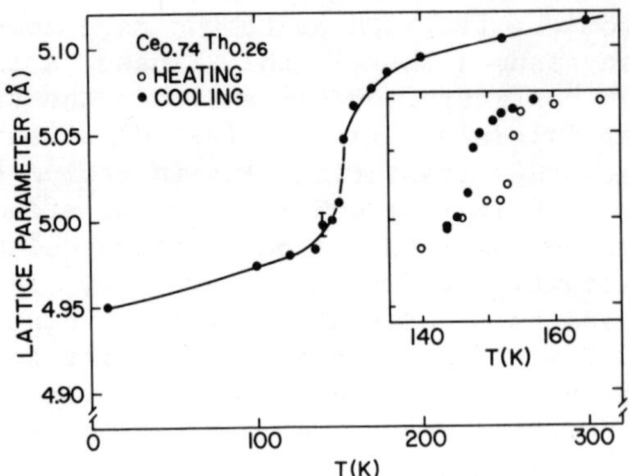

Fig. 4 - Lattice parameter versus temperature in $Ce_{.74}Th_{.26}$. The inset shows the observed hysterisis in the transition region. Figure taken from Ref. 9

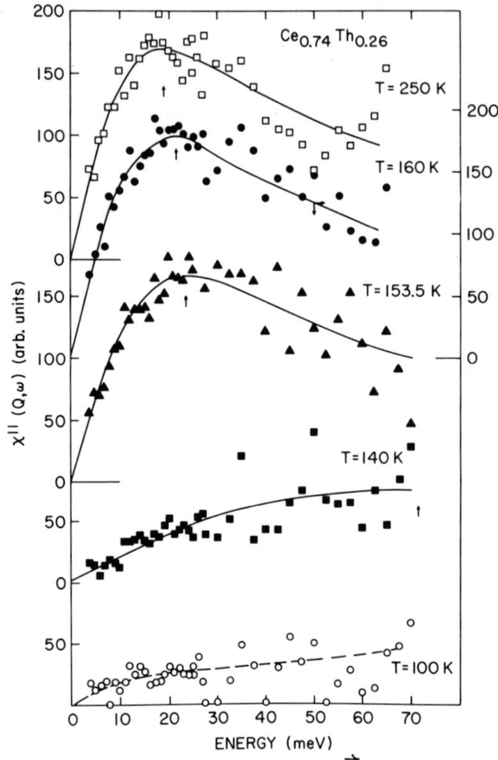

Fig. 5 - Temperature dependence of $\mathrm{Im}\chi(\vec{Q},\omega)$ for $Ce_{.74}Th_{.26}$. The low energy $\hbar\omega$<30 meV data were taken at Q=1.5 $Å^{-1}$ while for $\hbar\omega$> 25 meV data were taken at Q=3.0 $Å^{-1}$. The data were matched in the overlapping region with the scaling factor deduced from Fig. 6. This procedure is justified by the fact that the shape of Imχ is Q-independent in this Q-region. Figure taken from Ref. 9

occupancy thus seems to be non-integral valent at all temperatures. The inelastic measurements turn out to be rather complicated because of phonon effects and because of the polycrystalline nature of the sample. These complications are discussed in detail by Shapiro et al.[9] and we limit ourselves here to a brief presentation of their final results. The reader is referred to their paper for the details, especially vis-a-vis possible caveats. Fig. 5 gives the resultant $\mathrm{Im}\chi(\vec{Q},\omega)$ in $Ce_{.74}Th_{.26}$ at five temperatures from 100 K to 250 K obtained on heating. The associated form factor squared for E=30 meV is given in Fig. 6. The data in Fig. 5 correspond to an average over the Brillouin zone although, as discussed by Shapiro et al., $\mathrm{Im}\chi$ is at best weakly Q-dependent.

It may be seen instantly that $\gamma Ce_{1-x}Th_x$ differs significantly from semiconducting SmS. In a cubic system the $4f^1\ {}^2F_{5/2}$ manifold splits into a Γ_7 doublet and a Γ_8 quartet.[10] Point charge estimates indicate a possible splitting of 10 meV. Thus for a conventional integral valent Ce^{3+} system one would expect to see sharp Γ_7-Γ_8 crystal field excitons in direct analogy to the ${}^7F_0$-${}^7F_1$ excitons in SmS. Instead, even far from the critical point, one observes a purely Lorentzian response, that is, $\mathrm{Im}\chi(\vec{Q},\omega)\rightarrow\omega/(\omega^2+\Gamma^2)$. The solid lines represent a fit to this functional form. From Fig. 5 it may be seen that $\hbar\Gamma$ increases gradually from ~18 meV to ~23 meV as one approaches the first order line at 153.5 K. There is then a precipitous increase in the spin fluctuation energy to $\gtrsim$70 meV in the α phase. The form factor measurements in Fig. 6 demonstrate that the spin fluctuations are dominated by the $4f^1$

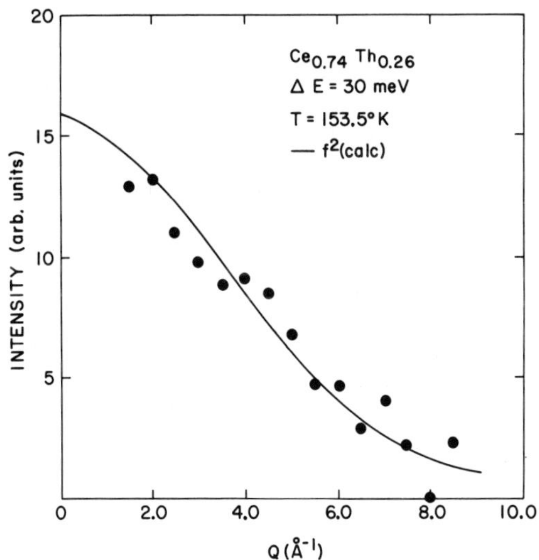

Fig. 6 - Intensity at an energy transfer of 30 meV vs. Q. The solid line is the $4f^1$ form factor squared with an arbitrary scaling factor. Figure taken from Ref. 9.

part of the wave function.

As discussed by Shapiro et al.[9] a more detailed examination of these results shows that $\mathrm{Im}\chi(\vec{Q},\omega)$ is at best weakly Q-dependent and further that $\chi(\vec{Q})$ deduced from Fig. 5 is completely consistent with the static susceptibility measurements of Parks and Lawrence.[3] Thus all of the available data on $Ce_{1-x}Th_x$ are internally consistent. They demonstrate then that: (a) this is a local transition; (b) $Ce_{1-x}Th_x$ is in some sense mixed-valent in both the γ and α phases; (c) the most dramatic difference between the γ and α phases is the spin fluctuation energy, ~20 meV in the former and >70 meV in the latter. These results are also quite similar to those in pure cerium in the γ and α phases although these latter experiments are rather less complete.[11] There is, as yet, no adequate theory for the mixed valent state.[12] However these neutron scattering measurements now give an energy scale for the spin fluctuation energies which can be related to f-band-widths, etc. in any proper theory. We refer the reader to the paper by Shapiro et al.[9] for some speculative ideas along these lines.

ACKNOWLEDGMENTS

We should like to thank Chandra Varma for his constant interest and advice on these experiments. We are also grateful to our collaborators Ernst Bucher, John Axe, Jon Lawrence and Ronald Parks, both for their role in stimulating these experiments and for their many positive contributions during the course of this work.

REFERENCES

1. W. Marshall and S. W. Lovesey, Theory of Thermal Neutron Scattering (Oxford U. P., New York, 1971).
2. E. Bucher, V. Narayanamurti and A. Jayaraman, J. Appl. Phys. 42, 1741 (1971) and references therein.
3. J. M. Lawrence, M. C. Croft and R. D. Parks, Phys. Rev. Lett. 35, 289 (1975); R. D. Parks and J. M. Lawrence, A. I. P. Conf. Proc. 29, 479 (1976) and to be published.
4. R. J. Birgeneau, A. I. P. Conf. Proc. 10, 1664 (1973).
5. S. M. Shapiro, R. J. Birgeneau and E. Bucher, Phys. Rev. Lett. 34, 470 (1975).
6. M. I. Nathan, F. Hotzberg, J. E. Smith, Jr., J. B. Torrance and J. C. Tsang, Phys. Rev. Lett. 34, 467 (1975) E 34, 1136 (1975).
7. R. J. Birgeneau, E. Bucher, L. W. Rupp, Jr. and W. M. Walsh, Jr., Phys. Rev. B 5, 3412 (1972); F. Mehran, K. W. H. Stevens, R. S. Title and F. Holtzberg, Phys. Rev. Lett. 27, 1368 (1971).

8. J. Donohue, The Structure of Elements (John Wiley, New York, 1975, pp 88-94).
9. S. M. Shapiro, J. D. Axe, R. J. Birgeneau, J. M. Lawrence and R. D. Parks (to be published).
10. K. R. Lea, M. J. M. Leask and W. P. Wolf, J. Phys. Chem. Solids 23, 1381 (1962).
11. B. D. Rainford, B. Buras and B. Lebech (to be published).
12. C. M. Varma, Rev. Mod. Phys. 48, 219 (1976) and the multitudinous theoretical papers in these proceedings.

COMMENTS (PAPER BY BIRGENEAU AND SHAPIRO)

CAMPAGNA: Is it possible to obtain the occupation number for the f-shell from the form factor measurement?

BIRGENEAU: I believe that it is not possible to go directly from the form factor to the occupation number without a theoretical model which relates the two. Others have attempted to show, using a local picture of the 4f-shell, that the integrated intensity scales like the occupation number. Our data are more consistent with the integrated intensity scaling with the occupation number squared, which is what one would expect if a hybridized wave function approach is used. The truth is probably somewhere in the middle.

CROFT: Interstitial hydrogen can be very important in lattice properties as we know from Wagner and Horner's work on the Nb-H system. Can you tell us how much interstitial hydrogen there is in your sample?

BIRGENEAU: My guess is about 1-2%. We can see a mode at 90 mev which corresponds to the local hydrogen mode. Also, the incoherent scattering cross-section for hydrogen is huge, viz., approximately 85 barns compared to a few barns for the coherent scattering cross-section for the rare earths.

WOHLLEBEN: It is difficult to imagine a temperature-independent band at relatively high temperatures where your experiment was done.

BIRGENEAU: All I can say is that there is a characteristic energy which for the γ phase is $\sim$25 mev.

ORBACH: Is it possible that the band is narrowing as the temperature increases, which might be interpreted as evidence for polaronic-like behavior?

BIRGENEAU: That is a possible interpretation since the lattice constant is changing. Also, one can dream up a self-consistent picture wherein the f-band overlaps the Fermi level and at the phase transition the band structure changes; i.e., as the temperature increases the band narrows since the lattice is expanding. There is no need to include polaronic effects.

ORBACH: But in fact, polaronic effects will give a band narrowing as you increase the temperature.

BIRGENEAU: Yes.

THE PHASE DIAGRAM AND SOME ASPECTS OF THE SUBSTITUTED SmS COMPOUNDS

A. Jayaraman, P. D. Dernier and L. D. Longinotti

Bell Laboratories
Murray Hill, New Jersey 07974

INTRODUCTION

The pressure induced electronic transition in SmS, involving a change in the valence state of Sm from +2 towards the +3 state, can also be induced by substituting a trivalent rare earth ion in the SmS lattice (1). The effects of such substitution with Gd and Y (2) has been rather extensively investigated, and for the $Sm_{1-x}Gd_xS$ system the phase boundaries in the temperature-composition plane (T-x) have been delineated (3). This phase diagram shows some unusual features in the temperature regime from 4.2°K to 900°K (3) and also some interesting metastability phenomena (3). In this paper we discuss these aspects.

It has been noted in earlier publications (1),(4) that in the chemically induced transition, the size of the substituted ion as well as its electronic structure is important. Recently we have studied under pressure the $Sm_{1-x}M_xS$ compounds with Eu, Yb and Ca for M and have obtained some new results. The results of this study and the role that the electronic band structure of the substituent plays will be discussed. In particular, we will show how important the 4f-5d gap and the stiffness of the lattice are in determining the character of the transitions.

THE PHASE DIAGRAM

Figure 1 shows the T-x diagram for $Sm_{1-x}Gd_xS$ and the P-T diagram for SmS. It is believed that the T-x diagram for other trivalent rare earth substitutions are similar, although the actual composition ranges for the boundary locations may differ somewhat

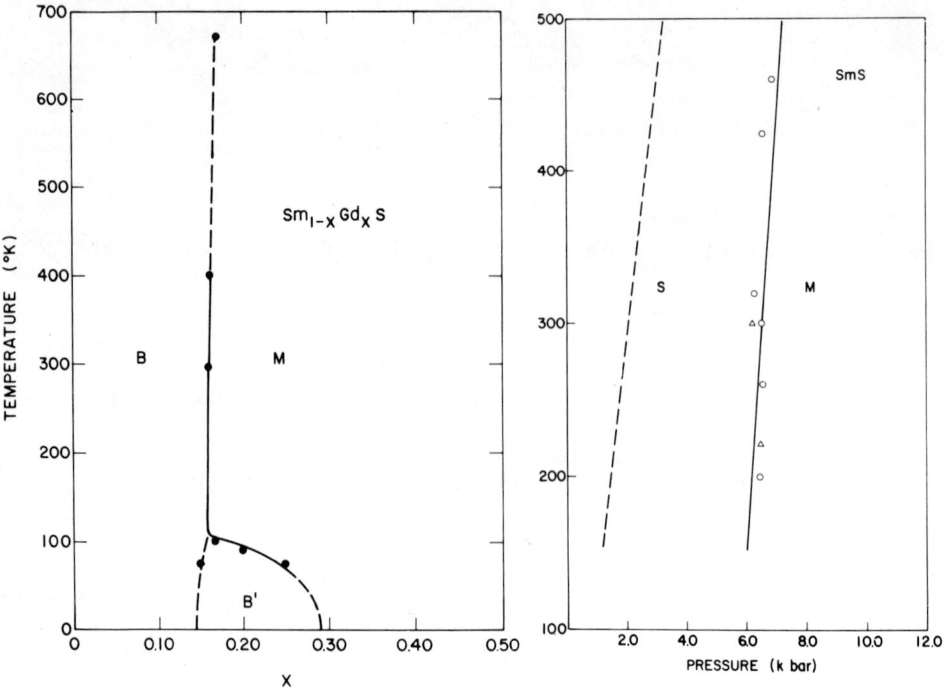

Fig. 1(a) The T-x diagram of $Sm_{1-x}Gd_xS$ system and Fig. 1(b) the P-T diagram of SmS

for different substitutions. The diagram has three phases, designated as B for the normal black phase, M for the metallic phase and B′ for the black phase resulting from the first-order transition observed with the M phase on lowering the temperature. We have also indicated the region where a continuous transformation to the B phase occurs, when the M phase is heated. All the phases in the diagram have the NaCl-type structure, but differ in their valence state. The shape of the metallic-nonmetallic phase boundary is a very interesting feature of the phase diagram. The re-entrant phase B′, we believe is different from the B-phase; the former has a lower entropy relative to the metallic phase M, while the B-phase has a higher entropy compared to the M-phase. The stability of phases in the sequence B′→M→B with increasing temperature is evidently due to the relative importance of the various terms that are contributing to the entropy At high temperatures the lattice contribution to the entropy is expected to dominate making the expanded phase stable, while at intermediate temperatures the electronic contribution dominates, making the M-phase stable. At low temperatures the lattice contribution is expected to be small and the system reduces its entropy by localizing the electrons.

In the diagram, part of the B-M phase boundary has been shown as a dashed line to indicate that a critical point is to be expected in this region. The upper limit of about 675°K has been arrived at from the thermal behavior of a nominally $Sm_{0.85}Gd_{0.15}S$ sample in which both the B and M phases appeared (3). On heating above room temperature, the M phase in the latter sample shows anomalous thermal expansion, while the B phase undergoes a slight contraction. The two lattice parameters merge at about 675°K (3). This data is shown in Fig. 2. The metallic golden sample containing 17 atomic percent Gd for instance exhibits a continuous but anomalously large thermal expansion (see Fig. 3). This we believe is due to closeness of the temperature traverse to the critical point. In the case of pure SmS, Tonkov and Aptekar (5) delineated the B-M boundary using the DTA method and have suggested ~950°K and 7 kbar as the probable coordinates for the critical point. The temperature coordinate of the critical point is probably very sensitive to the composition, in the case of substituted systems. Thus the existing data support the termination of the B-M phase boundary at a critical point.

PRESSURE, CHEMICAL SUBSTITUTION AND METASTABILITY PHENOMENA

That chemical substitution is equivalent to pressure has been demonstrated (3). On this basis we can expect the phase diagram of pure SmS in the P-T plane to correspond to the T-x diagram of the trivalent rare earth substituted SmS, such as the $Sm_{1-x}Gd_xS$ system. However, experimentally we have information only on the higher

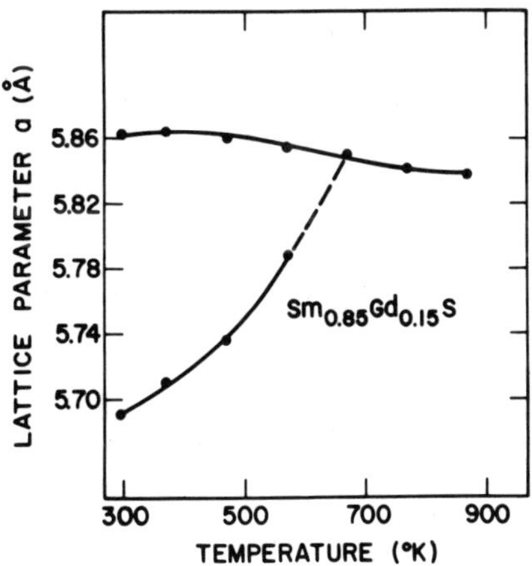

Fig. 2 Thermal behavior of the lattice parameter in a nominally $Sm_{0.85}Gd_{0.15}S$ sample. (From the data of Jayaraman et al. (3).)

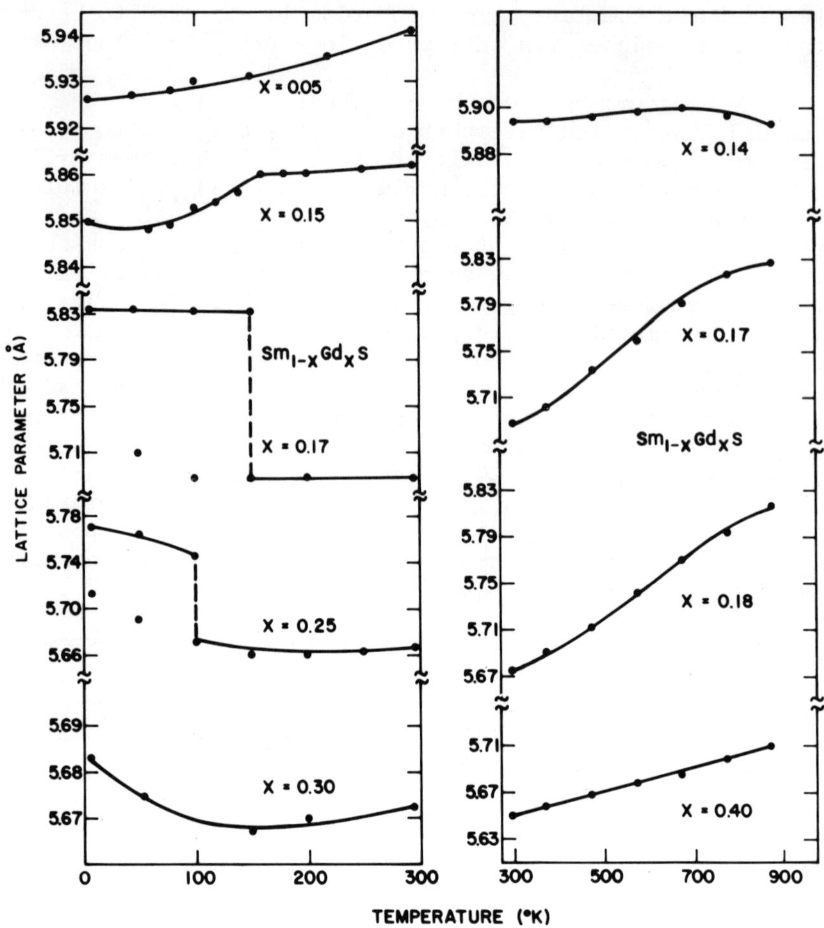

Fig 3 Anomalous thermal expansion behavior exhibited by $Sm_{1-x}Gd_xS$ samples. (From the data of Jayaraman et al. (3).)

temperature part of the B-M phase boundary for pure SmS (Fig. 1b), and the nature of the phase boundary at the low temperature part of the phase diagram has not been delineated. We will assume the similarities to exist and draw a generalized phase diagram to show graphically the effect of chemical substitution. In Fig. 4 the solid line represents the phase boundary drawn on the basis of the transformation data taken on increasing pressure. The dotted line is the boundary for the reverse transformation. The location of the different vertical lines are again based on the experimental data obtained with $Sm_{1-x}Gd_xS$ compounds. The compounds, whose composition lies within the solid and dotted curves, are black and

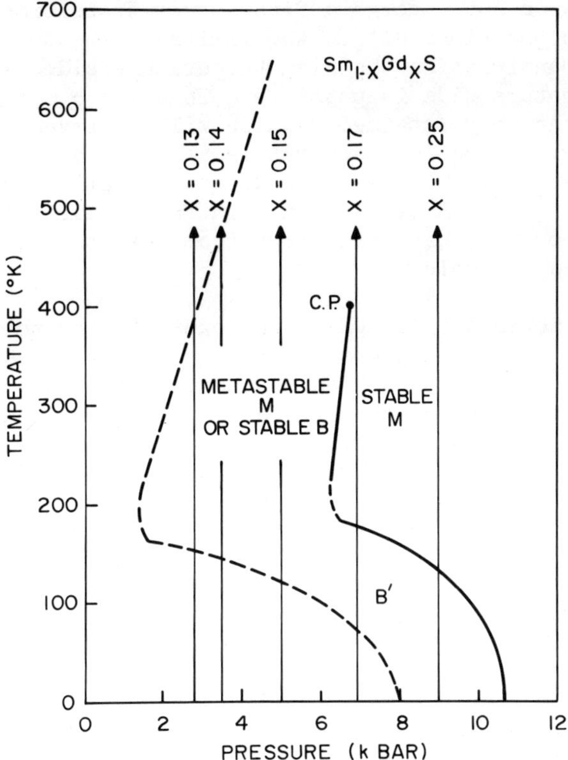

Fig. 4 Generalized phase diagram for the SmS doped systems. The effect of trivalent rare earth substitution is illustrated with the $Sm_{1-x}Gd_xS$ compounds The placement of the vertical lines correspond to the effective pressure the Gd substitutions represent.

undergo the B-M transition when pressure is applied. On releasing pressure the golden metallic phase is metastably retained at ambient pressure and temperature. The compositions which lie outside these limits do not exhibit metastability at ambient pressure and temperature. Thus the B-phase samples undergo the B-M transition, but revert when pressure is released. The compounds which are golden transform to the B′ phase on intersecting the dotted line, reverting back to the M-phase on warming. However, for the golden samples, a first-order M→B transition is not observed on the high temperature side, and only a continuous transition is observed. As mentioned earlier, this may be because of the critical termination of the B_M phase boundary. It is also to be noted in this connection that the boundary for the M→B transition (dashed line) lies too far from the temperature traverse and this can be another reason why the golden samples do not undergo a first-order transition on heating. However for compositions within the limits of the two lines, a first-order transition on the high temperature side, as well as on the low temperature side is possible. Thus for a $Sm_{0.86}Gd_{0.14}S$ sample the metastably retained M-phase will intersect the dashed phase boundary, on the low temperature as well as on the high temperature side. Further, the lower the Gd concentration the lower is the temperature for the M→B transition. In fact a large part of the dashed phase boundary has been drawn from results obtained on several $Sm_{1-x}Gd_xS$ samples.

Thus the metastability phenomena exhibited by the system is connected with the large pressure and thermal hysteresis associated with the valence transition. While thermodynamic arguments based on the relative positions of the free energy minima of the two phases can be advanced to explain the unusually large hysteresis, the microscopic mechanisms underlying this need to be understood. The phase diagram shown provides a basis for understanding the metastability, and retransformations of the retained phases.

THE INFLUENCE OF SIZE AND THE BAND STRUCTURE OF THE SUBSTITUENT ON THE VALENCE TRANSITION

To evaluate the influence of size and the electronic band structure in promoting the valence transition in SmS, we have investigated the $Sm_{1-x}Eu_xS$, $Sm_{1-x}Yb_xS$ and $Sm_{1-x}Ca_xS$ compounds under high pressure. The choice of these substitutions was dictated by the following considerations: (1) The lattice parameter of EuS and SmS are almost identical and hence in these mixed compounds, the size effect could be eliminated but the effect of larger 4f-5d energy gap can be studied. (2) If the size difference alone matters, the divalent Yb or Ca substitution may be expected to give a definite answer, because the lattice parameters of both YbS and CaS (about 5.68 Å) are much smaller compared to that of SmS (5.97 Å), and indeed are about the same as for some of the trivalent rare earth monosulfides, which are known to promote the valence transition (3). Both YbS and CaS are semiconductors but have different electronic structure; CaS has neither the 4f level nor the 5-d band. This circumstance may be expected to reveal the difference between having and not having the 4f-5d states in the electronic structure of the substituent monosulfide.

Some relevant data are presented in Table I for a few of the substituted compounds and for the pure components involved. The transition pressures in column 3 were determined from resistivity measurements performed under hydrostatic pressure. Qualitative observations on the reflectivity were also made on samples over a wider range of composition, to roughly determine the pressures of transition as well as its character.

TABLE I

DATA OF INTEREST IN CONNECTION WITH THE VALENCE TRANSITION

Compound	Lattice Parameter (a in Å)	Transition Pressure (kbar)	Energy Gap (eV)	$\frac{dE_g}{dP}$ (meV/kbar)	Bulk Modulus (kbar)
SmS	5.970	6.5	~0.10	10	~475 (6)
EuS	5.969	-	1.64	7.9(7)	600 (8)
CaS	5.683	-	2.5-3	-	730 (8)
YbS	5.677	150-200	~1.0	5-6	740 (8)
$Sm_{0.80}Eu_{0.20}S$	5.97	15	-	7 ?	500
$Sm_{0.80}Yb_{0.20}S$	5.906	7.5	-	3.5?	520
$Sm_{0.90}Ca_{0.10}S$	5.942	7.5	-	8 ?	501
$Sm_{0.80}Ca_{0.20}S$	5.907	12.4	-	13 ?	526
$Sm_{0.70}Ca_{0.30}S$	5.845	Continuous			551

The numbers in parenthesis are the references

The lattice parameter data reveal that the SmS lattice does not undergo the valence change when the substituent sulfide is a semiconductor with a large band gap, even though the size of the cation is favorable. This strongly suggests that the electronic structure plays a key role in the valence transition. From a comparison of our results on $Sm_{0.80}Eu_{0.20}S$, $Sm_{0.80}Yb_{0.20}S$ and $Sm_{0.80}Ca_{0.20}S$ compounds the following additional facts emerge:

With the addition of Eu, the transition pressure for the valence change moves to higher pressures. This must be because of the increase in the 4f-5d energy gap in the $Sm_{1-x}Eu_xS$ compounds with increasing Eu concentration. For Ca and Yb substitution a lowering of the pressure of transition may be expected, if the size of the substituted ion alone is important. This is not observed and in fact the pressure of transition increases with increasing Ca or Yb concentration, revealing thereby that the electronic band structure of the substituent sulfide plays a dominant role in promoting the valence change in SmS. The band gap of both YbS and CaS are an order of magnitude larger compared to SmS and hence more than compensate for the size effect which would tend to lower the transition pressure. The difference in the band gap between YbS and CaS reflects itself in the transition pressures of 7.5 and 12.5 kbar for the $Sm_{0.80}Yb_{0.20}S$ and $Sm_{0.80}Ca_{0.20}S$ respectively. This result suggests that larger is the band gap the higher should be the pressure of transition, provided the size effect is similar. In both Eu substituted and Yb substituted compound it would be of interest to find out if the Eu and Yb are still divalent when the valence transition in SmS has taken place. That there can be two f-levels in the Yb substituted SmS has been shown by Suryanarayanan et al. (15), which have been identified with $4f^6$ of Sm and $4f^{14}$ of Yb.

CONTINUOUS AND DISCONTINUOUS TRANSITION

Our resistivity studies on several Ca doped SmS compounds indicate that the transition remains first-order only at the Sm-rich end. Already, the sample containing 30 atomic percent Ca does not exhibit an abrupt decrease in resistance, characteristic of a first-order transition. This is also corroborated by the reflectivity changes that take place, when the samples are pressurized in a diamond anvil high pressure apparatus. In samples containing 30 atomic percent and higher amounts of Ca, the color progressively changes in the sequence, purple→bronze→copperish and finally to golden yellow, on increasing the pressure. The higher the concentration of Ca the higher is the required pressure to attain golden yellow color. In the sample containing 70 atomic percent Ca the metallic yellow color appears at about 200 kbar pressure. The change from first-order to a continuous transition has previously been noted in $SmS_{1-x}Se_x$ compounds by Bucher and

Maines (9) who have discussed their results in terms of the Falicov-Kimball (10) model. Since then several models have been proposed for the valence transition in Sm monochalcogenides and to explain the continuous and discontinuous nature (11). The strong electron-lattice coupling has been invoked, as also other decelerating mechanisms that can come into play (12). More recently Penney and Melcher (13) have proposed a model in which the bulk modulus has been brought in explicitly and shown to play an important role in the transition. Also in keeping with this Dernier, Weber and Longinotti (14) have found evidence for softening in lattice phonons which are intimately related to the bulk modulus. We believe that the change in the character of the transition from first-order to a continuous one in Ca-doped compounds and possibly for other dopings, is caused by the increase in the bulk modulus with Ca substitution. We believe that there are two factors which principally influence the transition. The first is the change in the 4f-5d gap of SmS with substitution and the second is the change in bulk modulus with substitution. The following conditions for a first-order transition can be stipulated. Firstly the 4f-5d energy gap (ΔE_g) should go to zero under the given constraint. Secondly the concentration dependent B(x), or the pressure dependent bulk modulus B(P) should not exceed a certain critical bulk modulus (B_c), when the first condition is satisfied. We can write these in the following simple form:

$$[\Delta E_g-(\Delta E_g/dX)X] \quad \text{or} \quad [\Delta E_g-(d\Delta E_g/dP)P] \rightarrow 0 \tag{1}$$

when

$$[B_c-B(X)] \quad \text{or} \quad [B_c-B(P)] \quad \text{is positive.} \tag{2}$$

When this condition is not satisfied only a continuous transition can occur. Increase either in the energy gap or in the bulk modulus will shift the transition to higher pressure.

In summary we find, (1) that the various metastability phenomena are related to the unusual features of the phase diagram, (2) that the anomalous thermal expansion in the high temperature region in mixed compounds is very likely to be due to the proximity of the critical point and, (3) that the electronic band structure plays a key role in the transition and (4) that the first-order transition is essentially controlled by the 4f-5d gap going to zero, before the bulk modulus reaches a critical value.

REFERENCES

(1) A. Jayaraman, P. D. Dernier and L. D. Longinotti (Review) in High Temperature-High Pressure 7, 1 (1975).

(2) L. J. Tao and F. Holtzberg, Phys. Rev. B 11, 3842 (1975).

(3) A. Jayaraman, P. D. Dernier and L. D. Longinotti, Phys. Rev. B 11, 2783 (1975).

(4) M. Campagna, G. K. Wertheim and E. Bucher "Structure and Bonding" under publication.

(5) E. Yu. Tonkov and I. L. Aptekar, Fiz. Tverd. Tela. 16, 1507 (1974) [Sov. Phys.-Solid State 16, 972 (1974)].

(6) T. Penney, R. L. Melcher, F. Holtzberg and G. Guntherodt, AIP Conf. Proc. 29, 392 (1975).

(7) P. Wachter, Sol. State Comm. 7, 693 (1969).

(8) A. Jayaraman, A. K. Singh, A. Chatterjee and S. Usha Devi, Phys. Rev. B 9, 2513 (1974).

(9) E. Bucher and R. G. Maines, Sol. State Comm. 11, 1441 (1972).

(10) L. M. Falicov and J. C. Kimball, Phys. Rev. Lett. 22, 997 (1969).

(11) C. M. Varma (Review) in Rev. Mod. Phys. 48, 219 (1976).

(12) L. L. Hirst, J. Phys. Chem. Solids 35, 1285 (1974).

(13) T. Penney and R. L. Melcher, Proc.Conf. on Metal-Nonmetal Transitions, Autrans, June 1976 (to be published in Journal de Physique).

(14) P. Dernier, W. Weber and L. D. Longinotti, Phys. Rev. 14, (1976).

(15) R. Suryanarayanan, J. A. Smirnov, G. Brun and S. G. Shulman. Proc. Conf. on Metal-Nonmetal Transitions, Autrans, June 1976 (to be published in Journal de Physique).

COMMENTS

TOURNIER: In the case of $Sm_{1-x}Eu_xS$ we have verified that Eu is always magnetic. I have a question about your SmS transition: were all your samples broken after the transition?

JAYARAMAN: Yes.

TOURNIER: The width of the transition probably reflects the fact that different parts of the sample have different critical pressures. Since you have shown that the critical pressure increases with the number of cycles, do you change the sample after each cycle?

JAYARAMAN: Yes. We can carry out only one meaningful experiment with one sample. You cannot trust the results on a previously cycled sample because of residual pressure effects and cracking of the sample. I would like to make another comment on these systems, particulary those containing additions of Eu or Yb. Are there two 4f levels in such systems; and, if there is, does the application of pressure, which collapses the Sm^{2+} level, also collapse the Eu or Yb^{2+} states?

ORBACH: Didn't Tournier just say that Eu always stays 2^{+}?

JAYARAMAN: Yes, he said it, but I don't know if he looked at the metallic phase.

A PHENOMENOLOGICAL MODEL OF THE PHASE TRANSITION IN SmS ALLOYS

Dieter W. Pohl

IBM Zurich Research Laboratory

(Säumerstrasse 4) 8803 Rüschlikon, Switzerland

ABSTRACT

A number of experimental properties of SmS alloys, e.g., transition pressure and lattice contraction *vs* dopant concentration or chemical collapsing, are described semi-quantitatively by means of a mean-field calculation. Our approach predicts a critical point in the pressure/concentration diagram near $X_{cr} = 27\%$ independent of the kind of dopant. We have strong experimental evidence that such a point exists in $Sm_{1-x}La_xS$ at about 2 kbar. In various other alloys the critical pressure gets negative and chemical collapsing occurs before the concentration reaches X_{cr}. Close to X_{cr} the La-dopant samples do not break when going through the transition which allows for more quantitative measurements than on other crystals of this system.

I. INTRODUCTION

A considerable amount of experimental information on SmS alloys and their valence transitions has been accumulated in the past three to four years.[1-6] The valence transition is connected with a strong compression of the lattice. The accompanying strain forces are of long range and for this reason, it can be well described by a mean-field theory although the underlying microscopic mechanism has not yet been understood completely. For instance, the phase diagram of pure SmS, SmSe, and SmTe has been successfully calculated in this way.[7-9]

II. MEAN-FIELD APPROACH

We shall now enlarge upon this approach in order to include the effect of alloying. In the derivation of the results, we shall follow, to a certain extent, the analysis of Alascio, Lopez, and Wio.[8]

Doping of SmS with $\underline{R}S$ influences the properties of the intrinsic material in a threefold way: lattice compression, band-gap shift, and conduction-band population (for trivalent cations). The stability of a system at a given pressure depends on the minima of the free enthalpy G rather than of the free energy. G is given by the well-known combination of internal energy U, entropy S, and volume V:

$$G = U - TS + pV . \qquad (1)$$

Both U and S depend on composition x in addition to the variables necessary to describe pure SmS, *viz.*, temperature T, pressure p, occupation number n or $\eta = n - 1/2$, and lattice compression $\Delta = -(R-R_0)/R_0$ (R_0 = cation-anion distance for x = 0, p = 0), i.e., pure $Sm^{2+}S$. The values of n, η and Δ can be derived from the minimum condition for G if explicit expressions for U and S are known, For this purpose, the following simplifying assumptions are made: a) U can be split into a lattice energy U_ℓ and an electronic energy U_e. U_ℓ may be a Born-Meier potential with n-dependent repulsive range.[8] It may also be expressed in terms of the bulk modulus and an n-dependent equilibrium volume,[9] or be a purely formal Taylor series expansion in Δ and n (or η). We shall use here the third approach. b) The lattice is considered to be rigid such that the distance R between neighboring ions is the same for both pairs of Sm-S and $\underline{R}$-S. U_ℓ is written as a linear combination of the lattice potentials of the pure SmS and $\underline{R}S$ both taken at the same value Δ:

$$U_\ell = (1-x)U_\ell^{Sm} + xU_\ell^{\underline{R}} \qquad (2)$$

with

$$U_\ell^{Sm} = f_{00} + f_{10}\,\alpha\, n + \frac{1}{2}(f_{02}\,\Delta^2 + 2f_{11}\alpha\, n\Delta + f_{20}\,\alpha^2 n^2). \qquad (3)$$

In Eq.(3), the terms f_{10}, f_{20} stand for the change of lattice energy when going from the di- to the tri-valent state. The change of energy is intimately related to the (relative) decrease α in equilibrium lattice constant; and this value has therefore been introduced as smallness parameter into Eq.(3). Neglecting variations in bulk modulus, we may estimate $f_{11} \simeq f_{02} \simeq f_{20}$. An analogous smallness parameter β can be defined for $\underline{R}S$ and replace αn in the expansion for $U_{\underline{R}}$ (α = 0.059; for β see Table I). c) Following Falicov and Kimball,[10] the electronic energy U_e consists of three parts:

$$U_e = \varepsilon_c N_c + (1-x)n\tilde{\Delta} - \tilde{G} N_c n. \tag{4}$$

$\varepsilon_c \simeq \tilde{B} N_c$ is the one-particle energy in the conduction band. $N_c = (1-x)n + x$ is the total number of conduction electrons. $\tilde{\Delta}$ is the gap between localized and band states,

$$\tilde{\Delta} = \tilde{\Delta}_0 + x\,(\partial\tilde{\Delta}/\partial x) - \gamma\, R_0 \Delta. \tag{5}$$

$\tilde{G}$ accounts for the Coulomb attraction between 4f holes and conduction electrons.

The expression for the entropy may be written as a sum of band (S_B), spin (S_S^{Sm}), and configuration (S_C^{Sm}) disorder as in Ref. 8. The disorder in the arrangement of Sm and $\underline{R}$ cations can be ignored since it is frozen-in and does not depend on η. It is also assumed that S does not depend directly on Δ. G can be expanded around its equilibrium value for $x_0 = \Delta_0 = n = 0$, i.e., pure SmS at ambient conditions. Differentiations with respect to η and Δ will then yield simple expressions for the equilibrium values of these quantities *vs* pressure and composition. (The temperature will be considered constant.) Before presenting results, it is convenient to introduce some illustrative abbreviations. Following Ref. 8, we define pressures

$$p_0 = [\gamma R_0 - \alpha f_{11}] / 6R_0^3, \tag{6}$$

$$p_1 = f_{02} / 6R_0^3. \tag{7}$$

Note that p_1 equals three times the bulk modulus of SmS ($\simeq$ 1400 kbar). Analogously, a "chemical" and an "electronic" pressure can be defined as

$$p_c = f_{11}\beta / 6R_0^3 \simeq p_1\beta \tag{8}$$

$$p_e = [(\partial\tilde{\Delta}/\partial x) + \tilde{B}]\, p_1/6p_0 R_0^3. \tag{9}$$

A few values for p_c are contained in Table I. p_e cannot easily be determined with the present incomplete knowledge on the band structure of rare-earth chalcogenides.

DOPANT	β	p_c [kbar]	x_{cc}(exp)	x_{cc}(calc) for p_0=151	90 kbar
Y	0.079	113	0.15	0.11	0.06
Gd	0.070	100	0.16	0.15	0.07
La	0.020	29	–	–	–
Nd	0.040	55	0.15	–	–
Yb	0.039	59	–	–	–

TABLE I. Smallness parameter β, chemical pressure p_c, experimental and calculated concentration for chemical collapsing x_{cc}.

The minimum condition $\partial G/\partial\Delta = 0$ yields the equilibrium values of $\Delta(\eta)$ in terms of these pressures:

$$\Delta = [\, xp_c + (1-x)(\eta + \tfrac{1}{2})\, p_0 + p\,] \,/p_1 . \tag{10}$$

Bearing in mind Eq.(10), the free enthalpy is written

$$G \simeq G_0(x) + (1-x)\ \{G_1\eta + G_2\ \eta^2 + G_4\ \eta^4\} \tag{11}$$

with the expansion coefficients

$$G_1 = F_1 - (6R_0^{\ 3}p_0/p_1)\ [x(p_c - p_e - p_0/2) + p + p_0/2] + 2xkT \tag{12}$$

$$F_1 = \tilde{\Delta}_0 + kT[\xi_1 + \ln\ (Q_M/2Q_I)] + \alpha f_{10} + \tilde{B} - \tilde{G} \tag{13}$$

$$G_2 = 3\alpha^2 R_0^{\ 3} p_1 + \xi_2(x)kT - (1-x)(3R_0^{\ 3} p_0^{\ 2}/p_1 + \tilde{G} - \tilde{B}). \tag{14}$$

The terms containing kT are contributions from the series expansion of the entropy; Q_M and Q_I are the spin multiplicities of the two Sm states; $\xi_1 = 1-\ln 2$ is a constant; $\xi_1 = (3+x)/(1+x)$. Expressions for G_3 and G_4 could formally be written down in a similar way; in view of the rough approximation made, however, such a procedure does not appear appropriate. It can be estimated that G_3 is small and may be ignored. G_4 shall be regarded as a phenomenological constant accounting for the existence of non-integral valences.

III. DISCUSSION

The type of transition - which should be either first-order or continuous according to experimental experience - depends on the sign of G_2. A first-order transition like that of pure SmS requires $G_2 < 0$. $G_2(x)$ changes sign at a critical concentration x_{cr}. The transition becomes continuous above this value. x_{cr} is independent of the type of material R, i.e., its "chemical" and "electronic" pressure.

Wio, Alascio, and Lopez[8] estimate $G_2(0) = -0.095$ eV on the basis of the transition pressure for pure SmS. With the known values of α, R_0, and p_1 we find $x_{cr} = 27\%$.

A careful survey of the literature data reveals that indeed no first-order transition has ever been found above $x = 0.3$. Jayaraman *et al.*[5] recently investigated the temperature-induced transition in $Sm_{(1-x)}Gd_xS$ and also came to the conclusion that a critical point might exist for this alloy close to 27%.

Closely related to the existence of a critical point is the "softening" of the first-order transition with increasing x. The lattice contraction $\delta\Delta$ at the transition may serve as a measure of the softening. Equation (10) yields

$$\delta\Delta/(1-x) = (p_0/p_1)\delta\eta(0)\sqrt{1-x/x_{cr}}. \tag{10a}$$

$\delta\eta(0)$ is the valence change at the transition in pure SmS (0.7). Equation (10a) is to be compared with the experimental lattice contraction plotted in Fig. 1. The calculated dashed and solid curves correspond to $p_0 = 151$ kbar as proposed in Ref. 8 and a least-square fit to the experimental points,[4,5] respectively. The fit yields the smaller value of $p_0 = 90$ kbar and $x_{cr} = 0.29$. The deviations of the calculated curves from the experimental points are considerable, but the decay towards x_{cr} is quite obvious.

With regard to the transition pressure $p_t(x)$ we shall ignore the experimental hysteresis. Being strongly influenced by G_4 it cannot be treated adequately within the framework of the present model. The limit of stability is then given by the condition that the minima of G be at equal height requiring (since we neglect third-order terms) $G_1(x,p) = 0$. Equation (12) shows that the transition pressure will decrease if $p_c > p_0/2 + p_e$; otherwise it will increase. The experimental transition-pressure curves shown in Fig. 2, indicate that this condition is satisfied for R = Y,Gd,[5] and La (see below). It is not satisfied for R=Yb[15] where is large and positive.

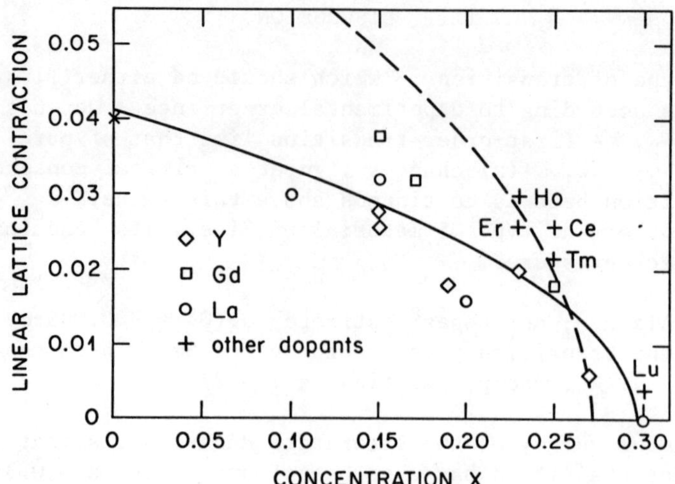

Fig. 1. Experimental and calculated relative linear lattice contractions at the phase transition. Except for the La alloys, the experimental data are taken from Refs. 4 and 5. The dashed curve is calculated from the data of Ref. 8; the solid curve is a best fit to the Y, Gd, and La points.

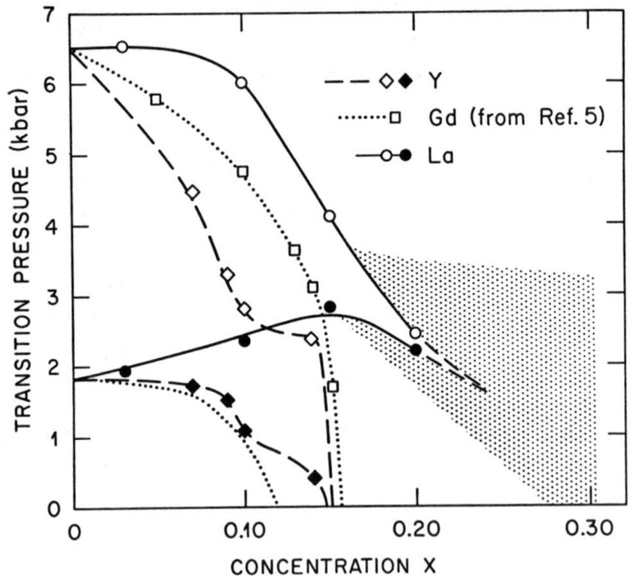

Fig. 2. Experimental transition pressures of $Sm_{(1-x)}\underline{R}_xS$ with $\underline{R}$ = Y, Gd (Ref.5) and La. The shaded area indicates the range of continuous transition. The curves are guidelines to the eye only.

For Y and Gd doping (and many other rare earths[5]) the effect of chemical pressure is so strong that the transition pressure p_t becomes zero far below x_{cr}. The condition for "chemical collapsing" is

$$x_{cc} = p_t(0)/(p_c - p_e - p_0/2). \tag{15}$$

For a crude estimate we assume $p_e \ll p_c$ (Table I). The quantitative agreement of the first value with the experimental data for Gd and the steep decay for Y at the calculated x_{cc} may be accidental but shows the qualitatively correct approach of Eq.(15). The recently observed softening of the bulk modulus in $Sm_{1-x}Y_xS$ near 30% is another evidence of the proximity of the critical point.[11]

IV. EVIDENCE FOR A CRITICAL POINT IN $Sm_{1-x}La_xS$

The lattice constant of LaS (5.85 Å) lies in between that of divalent (5.97 Å) semiconducting and trivalent (5.62 Å) metallic SmS.[3,12] Chemical collapsing does not take place in $Sm_{1-x}La_xS$[3] and a critical point with $p_{cr} > 0$ can therefore be expected. For this reason, we investigated transition pressure, reflectivity, and lattice compression as a function of composition and hydrostatic pressure.[13,14]

The phase transition was readily visible using a microscope focused onto the sample in the pressure cell. The well-known sudden change in color at the transition was observed for all our La-doped samples (x = 3,10,15,20%), except for $Sm_{0.7}La_{0.3}S$ which has a continuous transition.

The transition pressure of $Sm_{(1-x)}La_xS$ differs from the ones for Y and Gd (Ref. 5 and measurements of our own) in an interesting way (Fig. 2): the lower p_t *vs* x curve (back transformation) seems first to increase ($x < 0.1$) while the upper one remains constant. The latter then strongly decreases and the hysteresis gets very small. We can follow the transition up to $x = 0.2$, where the transformation pressures differ by no more than 0.2 kbar. No first-order transition seems to exist in the sample with the next-higher concentration, $x = 0.3$, at room temperature. This sample changes its properties continuously over a fairly large range of pressures. The strong response is typical for the proximity of a phase transition.

The lattice contraction as derived from our measurements has been plotted in Fig. 3(a) *vs* pressure. The linear contraction *at*

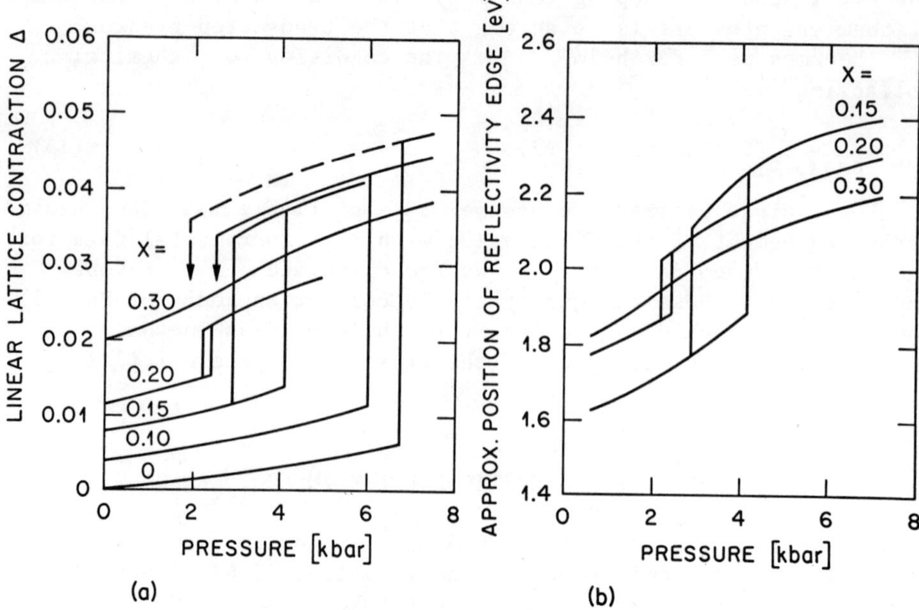

Fig. 3. Experimental lattice contraction (a) and position of reflectivity (b) for $Sm_{(1-x)}La_xS$.

the transition $\delta\Delta$ and its decay towards x_{cr} goes well with the previously shown results from other alloys (Fig. 1).

The reflectivities of different La-alloyed samples with the same x were found to vary considerably, which might be a result of variations in the surface preparation and crystal quality. The variation of the position of the reflectivity edge, however, could be obtained quite reliably. It gives some insight into the change of the Sm valence with pressure. Figure 4, shows the reflectivity edges for $x = 0.15$, 0.20 and 0.30. Crystals with 20% and 30% La *do not break* even after a large number of pressure cycles; those with $x = 0.15$ tend to break but the fracture pattern remains fairly stable after a few pressure cycles. In these samples, the variation of reflectivity *vs* pressure can be determined with good reproducibility. The reflection edge is located almost in the visible spectrum for the higher concentrations since each La ion gives one electron to the conduction band. The reflectivity edge shifts considerably already below the transformation pressure and continues to shift above it. This behavior is

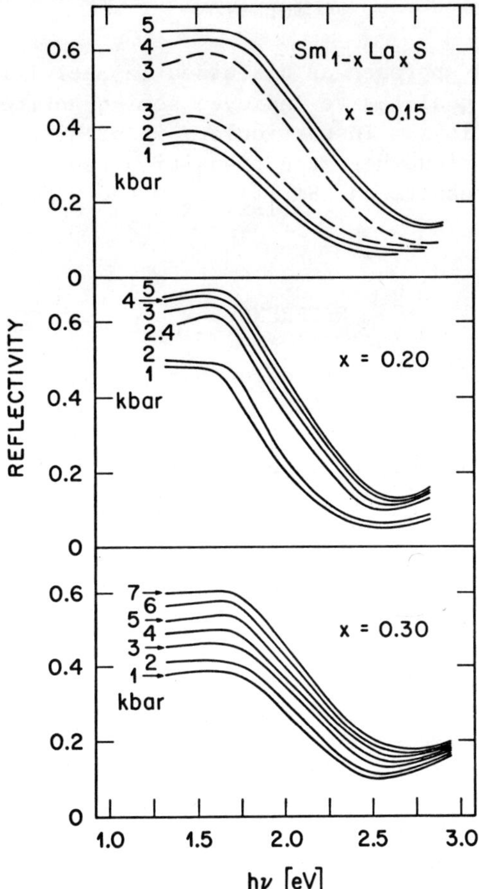

Fig. 4. Reflectivity curves for various pressures. The dashed curves, taken at the same pressure, demonstrate the effect of hysteresis.

typical for the softening of the first-order transition with increasing x.

The change in character of the transition is summarized in Fig. 3(b), where the frequency of a characteristic value of reflectivity (35%) close to the reflectivity edge has been plotted. The decreasing hysteresis and the strong pre- and post-transitional variations of reflectivity are clearly visible.

V. CONCLUSION

The mean-field approach as discussed in Sec. III, seems to be well suited for a qualitative and even semi-quantitative description of $Sm_{(1-x)}R_xS$ alloys. In particular, it predicts a "universal" critical concentration where the transition becomes continuous. The experimental results on $Sm_{(1-x)}La_xS$ and other alloys support this finding.

REFERENCES

[1]Fairly complete lists of references may be found in Refs.5 and 12 and in the review article of C.M. Varma, Rev. Mod. Phys. 48, 219 (1976).
[2]A. Jayaraman, E. Bucher, P.D. Dernier, and L.D. Longinotti, Phys. Rev. Lett. 31, 700 (1973).
[3]F. Holtzberg, *AIP Conference Proceedings* 18, 478 (1973).
[4]L.J. Tao and F. Holtzberg, Phys. Rev. 11, 3842 (1975); G. Güntherodt and F. Holtzberg, Solid State Commun. 18, 181 (1976).
[5]A. Jayaraman, P. Dernier, and L.D. Longinotti, Phys. Rev. B 11, 2783 (1975).
[6]T. Penney and F. Holtzberg, Phys. Rev. Lett. 34, 322.
[7]R. Ramirez, L.M. Falicov, and J.C. Kimball, Phys. Rev. B2, 3383 (1970); C.E.J. Conçalves da Silva and L.M. Falicov, J. Phys. C5, 906 (1972), and Solid State Commun. 17, 1521 (1975).
[8]B. Alascio and A. Lopez, Solid State Commun. 14, 321 (1974); H.S. Wio, B. Alascio and A. Lopez, Solid State Commun. 15, 1933 (1974).
[9]C.M. Varma and V. Heine, Phys. Rev. B11, 4763 (1975).
[10]L.M. Falicov and J.C. Kimball, Phys. Rev. Lett. 22, 997 (1969).
[11]T. Penney, R.L. Melcher, F. Holtzberg, and G. Güntherodt, *21st Conference on Magnetism and Magnetic Materials, Philadelphia* (1975), *AIP Conference Proceedings* (to be published).
[12]B. Batlogg, E. Kaldis, A. Schlegel, and P. Wachter, Phys. Rev. (to be published 1976).
[13]D.W. Pohl, to be published.
[14]The author is indebted to F. Holzberg for making the crystals available to him.
[15]M.Gronau and S.Methfessel,ACM Conf. Proc.,Amsterdam,1976 (to be published).
[16]M.Gronau, private communication.

Note added in proof:
After finishing the manuscript,the author got informed about a recent investigation of M.Gronau,Physics Department, University of Bochum,Germany on Pr-,Gd-,Tb-,Dy-,and Ho-doped SmS. The experimen-

tal data on the lattice contraction at the transition[15] are fairly large for Tb and Dy at about 22% concentration but on the whole fit well with the data of Fig. 1.

COMMENTS

WALSH: The work of Bucher and Maines on the Sm-S and Sm-Se series where they observed the transition going from discontinuous to continuous as a function of anion would seem to be relevant to your experiments. It would be interesting to consider these results in a theory of your kind.

POHL: I think one cannot straightforwardly extend these ideas to the case of anion sustitution (in SmS). The reason is that the compression of the cation results from the push of the anion without a change in the anion-cation separation. I think the situation is similar in the arsenides.

SHERRINGTON: There are two ways of viewing the Sm ion when one calculates energies due to it: (1) as having an average size lying between 2^+ and 3^+ à la Alascio and Lopez or (2) as having two different sizes as à la Anderson and Chui. Is it correct that you take a sort of average size model.

POHL: You're absolutely right. Yes.

SHERRINGTON: I would like to make a comment which is really related perhaps more to Lawrence's talk and that is to see which of the two pictures applies. Some time ago Riseborough and I tried to look at the model microscopically and found that the relevant criterion was the comparison of a fluctuation time (or fluctuation width if you like the hybridization thing) with the phonon frequencies. If one is on one side of Lawrence's criterion you would expect the Anderson and Chui picture to apply, whereas on the other side you would expect the Alascio and Lopez picture to hold. I think that the real world has a bit of both.

POHL: Our model has enough tolerance to accept both, because we have two parameters in it. One is the direct lattice compression; the other is the smallness parameter associated with the population of the upper band so you may say that if one of these parameters changes the other compensates for this change.

JEFFERSON: Don't you find with your model that the intermediate valence phase is thermodynamically unstable unless $\tilde{G}$ is very large? What arrests the valence transition. What stops it from going all the way to the pure valence phase of Sm^{3+}?

POHL: All that we have considered here is the expansion up to second order; and, of course, you cannot expect intermediate valences to be obtained at a first order transition process. I think it would be very interesting if one could go from 20% to 30% in smaller steps. Of course, it's a matter of materials preparation.

VALENCE INSTABILITIES IN Sm COMPOUNDS

T. Penney*

Centre de Recherches sur les Tres Basses Temperatures

C.N.R.S., BP 166 X, 38042 Grenoble-Cedex, FRANCE

SmS collapses to an intermediate valence state upon alloying with YS or SmAs in a manner somewhat similar to that of SmS under pressure. We shall first outline some of the conclusions reached concerning intermediate valence and the valence transition. Finally, we shall mention some local magnetic and electronic effects observed at small As concentrations in the SmS-SmAs system.

The current state of research and understanding in the field of valence instabilities is well described by the proceedings of this conference, the proceedings of the Autrans conference on metal-nonmetal transitions and a recent review article by Varma[1]. Extensive references may be found in these places. For this reason we will not here review or make reference to all of the work on Sm compounds.

A homogeneous density of states model for alloys of SmS with YS and SmAs has been described by von Molnar, Penney and Holtzberg[2]. It was suggested that the two six electron Sm configurations, f^6 and f^5d, can be approximated, in a quasi one electron picture, by a relatively narrow f level and a broader d band. The bands are not rigid, but move relative to each other with changes in composition pressure and temperature. In the integral valent Sm^{2+} or Sm^{3+} states the Fermi level, E_F, lies wholly above or wholly below the f level. In the intermediate valent state, E_F cuts both the f and d states. The quantum mechanical ground state is not described but it is assumed that all Sm are equivalent and in the same state. In the integral valent state, even though the f and d states may be degenerate, the Fermi level does not cut the f states. The physical properties are similar to those observed in other integral valence rare earth compounds.

On the other hand in the intermediate valence state, the Fermi level cuts both the d and the f states. In this case the conductivity of the d band is reduced due to scattering to the f states and the Hall effect is no longer interpretable in a one band model.[3] The linear term, γ, in the low temperature specific heat of the SmS-YS system is an order of magnitude larger in the case of intermediate valence than for integral valence, indicating a much larger density of states at the fermi level.[4] In this picture the intermediate valence compounds will have a soft bulk modulus, as observed[5], if the f and d states move relative to one another under volume strain. In this case since E_F cuts both f and d states the electrons can move from one to the other. This electronic redistribution makes it easier to strain the crystal.

The bulk modulus (almost an order of magnitude smaller than SmS) is a good example of the strong interaction between the electronic configuration and the lattice. The other example is, of course, the phase transition which occurs in SmS with pressure or doping. At this phase transition there is a change in color from black to gold, a Sm valence change from divalent to intermediate valence and a 10% lattice collapse. The reason for the strong coupling between electronic and lattice properties appears to be simply that $Sm^{3+}S$ (like $Gd^{3+}S$) is very much smaller than $Sm^{2+}S$. The rare earths are very well known as a transition series with strong ionic behavior, because the f electrons are well shielded from their environment by the valence electrons. A related effect is that if an electron is removed, the valence electrons collapse since they are screened from the nucleus by one less f electron. This change in ionic size with valence is part of the electron lattice coupling and contributes to both the phase transition and the soft bulk modulus. In addition the electron removed from the f level will couple to the lattice through metallic bonding and the crystal field interaction.

In order to examine the phase transition and the soft bulk modulus in a semiquantitative way, a simple theory treating electronic configuration and lattice collapse, based on the density of states model, has been worked out[6]. One assumes first that the f^6 and f^5d configurations are nearly degenerate, due to the crystal field splitting and band width of the d electron. Since this effect is volume dependent the second assumption, that the f^5d energy decreases with respect to the f^6 with volume decrease, immediately follows. When calculating the elastic energy the change in the Sm ionic volume must be considered and also the fact that the bulk moduli of YS and integral valent $Sm^{3+}S$ are observed to be about twice as large as for divalent $Sm^{2+}S$.

The theory gives a reasonably good fit to the collapse of SmS with substitution of YS or SmAs and shows the similarity of these compositional changes to the pressure collapse of SmS. It also shows how a soft bulk modulus, results from intermediate valence. The intermediate valence state is stable because the bulk modulus of integral valent $Sm^{3+}S$ and YS are about twice that for $Sm^{2+}S$. The resulting large anharmonic term in the lattice energy resists conversion to pure Sm^{3+}. It is amusing to note that it is by taking account of the large $Sm^{3+}S$ bulk modulus that the intermediate valence is stablized and a small bulk modulus produced.

In the SmS-YS system the observed decrease of volume with YS concentration, x, is slightly faster than the theory for the black region ($0 < x < 15\%$) before the phase transition. This difference can be rationalized because the theory is crude and for zero temperature only. Whereas the theory predicts pure Sm^{2+} up to the phase transition, we expect some thermal occupation or hybridization with the f^5d state as it approaches the f^6, just before the transition. However, perhaps it is not so simple, for the agreement is good in the case of black SmS under pressure ($p < 4.5$ kbar). In the case of the SmS-SmAs system, in the black region ($0 < x < 5\%$) before the phase transition, the agreement between theory and experiment is very bad. With increasing x, the lattice constant decreases much faster than the theory. The problem is that this theory is homogeneous, treating all Sm equivalently, whereas the Sm neighbors of Y or As may be very different from Sm with all sulfur neighbors.[7]

A difference between the effect of pressure and Y on the black phase of SmS has been observed in the Mössbauer isomer shift[8], which is a measure of the valence. In the case of Y substitutions, the isomer shift (valence) has moved 10-15% of the way to the Sm^{3+} value at one half the transition concentration,while in the case of pressure there is no change in isomer shift at one half the critical pressure, indicating that all Sm remains divalent. For these small Y concentrations it is not possible to distinguish between intermediate valence and a mixture of Sm^{2+} and Sm^{3+}, as it is at larger concentrations.

In the case of the SmS-SmAs system there are certainly local, non-homogeneous effects at small concentrations. A detailed description of the effects of As on the lattice constant, magnetization, and resistivity are given elsewhere in these proceedings[9]. It is shown that for $x < 5\%$ each As causes its 6 nearest neighbor Sm to collapse to the 3+ state. Such strong local environmental behavior is far removed from the homogeneous approaches that have been used to describe the intermediate valence state. It appears that in the case of Sm compounds, pressure and composition may produce similar effects at large concentrations but may be completely different at small concentrations. From this point of view SmS with pressure may appear to be a cleaner system but it should be noted that SmS often contains Sm^{3+} impurities[10] due to deviations in the stoichiometry and other defects. A final question is whether at large As concentrations ($10\% < x < 50\%$) there are remanents of the strong inhomogeneous behavior observed at low concentration (x < 5%) or whether there is a uniform intermediate valence. In the case of the SmYS system the isomer shift measurements clearly show intermediate valence at large Y concentrations but would not detect either Sm^{2+} or Sm^{3+} up to about ten percent[8].

REFERENCES

1) C. M. Varma, Rev. Mod. Phys. 48, 219 (1976).

2) S. von Molnar, T. Penney and F. Holtzberg, J. Physique Coll. 37, C4- (1976). Proceedings of the Autrans conference on the Metal-Non Metal Transitions.

3) T. Penney and F. Holtzberg, Phys. Rev. Letts. 34, 322 (1975).

4) S. von Molnar and F. Holtzberg, AIP Conf. Proc. 29, 394 (1976).

5) R. L. Melcher, G. Güntherodt, T. Penney, and F. Holtzberg, 1975 IEEE Ultrasonics Symposium Proc. IEEE Cat. #75 CHO 994-4SA, pg. 16. T. Penney, R. L. Melcher, F. Holtzberg and G. Güntherodt, AIP Conf. Proc. 29, 392 (1976).

6) T. Penney and R. L. Melcher, J. Physique Coll. 37, C4- (1976).

7) T. Kasuya, J. Physique Coll. 37, C4- (1976).

8) J. M. D. Coey, S. K. Ghatak, F. Holtzberg, AIP Conf. Proc. 24, 38 (1975) and J. M. D. Coey, S. K. Ghatak, M. Avignon and F. Holtzberg, Phys. Rev. (1976) in press.

9) F. Holtzberg, O. Pena, T. Penney and R. Tournier, this conference.

10) G. Chouteau, F. Holtzberg, O. Pena, T. Penney, R. Tournier, S. von Molnar, J. Phys. Coll. 37, C4-001 (1976) and A. Chenevas-Paule, P. Haen, F. Lapierre, R. Tournier and L. Vieux-Rochas, this conference.

* IBM Watson Research Center, Yorktown Heights, New York, 10598 U.S.A.

PHYSICAL PROPERTIES AND BEHAVIOR OF ALLOTROPICALLY PURE α-, β- AND γ-Ce

K. A. Gschneidner, Jr.

Ames Laboratory-ERDA and Department of Materials Science and Engineering
Iowa State University, Ames, Iowa 50011

INTRODUCTION

Cerium was the first material found to undergo a valency change. X-ray studies in 1949 – 1950 revealed that when Ce was compressed to modest pressures (> 8 kbar at 300 K)[1] or cooled to low temperature (< 100 K at 1 atm)[2] a large volume contraction occurred (> 12%) while the crystal structure remained the same. When these authors discussed this unusual behavior with their colleagues, both Zachariasen[3] and Pauling[4] independently suggested that this volume contraction corresponded to a valence change of 3 for γ-Ce to 4 for α-Ce, i.e. the 4*f* electron of γ-Ce was promoted to the valence band, leaving the 4*f* level of α-Ce empty. Several other authors subsequently reanalyzed the physical properties of the phases involved in the valence change and they concluded that about 0.5 to 0.67 of an electron per atom is transferred from the 4*f* level to the valence band.[5] Today the consensus of scientists favors the 0.67 value. Furthermore, Gschneidner and Smoluchowski[5] point out that the valences of the two phases involved (γ- and α-Ce) change with temperature and pressure. As a result of this valence change, the γ-α pressure-temperature phase boundary ends at a critical point. Ce is the only solid known to have a critical point.

More recent investigations have found that in addition to the large volume contraction associated with the $\gamma \rightarrow \alpha$ transformation, Ce, under various conditions of temperature and pressure, is an antiferromagnet,[6,7] a superconductor[8] and the only pure element to exhibit Kondo scattering.[9-11] All of the phenomena occur because the 4_f_ level lies near or at the Fermi level. These behaviors make Ce in its elemental form the most fascinating member of the periodic table; and the knowledge and understanding of this element will bring much insight concerning valence instabilities and the electronic nature of other materials.

The major emphasis on our work on cerium in the past 6 or 7 years has been to: (1) prepare well characterized samples with respect to the presence of other elements, both interstitial and metallic, and phase purity; and (2) to make reliable physical property measurements on these materials. The results of these efforts are summarized below.

EXPERIMENTAL RESULTS

Preparation of Pure Allotropes

Three different Ce stocks which have been used throughout this study were prepared at the Ames Laboratory-ERDA by Ca reduction of CeF_3.[12] The chemical analyses showed these three materials are the highest purity Ce ever prepared (> 99.8 at.% pure with respect to 82 elements, including the interstitials).[11] Although to date there has been no evidence for the influence of impurities (< 0.1 at.%) on the properties of the various Ce allotropes, this author believes that it is quite likely that impurities may have pronounced effects, especially on α-Ce. For this reason it is urged that future work be done on materials which have been well characterized both chemically (impurity content) and with respect to phase purity.

Although the existence of α-, β- and γ-Ce was well established by the mid 1940's it was not until the early 1970's that allotropically pure α- and β-Ce were prepared for studies at low temperatures and atmospheric pressure. The preparation of α-Ce involves a high pressure – cooling – decompression cycle,[13-15] while that of β-Ce involves the cooling down to 4.2 K, warming to 298 K ten times at 1 atm followed by a week's anneal at 348 K, and then repeating this procedure three more times.[15]

Properties of α-Cerium

Low temperature heat capacity (1 - 20 K) and magnetic susceptibility (1 - 150 K) have been measured by Koskimaki and Gschneidner[16] on an allotropically pure α-Ce sample. The C/T vs T^2 plot was linear from the lowest temperature (1.6 K) to about 5 K and the resultant γ and θ_D values are 12.8 mJ/g-at. K^2 and 179 K, respectively. The electric specific heat constant, γ, for α-Ce is the highest known value for pure metals (excluding values reported for magnetically ordering metals and Pu). It is 20 to 100% higher than the values of closely related rare earth metals Sc, Y, α-La and Lu,[17-19] and ~ 3 times larger than that of Th.[20] This high value is consistent with a 4f level at the Fermi surface. High pressure heat capacity measurements of α-Ce by Phillips et al.[21] yielded a γ of 11.3 mJ/g-at. K^2 and a θ_D of 200 K at 10 kbar. These data suggest that the density of states at the Fermi surface decreases and the lattice stiffens with increasing pressure.

The magnetic susceptibility of α-Ce is shown in Fig. 1 and is unusual in that there is a rapid increase in the susceptibility below 20 K. This rise could not be accounted for by the presence of magnetic impurities in the Ce sample. The other results shown in Fig. 1 were taken on less reliably characterized cerium samples — that of Grimberg et al.[22] contained some β-Ce, while both the MacPherson et al.[23] and Grimberg et al. samples showed a field dependence indicating the presence of magnetic impurity atoms. In addition to the unique temperature dependence, the magnetic susceptibility is 4.5 times the Pauli contribution calculated from the electronic specific heat constant. This large discrepancy suggests that α-Ce is an exchange enhanced paramagnet.[23] The

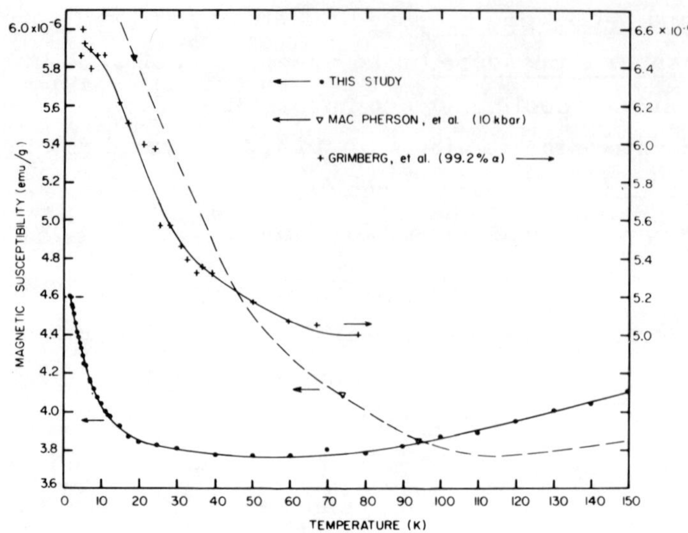

Fig. 1. The magnetic susceptibility of α-Ce after Koskimaki and Gschneidner.[16] Note the shift in scale for the results of Grimberg et al.

third unusual feature of the α-Ce susceptibility is the slow rise with temperature above 60 K.

The resistivity of α-Ce has been measured by a number of workers[14,22,24,25] who used various pressure methods to prepare their samples. There is some disagreement in results, mainly whether or not a T^2 dependence exists at low temperatures, which is expected for metals with large exchange enhancements. Katzman and Mydosh[24] found such a dependence but none of the other workers have, except, as might be expected, in the cases where β-Ce or impurities were present. They also found the T^2 dependence to decrease with increasing pressure, while Nicolas-Francillon and Jerome[14] found no change in resistivity up to 8.5 kbar. A second effect expected for spin fluctuation resistivity is a negative departure from a linear T law at higher temperatures, but the results of Brodsky and Friddle[25] are linear to 298 K for α-Ce at 18 kbar. Thus the departure from linearity, if it exists, occurs above 300 K, which does not necessarily preclude the existence of spin fluctuations in α-Ce.

Properties of β-Cerium

Studies on dhcp β-Ce have been quite sparce primarily because of the difficulty in preparing an allotropically pure sample. In the past few years great strides have been made since Koskimaki et al.[15] developed a technique for making such samples. Recent studies include heat capacity (1 - 20 K) measurements,[26] electrical and magnetic susceptibility studies,[9,11] elastic and inelastic neutron scattering,[11,26] and magnetoresistance[27] measurements.

One of the problems with β-Ce is that if slowly cooled below 50 K it will partially transform to α-Ce. If, however, one quenches β-Ce to 4.2 K it will not transform to α-Ce. On warming β-Ce will begin to transform to α-Ce at 15 K, but again rapid heating to above 50 K enables one to prevent any transformation from occurring. This is illustrated in Fig. 2, where it is seen that β-Ce begins to transform to α-Ce at point A upon cooling and at point B upon

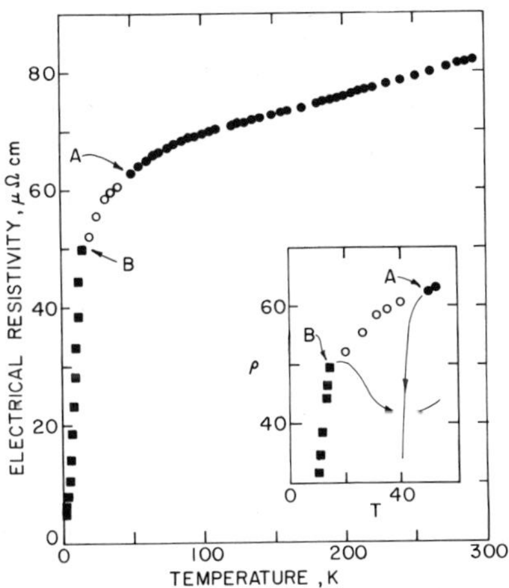

Fig. 2. The electrical resistivity for β-Ce, after Burgardt et al.[11] The solid circles and squares are data points which were obtained from samples at thermal equilibrium. The open circles were measured by recording the resistivity while warming the samples in 5 K steps every 15 to 30 seconds.

warming. Burgardt et al.[11] also showed that the $\beta \rightarrow \alpha$ transformation takes place at the surface and does not occur in the sample below a depth of 0.3 mm. Furthermore, only ~10% of the β-Ce in this outer 0.3 mm portion of the sample transforms to α-Ce.

The low temperature heat capacity of β-Ce[26] showed a large magnetic contribution to the specific heat due to the antiferromagnetic ordering of β-Ce. Indeed, for the first time these data showed that the magnetic ordering occurs at 12.45 and 13.7 K for the two types of site symmetry (the upper temperature was thought to be due to ordering on the hexagonal sites, and the lower one corresponds with cubic site ordering). The lattice and electronic contributions were assumed to be the same as for α-La (dhcp), and the Schottky contribution was calculated from the energy levels of the crystal fields, which were determined from neutron scattering results. The magnetic contribution was obtained when the first three contributions were subtracted from the measured heat capacity. The calculated magnetic entropy was within 2% of the theoretical R ln 2 value expected for a doublet ground state. The magnetic heat capacity below 1.33 K was found to exhibit an exponential behavior, yielding a spin wave energy gap of 1.41 K. Between 3.5 and 6.5 K the magnetic heat capacity exhibited a T^3 behavior. This observation of two different temperature behaviors above and below the spin wave gap temperature is consistent with Cooper's theory[28] and is the first time it has been found in a rare earth metal.

The temperature dependence of the electrical resistivity of β-Ce is unique among the pure elements – the resistivity is anomalously large above 60 K and exhibits an order of magnitude drop below 20 K, Fig. 2.[9,11] The magnetic susceptibility shows a Curie-Weiss behavior characteristic of a $4\underline{f}^1$ configuration down to ~20 K, just a few degrees above its Néel temperature.[9,11] In order to determine the anomalous part of the resistivity of β-Ce, the phonon contribution was assumed to be the same as that of α-La (both

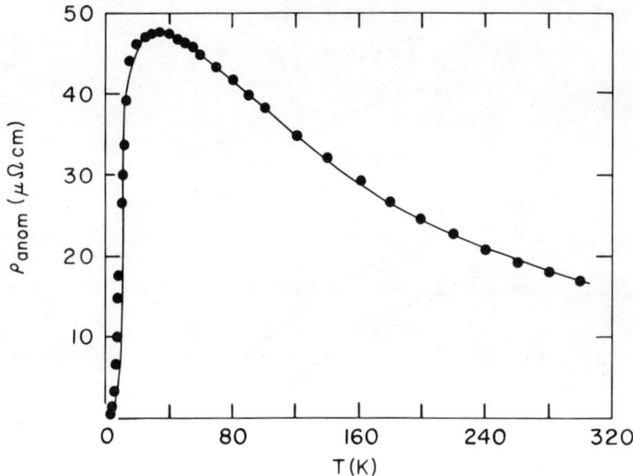

Fig. 3. The anomalous resistivity of β-Ce (solid points) and theoretical resistivity (solid line) after Liu et al.[10]

are dhcp). The resultant ρ_{anon} is shown in Fig. 3. The authors[9-11] showed that this curve could not be explained by existing theories including those which take into account crystal fields. They suggest that the rise in ρ_{anon} above 40 K is due to Kondo scattering, which at low temperature is quenched by the internal field associated with magnetic ordering. A more quantitative model was developed by Liu et al.[10] This model accounts for the temperature dependence of the resistivity between 20 and 80 K by using a mean-field calculation of the internal field in an antiferromagnet, and for T < 20 K by using a two-spin cluster model. Above 80 K the resistivity follows the well known formula for Kondo scattering $\rho(T) = \rho_o[1 - a \ln T/T_F]$. The results of fitting this model to the experimental data by normalizing the theory to experiment at 50 K are shown in Fig. 3, where the agreement over the whole temperature range is excellent.

Data from magnetoresistivity measurements on β-Ce are shown in Fig. 4 where the normal negative slope is seen for a field suppression of Kondo scattering. The field dependence of the ρ_{spin} contribution was found to follow a $-AH^n$ law where A and n are constants and H is the applied field in kOe. The value of n was found to be constant (1.61) for all temperatures above the Néel

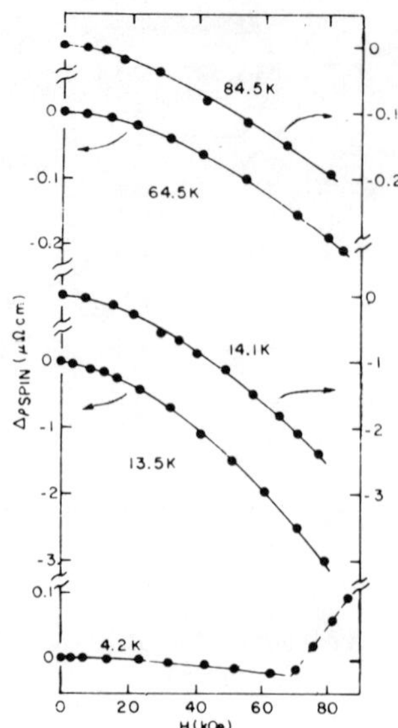

Fig. 4. The spin resistivity of β-Ce as a function of field at various temperatures. The solid points are experimental values and the solid lines represent the fits of the data, after Burgardt et al.[28]

temperature, while A increases with decreasing temperature from 1.4×10^{-4} at 64 K to 32.2×10^{-4} at 13.5 K.

The magnetic susceptibility of β-Ce near the Néel temperature was found to vary with the applied magnetic field below 1 kOe.[11] High field measurements (> 10 kOe) show that β-Ce follows the Curie-Weiss law, with an effective moment of 2.61 μ_B and a paramagnetic Curie temperature of -41 K.[11] The sensitivity of the susceptibility to the applied field is similar to, but about an order of magnitude smaller than, that observed in α-Nd (dhcp) and α-Sm (rhombohedral). We expect that the magnetic behavior of β-Ce is complex, like that observed in Nd and Sm, but this cannot be confirmed until single crystals of β-Ce are available for study.

Properties of γ-Cerium

Little is known about γ-Ce below room temperature because it begins to transform to β-Ce at ~ 273 K, and whatever γ-Ce that remains at ~ 110 K will transform to α-Ce. High temperature magnetic susceptibility data (> 298 K) show that γ-Ce obeys a Curie-Weiss law, with an effective moment of ~ 2.45 μ_B and a paramagnetic Curie temperature which lies between -9 and -50 K.[29,30] The room temperature (300 K) electrical resistivity of γ-Ce is 74.4 μΩ cm, approximately 8 to 15 μΩ cm smaller than that of β-Ce. Upon cooling it appears that the temperature dependence of the electrical resistivity is similar to that of β-Ce, which suggests that γ-Ce also exhibits Kondo scattering which is destroyed when γ-Ce transforms to α-Ce.[11] An analysis of the room temperature heat capacity[31] yields an electronic specific heat constant of 7.5 mJ/g-at. K^2, which is considerably lower than that of α- or β-Ce. This suggests that the electronic density of states at the Fermi surface of γ-Ce is significantly less than that of β-Ce (~20%) and considerably less than that of α-Ce (~ 40%).

CONCLUSIONS

In this paper I have tried to summarize most of the reliable information available on the physical properties of the three cerium allotropes. But no attempt has been made to support or disprove any of the current theories concerning the nature of α- or γ-Ce and the α-γ transformation. A large part of these theories have been based on unreliable physical property data (which were the only data available at the time), but now with the availability of good data on allotropically pure cerium phases, these theories should be re-examined.

Still, much experimental work needs to be done, especially on well characterized, high chemical purity and allotropically pure samples. For example, electrical resistivity and transport properties need to be remeasured on α-Ce; the influence of small concentrations on the physical properties, especially the susceptibility and resistivity, of α-Ce should be started; and the preparation of single crystal α- and β-Ce needs to be accomplished. The later would open many doors for further investigations which are needed to fully understand Ce– the keystone element of the periodic table.

REFERENCES

1. A. W. Lawson and T.-Y. Tang, Phys. Rev. 76, 301 (1949).
2. A. F. Schuck and J. H. Sturdivant, J. Chem. Phys. 18 145 (1950).
3. W. H. Zachariasen, unpublished information quoted by Ref. 1.
4. L. Pauling, unpublished information quoted by Ref. 2.
5. K. A. Gschneidner, Jr. and R. Smoluchowski, J. Less-Common Metals 5, 374 (1963); and see citations therein to earlier papers.
6. J. M. Lock, Proc. Phys. Soc. 70B, 566 (1957).
7. M. K. Wilkinson, H. R. Child, C. J. McHargue, W. C. Koehler and E. O. Wollan, Phys. Rev. 122, 1409 (1961).
8. J. Wittig, Phys. Rev. Lett. 21, 1250 (1968); and C. Probst and J. Wittig p. 453 in Proc. 14th Intern. Conf. Low Temp. Phys., Vol. 5, M. Krusius and M. Vuorio, eds. North Holland/American Elsevier, Amsterdam/New York (1975).
9. K. A. Gschneidner, Jr., P. Burgardt, S. Legvold, J. O. Moorman, T. A. Vyrostek and C. Stassis, J. Phys. F 6, L49 (1976).
10. S. H. Liu, P. Burgardt, K. A. Gschneidner, Jr. and S. Legvold, J. Phys. F 6, L55 (1976).
11. P. Burgardt, K. A. Gschneidner, Jr., D. C. Koskenmaki, D. F. Finnemore, J. O. Moorman, S. Legvold, C. Stassis and T. A. Vyrostek, Phys. Rev. B 14, 2995 (1976).
12. B. J. Beaudry and P. E. Palmer, p. 612 in Proc. 11th Rare Earth Research Conf., J. M. Haschke and H. A. Eick, eds. (CONF-741002, Part 2) National Technical Information Service, Springfield, VA 22151.
13. N. T. Panousis and K. A. Gschneidner, Jr., Solid State Comm. 8, 1779 (1970).
14. M. Nicolas-Francillon and D. Jerome, Solid State Comm. 12, 523 (1973).
15. D. C. Koskimaki, K. A. Gschneidner, Jr. and N. T. Panousis, J. Cryst. Growth 22, 225 (1974).
16. D. C. Koskimaki and K. A. Gschneidner, Jr., Phys. Rev. B 11, 4463 (1975).
17. T.-W. E. Tsang, K. A. Gschneidner, Jr. and F. A. Schmidt, Solid State Comm. 20, 737 (1976).
18. P. Wells, P. C. Lanchester, D. W. Jones and R. G. Jordan, J. Phys. F 6, 11 (1976).

19. D. L. Johnson and D. K. Finnemore, Phys. Rev. 158, 376 (1967)

20. N. E. Phillips, Crit. Rev. Solid State Sci. 2, 467 (1972).

21. N. E. Phillips, J. C. Ho and T. F. Smith, Phys. Lett. A 27, 49 (1968).

22. A. J. T. Grimberg, C. J. Schinkel and A. P. L. M. Zandee, Solid State Comm. 11, 1579 (1972).

23. M. R. MacPherson, G. E. Everett, D. Wohlleben and M. B. Maple, Phys. Rev. Lett. 26, 20 (1971).

24. H. Katzman and J. A. Mydosh, Phys. Rev. Lett. 29, 998 (1972).

25. M. B. Brodsky and R. J. Friddle, Phys. Rev. B 7, 3255 (1973).

26. D. C. Koskimaki and K. A. Gschneidner, Jr., Phys. Rev. B 10, 2055 (1974); and T.-W. E. Tsang, K. A. Gschneidner, Jr., D. C. Koskenmaki and J. O. Moorman, Phys. Rev. B 14, (to be published in Nov. 15, 1976 issue).

27. P. Burgardt, S. Legvold, J. H. Queen and K. A. Gschneidner, Jr., p. 527 in Magnetism and Magnetic Materials - 1975, J. J. Becker, G. H. Lander and J. J. Rhyne, eds., AIP Conf. Proc. No. 29, Amer. Institute of Phys., New York (1976).

28. B. R. Cooper, Proc. Phys. Soc. London 80, 1225 (1962).

29. R. V. Colvin, S. Arajs and J. M. Peck, Phys. Rev. 122, 14 (1961).

30. C. R. Burr and S. Ehara, Phys. Rev. 149, 551 (1966).

31. K. A. Gschneidner, Jr., p. 153 in Rare Earth Research III, L. Eyring, ed., Gordon and Breach, New York (1965).

ACKNOWLEDGMENT

This work was partially supported by the U.S. Energy Research and Development Administration under Contract Number W-7405-eng.-82. The author also wishes to thank O. D. McMasters for his critical comments and valuable discussions.

COMMENTS

ANDERSON: Is there no anomaly in the resistivity of β-Cerium at the 12°K transition temperature, as far as you can see?

GSCHNEIDNER: No. However, even if it were present it may be difficult to see because the spin-disorder resistivity (∿2 μΩcm) is so small compared to the total resistivity change (∿50 μΩcm) which is dropping rapidly with temperature in that region (see Fig. 2).

DONIACH: Is anything known about the sublattice moment in the antiferromagnetic-like neutron diffraction?

GSCHNEIDNER: No. That's one of the things we would like to determine, but we need single crystals and that's another story.

ORBACH: So what evidence is there that it's an antiferromagnet?

GSCHNEIDNER: First of all the susceptibility, and also the heat capacity.

ANDERSON: There are lots of reasons for (anomalous) susceptibilities and heat capacities other than antiferromagnetism.

GSCHNEIDNER: There are the neutron scattering results obtained many years ago (1961) which showed satellites next to the nuclear peak?

TOURNIER: The properties of this system seem analogous to $CeAl_3$ and $CeIn_3$. In each case there is essentially nothing in the resistivity except perhaps a small change in slope. The specific heat has a peak and in $CeIn_3$ there is a very large γT term in the specific heat. I think all of these substances are exchange-induced antiferromagnets in which there is competition between the Kondo effect and magnetic ordering.

Gd^{3+} EPR PROBE OF THE DETERMINATION OF VALENCE STATES+

C. Y. Huang*

*Los Alamos Scientific Laboratory, Los Alamos, NM 87545

ABSTRACT

The theory of the Gd^{3+} probe EPR technique is reviewed. This technique has been employed to investigate at 9.2 GHz the Gd^{3+} EPR in 0.1 at. % GdS in TmS, 1 at. % GdTe in TmTe, and 1 at. % GdSe in TmSe. From the temperature dependence of the Gd^{3+} EPR linewidth, we have concluded that Tm in TmS is trivalent with a singlet ground-state. The temperature variation of the Gd^{3+} linewidth in TmTe has been interpreted in terms of the "delocalization" of the extra electron "localized" around Gd^{3+}. From the linewidth and g-value data, we have shown that Tm in TmTe is divalent. The linewidth and the g-value of the Gd^{3+} resonance in TmSe have been found to be independent of temperature. These striking results have been explained in terms of the "motional narrowing" effect attributed to the interconfigurational fluctuations of Tm in TmSe.

INTRODUCTION

There has been much interest in recent years in the study of the valence states of the thulium ions in thulium monochalcogenides.[1-9] In this paper we report our results for TmSe, TmS and TmTe determined by the Gd^{3+} impurity EPR (electron paramagnetic resonance). It has been shown that TmSe is an unusual mixed-valence-state material. According to Refs. 3 and 5, the thulium ions in TmSe are

+Supported in part by NSF.
*Work performed under the auspices of ERDA.

predominantly trivalent; the Mössbauer results have revealed[2] that the correlation time of the interconfigurational fluctuations between the Tm^{2+} and Tm^{3+} electronic configurations falls within 10^{-10} - 10^{-12} second. On the other hand, from their reflectivity data, Batlogg, et al.[6] concluded that the interconfigurational fluctuation does not exist in TmSe. However, the valence states of the Tm ions in TmS and TmTe are definite. From the lattice-constant[1] and the susceptibility measurements, Bucher, et al.[3] have determined that the Tm ions in TmS are almost trivalent, and from the optical and photoemission experiments, Suryanarayanan, et al[4] have concluded that Tm in TmTe is essentially divalent.

Since TmSe, TmS and TmTe are all cubic with the NaCl structure,[3] in comparison with the rare-earth monopnictides,[10] the cubic crystal-field is expected to split the Hund's rule ground-state of Tm^{3+} (J=6) into various crystal-field states. This expectation has led to several unsuccessful attempts[3] to measure the crystal-field splittings in these compounds by neutron scattering, a technique which was successfully applied to rare-earth monopnictides.[11] According to these observations and the wide disparities between estimates of the Tm^{2+} and Tm^{3+} concentrations, it is evident that the thulium monochalcogenides form one of the most unique series of compounds for use in the study of unstable valence systems. Encouraged by the successful applications of the Gd^{3+} probe EPR technique in the study of the crystal-field and exchange effects in thulium monochalcogenides,[12,13] we have investigated the Gd^{3+} impurity EPR in TmS, TmTe, and TmSe. Our results[14] have confirmed that Tm is trivalent, divalent, and mixed-valent in TmS, TmTe, and TmSe, respectively. Hence this Gd^{3+} probe technique can be used for studying the material with mixed valence as well.

In the next two sections we will review the theories pertinent to explain our experimental technique and data, and then will present the experimental results concerning the valence states of Tm in our Tm monochalcogenide samples.

REVIEW OF THEORIES

In a recent paper, Sugawara, et al. have reported[12] their studies of some singlet-ground-state systems (Tm - and Pr - group VA intermetallic compounds). They obtained the exchange and crystal-field effects of these systems by observing the paramagnetic resonance of dilute Gd^{3+} impurities used as a probe. (For succinctness, we term this method the "Gd^{3+} probe EPR technique".) In this Sec., we will review the theories associated with this technique. A brief discussion of the narrowing of spectral lines by motions will be also presented.

A. g-Shift Due to Gd-Host Exchange

We consider a paramagnetic system in which there are dilute Gd^{3+} impurities (probes). If the exchange interaction between a Gd^{3+} ion and the surrounding host magnetic ions is isotropic, then the Hamiltonian of the S-state Gd^{3+} is written

$$\mathcal{H} = g_o \beta \vec{S} \cdot \vec{H} + \sum_j J_j \vec{S} \cdot \vec{S}_j, \quad (1)$$

where the first and second terms represent the Zeeman and exchange energies, respectively. g_o is the g-value of the isolated Gd^{3+} ion, β is the Bohr magneton; $\vec{S}$ and $\vec{S}_j$ represent the spin of Gd^{3+} and the j-th host magnetic ion surrounding the Gd^{3+} ion in question, respectively; and J_j is the exchange interaction coefficient of Gd^{3+} and the j-th host ion. Hutchings, et al.[15] have treated this problem, and they have shown that the Gd^{3+} ion will exhibit a spin resonance with a g-value shifted from g_o. As long as the exchange energy is small compared to the crystal-field splittings, the g-shift can be expressed as[12,13,15]

$$\Delta g = \left(\frac{\lambda-1}{\lambda} \right) \frac{J(0)}{\beta^2} \chi_{CF}, \quad (2)$$

with

$$J(0) \equiv \sum_j J_j. \quad (3)$$

Here λ and χ_{CF} are the Landé g-factor of the host magnetic ion and the crystal-field-only isothermal susceptibility of the host, respectively. When there are exchange interactions between the host magnetic ions Eq. (2) is modified.[16,12] However, for our purpose, Eq. (2) is a good approximation. Accordingly, the g-shift of Gd^{3+} is proportional to the isothermal susceptibility of the host magnetic ion; and the exchange interaction coefficient $J(0)$ can be obtained if we measure the g-shift of Gd^{3+} at zero temperature, and if the isothermal susceptibility of the host, χ_{CF}, is known. To date, Eq. (2) has been successfully utilized to interpret the temperature-dependent data of the wide variety of paramagnetic hosts (Tm and Pr monopnictides,[12] $PrPb_3$, TmCd, $PrIn_3$, and $PrAl_2$ compounds[17,18]).

B. Theory of the Linewidth

In addition to the static magnetic field, $\vec{H}$, a Gd^{3+} probe ion also experiences the exchange fields originating from the exchange interactions between the Gd^{3+} ion and its neighboring host magnetic ions. When the spins of the host ions fluctuate, a fluctuating field at the site of the Gd^{3+} ion results. This fluctuating field gives rise to an additional linewidth of the Gd^{3+} EPR. Moriya and Obata[19] have investigated this problem in detail. Their theory predicts that the EPR linewidth of the magnetic impurity is proportional to the auto-correlation function of the spin fluctuations of the host magnetic ion $\langle S_j(t)S_j(0)\rangle$. They concluded that the linewidth is approximately given by[19,20]

$$\Delta H = AkT\ (\chi_T - \chi_{iso}). \tag{4}$$

Here A is a proportionality constant, and χ_T and χ_{iso} are the isothermal and isolated susceptibilities of the host magnetic ion, respectively. (The isothermal susceptibility is the susceptibility measured under circumstances where the magnetic system remains in thermal equilibrium with the lattice during the measured period. This means that the thermal populations of the various crystal-field levels adjust themselves in a way according to the change in Boltzmann factors as the energies of the levels change with applied static magnetic field. The isolated susceptibility is that measured when the thermal populations of the levels do not change. The effect measured is due to the change in wavefunctions, and hence the moments, with the applied magnetic field, i.e. the polarization effect.) In case of the singlet-ground-state system, Eq. (4) signifies that the population fluctuation in the excited crystal-field states of the host magnetic ions contributes to the line broadening of the Gd^{3+} probe. This means that we expect the linewidth of Gd^{3+} in a singlet-ground-state system to increase sharply as the temperature is raised to the order of the crystal-field splitting to the first excited state of the host magnetic ions; and if the temperature is high compared to the overall crystal-field-splitting, we expect the linewidth to be temperature independent. Since the isothermal and isolated susceptibilities are directly related to the crystal-field splittings, the measurements of the EPR linewidth of Gd^{3+} can provide very useful information about the crystal-field splittings of the host non-Kramers ions. As demonstrated in Refs. 12 and 13 the crystal-field splittings in Tm and Pr monopnictides agree with those obtained by neutron scattering. However, if the ground-state of the host magnetic ion is magnetic, the Gd^{3+} probe linewidth is expected to be very broad and nearly temperature-independent. Therefore, by studying the temperature dependence of the Gd^{3+} probe linewidth we may determine whether the ground-state of the host magnetic ion is magnetic or not. Therefore, this technique can be used to identify the valence state of the host magnetic ions.

C. Motional Narrowing

Motional narrowing was first discussed in the classic paper[21] by Bloembergen, Purcell, and Pound (BPP). Anderson's explanation[22] of the phenomenon will be presented here. It is clear that non-magnetic interactions cannot affect the magnetic resonance line directly. The magnetic interactions responsible for the linewidth depend on the positions of the magnetic ions. Some non-magnetic interactions cause the actual random motions of the ions, and the magnetic interactions will then vary in time in some way controlled by the ionic motions. As a result, the magnetic resonance line experiences a time averaged-effect of the magnetic interactions. When the ionic motions are sufficiently rapid, this average effect may be much smaller than otherwise. Therefore the broadening effect of the magnetic interactions is reduced; the resonance line is narrowed and this is called motional narrowing.

Actually, before BPP, the phenomenon of exchange narrowing was discussed by Gorter and Van Vleck.[23,24] Since the mechanism is in principle the same in both motional and exchange narrowing, we will first discuss the phenomenon of exchange narrowing in $CuSO_4 \cdot 5H_2O$, and then apply this result to our case. In $CuSO_4 \cdot 5H_2O$, there are two inequivalent Cu^{2+} sites in a unit cell. Consequently, there are two g-values, g_1 and g_2, and two corresponding resonance absorption lines. This spectrum was observed[25] at 30 GHz. However, when the microwave frequency was lowered to 10 GHz, only one narrow line was observed.[25] This phenomena may be explained in the following classical language.

Since J/h where J is the exchange interaction coefficient between the spins on sites 1 and 2, may be taken to be the rate at which the resonant spin on sites 1 and 2 "exchange", the spins precess at frequency $g_1\beta H_o/h$ for a while and then at $g_2\beta H_o/h$. If the rate at which the exchange between these two Larmor frequencies is faster than $|g_2 - g_1|\beta H_o/h$, then the line is "exchange narrowed". A theoretical model calculation appropriate to this problem has been made by Anderson.[22]

The situation for TmSe is similar to this example. In TmSe, Tm fluctuates between Tm^{2+} and Tm^{3+} at the rate of $\sim 10^{11}$ per sec.[2] If g_1 and g_2 are the g-values of the Gd^{3+} impurity probe when all the surrounding Tm are Tm^{2+} and Tm^{3+}, respectively, the fluctuating Tm ions will cause the Gd^{3+} g-value to fluctuate between g_1 and g_2 at the rate ($\sim 10^{11}$ per sec.) much faster than $|g_2 - g_1|\beta H_o/h$. In this case, the line is "motionally narrowed" by the valence fluctuations, and a single narrow line is expected.

EXPERIMENTAL RESULTS

The powder samples[14] of TmS, TmTe, and TmSe doped with Gd^{3+} were used in this investigation. They were prepared by a method similar to that described in Refs. 12 and 13. All these samples have the cubic NaCl structure,[3] and hence the g-values are isotropic. The EPR spectometer used is a conventional one operating at 9.2 GHz.

A. TmS

Fig. 1(a) displays the temperature dependence on the Gd^{3+} EPR linewidth of 0.1 at. % GdS in TmS. The linewidth increases with temperature. Following our theoretical discussion in the last section, this temperature dependence suggests that the Tm ions in TmS are trivalent with non-magnetic ground-states. This conclusion is in agreement with the energy level scheme of Tm^{3+} in TmS assumed by Bucher, et al.[3] For Tm monopnictides,[10,12,13] $x \simeq -1$, where x is

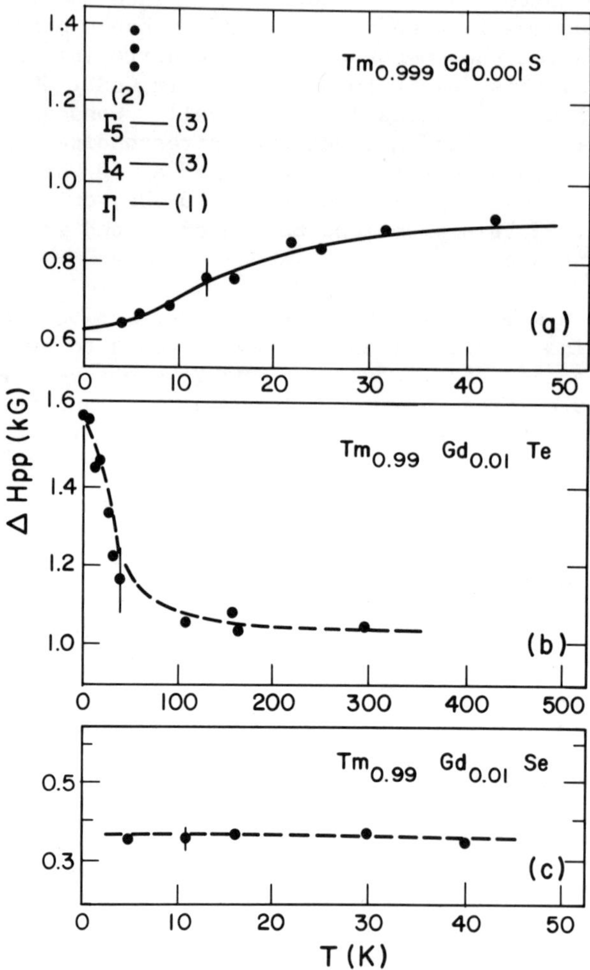

Fig. 1. Gd^{3+} EPR linewidth versus temperature for (a) 0.1 at. % GdS in TmS, (b) 1 at. % GdTe in TmTe, and (c) 1 at. % GdSe in TmSe.

the crystal-field parameter as defined by Lea, Leask, and Wolf.[26] If we also assume $x \simeq -1$ for TmS, the crystal-field level scheme of Tm^{3+} is as depicted in the inset of Fig. 1(a), and TmS is a singlet-ground-state system. By fitting Eq. (4) with the wavefunctions provided in Ref. 26 to the data, we obtain the solid curve in Fig. 1(a) with the crystal field splitting $E(\Gamma_4) - E(\Gamma_1) = 15 \pm 3$ K. This value of the splitting is in close agreement with that (16 K) estimated by Bucher, et al.[3] from their specific heat data. It is interesting to note that there is no linewidth anomaly near the Néel temperature[3] of TmS (5.18 K) due to the critical slowing-down as is predicted and observed in the pure antiferromagnet.[27] The absence of this anomaly is attributed[28] to the weak exchange interaction between Gd^{3+} and Tm^{3+}.

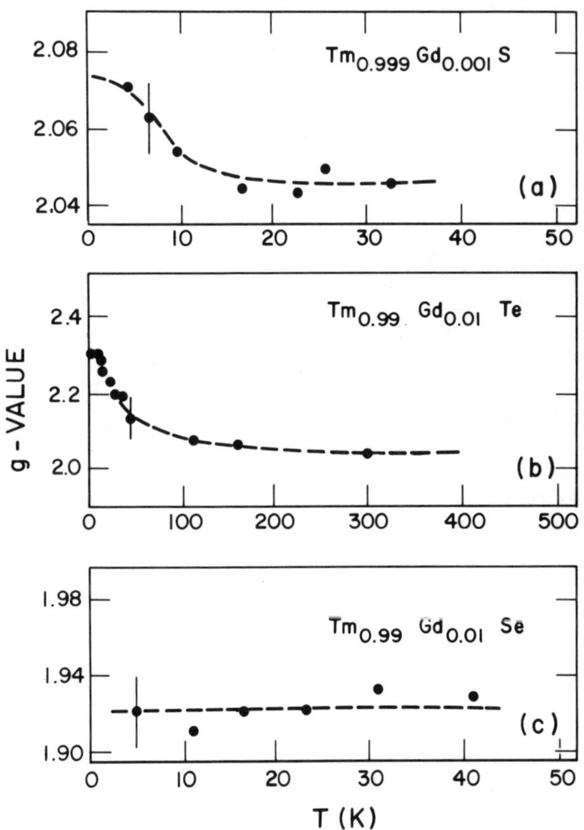

Fig. 2. Gd^{3+} resonance g-value versus temperature for (a) 0.1 at. % GdS in TmS, (b) 1 at. % GdTe in TmTe, and (c) 1 at. % GdSe in TmSe.

Fig. 2(a) depicts the temperature dependence of the Gd^{3+} g-value. The temperature dependent part of the g-value is similar to those of other singlet-ground-state systems.[12,13] Accordingly, the data of the g-measurement also support the idea that Tm in TmS is trivalent. Using Eq. (2), we obtain the exchange interaction coefficient $\mathcal{J}(0) = +0.74 \pm 0.07$ K, in which the positive sign denotes that Gd^{3+} and Tm^{3+} are ferromagnetically coupled.

B. TmTe

First, let us examine the consequences of the assumption that Tm in TmTe is divalent. According to Lea, Leask, and Wolf,[26] the ground-state is the Γ_6 doublet and the excited states are the Γ_7 doublets and the Γ_8 quartet. Since the Γ_6 ground-state is magnetic, and by using Eq. (4) and the wavefunctions provided by Lea, Leask, and Wolf,[26] the linewidth of the Gd^{3+} probe in TmTe is expected to increase slightly with increasing temperature. Contrarily, as shown in Fig. 1(b), the data of 1 at. % GdTe in TmTe demonstrate that the Gd^{3+} linewidth decreases drastically from 5 K to 40 K. This surprising temperature dependence is believed to originate from the presence of the "extra" electrons contributed by the doping of the trivalent Gd ions into a host of the divalent Tm ions as required by charge compensation. In analogy to the case[16] of Gd^{3+} in Sm^{2+} monochalcogenides, we assume that there is an "extra" electron localized around the Gd^{3+} ion in a cubically symmetrical orbital. The presence of the "extra" electron contributes directly to the exchange interaction between the Gd^{3+} ion and its surrounding Tm ions. This "extra" exchange interaction in turn contributes to the line broadening. At high temperatures, the "extra" electron is delocalized, resulting in the reduction of the $Gd^{3+} - Tm^{2+}$ exchange interaction and, hence, giving rise to the decrease of the Gd^{3+} probe linewidth. From the temperature dependence of the linewidth the activation energy needed to delocalize the "extra" electron has been estimated to be $\sim$ 50 K.

According to Eq. (2) the temperature variation of the exchange interaction should be also reflected in that of the g-value. As shown in Fig. 2(b), the g-value also decreases with increasing temperature. Utilizing Eq. (2), $\mathcal{J}(0) = 1.3 \pm 0.1$ K at T = 5 K.

Walsh, et al.[29] pointed out that the higher concentration of Gd^{3+} in Sm monochalcogenides could result in the delocalization of the "extra" electrons and the decrease of the g-shift. For this reason we have investigated 3 at. % GdTe in TmTe. As expected, the g-shift observed has been found to be much smaller than that of the 1 at. % samples. In addition, the linewidth of our 3 at. % samples is nearly temperature independent, indicating that the ground-state of Tm^{2+} is magnetic. In order to further support the idea of "extra" electrons, we have measured the electrical conductivity. It is known that pure TmTe is a semiconductor.[3] However, all our Gd doped samples exhibit that the electrical conductivity decreases with increasing temperature, and hence they are metals. This observation suggests that Gd are donors, and when the Gd concentration is high enough, the sample becomes metallic.

C. TmSe

The interconfigurational fluctuations of a Tm ion arises from an interplay of conduction electrons with the unstable 4f shell of the Tm ion.[30] Accordingly, we expect the exchange interactions between a Gd^{3+} probe and its neighboring Tm ions, J, to fluctuate in addition to the valence fluctuations of Tm. Even though the detailed nature of the exchange interactions is not known, we would anticipate that J fluctuates, more or less, at the same rate as that of the valence fluctuation of Tm. Consequently, in computing the linewidth following the theory of Moriya and Obata,[19] we should first calculate the time auto-correlation function of the form:

$$\langle J(t)\, S_j(t)\, J(0)\, S_j\, (0)\rangle,$$

where $J(t)$ is the exchange interaction coefficient between the Gd^{3+} ion and its j-th neighboring Tm ion at time t. At present, there is no computed result of this joint auto-correlation function including the fluctuations of both J and S_j. However, according to the Mössbauer data,[2] the valence of Tm and hence S_j fluctuates at the rate[31] of $\sim 10^{11}$ sec^{-1}, which is somewhat greater than our microwave frequency (9.2×10^9 sec^{-1}). Hence, it is reasonable to assume that the correlation time of the joint time auto-correlation function to be $\tau_c \lesssim 10^{-11}$ sec. In the paramagnetic regime, if the temperature is not very close to the critical temperature, according to our data presented above and those in Refs. 12, 16, and 18, the exchange g-shift is $\lesssim 0.4\, g_o$. Thus, at our microwave frequency (9.2 GHz), the separation between a Gd^{3+} line due to the influence of Tm^{2+} and that due to Tm^{3+}, $|g_2 - g_1|\beta H_o/h$, is on the order of 10^9 sec^{-1}. This separation is much less than the fluctuation rate of $\sim 10^{11}$ sec^{-1}. In this extreme "motional narrowing" case, based on our discussion presented in the last section, the resonance absorption is expected to be a narrow line. This is exactly what was observed. Fig. 1(c) shows the linewidth data for 1 at. % GdSe in TmSe. The linewidth is temperature independent within our experimental accuracy and is only $\sim$ 350 Gauss. This narrow linewidth is believed to be partly due to the zero-field splitting of the S-state Gd^{3+} (S = 7/2) and partly due to the Gd^{3+} - Gd^{3+} dipole interactions. Fig. 2(c) depicts our g-value data from 5 K to 40 K. Within our experimental errors, the g-value is independent of temperature, and is 1.92 ± 0.02. This g-value yields that the g-shift, $\Delta g \simeq -0.07$, where the minus sign indicates that the Gd^{3+} probe is "coupled" to the neighboring Tm ions antiferromagnetically.

CONCLUDING REMARKS

It is evident that the application of the Gd^{3+} probe EPR technique to the study of mixed-valence systems is still at an early stage and much more work has to be done in order to understand the physics associated with the technique with confidence. Several problems of interest are: (1) The theory of the fluctuating exchange interaction, $J(t)$, between Gd^{3+} and the fluctuating host magnetic ion. (2) The theory of the joint auto-correlation function including $J(t)$ and $S_j(t)$. (3) The theory of the zero-field splitting of Gd^{3+} in a valence-fluctuating host. (4) The theory of the Gd^{3+} probe lineshape and linewidth, including the effects of conduction electrons.

ACKNOWLEDGEMENT

The author is grateful to R. P. Guertin, K. Sugawara, C. M. Varma, and W. M. Walsh, Jr. for helpful discussions.

REFERENCES

1. M. Campagna, E. Bucher, G. K. Wertheim, D. N. E. Buchanan, and L. D. Longinotti, Phys. Rev. Lett. 22, 885 (1974).

2. B. B. Triplett, N. S. Dixon, P. Boolchand, S. S. Hanna, and E. Bucher, J. Phys. (Paris), Collog. 35, C6 - 653 (1974).

3. E. Bucher, K. Andres, F. J. diSalvo, J. P. Maita, A. C. Gossard, A. S. Cooper, and G. W. Hull, Jr., Phys. Rev. B11, 500 (1975), and the references therein.

4. R. Suryanarayanan, G. Güntherodt, J. L. Freeouf, and F. Holtzberg, Phys. Rev. B 12, 4215 (1975).

5. R. W. Ward, B. P. Clayman, and T. M. Rice, Sol. Stat. Commun. 17, 1297 (1975).

6. B. Batlogg, E. Kaldis, A. Schlegel, and P. Wachter, Phys. Lett. 56A, 122 (1976); B. Batlogg, A. Schlegel, and P. Wachter, to be published in J. Phys. (Paris).

7. A. Jayaraman, P. D. Dernier, and L. D. Longinotti, High Temp. High Pressure 7, 1 (1975).

8. C. M. Varma, Rev. Mod. Phys. 48, 219 (1976).

9. R. P. Guertin, S. Foner, and F. P. Missell, Phys. Rev. Lett. 37, 529 (1976); R. P. Guertin, to appear in Proc. 2nd Int. Conf. on Crystal Field Effects in Metals and Alloys, Zurich, Switzerlnad (1976).

10.. B. R. Cooper, CRC Crit. Rev. Sol. State Sci. 3, 83 (1972) and references therein.

11. R. J. Birgeneau, E. Bucher, L. Passell and K. C. Turberfield, Phys. Rev. B4, 714 (1971); R. J. Birgeneau, E. Bucher, J. P. Maita, L. Passell, and K. C. Turberfield, Phys. Rev. B8, 5345 (1973); H. L. Davis and H. A. Mook, AIP Conf. Proc. 18, 1068 (1974).

12. K. Sugawara, C. Y. Huang, and B. R. Cooper, Phys. Rev. B11, 4455 (1975).

13. K. Sugawara, Ph.D. thesis (Case Western Reserve Univ., 1974) (unpublished).

14. C. Y. Huang and K. Sugawara, to be published.

15. M. T. Hutchings, C. G. Windsor, and W. P. Wolf, Phys. Rev. 148, 444 (1966).

16. R. J. Birgeneau, E. Bucher, L. W. Rupp, Jr., and W. M. Walsh, Jr., Phys. Rev. B5, 3412 (1972).

17. K. Sugawara and C. Y. Huang, J. Phys. Soc. Japan 41, No. 5 (1976).

18. K. Sugawara, submitted for publication.

19. T. Moriya and Y. Obata, J. Phys. Soc. Japan 13, 1333 (1958).

20. Eq. (4) can be also expressed as $\Delta H = B[\langle(\partial E/\partial H)^2\rangle - \langle\partial E/\partial H\rangle^2]$, where B is a constant, E is the energy of the host magnetic ion, and $\langle\cdots\rangle$ denotes the thermal average.

21. N. Bloembergen, E. M. Purcell, and R. V. Pound, Phys. Rev. 73, 679 (1948).

22. P. W. Anderson, J. Phys. Soc. Japan 9, 316 (1954).

23. C. J. Gorter and J. H. Van Vleck, Phys. Rev. 72, 1128 (1947).

24. J. H. Van Vleck, Phys. Rev. 74, 1168 (1948).

25. D. M. S. Bagguley and J. H. E. Griffiths, Proc. Roy. Soc. (London) A201, 366 (1950).

26. K. R. Lea, M. J. M. Leask, and W. P. Wolf, J. Phys. Chem. Solids 23, 1381 (1962).

27. M. S. Seehra and D. L. Huber, AIP. Conf. Proc. No. 24, 261 (1974).

28. K. Sugawara, C. Y. Huang, and B. R. Cooper, in preparation.

29. W. M. Walsh, Jr., L. W. Rupp, Jr., E. Bucher, and L. D. Longinotti, AIP Conf. Proc. No. 18, 535 (1973).

30. L. L. Hirst, Phys. Kondens. Matter. 11, 255 (1970); J. Phys. Chem. Solids 35, 1285 (1974).

31. P. Boolchand, private communication. He has informed us that the fluctuation rate is faster than 10^{10} sec^{-1} in the temperature range we did our measurements. We would like to thank him for informing us of this unpublished result.

COMMENTS

VARMA: The fact that the bottleneck effect as worked out by Hasegawa, Orbach and others was not observed is very interesting indeed.

HIRST: No. A bottleneck effect occurs only when a relatively narrow EPR line is possible in the host material, which is the case only for host materials with spin-only magnetism.

WOHLLEBEN: Can you tell me the lower limit of the fluctuation rate?

HUANG: For TmSe, the fluctuation rate should be faster than 10^9 sec^{-1}, but let me remind you that my samples are not pure TmSe; they are 1% Gd-doped samples. So, even if my samples demonstrated that the Tm ions fluctuate, this does not necessarily imply that the Tm ions in pure TmSe fluctuate.

KASUYA: I would like to comment that we have found in our own laboratory that in another typical valence fluctuation material SmB_6 the linewidth depends on the Gd impurity concentration.

HUANG: It is true that the linewidth does in fact depend on the Gd concentration. But, as I showed earlier, for 0.1, 3 and 10 at.% GdP in TmP and 0.1 and 3 at.% GdSb in TmSb, the temperature dependent parts of the linewidths are independent of the Gd concentration. Therefore, our conclusions would not be altered by making the concentration dependence measurements. Of course, the Gd concentration should be kept low enough to avoid complications.

APPLICATION OF DISORDER TO PROBE THE GROUND STATE OF $CeAl_3$*

A. S. Edelstein and R. E. Majewski

Department of Physics, University of Illinois at Chicago Circle, Argonne National Laboratory

T. H. Blewitt

Argonne National Laboratory

The magneto-resistance of $CeAl_3$ decreases rapidly below 2K. This may happen because the 4f electrons make a transition into an itinerant narrow band. Disorder created by neutron irradiation or chemically by alloying with La tends to suppress this decrease and increases the resistivity ρ much more at low temperatures than for $T > 100$ K. For disorder created by neutron irradiation, the change in resistivity $\Delta \rho \propto \ln T$ for $1.5 \text{ K} \leq T \leq 10$ K. Disorder inhibits the transition.

The intermetallic compound $CeAl_3$, which has been classified[1] as a valence fluctuation system, has remarkable properties. Susceptibility measurements[2-4] show no evidence of magnetic ordering, yet the effective moment $\mu_{eff} \equiv (3kT\chi)^{\frac{1}{2}}$, where χ is the susceptibility, goes to zero at low temperatures ($T<10$ K). Despite the absence of magnetic ordering, specific heat measurements[4,5] account for essentially all the spin entropy $R \ln (2J+1)$, $J=5/2$. The broad specific heat anomaly[5] centered at 5 K and which has two sharp features is not understood. There is an enormous γT contribution ($\gamma = 1620$ mJ/mole K^2) to the specific heat for $T < 0.3$ K. Unlike $Ce_{.8}Th_{.2}$, which makes a first order phase transition and shows a large decrease below the transition in the fractional occupancy[6] η of the localized $4f_1$ configuration, η only decreases[6] slightly in $CeAl_3$ as the temperature is decreased to 11 K. As we shall discuss below, η may decrease considerably more below 11 K. Below 130 K the resistivity[3,7] increases logarithmically with decreasing temperature, shows a broad maximum at $T \approx 37$ K and decreases gradually. The resistivity increases[4] quadratically with increasing T for $T < 0.2$. The negative magneto-resistance[3] typical of local moment systems decreases in magnitude for $T < 2$ K and changes sign at $T \approx 0/5$ K. Andres et al.[4] have used a virtual bound state model to interpret much of the above data. At high temperatures they suggest the behavior is like that of local moments, but at low temperatures coherence develops between the virtual bound states. A band of collective states is formed that has a very high density at the Fermi energy.

Since $\mu_{eff} \to 0$ and essentially all the spin entropy has been removed by 1 K, the ground state is a singlet. In this case, for a well ordered compound like $CeAl_3$, the only scattering left at low temperatures is due to defects, impurities, and other things that break the symmetry. In order to investigate the scattering off the Ce atoms, it is necessary to disrupt the perfect periodicity. Here we report resistivity and magneto-resistivity measurements on $CeAl_3$ which has been disordered by neutron or chemically by alloying with La. In each case we have found that disorder increases the resistivity much more below 100 K than above. Further, the magnitude of the negative magneto-resistance of the disordered samples decreases less. These results probably imply that the disorder breaks the coherence necessary for the formation of a narrow band and tends to inhibit the transition of $CeAl_3$ into its ground state.

The compound $CeAl_3$ has a hexagonal DO_{19} structure (Ni_3Sn-type). Our samples of $Ce_{1-x}La_xAl_3$ ($x = 0$ and 0.01) were prepared by arc melting and subsequently annealed in vacuum at 1100°C for 2 weeks. X-ray analysis indicates that they have the correct structure and the same lattice constants to 1 part in 10^3. The crystal field splits the energy levels of the three doublets. Recent measurements[8] indicate that $J_z = \pm 3/2$ is lowest and $J_z = \pm 5/2$ and $J_z = \pm 1/2$ are respectively 5.3 and 7.2 meV higher.

A sample of $CeAl_3$ was exposed to a dose of 3.3×10^{18} n/cm^2 with an energy greater than 0.1 meV with the degraded fission spectrum in the Argonne low temperature ($T = 5$ K) CP-5 fast flux facility.[9] The resistivity was measured _in situ_. It was found that the change in the resistivity $\Delta\rho$ was nearly proportional to the radiation dose. Each "knock-on" collision with a newtron disorders approximately 5000 atoms within a displacement cascade[10] which contains 10^5 atoms. From a consideration of the neutron cross-section and the dose we estimate that about 8% of the atoms were disordered. Resistivity measurements below 1 K and higher precision measurements above 1 K were made by transferring the sample at room temperature to a different cryostat which was external to the reactor. The resulting annealing decreased the change in resistivity by about a factor of 2, but no further annealing occurred even when the sample was kept at room temperature for months. The cross-sectional area of this sample was approximately 0.02 cm^2 and the separation between the voltage leads was 1.32 cm.

Figure 1 shows a plot of the resistivity of the irradiated sample, ρ_{irrad}, the unirradiated sample, $\rho_{unirrad}$, and the difference, $\Delta\rho = \rho_{irrad} - \rho_{unirrad}$. The solid curves are spline fits to the data with the stiffness parameter equal to the number of data points. For the curves representing ρ_{irrad} and $\rho_{unirrad}$ the error used in the spline fits was 0.3 μΩ cm. To indicate the sensitivity of the spline curve to the error assumed for ρ, both 0.3 and 0.75 μΩ-cm errors were employed. The larger error for Δρ is more consistent with the error employed in ρ and the curve for Δρ using this larger error is more likely to represent the correct behavior of Δρ. The difference Δρ has the following properties: 1) Δρ is increased more at low temperatures than at high temperatures. 2) $\Delta\rho \propto \ln T$ for $1.5\ K \leq T \leq 10K$. 3) $\Delta\rho = 27.5 [1-(T/4.58)^2]$ μΩ-cm for $0.35 \leq T \leq 1.5\ K$. 4) Δρ is minimum where ρ is maximum.

We have investigated the effect of chemical disorder by measuring ρ and hence Δρ of $Ce_{.99}La_{.01}Al_3$. The results of these measurements are shown in Figure 2. The dashed curve, using an error of 0.75 μΩ-cm for Δρ, is more likely to be correct. Alloying with La has shifted the temperature at which the maximum

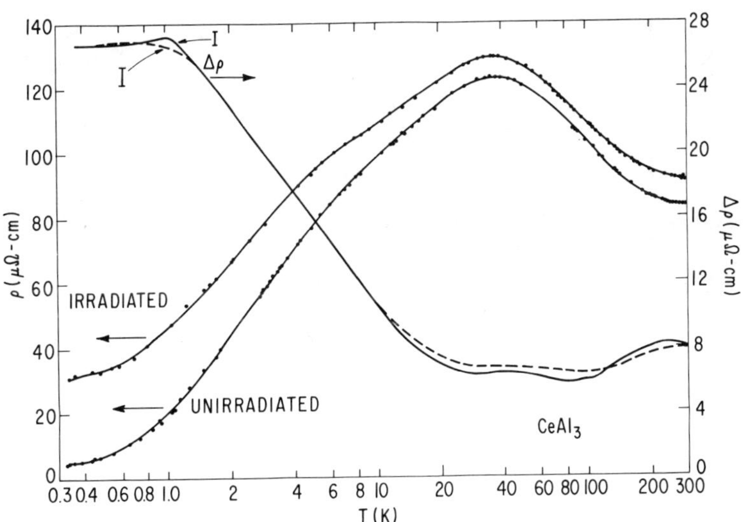

Fig. 1 Plots of the resistance of unirradiated $CeAl_3$, neutron irradiated $CeAl_3$ and the spline fitted difference Δρ versus temperature.

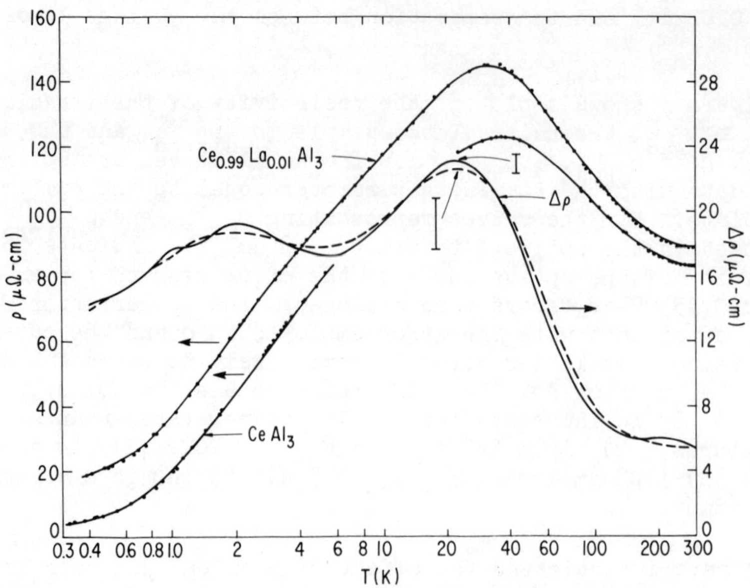

Fig. 2 Plots of the resistance of $Ce_xLa_{1-x}Al_3$ (x = 0.0, 0.01) and the spline fitted difference $\Delta\rho$ versus temperature.

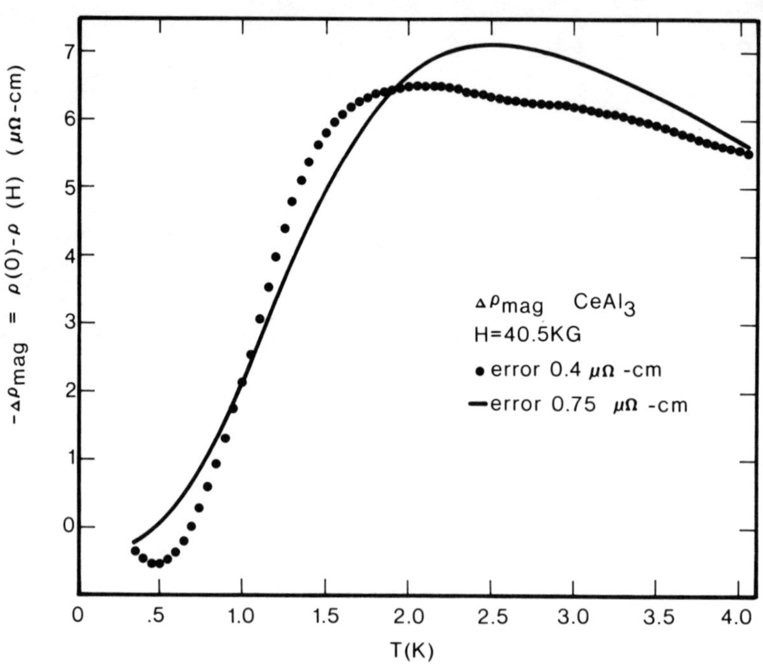

Fig. 3 Plot of the spline fits to $-\Delta\rho_{mag} \equiv \rho(0)-\rho(H)$ for $CeAl_3$ versus T.

in ρ occurs and has increased its magnitude. The difference reflects this shift but in addition shows additional structure at 2K. This additional structure may result from a contribution like that shown in Fig. 1. The disorder is, as discussed above, larger in the neutron irradiated sample than 1%. Hence, we expect Δρ to be larger in Fig. 1 than in Fig. 2 and it is.

Magneto-resistivity measurements have proved very valuable in understanding the effects of disorder. In Fig. 3 the negative of the magneto-resistance ρ_{mag} of $CeAl_3$ is allotted as a function of T. As usual we have defined $-\rho_{mag}$ as ρ(0)-ρ(H) where ρ(H) where ρ(H) is the resistivity measured in an external magnetic field H. A field of 40.5 KG was employed. At 4.2 K, ρ(H) decreased approximately as H^2. The small deviations from an H^2 dependence were like those observed in the magnetization M which showed some tendency toward saturation. From the coefficient of the H^3 term in M a characteristic temperature of 5.1K was estimated.[11] The solid spline curve shown in Fig. 3, based on an error of 0.75 μΩ-cm, is believed the more correct one. One sees that the magnitude of the magneto-resistance has a maximum at 2.5 K and decreases

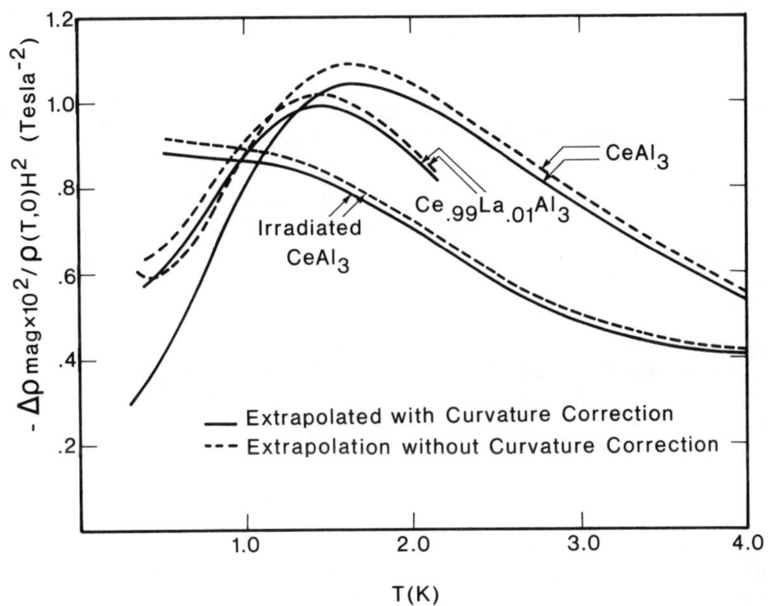

Fig. 4 Plot of the spline fits to $-\Delta\rho_{mag}/\rho H^2$ versus T. The quantity $-\Delta\rho_{mag} \equiv \rho(0)-\rho(H)$ and H is the external magnetic field in tesla. H = 4.5, 4.507, 4.505 tesla for $CeAl_3$, the irradiated sample of $CeAl_3$ and $Ce_{.99}La_{.01}Al_3$ respectively. The correction for the positive magneto-resistance is discussed in the text.

rapidly for T < 2 K. The magneto-resistance changes sign at 0.5 K. A negative magneto-resistance is characteristic of local moment systems and follows from perturbation theory calculations.[12,13] For dilute alloy Kondo systems, the magneto-resistance has the form:[13]

$$\rho(H) = \rho_o\{1 + B\ln[(T^2 + \theta^2)^{\frac{1}{2}}/\varepsilon_F]\} \quad (1)$$

where ρ_o is the first order perturbation theory result, ε_F is the Fermi energy, $\theta^2 = \tau^{-2} + \omega_o^2$, ω_o is the molecular field and τ is the spin relaxation time. For $\omega_o << T$ or τ^{-2} one can write

$$\frac{\rho(H)-\rho_o}{\rho_o} = \Delta\rho_{mag}^{(H)}/\rho_o = \tfrac{1}{2}B\omega_o{}^2/(T^2+\tau^{-2}) \quad (2)$$

This result or that of Beal-Monod and Weiner[14] fits the data[15] on Cu Fe. In the absence of local fields $\omega_o \rightarrow H$ and for $T >> \tau^{-1}$ the right hand side of Eq.(2) will be proportional to the square of the magnetization M. At low temperatures $T << \tau^{-1}$, $M \propto \tau$, and again the right hand side of Eq.(2) is proportional to $M.^2$ We believe that it is reasonable to suppose, for a concentrated local moment system, above its ordering temperature that

$$-\rho_{mag} \propto \rho M^2 \quad (3)$$

To test this relation we have plotted in Fig. 4, $-\Delta\rho_{mag}/(\rho H^2)$ as a function of T for $CeAl_3$, the irradiated sample of $CeAl_3$ and $Ce_{.99}La_{.01}Al_3$. We divided by H^2 to take account of the fact that somewhat different fields were employed in the measurements on different samples. Since the magneto-resistance is the sum of two terms, the negative magneto-resistance $\Delta\rho^-_{mag}$ which we have been discussing and the usual positive magneto-resistance $\Delta\rho^+_{mag}$ observed in the absence of local moments, we have to correct our data to take account of the positive magneto-resistance. We assume $\Delta\rho^+_{mag}$ is temperature independent for T < 4 K and that $\Delta\rho^-_{mag}$ of $CeAl_3$ goes to zero as $T \rightarrow 0$. This latter assumption is not unreasonable since $\Delta\rho_{mag}$ changes sign and ρ becomes very small at low temperatures. By extrapolating the solid curve shown in Fig. 3 and using these assumptions we have been able to estimate $\Delta\rho^+_{mag}$ for all samples by adding $\Delta\rho^+_{mag}$ to it. To demonstrate the sensitivity of our result to the extrapolation procedure, Fig. 4 shows the result of two different extrapolation procedures, one based on a linear extrapolation and the other taking the obvious curvature into account. The latter extrapolation should be better. In Fig. 4 one sees that $-\Delta\rho_{mag}/(H^2\rho)$ of $CeAl_3$ is in qualitative agreement with Eq.(3) for T > 1.9 K; i.e., both $-\Delta\rho_{mag}/\rho$ and M^2 are monotonically increasing with decreasing T. For $0.5 < T \leq 1.9$, M continues to increase[3,4] with decreasing T while in contrast $-\Delta\rho_{mag}/\rho$ decreases. For dilute systems[15] there is no such decrease in the magnitude of the magneto-resistance, even at low temperatures where perturbation theory is invalid. Hence, the data for $CeAl_3$ shown in Fig. 4 is not easily understood in terms of local moment behavior. The decrease in $-\Delta\rho_{mag}/\rho$ is consistent with the description by Andres et al.[4] of virtual bound states overlapping at low temperatures to form coherent states in a narrow band located at ε_F. Our data suggests that this band is formed at $T \approx 2$ K. Though the broad specific heat anomaly[5] is centered at a somewhat higher temperature ($T \approx 5$ K), it still may be associated with the formation of this band.

Now we shall consider the effect of disorder on the magneto-resistance. One sses in Fig. 4 that disorder inhibits the decrease in $-\Delta\rho_{mag}/\rho$ and that this tendency increases as the disorder increases. It should be remembered that the neutron irradiated sample is more disordered than the $Ce_{.99}La_{.01}Al_3$ sample. One might question using the value of $\Delta\rho^+$ for the disordered samples that was estimated for $CeAl_3$, but one can see from Fig. 4 that our results for the disordered samples are not overly sensitive to the estimate of $\Delta\rho^+_{mag}$. These effects can be understood in terms of the description of Andres et al.[4] Disorder will disrupt the coherence, inhibit the formation of a narrow band, and consequently inhibits the transition into the ground state.

In Fig. 5, we plot something akin to the fractional change in the magneto-resistance due to disorder. We define f as $-\Delta\rho_{mag}(T)$

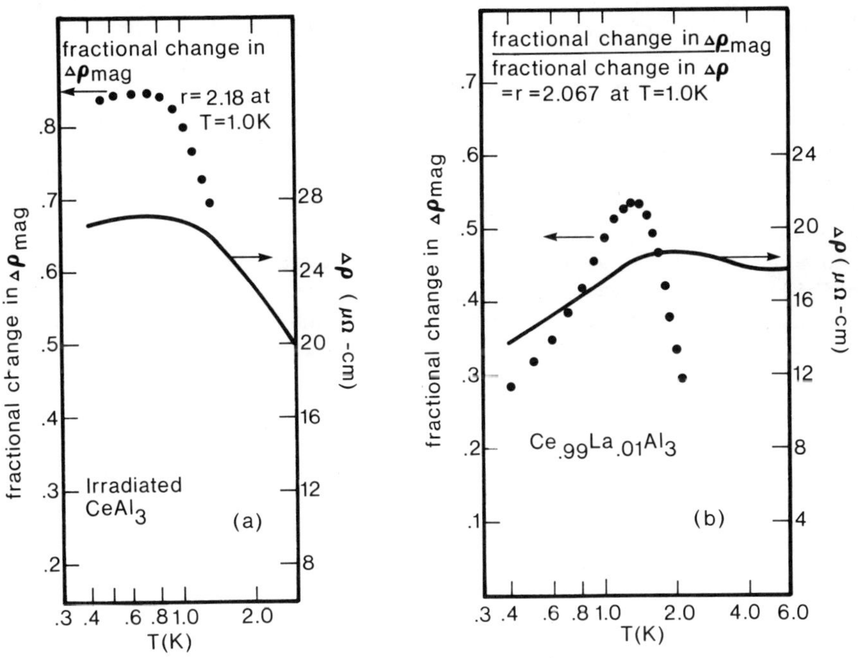

Fig. 5 Comparison of f and $\Delta\rho$ as a function of temperature for a) the neutron irradiated sample and b) $Ce_{.99}La_{.01}Al_3$.

in the disordered sample minus $-\Delta\rho_{mag}(T)$ of $CeAl_3(T)$ divided by $-\Delta\rho_{mag}(4)$ of $CeAl_3$. Also shown is $\Delta\rho$. One observes, that f and $\Delta\rho$ possess some similarities in their temperature dependence. In the case of the neutron irradiated sample, Fig. 5a, both become temperature independent below 1 K. For $Ce_{.99}La_{.01}Al_3$ both peak at roughly the same temperature. The magnitudes of f and $\Delta\rho$ are correlated in similar ways for both kinds of disorder. We define $r = f/(\Delta\rho/\rho_N)$ where ρ_N is the resistivity of $CeAl_3$ at 4 K. The value of r at 1 K is 2.18 for our neutron irradiated sample and 2.07 for $Ce_{.99}La_{.01}Al_3$.

It is worth noting that chemical disorder at high concentrations of $Ce_{1-x}La_xAl_x$ (x = 0.1 and 0.2) drastically affects the specific heat.[16] As one alloys with La the specific heat anomaly is shifted to lower temperatures and the higher sharp feature in the specific heat is suppressed.

We have demonstrated that disorder produced by neutron irradiation or chemically by alloying with La drastically affects the resistivity and magneto-resistance. The effect on the magnetoresistance is consistent with the disorder disrupting the coherence and suppressing the formation of a narrow band near ε_F. It is not clear over what spatial range the disorder disrupts the coherence. Near the disordered region one probably would expect the coherence to be disrupted even at T = 0. The results shown in Fig. 4 are consistent with this view. An extrapolation of the curve of $-\Delta\rho_{mag}$ for the irradiated sample does not go to zero. Neutron measurements on $CeAl_3$ at lower temperatures than those previously employed are planned to investigate the magnetic scattering to see whether we can observe any effects that may be associated with the formation of a narrow band near ε_F. For example, we might see η, the factional occupancy of the localized $4f^1$ configuration, decrease. Our work raises several further questions, some of which are easily answered by further experiments. Is the susceptibility increased within the disordered regions? Why does $\Delta\rho$ have the temperature dependence we have observed? We plan a systematic investigation of this system and a determination of whether the effects we have observed are important in other systems.

We wish to thank T. Scott for his technical assistance and T. Arai and J. E. Robinson for helpful conversations.

*Work performed under the auspices of the U.S. Energy Research and Development Administration.

1. B. Batlogg, E. Kaldis, A. Schlegel, G. von Schultess and P. Wachter, Solid State Comm. 19, 673 (1976).
2. K. H. Mader and W. M. Swift, J. Phys. Chem. Solids 29, 1759 (1968).

3. A. S. Edelstein, C. J. Tranchita, O. D. McMasters and K. A. Gschneidner, Jr., Solid State Comm. 15, 81 (1974).
4. K. Andres, J. E. Graebner, and H. R. Ott, Phys. Rev. Letters 35, 1779 (1975).
5. J. V. Mahoney, V.U.S. Rao, W. E. Wallace, R. S. Craig, N. G. Nereson, Phys. Rev. B 9, 154 (1974).
6. A. S. Edelstein, H. R. Child, and C. J. Tranchita, Phys. Rev. Letters 36, 1332 (1976).
7. K. H. J. Buschow, H. J. van Daal, F. E. Maranzana, and P. B. van Aken, Phys. Rev. B 3, 1662 (1971).
8. A. S. Edelstein, R. Majewski, S. K. Sinha, T. Brun, C. A. Pelizzari, Proc. Neutron Scattering Conf., Gatlinburg, Tenn., 1976; ed. R. M. Moon; A. P. Murani, K. Knorr, and K. H. J. Buschow, Second International Conf. on Crystal Field Effects in Metals and Alloys, Zurich, 1976.
9. T. H. Blewitt and T. J. Koppennall, Irradiation Hardening in Copper and Copper Alloys, IMD-AIME Symp. on Radiation Effects, W. F. Shelly, ed., Gordon and Breach, New York, 37, 561-625 (1967).
10. Defects and Radiation Damage in Metals, by M. W. Thompson, Chap. 5 (Cambridge Un. Press, Cambridge, 1969).
11. A. S. Edelstein, T. O. Brun, G. H. Lander, O. D. McMasters, and K. A. Gschneidner Jr., in Magnetism and Magnetic Materials-1974, AIP Conference Proceedings No. 24, edited by C. D. Graham, Jr., G. H. Lander, and J. J. Rhyne (American Inst. of Phys., New York, 1975), p. 428.
12. K. Yosida, Phys. Rev. 106, 893 (1957).
13. J. W. Garland, K. H. Bennemann, A. Ron, and A. S. Edelstein, J. Appl. Phys. 41, 1148 (1970).
14. M. T. Beal-Monod and R. A. Weiner, Phys. Rev. 170, 552 (1968).
15. P. Monod, Phys. Rev. Letters 19, 1113 (1967).
16. M. H. van Maaren, K. H. J. Buschow and H. J. van Daal, Solid State Comm. 9, 1981 (1971).

COMMENTS

ANDRES: Would you not expect the virtual bound states to form a narrow f band at low temperature? Have you observed any change in the coefficient of the T^2 term in the low temperature resistance of your irradiated samples?

EDELSTEIN: The answer to the first question is perhaps and the answer to the second question is no.

TOURNIER: I would like to make a few comments about the experiments we have done on $CeAl_3$. We observe that the peak in specific heat (at normal pressure) at 4°K sometimes is not present, suggesting that this peak is not an intrinsic property of the system but is due to the presence of another phase, e.g., $CeAl_2$. However, it seems that the specific heat peak at 6°K is always present. When we apply pressure the linear term in the specific heat decreases as much as a factor of 3, but there is no change in the position of the 6°K peak. We also observe a decrease in susceptibility under pressure as in all Kondo systems. In some samples we observe remnant magnetization which disappears above 6°K. Since the susceptibility follows the Curie-Weiss law with a large negative Curie temperature, we attribute the remnant magnetization to uncompensated antiferromagnetic ordering below 6°K, this in turn resulting from an unequal number of spin-up and spin-down spins caused by defects. This ordering process takes place in the Kondo state and it is similar in nature to the one that takes place in singlet ground state systems like Praseodymium.

KAPLAN: Have you seen magnetic ordering in your neutron scattering experiments?

EDELSTEIN: No, at least not down to the temperatures we have looked at. The question of the 6°K specific heat anomaly is open. Higher order derivatives of the resistance do show some structure at 6°K.

FONER: When you make corrections to the magnetoresistivity do you take into account that the magnetic moment is temperature and field dependent?

EDELSTEIN: That is an open problem that has to be looked into.

VALENCY CHANGES OF Ce, Yb AND U IN INTERMETALLIC COMPOUNDS WITH Ni AND Cu

K. H. J. Buschow, M. Brouha, H. J. van Daal and
A. R. Miedema
Philips Research Laboratories
Eindhoven, The Netherlands

1. INTRODUCTION

It is well known that in intermetallic compounds Yb as well as Ce can have two different valence states. Recently indications were obtained that this is also the case for U. The valence adopted by these elements depends in general on the composition of the particular intermetallic compound, on the temperature and on the applied pressure. In some cases the valence state seems to be less clearly defined in that the elements mentioned show valences apparently intermediate between two integral values.

In this investigation we have studied the valence states of Yb, Ce and U in pseudo-binary compounds with Ni and Cu. The differences in electronegativities and electron densities between Yb, Ce or U on the one hand and Ni or Cu on the other are such that in compounds with Ni the higher valence state prevails whilst in compounds with Cu the lower valence state is dominant (1). Gradual replacement of Cu by Ni is expected therefore to induce a valence change. The nature of this valence change has been studied in more detail by measurements of the concentration dependence of the lattice constants, while further information of the progress of the valence change was derived from the results of magnetic measurements and incidentally also from the results of resistivity measurements under applied pressure.

2. EXPERIMENTAL

The samples $Ce(Ni_{1-x}Cu_x)_5$, $Yb(Ni_{1-x}Cu_x)_5$ and $U(Ni_{1-x}Cu_x)_5$ and $YbNi_{1-x}Cu_x$ were prepared by arc melting followed by vacuum annealing for the first two series. The compounds $YbNi_{1-x}Cu_x$ were prepared by sealing the appropriate amounts of starting materials into a molybdenum container and heating for 1 hr. at 1000°. X-ray diffraction was performed on a Philips X-ray powder diffractometer with CuKα radiation in combination with an X-ray monochromator. The magnetic measurements were made on polycrystalline single-phase samples using an adaptation of the Faraday method. The resistivity of $YbNi_{1-x}Cu_x$ (at 100 K) as a function of the applied pressure was measured up to 150 kbar between Bridgman anvils. As calibration points for the pressure applied the phase transitions of Bi(I → II, III → V), Ba(I → II) and Pb were used (2,3).

3. RESULTS AND DISCUSSION

The magnetic properties of the system $U(Ni_{1-x}Cu_x)_5$ have already been reported before (4). The compound UNi_5 has a perfectly constant magnetic susceptibility ($\chi_M = 0.8 \times 10^{-4}$ cm^3/mole) up to temperatures higher than 1000 K. This behaviour has been interpreted as being due to uranium ions in the tetravalent state (U^{4+} with $5f^2$, comparable to Pr^{3+} with $4f^2$). The expected high crystal field splitting of the $5f^2$ state leads to a non-magnetic ground state. The magnetic susceptibility is then almost exclusively determined by the temperature independent Van Vleck contribution. Upon replacement of Ni by Cu the susceptibility increases gradually and becomes temperature dependent. The magnetic susceptibility of UCu_5 does not show Curie Weiss behaviour. A Neel type paramagnetic - antiferromagnetic transition is observed at 15 K. The transition from nonmagnetic UNi_5 to magnetic UCu_5 has been interpreted in terms of a decreasing valency of the uranium atoms from 4+ to 3+. In the 3+ valence state ($5f^3$, comparable to $4f^3$ of Nd) there will be a magnetic ground state for any crystal field (Kramers ion). Indications of a valence change in $U(Ni_{1-x}Cu_x)_5$ were furthermore obtained from measurements of the lattice constant, the specific heat, electrical resistivity and absolute Seebeck coefficient (see Ref. 4).

The magnetic properties of $Ce(Ni_{1-x}Cu_x)_5$ are shown in Fig. 1. For $CeNi_5$ Curie Weiss behaviour is followed only approximately at temperatures in excess of 130 K. The effective moment (2.6 μ_B/Ce) is close to the value expected for Ce^{3+} (2.54 μ_B/Ce). The asymptotic Curie temperature is strongly negative (-120 K). The most prominent feature in the χ^{-1} versus T curve of this compound is the occurrence of a minimum close to

T = 90 K. Below this temperature the reciprocal susceptibility first increases and at the lowest temperatures decreases again. This latter decrease is probably not an intrinsic property of $CeNi_5$ but has to be accounted for by the presence of a few tenths of at .% of Ce_2O_3. The minimum in the χ^{-1} versus T curve, or the maximum in the corresponding χ versus T curve, strongly suggests that Ce is present in $CeNi_5$ in an intermediate valence state. In a previous paper it was shown already that for a situation where an excited valency state with an effective moment μ and degeneracy 2J+1 is lying at an energy ΔE above the ground state, having an effective moment equal to zero and a degeneracy equal to 1 the susceptibility is given by (5)

$$\chi_f(T) = \frac{N\mu^2}{3k_B} \frac{c(T)}{(T+\theta)} \tag{1}$$

where $C(T) = (2J+1)\exp(-\Delta E/k_BT)/[1+(2J+1)\exp(-\Delta E/k_BT)]$ and N denotes the concentration of Ce ions. According to this expression Curie Weiss behaviour should be observed at high temperatures with an effective moment close to the moment μ of the excited state. Upon lowering of the temperature the exponential decrease of the population of the excited state outweighs the Curie Weiss like increase in magnetization and this results in a maximum of $\chi(T)$ or a minimum of $\chi^{-1}(T)$. Qualitatively expression (1) describes the experimentally observed temperature dependence of the susceptibility rather well if one uses $\mu = 2.54\ \mu_B$/Ce, and J = 5/2 appropriate to Ce^{3+} and θ = 120 K.

It is to be noted that a similar temperature dependence of χ^{-1} was observed for $YbAl_3$. A quantitative fit with expressions like (1) was obtained in two different occasions. Havinga et al. (5) used the (Friedel-Anderson) virtual bound type approach. Essential in this approach is, that in order to describe the observed $\chi(T)$, the contribution χ_f in expression (1) has to be combined with a strongly temperature dependent Pauli paramagnetic contribution χ_p due to the valence state prevailing at the lowest temperatures. In this description θ reflects the more or less enhanced interaction θ_p of the magnetic moments corresponding to the excited state, implying that this quantity is temperature dependent (proportional to the magnetic moment concentration, $\theta = c\theta_p$). Sales and Wohlleben (6), using the ionic based approach of Hirst (7), were able to fit the data with a constant θ. In this latter approach c(T) is replaced by c(T+θ) and θ represents a characteristic temperature T_{sf}, being a measure of the transition rate ω_f between the two valency states E_n and E_{n-1} ($k_B\theta = k_BT_{sf} = h\omega_f$). Unfortunately a quantitative fit of all the data of $CeNi_5$, which could decide between the two models, is somewhat hampered by the presence of Ce_2O_3 (or better $CeO_{1.5\pm\delta}$) impurities for which the magnetic properties (unlike Yb_2O_3) are not yet well

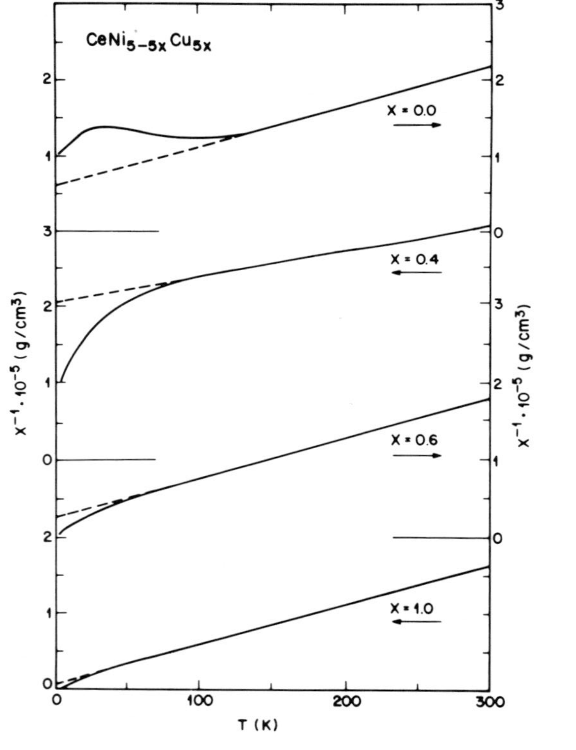

Fig. 1.

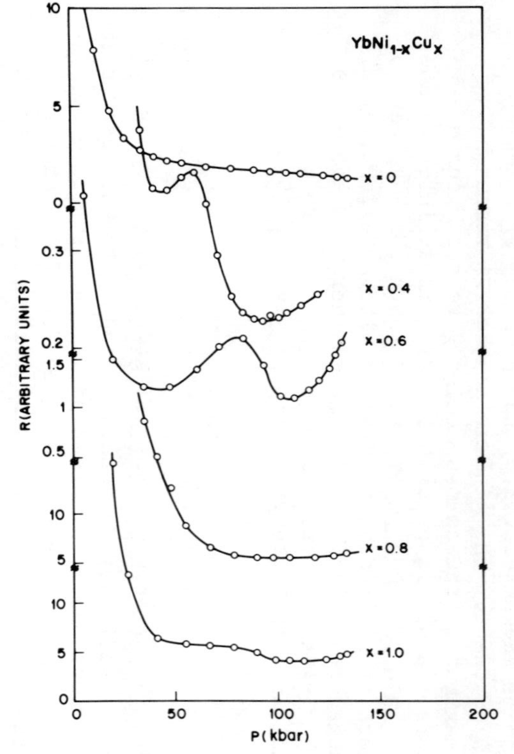

Fig. 2.

Fig. 1. Temperature dependence of the reciprocal susceptibility of compounds of type $CeNi_{5-5x}Cu_{5x}$.

Fig. 2. Pressure dependence of the electrical resistivity of compounds of type $YbNi_{1-x}Cu_x$, measured at 100 K. The initial decrease of the resistivity is due to the brittle nature of the samples. The pressure applied will reduce the empty space associated with cavities and microcracks and lead to a substantial decrease of the electrical resistivity.

defined. The most critical fit, undoubtedly, is that involving epxression (1) with $c(T)$ replaced by $c(T+\theta)$ and with θ fixed ($\theta = T_{sf}$). Since θ is given by the Curie Weiss intercept it contains only a single adjustable parameter, ΔE. Attempts have therefore been made to attain such a fit by restricting ourselves to the high temperature data ($T > 50$), where the impurity contribution is less important. A fit could be obtained, however, only by using $\chi(T) = N(2.54)^2\mu_B^2 c(T+\theta)/3k_B(T+\theta) + \chi_p(T)$, with $\Delta E \approx 375$ K, where $\chi_p(T)$ represents a temperature dependent paramagnetic contribution ($\chi_p = 1.6\times10^{-6}$ emu/g at 300 K increasing to $\chi_p = 3\times10^{-6}$ emu/g at 50 K). The necessity of the $\chi_p(T)$ term does not disprove the validity of the ion-based approach, however, since such a contribution is expected to arise from the Ni 3d band. When R is trivalent in RNi_5 the Fermi level resides in a minimum of the density of states of the 3d band (8), implying a low value of χ_p. The valence change involved when $CeNi_5$ is cooled, very probably will increase the Fermi level and hence lead to the required increase of χ_p. Upper limits of the values of χ_p are set by the susceptibilities determined by bulk measurements on poly- and single-crystalline materials like $LaNi_5$ or YNi_5 ($\chi \approx 5\times10^{-6}$ emu/g) (8). The values of χ_p applied certainly fall within this range.

We mentioned already that upon continued Cu substitution in $CeNi_5$ one expects a stabilization of the trivalent state with respect to the tetravalent one. If the trivalent state becomes the ground state (upon increasing x) one can equally well use expression (1) to describe the temperature dependence of the susceptibility, noting however, that ΔE is now a negative quantity. For a large enough energy separation, Curie Weiss behaviour is expected with an effective moment close to 2.54 μ_B. All of the magnetic data shown in Fig. 1 for the compounds containing Cu can indeed be explained in this way: A good fit with expression (1) could be reached by using again only the data obtained above about 50 K where the oxyde impurity contributions are of little importance. Furthermore the data had to be corrected a priori by a temperature independent Pauli paramagnetic contribution approximately given by $(1-x)\times1.6\times10^{-6}$ emu/g, where 1.6×10^{-6} emu/g is the value used in $CeNi_5$. The temperature indendency is indeed expected now since in the temperature range considered, temperature variation will not lead to significant changes in the relative occupation of the two valence states so that the temperature induced electron transfer Ce → Ni can be disregarded. It can be concluded therefore that in all the compounds of Fig. 1 containing Cu one has a ground state corresponding to the lower valency state. It is interesting to compare the values of θ of these compounds. After applying the correction due to the Pauli paramagnetic term the Curie Weiss intercepts are:

$CeNi_3Cu_2$: θ = 440 K
$CeNi_2Cu_3$: θ = 30 K
$CeCu_5$: θ = 17 K

In terms of the ion-based model the transition rate ω_f between the valence states is proportional to $\exp-(|\Delta E|/\Delta)$, where Δ represents the 4f conduction electron mixing strengths. Since in this model $T_{sf} = \theta$ and $h\omega_f = k_BT_{sf}$ the observed decrease of the Curie Weiss intercepts, in going from $CeNi_3Cu_2$ to $CeCu_5$, reflects an increasing value of ΔE. This trend agrees closely with the above-mentioned preference of the lower valence in compounds with large amounts of Cu. The observed change in the θ_p values is much more difficult to interpret in terms of the virtual bound state approach. For all three compounds the values of **c** in expression (1) come close to 1 in the temperature range considered (at all temperatures one has $6/7 < c < 1$). Hence $\theta(c) = c\theta_p \simeq \theta_p$, where θ_p is a measure of the strength of the magnetic coupling between the localized moments. In contradistinction to $CeNi_5$ and $YbAl_3$ the magnetic state is now the lowest state in the above compounds. Within the virtual bound state model it is therefore difficult to understand why a mean interaction strength of θ_p = 400 K (in $CeNi_3Cu_2$) does not lead to magnetic ordering upon cooling.

Finally we would like to mention that the deviation of the copper rich compounds from Curie Weiss behaviour at low temperatures may not only be due to oxyde impurities but could equally well be due to crystal field effects.

For reasons that will become clear below the system $YbNi_{1-x}Cu_x$ does not lend itself to a study of the concentration dependent changes of the magnetic susceptibility. Instead we have studied the changes in the curves of the electrical resistivity measured as a function of applied pressure. The results are shown in Fig. 2. From magnetic data published in literature it follows that Yb in YbNi has adopted the higher valence state, i.e. Yb^{3+} (YbNi: μ_{eff} = 4.5 μ_B/Yb; θ = 15 K) (9). No Curie Weiss behaviour is observed for YbCu (10). In this compound χ rises slightly with temperature. According to the analyses given by Debray et al. (10) Yb has a mixed valency here with $E_n(4f^{14})$ lying 4500 to 6000 K below $E_{n-1}(4f^{13})$. Since applied pressure stabilizes the E_{n-1} state relative to the E_n state one can expect a pressure induced valence change only for those compositions where the E_n state lies lowest. Applied pressure will cause then a crossing of the E_n and E_{n-1} states. These two states will meet each other necessarily at the Fermi level so that the level crossing will be accompanied by a marked effect in the resistivity. In the two compounds YbCu and $YbNi_{0.2}Cu_{0.8}$ the energy separation between the two states is apparently too high to give rise to the effect mentioned. The maxima in the p(P) curves of $YbNi_{0.4}Cu_{0.6}$ and $YbNi_{0.6}Cu_{0.4}$ show that here the requirements for level crossing are met. It follows from the results described above that in $YbNi_{1-x}Cu_x$ the higher valence state is lowest in energy only when x is small or zero. For $x > 0.4$ the lower valence state is lowest in energy which means that a concentration induced level crossing occurs in the range $0 < x < 0.4$. This is the same concentration range as in $Ce(Ni_{1-x}Cu_x)_5$.

As a last example we would like to mention the valence changes in $Yb(Ni_{1-x}Cu_x)_5$. Due to the presenceof a ferromagnetic impurity (Yb_2Ni_{17}) in addition to Yb_2O_3 a meaningful interpretation of magnetic data of the Ni rich compounds can not yet be given. It follows, however, from a comparative study of the lattice constants of the RNi_5 compounds (R = La to Lu) that Yb in $YbNi_5$ is trivalent (11). The compound $YbCu_5$ (or better $YbCu_{5+\delta}$, see below) has a more or less temperature independent magnetic susceptibility ($\chi = 1\times10^{-6}$ emu/g) and this points to Yb in the divalent state. Increasing x in $Yb(Ni_{1-x}Cu_x)_5$ therefore is expected to involve a change in valency, although details of this valence change have to await further experimental studies.

We will now discuss the changes in the crystal lattice that accompany the above-mentioned valency changes. The experimental results are depicted in Figs. 3-6. In general the lower valency state entails a larger metallic radius than the higher valency state so that, with increasing x, one expects an increase in lattice constants. In the system $U(Ni_{1-x}Cu_x)_5$ and $Ce(Ni_{1-x}Cu_x)_5$ this increase in lattice constant seems to proceed in a more or less continuous fashion, although it would seem that the largest effects occur at the Cu rich ends. A quite different situation is encountered in the system $YbNi_{1-x}Cu_x$ (Fig. 5). Here the change in lattice constants is discontinuous. Different heat treatments did produce the same results so that effects due to clustering can largely be excluded. An even more complicated situation arises in $Yb(Ni_{1-x}Cu_x)_5$. Here a structure different from the hexagonal $CaCu_5$ type is adopted in the region $0.57 < x < 0.96$. In this region the structure has become cubic, $AuBe_5$ type (the cubic region is separated from the hexagonal $CaCu_5$ type region by two-phase regions). This complicated behaviour needs further explanation: All rare earth compounds RNi_5 (R = La to Lu; Ca) are hexagonal ($CaCu_5$ type). In terms of metallic radii and valence electron concentration this structure can accommodate Yb in the divalent as well as in the trivalent state. The situation for RCu_5 is different: The elements R = La to Gd and R = Ca also adopt the $CaCu_5$ type but the elements R = Tb to Lu give rise to the cubic $AuBe_5$ type. Consequently, if Yb would stay trivalent or predominantly trivalent at all x in $Yb(Ni_{1-x}Cu_x)_5$ one would expect only one structure change, i.e. from hexagonal ($CaCu_5$) to cubic ($AuBe_5$).

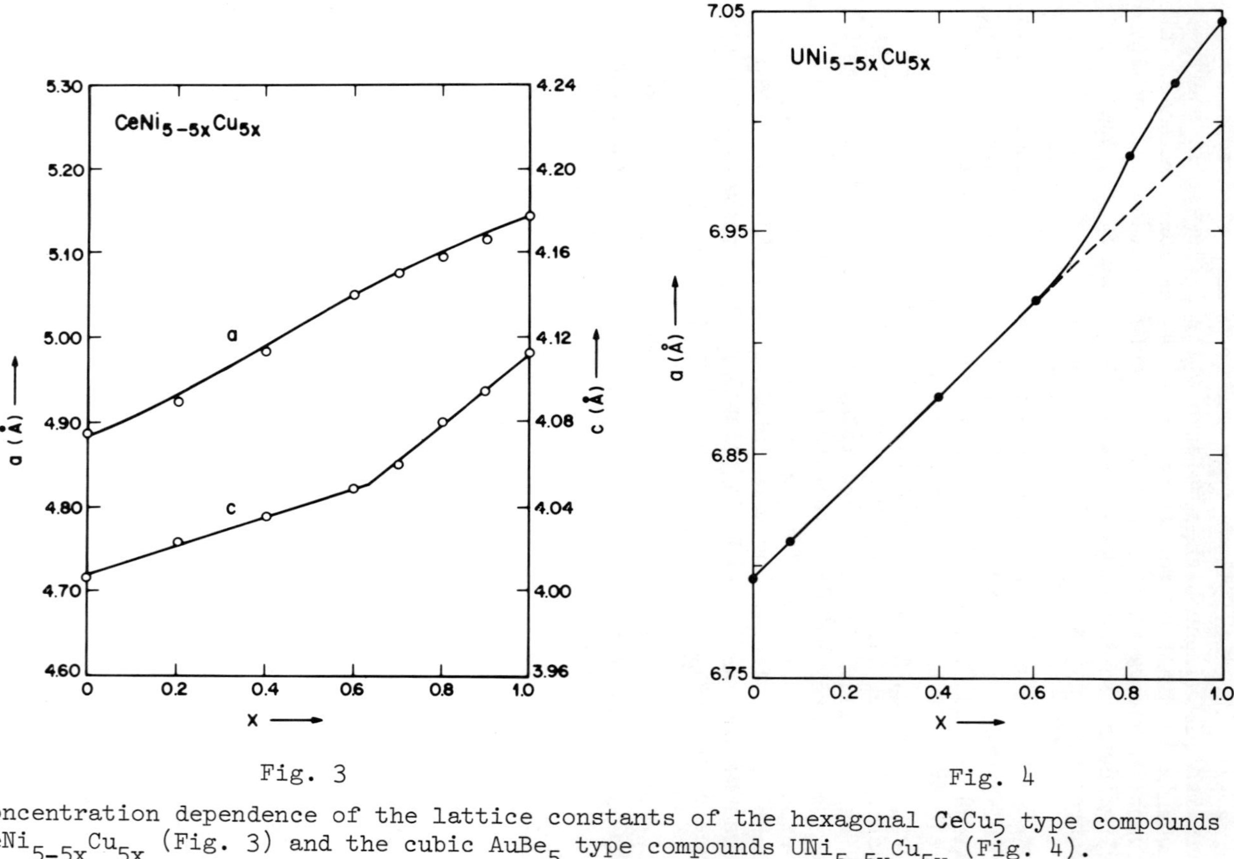

Fig. 3

Fig. 4

Concentration dependence of the lattice constants of the hexagonal $CeCu_5$ type compounds $CeNi_{5-5x}Cu_{5x}$ (Fig. 3) and the cubic $AuBe_5$ type compounds $UNi_{5-5x}Cu_{5x}$ (Fig. 4).

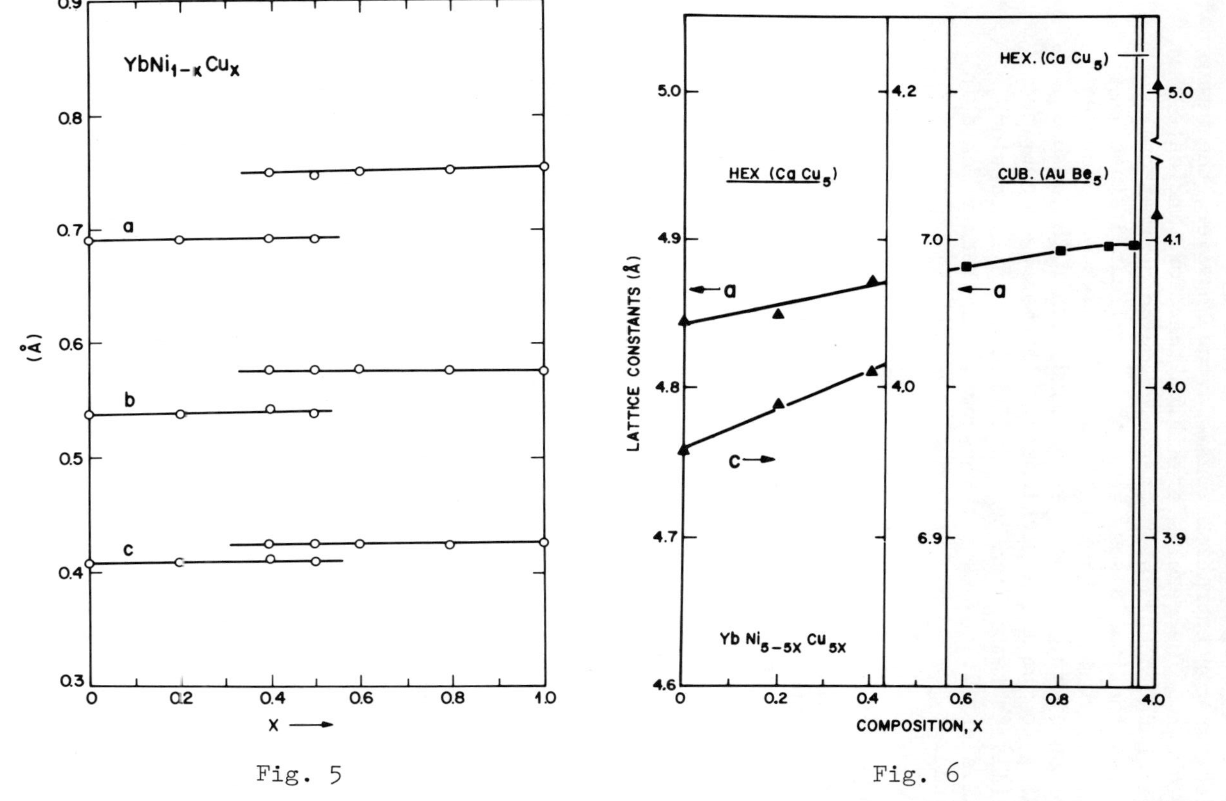

Fig. 5

Fig. 6

Fig. 5. Concentration dependence of the lattice constants of the compounds $YbNi_{1-x}Cu_x$ (orthorhombic FeB type).

Fig. 6. Concentration dependence of the crystal structures and the lattice constants of the compounds $YbNi_{5-5x}Cu_{5x}$.

As soon as Yb in these compounds would lower its valence with increasing x, the larger metallic radius and the decreasing valence electron concentration would favour, however, the hexagonal $CaCu_5$ type again. The results of Fig. 6 therefore indicate that the valence change occurs at rather high Cu concentration. Actually the situation is somewhat more complicated since in the $CaCu_5$ type lattice of $YbCu_5$ part of the Yb sites are occupied by pairs of Cu atoms, making the compound richer in Cu than corresponding to the 1:5 stoichiometry. It has been outlined elsewhere (11) that this can be considered as an indication that the divalent state has not yet completely been reached in this compound.

4. CONCLUDING REMARKS

We have studied the valence changes of U, Ce and Yb in pseudo-binary compounds of the type $R(Ni_{1-x}Cu_x)_n$. It is found that the lattice can respond in three different ways to the change from the higher to the lower valence when x is increased: (i) the lattice constants increase continuously, as is observed in $U(Ni_{1-x}Cu_x)_5$ and $Ce(Ni_{1-x}Cu_x)_5$; (ii) the lattice constants increase discontinuously, as is observed in $YbNi_{1-x}Cu_x$; (iii) apart from a changing lattice constant the valence change can also give rise to changes in crystal structure and slight deviation from the stoichiometric composition ($Yb(Ni_{1-x}Cu_x)_5$).

A physically meaningful interpretation of the susceptibility data of compounds of the series $Ce(Ni_{1-x}Cu_x)_5$ with $x \geq 0.4$ could only be given in terms of the ion-based model of interconfiguration fluctuations and make the applicability to these compounds of the resonant scattering theory based on virtual bound states less likely. In $CeNi_5$ the ground state is nonmagnetic; the susceptibility arises through thermal excitation to the magnetic state.

REFERENCES

1. A. R. Miedema, R. Boom and F. R. de Boer, J. Less-Common Met. 41, 283 (1973); 46, 67 (1976); 46, 167 (1976).
2. E. C. Lloyd, "Accurate Characterization of High Pressure Environment", Proc. Symp. NBS Gaithersburg, 1971, pp. 1.
3. H. Mii, I. Fujishiro, M. Senoo and K. Ogawa, High Temp.-High Pressures 5, 155 (1973).
4. H. J. van Daal, K. H. J. Buschow, P. B. van Aken and M. H. van Maaren, Phys. Rev. Lett. 34, 1457 (1975).
5. E. E. Havinga, K. H. J. Buschow and H. J. van Daal, Solid State Comm. 13, 621 (1973).
6. B. C. Sales and D. K. Wohleben, Physical Rev. Letters 35, 1240 (1975).

7. L. L. Hirst, Phys. Kondens. Mater. 11, 255 (1970).
8. D. Gignoux, D. Givord and A. Del Moral, Solid State Comm. 19, 891 (1976).
9. J. P. C. Klaasse, W. C. M. Mattens, F. R. de Boer and P. F. de Chatel, AIP-Proc. (in the press).
10. D. Debray, B. F. Wortmann and S. Methfessel, Phys. Rev. (B) (in the press).
11. K. H. J. Buschow, J. Less-Common Met. 26, 329 (1972).

COMMENTS

NOWIK: Recent Mössbauer experiments on $EuNi_{5-x}Cu_x$ by Bauminger et al in Jerusalem indicate the presence of both Eu^{2+} and Eu^{3+} and that the relative intensities of these components change with concentration, demonstrating the existence of strong local environment effects.

BUSCHOW: Yes. I more or less expected this question, because data of a similar nature was presented some time ago by Ross et.al from Manchester and they also concluded that there were environmental effects. In the model of Sales and Wohlleben the transition rate, or the temperature associated with it, takes the form of a Curie-Weiss intercept. Hence, one can compare this type of analysis with the normal magnetic type of Curie-Weiss behavior. In this regard, consider the case of $GdAl_2$. Magnetic dilution of this compound results in a whole gamut of internal fields. Nevertheless, one finds a nice straight line Curie-Weiss behavior--the distribution of internal fields enters as an average effect. Similarly, in the Sales-Wohlleben formalism the Curie-Weiss intercept will be presumably an average of the different transition rates.

WOHLLEBEN: It seems to me much harder to make Eu intermediate-valent than anything else. If that is so then one might understand from that tendency the fact that Eu wants to be either-or and will not easily fluctuate at at high rate. If you have it fluctuating, a small change of parameters, e.g. a change in local environment, will easily drive the Eu ion out of the region of instability again.

NOWIK: I would like to add that in the Mössbauer spectra one sees, in addition to a divalent line and a trivalent line, also a line that moves as a function of temperature which really represents a fluctuating state. It means that for a certain configuration you can create Eu ions which are in the fluctuating state.

THEORY AND EXPERIMENT ON SmB_6

Tadao KASUYA, Kenichi KOJIMA[+] and Mitsuo KASAYA

Department of Physics, Tohoku University

Sendai, Japan

Experiments and theory on SmB_6 are reviewed critically. Then new experimental results for $Sm_{1-x}V_xB_6$ with vacancies V, $0 < x < 1/3$, are shown. A model in which 4f hole states of Sm are described by an extended Hubbard model with collective spin fluctuation modes is proposed. The theory explains neary all the anomalous properties self-consistently.

I. Introduction

Various Sm-compounds show the so called valence fluctuating state either under pressure or in alloying[1]. Among them, SmB_6 is unique in the point that it shows the valence fluctuating state in the atmospheric pressure without alloying. Therefore, it is most suitable to study the valence fluctuating state in detail.

A lot of experimental work has been done already and the following facts are now clear. (1) The ratio of Sm^{+++} and Sm^{++} is about 7:3 and is nearly temperature independent. This is checked by the measurements of X-ray absorption[2], lattice constant[3], X-ray photo emission[23] and Mössbauer isomer shift[4]. Note that the Mössbauer measurement can see only the time average of Sm^{++} and Sm^{+++}, which indicates that even at lowest temperature the valence of each Sm atom is fluctuating faster than the nuclear life time, about 10^{-9} sec. (2) Even if the ground state of $Sm^{+++}(4f^5)$ has magnetic moment, $S = 5/2$, $L = 5$, $J = 5/2$, $g = 2/7$, this is not observed in any experiment. The temperature dependence of the susceptibility[5] is fairly well fitted by that of 40% Sm^{++} and the Mössbauer measurement[4] can not see any spin ordering down to 1°K. This should have strong correlation with the valence fluctuation mentioned in (1). (3) All divalent rare earth hexaborides, $R^{++}B_6$, except EuB_6, which exists always as the degenerated semiconductor[6], are insulator, while all trivalent compounds $R^{+++}B_6$, are good metal with one conduction electron per R^{+++} atom, which is checked by the Hall constant measurement. In particular, in LaB_6, even the detailed information on the Fermi surface and the effective masses is obtained[7]. Therefore, SmB_6 should have conduction electrons of about 0.7 per Sm atom. The resistivity measurement shows[5,8], however, rather a n-type semiconductor like behavior with an impurity band, which is however, different from the usual semiconductors in the point that no essential change is observed even if the donor Gd^{+++} or the accepter Eu^{++} is doped[9]. (4) The thermoelectric power[10] has a small positive value below 2°K, changes the sign rapidly having a minimum value -350 μv/deg at 10°K. Then again changes the sign at 200°K, above which it has a small positive value. The Hall constant measurement shows the similar behavior, changing the sign twice[8]. (5) Compared with LaB_6, SmB_6 has a large anomalous specific heat at low temperature[8]. A simple calculation shows that nearly all the freedom of 4f states in Sm^{+++} is released by 50°K. (Fig. 4)

Based on their Hall constant measurement mentioned above, Nickerson et.al. proposed a very complicated band structure model[8]. It is, however, hardly acceptable by the following reason. First of all, extrapolation from the Hall constant measurement on EuB_6[11] indicates a considerable contribution of the anomalous Hall effect even in SmB_6. Furthermore, the Hall effect in the complicated bands and mechanisms can not be treated by a simple band model. Therefore, it is dangerous to guess the band structure based on the simplest band model without the anomalous Hall effect. In the second, they assumed real localization of a 5d conduction electron on each Sm^{+++} site to explain no observation of the naked Sm^{+++} moment. This is highly unacceptable from the fact of the observed really wide 5d band as mentioned in (3). The more natural model should be as follows. The $5d_{eg}$ bands are occupied by the conduction electrons of 0.7 per Sm atom. Occupied and unoccupied 4f levels in Sm are gathering near the Fermi energy E_f. They are the main source of the anomalous properties. The problem is how to explain the all anomalous properties mentioned above consistently on this model.

Recently, it was shown that SmB_6 can exist with a lot of vacancy randomly situated on Sm sites[12], namely, $Sm_{1-x}V_xB_6$, in which V means the vacancy and x can cover the range $0 < x < 1/3$. Note that $x \sim 1/3$ means that nearly all Sm should be Sm^{+++} and only very small amount of the conduction electrons can exist. Based on this fact, we studied various properties of $Sm_{1-x}V_xB_6$ experimentally to get more information on the valence fluctuating state of SmB_6. Some of them are shown in the next section. Based on the new informations together with the existing experimental results mentioned before, a new theoretical model is proposed to explain the experiment consistently. This is shown in Section III.

II. Experiment

Measurements were done on single crystals. The detail of the sample preparation is given in another paper[13]. The sample composition was determined both by the mass density measurement and the chemical analysis. Change of color from metallic to less metallic with increasing number of the vacancy, or the decreasing number of conduction electrons, is also significant. Note, however, that due to small sizes of samples used, there are considerable errors for the estimation of the composition, may by 5%, or more in some cases, in the Sm concentration.

Temperature dependence of resistivity is shown in Fig.1a for three samples, nearly stoichiometric, nearly $x \sim 1/3$, and intermediate. It is remarkable that, even though the high temperature resistivity increases as x increases, or the number of the conduction electrons decreases, the low temperature resistivity decreases as x increases. Below 4°K, increasing of resistivity slows down similar to the impurity conduction in the usual semiconductors. To check it more carefully, log ρ are plotted both against T^{-1} and $T^{-1/4}$ as shown in Fig.1b for the sample of $x \sim 0$. It is clear that, even though the temperature range is not so wide, $1.6 \sim 4$°K, the $T^{-1/4}$ low is definitely better than T^{-1}, showing existing of the so called variable range hopping[14] among the Anderson type localization[15]. Note that we do not expect the variable range hopping among the atomic localized 4f states. In the lower temperature, SmB_6 shows negative magnetoresistance given by $\Delta\rho/\rho = 6\times10^{-6}H^2$, in which H is given by the unit of kOe.

Temperature dependence of the magnetic susceptibility χ is shown in Fig.2a for the same samples as the resistivity measurement. The following features are remarkable. First, as x increases, the Curie-weiss type part of the susceptibility appears and grows indicating that some Sm^{+++} are trapped near at the vacancies and show atomic behavior. Secondly, even though the number of Sm^{++} decreases as x increases, the susceptibility per Sm atom in high temperature increases slightly as x increases. The atomic susceptibilities of Sm^{++} and Sm^{+++} are also shown in Fig.2a for comparison. It should be noted that the ground state of Sm^{++} is the singlet, S = 3, L = 3 and J = 0, while that of Sm^{+++} is magnetic J = 5/2. However, because of its small value of g-factor, g = 2/7, the temperature independent part is comparable with the Curie part above T = 100°K, or it is given below room temperature by[16].

$$\chi_{+++} = 0.124\left(\frac{0.713}{T} + 0.00556\right) = \frac{0.0885}{T} + 0.00069\ \frac{\text{emu}}{\text{mole}}\ . \qquad (1)$$

Furthermore, under the cubic field, the six folds J=5/2 states split into the doublet and the quartet, which is split further by a local field due to vacancies. Hereafter, we simply neglect the small local field effect. Then in temperature well below the cubic crystal field

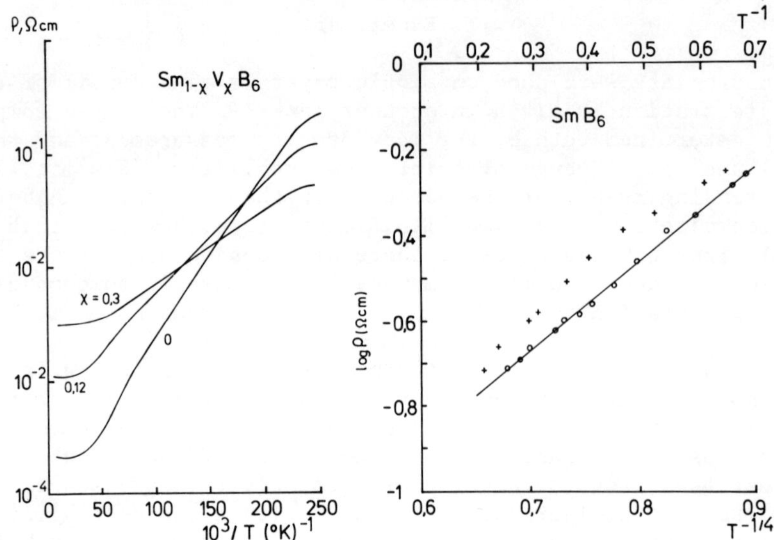

Fig.1a. The electrical resistivity ρ vs T^{-1} for $Sm_{1-x}V_xB_6$ with x = 0, 0.12, and 0.3.
Fig.1b. The low temperature part of SmB_6 in an enlarged scale. Both T^{-1} plot, + in the upper scale, and $T^{-1/4}$ plot, 0 in the lower acale, are shown. The solid line is the best fitted line for the $T^{-1/4}$ plotted experimental values.

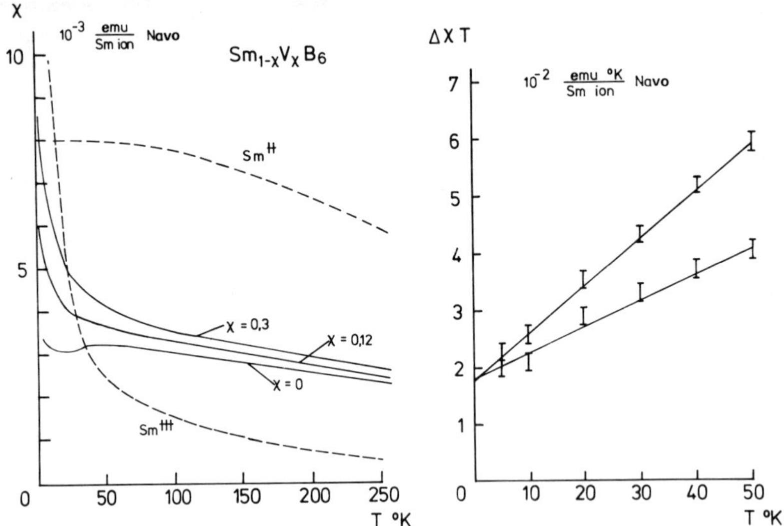

Fig.2a. The magnetic susceptibility χ per Sm ion in $Sm_{1-x}V_xB_6$ for x = 0, 0.12 and 0.3 vs T. For comparison, the values for free Sm^{++} and Sm^{+++} are also shown by broken lines.
Fig.2b. ΔχT vs T, in which $\Delta\chi = \chi_{0.3} - \chi_0$ for the upper values and $\Delta\chi = \chi_{0.3} - 1.1\chi_0$, for the lower values. The bars, I, show experimental values and the solid lines are the best fitted lines for both cases.

splitting Δ, the first term in the bracket in Eq.(1) should be replaced by

$$\frac{0.170}{T} + \frac{1.09}{\Delta}, \qquad \text{or} \qquad \frac{0.442}{T} + \frac{0.54}{\Delta}, \tag{2}$$

for the lowest doublet or the lowest quartet, respectively, in which Δ is given in the unit of °K. In the simple point charge model, the quartet is lower than the doublet. However, in the system such as SmB_6, the reversed case is also possible, as is observed in CeB_6[17]. To see the contribution of the atomic Sm^{+++} experimentally, we should subtract the background. In Fig.2b, $T(\chi_{0.3}-\chi_0)$ and $T(\chi_{0.3}-1.1\chi_0)$ are shown, in which $\chi_{0.3}$ and χ_0 show the susceptibilities of x = 0.3 and x = 0 samples, respectively, and these seem to give the upper and lower limits of the contribution of Sm^{+++}. These data are fairly well fitted by the formula, $\Delta\chi T=(0.018+0.00081T)=0.124(0.145+0.00653T)$ and $\Delta\chi T=0.124(0.145+0.00363T)$, respectively. The first term is fairly accurately determined. Then, by comparing with Eqs.(2), we can estimate the ratio of the atomic localized Sm^{+++} to be about 85% for the case of the doublet lower and 33% for the quartet lower. On the other hand, the second term has much more ambiguity, which offers a less accurate information on the crystal field splitting Δ. For the doublet lower, Δ should be very large, larger than several hundreds °K, while for the quartet lower, Δ may be around 40~100°K. We can not reject any one definitely on the present analysis, but the case for the quartet lower looks more reasonable. Note that the case for a very small Δ, that is the free Sm^{+++}, seems to be not consistent with the present analysis.

The paramagnetic spin resonances for dilutely doped Gd^{+++} and Eu^{++} are also measured for samples with various stoichiometries[13]. The g shift is negative and nearly temperature and sample independent, g = 1.92~1.94 for all cases. On the otherhand, the temperature dependence of the line width ΔH behaves very differently for Gd and Eu spins[13,18].

III. Theory

The fundamental model was already described in another paper[18]. The $5d_{eg}$ bands are filled with about 0.7 conduction electrons per Sm atom in pure SmB_6, while the 4f levels are situated at near the Fermi energy E_f. The effective transfer energy between the neighbour 4f states is induced by the mixings of the 4f states with the $5d_{eg}$ conduction electrons and the valence electrons on Borons. As is the case in Eu-chalcogenides[19], the 4f band width due to the d-f mixing may be of several hundreds °K, while that due to the antibonding effect from the 4f-valence bands mixing may be the order of thousand °K. These values are, however, those without the correlation effect. Correlation effect reduces the band width further. We applied the simplified Hubbard approximation to the six folds degenerated Sm^{+++} 4f bands assuming the singlet ground state of Sm^{++} as the vacuum[18]. Then, under the assumption that the band shape shrinks uniformly, the band width for the v-th Sm^{+++} band is reduced by a factor (C_2+C_v) while keeping the hight of the density of state constant, in which C_2 and C_v are the atomic ratio of the ground state Sm^{++} and that of the v-th Sm^{+++} state, respectively. In the paramagnetic state, assuming $C_2 = 0.3$, and thus $C_v = 0.117$, $C_2+C_v = 0.417$, the Sm^{+++} 4f hole state occupies about 28% of the each available Hubbard band states. This is shown in Fig.3a.

Because of its small band width, the polaron effect becomes important[20]. In the ionic crystal, the optical phonon polaron is most important. However, in metals such as SmB_6, in which the number of Sm^{+++}, n_3, is equal to that of the conduction electron n_c, the 4f charge fluctuation is screened very well and thus the optical phonon polaron is no more important. Then, the acoustic phonon polaron becomes more important. The band narrowing factor is given for each case by (see Appendix I)

$$\varepsilon_{nm}^{eff} = \varepsilon_{nm} \exp\left[-\zeta_{nm}\right] \tag{3}$$

$$\zeta_{nm}^{0} = \frac{\Delta E_0}{\hbar\omega_0}\left(1-\frac{1}{y}\int_0^y \frac{\sin x}{x}\,dx\right) \tag{4}$$

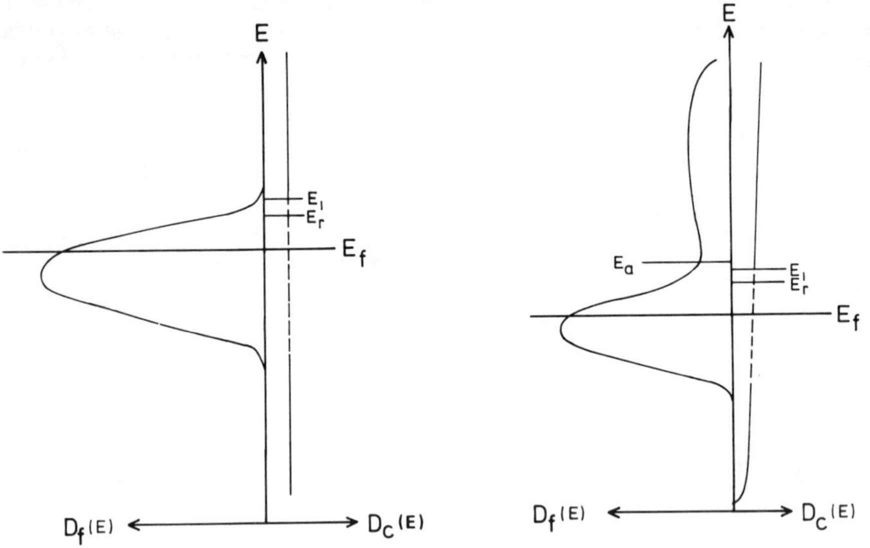

Fig.3a. Density of state for the 4f hole, $D_f(E)$, and that of $5d_{eg}$, $D_c(E)$ are shown schematically as functions of energy E. E_1 and E_r are the mobility edges for the resistivity and the ESR relaxation, respectively. Note that, similar to the impurity state, the 4f hole state density is not the usual one particle meaning.

Fig.3b. The same scheme for the sample with many vacancies. Above E_a, the 4f hole states are given by that of atomic localized Sm^{+++}.

$$\Delta E_0 = \frac{e^2 k_0}{\pi \varepsilon_p} \tag{5}$$

$$\zeta^a_{nm} = \frac{3\Delta E_0}{2\hbar u_a k_0} \left(1-2\frac{1-\cos y}{y^2}\right) \tag{6}$$

$$\Delta E_a = D^2/2Mu_a^2 \tag{7}$$

where ε_{nm} is the transfer energy without the polaron effect, ζ^0_{nm} and ζ^a_{nm} are the band narrowing factors for the longitudinal optical phonons and the longitudinal acoustic phonons, respectivilely, k_0 the Debye cut off wave vector, ε_p the polaron dielectric constant given by $\varepsilon_p^{-1} = \varepsilon_\infty^{-1}-\varepsilon_0^{-1}$, $\hbar\omega_0$ the optical phonon energy, u_a the acoustic phonon velocity, D the electron-phonon coupling constant, M the molecular mass and $y = k_0R_{nm}$. Similar to the SmS[18,21] the narrowing factor ζ^a_{nm} for the nearest neighbour transfer may be estimated from the electrical resistivity[6] of LaB_6 to be the order of unity, which reduces the effective 4f band width of SmB_6 further to a few hundreds °K.

Now, we should check whether the paramagnetic state is really stable compared with the ferromagnetic state in our model and the approximations. This is easily done and is given in Appendix II for the density of state of the 4f hole bands given by

$$D(E) = B|E - E_0|^P \tag{8}$$

for the region $0 \le E \le E_0$, or only for the edge region, then the middle region is represented simply by the parameter $\xi = N(P+1)/2BE_0^{P+1}$, the ratio of the real total state and the hypothetical one in which Eq(8) is assumed to be applicable to the whole region. The following conclusions are obtained. At first, for $\xi = 1$: (i) For $p = 0$, the rectangular density of state, the ferromagnetic energy $-U_f$ and the paramagnetic energy $-U_p$ are always same. (ii) For $p>0$, $U_p>U_f$ for $C_3 \le 0.5$ and $U_p<U_f$ for $C_3 \ge d/(d+1) = 6/7$, in which d is the multiplicity of Sm^{+++}, that is, $d = 2J+1 = 6$. The sign changes in a value of C_3 in between $6/7>C_3>0.5$. (iii) For $\xi \neq 1$, the paramagnetic state becomes more stable for $\xi>1$ but unstable for $\xi<1$. The values for some typical cases are shown in Table I.

The paramagnetic susceptibility is also easily calculated in the same way as done in Appendix II, and the value $AC_3/3$ is also shown in Table I. See Appendix II for the definition of A. Note that the susceptibility χ_b per Sm^{+++} is given by $(3T/AC_3E_0)\chi^c_{+++}$, in which χ^c_{+++} is the Curie part of the free Sm^{+++}, that is, $g^2\mu^2J(J+1)/3T$. When we assume that the temperature independent term, that is the mixing term of the susceptibility of Sm^{+++} and that of Sm^{++}, are same as those of atomic state, we can evaluate the total susceptibility χ_0 of the system. Note that when we assume the usual type of the spin-spin exchange coupling given by $I < gJ_z >^2$, the total susceptibility χ is given by

$$\frac{1}{\chi} = \frac{1}{\chi_0} + \frac{2I}{\mu^2} \quad . \tag{9}$$

For SmB_6, however, the most contribution for χ comes from Sm^{++}. Because the number of Sm^{++} has considerable error, may be 20% of error, it is nearly impossible to evaluate the contribution of the Sm^{+++} band susceptibility experimentally. We can only say very roughly that it will be less than 20% of the total susceptibility. On the theoretical side, too, there are still large unknown factors due to the crystal field effect and the spin-spin exchange interaction.

Table I

C_3	0.6	0.6	0.7	0.7+	0.7	0.8	0.8
P	0.5	0.8	0.5	0.5	0.8	0.5	0.8
U_p/NE_0	0.202	0.184	0.173	0.136	0.156	0.129	0.115
U_f/NE_0	0.193	0.173	0.172	0.132	0.155	0.135	0.123
$\Delta U/NE_0$	0.009	0.011	0.001	0.004	0.001	-0.006	-0.008
A	1.79	1.41	1.59	2.29	1.20	(1.41)	(1.03)
$AC_3/3$	0.36	0.28	0.37	0.53	0.28	(0.38)	(0.28)
2A/35	0.10	0.08	0.09	0.13	0.07	(0.08)	(0.06)

+ $\xi = 1.5$. For all other cases, $\xi = 1$.
The value in the bracked means an unstable mode.

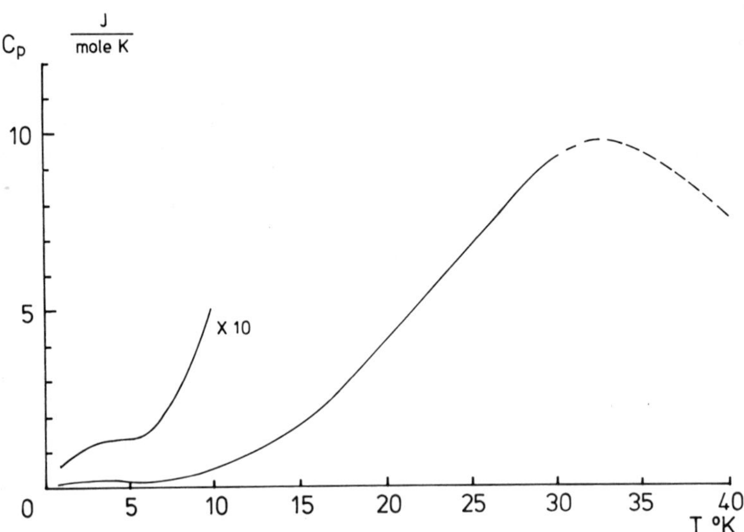

Fig.4. The magnetic part of the specific heat C_p in SmB_6 estimated from the data of Nickerson et.al.[8] Below 10°K, the ten times enlarged curve is also shown. The dotted curve above 30°K is not observed but expected from the entropy calculation.

Therefore, for SmB_6, we can only say very roughly that among the examples listed in Table I, the case for p = 0.5 and ξ = 1.5 may be the best. For samples with a large number of vacancies, we can say more definitely. Consider the case that one vacancy creates two Sm^{+++}. When we assume that, for the remaining (1-3x) Sm ions, 30% is Sm^{++}, atomic ratio of Sm^{++} is only $C_2 = 0.3(1-3x)/(1-x)$. Nevertheless, the susceptibility of the compound increases as x increases even at room temperature, where the contribution of the localized Sm^{+++} is very small. This is explained as follows. As was studied before for the sample with x = 0.3, it is more reasonable to accept the model that the quartet is lower and then the number of the atomic localized Sm^{+++} is less than 2x. This means very possibly that in the band, or non atomic localized, phase, the ratio C_3 is larger than 0.7. This means that, as shown before in our calculation, the system is more ferromagnetic like and thus the paramagnetic susceptibility is enhanced more. This enhancement for the Sm^{+++} band susceptibility χ_b overcomes the decreasing due to the decreasing number of Sm^{++}. On the other hand for the doublet lower, the number of the atomic localized Sm^{+++} is larger than 2x, which causes C_3 to be smaller than 0.7, that is more stable paramagnetism and thus sharp decreasing in the susceptibility. This is completely against the experimental result.

So far, we have considered the single particle excitation only. However, as well known, the single particle excitation in the Hubbard model is not enough to describe the elementally excitations in the system and the collective excitation becomes more important as the Hubbard bands are filled more. This is easily seen in the limit that all Sm are Sm^{+++}. Then there are no single particle excitation and only the spin fluctuation, or the paramagnon excitation, can exist. In a general configuration of Sm^{+++} and Sm^{++}, there are two modes, the 4f charge fluctuation and the 4f spin fluctuation. The charge fluctuation mode has, however, much higher energy than that of the spin fluctuation, so that this mode is neglected here after. On the other hand, the spin fluctuation mode requires much lower energy because the conduction electrons are not asked to follow it. Two kinds of forces are responsible for the paramagnon energy. One is the usual spin-spin exchange interaction as was discussed by Kasuya[19] and another is due to the band Sm^{+++} state. The latter is in some sense similar to the paramagnon in the d band and may be evaluated in the present model as the energy to induce $\langle J_z \rangle = 1$ in each lattice site, which is $2AE_0/35$ as shown in Appendix II, and the values for some cases are listed in Table 1. The paramagnon energy should be compared with the specific heat measurement. The magnetic part of the specific heat obtained by the experiment of Nickerson et.al.[8] is shown in Fig.4. The main peak around 35°K should be due to the paramagnon and thus be compared with the theory. Again, the case for C_3=0.7, p=0.5 and ξ=1.5 seems best in the Table I. Note that in general the calculated paramagnon energy is much smaller than the band width E_0. The first small peak around 5°K should be due to the atomic localized Sm^{+++} because its magnitude depends on sample[8] and is explained by about 1% of the free Sm^{+++}. It is difficult to evaluate the T-linear term due to the 4f Hubbard band definitely. It's coefficient seems to be around 0.01~0.02 joule/Mole deg^2, from the data below 10°K. This gives the density of state of 4~8 states/eV. Sm ion. This is, however, too small compared with what we expect from the six-fold Hubbard bands with the band width of a few hundreds °K. The main discrepancy seems to come from the fact that, as shown clearly in the transport experiment, the 4f hole state in the real SmB_6 crystal is not an extended band state but the Anderson's localized state. However, once we have a localized state, the single particle picture is not applicable in particular at very low temperature and for a very small energy excitation.

Due to its small band width, the potential fluctuation due to the vacancies can easily overcome the band width and thus causes the Anderson's localized state extending over many lattice sites. This is observed in the resistivity of the lowest temperature range. Furthermore, due to the large difference in the density of state, the $5d_{eg}$ states near the Fermi level can not exist as the Bloch band states but are mixed completely to the 4f states forming small tail parts. Therefore, for the $5d_{eg}$ bands, there are pseudo mobility edges as shown in Fig.3. This is the origin of the semiconductor like behavior of the resistivity above 4°K. In the room temperature, the Fermi level is shifted and situated completely above the mobility edge. Therefore, the situation is nothing different from the usual magnetic metals. The signs and the magnitudes of the thermoelectric power are also very well explained on the present model. Note that the sign of the thermoelectric power of LaB_6 varies in samples[6,10] showing a sensitive dependence on the scattering mechanism.

The different behaviors of Gd^{+++} and Eu^{++} in the line width of the paramagnetic ESR are also explained very well[18] based on the fact that the d-f exchange mechanism is, as usual, most important in Gd^{+++} but the f-f exchange is more imprtant in Eu^{++} because of the small $5d_{eg}$ amplitude on Eu^{++}.

IV. Conclusion

Various anomalous properties in SmB_6 are attributed to the strongly correlated, narrow 4f polaron bands. The extended Hubbard model with collective modes was used to describe the 4f state. Nearly all anomalous properties are explained consistently on this model. The model should be applicable to all other materials showing the similar valence fluctuating properties.

Finally, one of the author. T.K., expresses his thanks to Professor S. Methfessel for his kind hospitality and interesting discussions during the author's stay in Bochum, where the manuscript has been prepared.

Appendix I

The effective mass of the small polaron is calculated. The Hamiltonian is given by

$$\mathcal{H} = -\sum_{n,m} \varepsilon_{nm} a_n^+ a_m - \sum_{v,k} \hbar\omega_{vk} b_{vk}^+ b_{vk} - \sum_{v,k,n} V_{vkn} a_n^+ a_n b_{vk} + c.c. \quad \text{(AI1)}$$

where a_n^+, a_n, b_{vk}^+ and b_{vk} are the creation and anihilation operators for the Wanier state at site R_n and the phonon with wave vector k of v-th mode, ε_{nm} the transfer energy, $\hbar\omega_{vk}$ the phonon energy and V_{vkn} the simplified electron phonon coupling constant. The last two terms are diagonalized by the following Unitary transformation.

$$U_n = \exp\left[\sum_{vk} (f_{vkn} b_{vk}^+ - f_{vkn}^* b_{v-k})\right] , \quad \text{(AI2)}$$

$$f_{vkn} = (V_{vkn}^* / \hbar\omega_{vk}) a_n^+ a_n , \quad \text{(AI3)}$$

where the electron is localized on n-site, $|\rangle = a_n^+|0\rangle$. The last term of Eq.(AI1) is now diagonalized as

$$\Delta E = -\sum_{vk} (|V_{vkn}|^2 / \hbar\omega_{vk}) a_n^+ a_n , \quad \text{(AI4)}$$

which is due to the gain in the interaction term, $-2\Delta E$, and the loss in the phonon term, ΔE. The first term of Eq.(AI1), the transfer term, is now transformed as

$$\langle 0| a_n U_n^+ a_n^+ \varepsilon_{nm} a_m U_m a_m^+ | 0 \rangle \quad \text{(AI5)}$$

giving the result

$$\varepsilon_{nm}^{eff} = \varepsilon_{nm} \langle 0| U_n^+(1) U_m(1) | 0 \rangle \equiv \varepsilon_{nm} e^{-\zeta nm}$$

$$= \varepsilon_{nm} \exp\left[-\frac{1}{2} \sum_{vk} |f_{vkn} - f_{vkm}|^2\right] , \quad \text{(AI6)}$$

in which $U_n(1)$ means to take $a_n^+ a_n = 1$. Note that the present treatment is easily extented to the many electron system by taking $U = \Pi U_n$.

Appendix II

At first, let us consider the case of small Sm^{+++} concentration and assume the following shape of density of state without correlation effect near the edge,

$$D(E) = B|E_0 - E|^P . \quad \text{(AII1)}$$

Then, for both the paramagnetic and perfect ferromagnetic states, the binding energies at T = 0 are given respectively by

$$\frac{U_P}{C_3 N E_0} = 1 - \frac{d-1}{d} C_3 - \frac{p+1}{p+2} \left(1 - \frac{d-1}{d} C_3\right)^{P/(p+1)} \left(\frac{2\xi C_3}{d}\right)^{1/(p+1)} , \quad \text{(AII2)}$$

$$\frac{U_f}{C_3NE_0} = 1-\frac{p+1}{p+2}(2\xi C_3)^{1/(p+1)} \quad , \tag{AII3}$$

in which N is the total number of Sm ion, d is the manifold of Sm^{+++}, six in the present case, and $\xi=N(p+1)/2BE_0^{p+1}$ is the ratio of the actual total states and that of the hypothetical one in which the formula (AII1) is applicable as far as E=0. It is clear that, except p=0, the paramagnetic state is more stable in the limit of $C_3<<1$, while for p=0 the ferromagnetic or the paramagnetic state is more stable depending on $\xi<1$ or $\xi>1$. For a finite C_3, the situation depends complicatedly on the shape of D(E), but the general feature is the same as the case of p=0 and for p>0 the paramagnetic state is more stable than p=0.

For $\xi=1$, Eq.(AII1) is applicable for the whole region. For p=0, it is easily seen that $U_p=U_f$ for the whole region of C_3. For p>0, Eq.(AII3) is applicable only for $C_3\leq 1/2$. For $C_3>1/2$, C_3 should be replaced by $(1-C_3)$. Eq.(AII2) is applicable as far as $C_3\leq d/(d+1)$. For $C_3>d/(d+1)$, U_p is given by

$$\frac{U_p}{NE_0} = d(1-C_3)\left[1-\frac{d-1}{d}C_3-\frac{p+1}{p+2}\left(1-\frac{d-1}{d}C_3\right)^{p/(p+1)}\{2(1-C_3)\}^{1/(p+1)}\right]. \tag{AII4}$$

it is easily seen that for $C_3\leq 1/2$, the paramagnetic state is more stable, but, in between $1/2<C_3<d/(d+1)$, the sign is reversed and for $d/(d+1)\leq C_3$ the ferromagnetic state is more stable.

The magnetization under the magnetic field is also easily calculated. We treat here only the case for the paramagnetic state and the linear susceptibility. Assuming that the increasing number in each subband is given by αJ_z for each subband, we can calculate the increasing of the band energy to be

$$NE_0\alpha^2\,\frac{(2J+1)J(J+1)}{3}\left[\frac{p+2}{(p+1)^2}\left(1-\frac{d-1}{d}C_3\right)^{p/(p+1)}\left(2\xi\frac{C_3}{d}\right)^{1/(p+1)}\right.$$

$$\left.\left\{\frac{d(p+2)}{2C_3(p+1)}+\frac{p}{\left(1-\frac{d-1}{d}C_3\right)^2}\left(1-C_3+\frac{(2p+1)}{2(p+1)}\frac{C_3}{d}\right)\right\}-1\right] \quad , \tag{AII5}$$

while the gain in the Zeeman energy is given by $-Ng\mu H\alpha(2J+1)J(J+1)/3$. Now, α is given so as to minimize the above two terms, $\alpha=g\mu H/2E_0A$, and thus the susceptibility per Sm^{+++} ion is

$$\chi_b = \frac{g^2\mu^2(2J+1)J(J+1)}{6E_0AC_3} \quad , \tag{AII6}$$

in which A means the term within the large bracket in Eq.(AII5). Note that the sign of A gives a local stability of the paramagnetic state for an infinitely small uniform polarization. For example, for $p\to 0$, $A=4\xi-1$, which means the local stability is much better. Calculated numerical values for some typical cases are given in Table I. It is clearly seen that the paramagnetic state becomes more stable as C_3

decreases, p increases and ξ increases.

The paramagnon exitation energy may be evaluated to be the band energy increase to create $\langle J_z \rangle = 1$ at each site. Then, it is given by $2AE_0/35$. The value for $2A/35 = 0.057A$ is also given in Table I.

References

[+]Present address: Faculty of Integrated Arts and Science, Hiroshima Univ.

1. See, for example, S. von Molnar, T. Penney and F. Holtzverg, Proc.Int.Conf. Metal-Nonmetal Trans., Autrant 1976.
2. E.E. Vainshtein, S.M. Blokhin and Yu.B. Paderno, Sov.Phys.Sol. Stat. 6, 2318 (1965)
3. A. Menth, E. Buehler, H.J. Levinstein and T.H. Geballe, J.Appl. Phys. 40, 1006 (1969), 22, 295 (1969).
4. R.L. Cohen, M. Eibschütz and K.W. West, Phys.Rev.Letters 24, 383 (1970); J.Appl.Phys. 41, 898 (1970).
5. A. Menth, E. Buehler and T.H. Geballe, Phys.Rev.Letters 22, 295 (1969).
6. J.P. Mercurio, J.E. Etourneau, R. Naslain, P. Hagemuller and J.B. Goodenough, J.Sol.State Chem. 9, 37 (1974).
7. Y. Ishizawa, J.Phys.Soc. Japan 35, (1977).
8. J.C. Nickerson, R.B. White, K.N. Lee, R. Bachmann, T.H. Geballe and G.W. Hull jr., Phys.Rev. B3, 2030 (1971).
9. T.H. Geballe, A. Menth, E. Buehler and G.W. Hull, J.Appl.Phys. 41, 904 (1970).
10. Yu.B. Paderno, V.I. Novikoyand, E.S. Garf, Powder Metallurgy (USSR) 83, 70 (1969).
11. Y. Ishikawa, private communication, and T. Kasuya, Proc. of the Intern.Conf. Metal-Nonmetal Trans., Autrant, 1976.
12. K. Niihara, Bull.Chem.Soc. Japan 44, 963 (1971).
13. K. Kojima, Dr. Thesis, Tohoku-University 1976, to be published.
14. N.F. Mott and E.A. Davis, Phil.Mag. 17, 1269 (1968).
15. P.W. Anderson, Phys.Rev. 109, 1492 (1958).
16. J.H. van Vleck, Theory of Electric and Magnetic Susceplibity, Oxford, 1936.
17. J.C. Nickerson and R.M. White, J.Appl.Phys. 40, 1011 (1969).
18. T. Kasuya, Proc.Intern.Conf. Metal-Nonmetal Trans., Aurtant, France, 1976.
19. T. Kasuya, IBM J. of Research and Development 14, 214 (1970).
20. T. Kasuya, Critical Rev. in solid State Sciences 3, 131 (1972).
21. T. Kasuya, to be published.
22. P.G. de Gennes. Phys. Rev. 118, 141 (1960).
23. J.N. Chazalviel, M. Campagna, G.K. Wertheim and P.H. Schmidt, Solid State Commun. 19, 725 (1976).

COMMENTS

ORBACH: If you have magnon excitations, this should show up in the relationship between linewidth and g-shift in the resonance. Why is there no effect of the paramagnon susceptibility on the magnetic resonance? Specifically, if you had a paramagnon piece at 35 or 40°K you would not expect a linear T-dependence for the Korringa relationship.

KASUYA: It is difficult to find the resonance signal beyond 20°K, even at which temperature there is a large ambiguity.

KAPLAN: Are you saying that the $T^{1/4}$ resistivity law is holding down to the smallest vacancy concentration available:

KASUYA: Yes.

KAPLAN: Is the U_{ff} in your Hubbard model much larger than the width of the narrow band?

KASUYA: Yes. I just use the strong correlation limit.

KAPLAN: What is the model's magnetic character?

KASUYA: Just the usual paramagnet.

WACHTER: With these kinds of materials the energy scale (or time scale) of the measurement, as it relates to the energy of the virtual bound state, is important. In, e.g., electrical conductivity measurements the important energies are small and one finds a large temperature dependence. On the other hand in an optical experiment involving high energies like optical reflectivity one sees the plasma mode of the 5d-electrons, which is about 2 ev. In the latter case one finds the number of free electrons, which corresponds to about 0.7 as you said. However, this plasma edge does not change with temperature.

KASUYA: This is also consistent with our model.

CAMPAGNA: I would like to ask what is the exact origin of the conduction band you were talking about? Secondly, I did not completely understand why you expect the susceptibility to decrease rather than increase when you create vacancies. I think that convincing arguments could be made for the opposite result.

KASUYA: The answer to the second point is very simple. The susceptibility of Sm^{3+} at room temperature is an order of magnitude smaller than Sm^{2+} and vacancies transform Sm^{2+} to Sm^{3+}. Therefore, we would expect with increasing temperature (increasing number of vacancies) a decrease in susceptibility but experiment shows just the exact opposite. For the first question, the 4f band width is made by the anti-bonding effect with the valence band mixing.

WILSON: Dennis McWhan sees[+] a straight line extending down to 10°K on a $T^{-1/4}$ plot for SmS at 20 kbar (fluctuating metallic phase). The fit is slightly better than that obtained by forcing a T^{-1} fit, but the difference is mostly in philosophy. I tend to believe, as Professor Kasuya mentioned, that this is disorder localization. At 10 kbar the McWhan data yields around 30°K a kink in the $T^{-1/4}$ plot which may indicate that one is entering the B' phase in SmS as in the $Sm_{1-x}Gd_xS$ system. The turnover in ρ observed below 10°K could possibly result from surface conductivity of fully collapsed non-fluctuating material.

TOURNIER: We have seen a beautiful T^2 behavior in the resistivity of SmS thin films at low temperature. We believe it's a Kondo effect of Sm^{3+} which is present in the collapsed phase of SmS. I think you have to plot your resistivity down to 0°K.

[+]Plots in this form of McWhan's data appear in J. A. Wilson, "Structure and Bonding" - to appear late 1976.

THEORETICAL APPROACH TO THE CONFIGURATION MIXING IN Sm CHALCOGENIDES*

T.A. Kaplan, S.D. Mahanti, and Mustansir Barma

Department of Physics, Michigan State University

East Lansing, Michigan 48824

ABSTRACT

We discuss the approach to the properties of the Sm chalcogenides based on an essentially localized or excitonic picture of most of the Sm 5d-electrons in the collapsed phase, with a small number (≈ 0.1 electron per Sm) occupying the free-electron-like states at the bottom of the 5d-band. The way this picture can lead to the observed results on volume and degree of mixing vs. pressure, as well as electrical conductivity, magnetism, and specific heat is reviewed. In addition, it is shown that this model, in the case of SmS can lead to the observed plasma edge frequency and furthermore predicts strong optical absorption at low energies (≈ 10^{-1} ev.). We give a comparison of the above picture with the more common models where all the 5d-electrons (~.7 per Sm in SmS) occupy free-electron-like states, and a critique of the latter.

I. INTRODUCTION

Samarium chalcogenides have attracted considerable experimental [1-7] and theoretical [8-19] attention in the last few years. Some of these compounds undergo a transition to a mixed-configuration (or "mixed-valence") phase on applying pressure. At temperature $T \cong 300°K$, SmS undergoes [1-3] a first-order phase transition at pressure $p \sim 6$ kbar, with a decrease in volume of ~10%. In the mixed phase, which is apparently nonmagnetic [3], each Sm ion is in a mixture of f^6 and f^5d (or f^5s) configurations. The amount of admixture has been determined from the lattice constant [1], isomer shift [2], magnetic susceptibility [3], XPS data [5]; a ratio of f^6 to f^5d of about 3:7 has been found. While the low-p phase is a semiconductor, the mixed phase is a peculiar metal, its resistivity ρ at room temperature being ~100 times larger than that of Cu, increasing with decreasing T, and the low-T specific heat being $\propto T$ and anomalously high [6]. Further, a plasma edge in the optical absorption was found to occur at a bare value (after correction for the high-frequency dielectric constant) corresponding to $n \cong .9$ electrons/Sm in the formula $\omega_p^2 = 4\pi ne^2/m^*$. [7] Finally, SmSe and SmTe change smoothly from the semiconducting to the "metallic" phase [1].

In all but one [8,9,10,20] of the models of the Sm-chalcogenide mixed phase, all the d-electrons (~0.7/Sm atom) are in a broad one-electron type of band. We [8,10] were led to a sharply contrasting picture, in which only about .1 electrons/Sm atom are of this "band-like" character, the remaining d-electrons being essentially localized. One of our motivating considerations may be discussed with the help of Fig. 1, where (a) and (b) show a schematized picture of the observed optical absorption (abscissa) vs. energy (ordinate) in the semiconducting phase of SmS. In this phase the ground state has each Sm^{2+} ion in $4f^6$, $^7F_0$. The pressures in (a) and in (b) are atmospheric and a value just smaller than the critical pressure p_c, respectively. In the data [7,21], there are two sharp and intense low-lying peaks at ~.8 and 1.6 ev, each with a width of ~.5 ev. These were interpreted [21] as intraatomic transitions, $f^6 \rightarrow f^5 d_\ell$, where d_ℓ is a localized t_{2g}-state, and the doublet originates in the ^6H_J and ^6F_J terms of $4f^5$; furthermore the 1/2-volt width is attributed to the fine structure of these terms [22]. Using this interpretation, but simplifying, we have chosen to schematize the actual data by showing a single peak, considerably narrower than .5 ev, at .6 ev, which would be about the energy of the lowest state of ^6H_J given this interpretation. The broader background is also a schematized version of the absorption thought to occur [7] in transitions $f^6 \rightarrow f^5 d_k$ where d_k is a conduction-band Bloch state. G_0 is the band gap thought to be about .1 to .2 ev. [7,23].

The nature of the sharp peaks deserves further comment. A band calculation [24] for semiconducting SmS shows a band gap (4f-5d) at about .1 ev, with the lower part of the conduction band being broad, with a structureless density of states starting free-electron-like at the bottom and extending to about 2.5 ev. above the bottom of the band. (Above this, more complicated 5d structure occurs.) In other words, band theory does not account for these prominent features. The situation is quite similar to that in EuO (see Methfessel et. al., footnote 22), where as noted by Lendi [25], the very prominent peak in the optical absorption near the bottom of the conduction band could not be explained, according to his band calculation, by excitation to the one-electron Bloch functions. He [25] therefore suggested an excitonic mechanism, apparently unaware of the earlier quantitatively rather successful, interpretation [22] on the basis of a Frenkel exciton. There is a notable difference between the "extra" peak in EuO and those in SmS, namely,

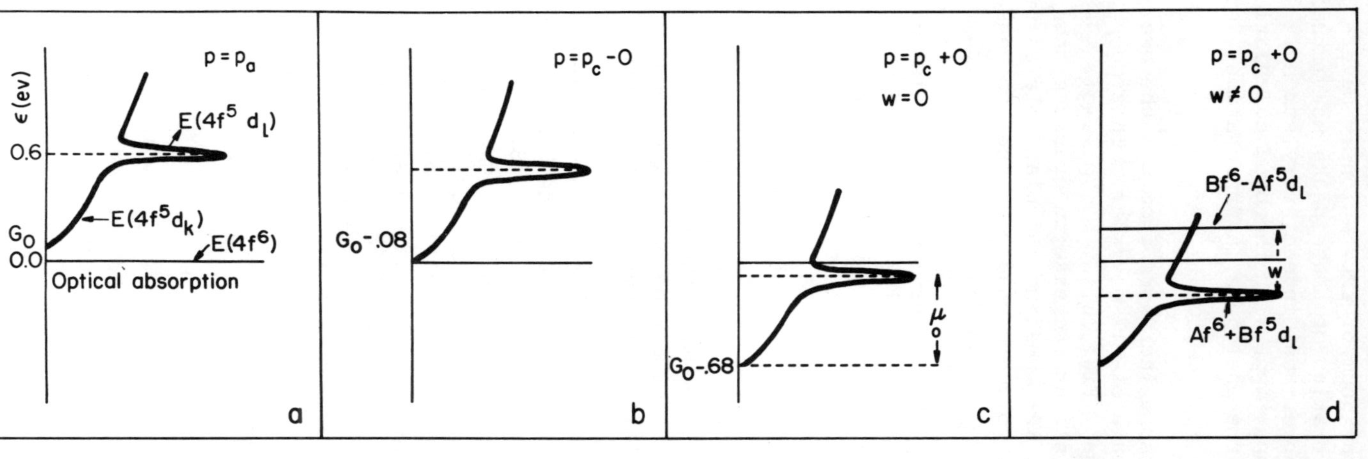

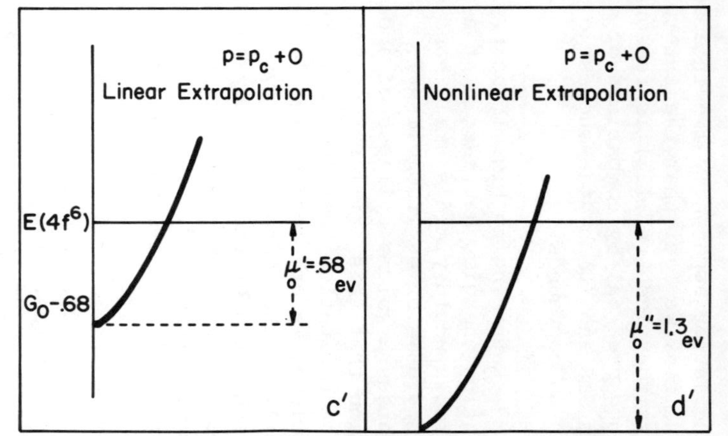

Figure 1. Electron-hole excitations from semiconducting ground state ($4f^6$)
(a) and (b): Semiconducting phase.
(c) and (d): Extrapolation to collapsed phase (present picture).
(c') and (d'): Extrapolation to collapsed phase (common picture).

the latter apparently lie in a continuum of excitations. Nevertheless, we feel that the interpretation of these peaks as virtual excitons (see ref. 26) is reasonable, and, indeed, very strongly suggested (a) by comparison of experiment with the band calculation on SmS and (b) by the success of the analogous interpretation [22] in Eu compounds.

Returning to Fig. 1, as pressure increases from p_a, the excitation energies decrease [7] at a rate of ~12 mev/kbar. This decrease would yield a small band gap (Fig. 1b) at $p = p_c - 0$ of $\sim G_o - .08$ $(= .02$ ev if $G_o = .1$ ev$)$. A slight increase in pressure then causes a volume collapse of about 12%. Making the plausible assumption that this still rather small volume change continues to cause the excitation energies to decrease approximately linearly with volume, we are led to Fig. 1c [27]. But this represents an instability: the state of the system in which every Sm is in f^6 is no longer the ground state, system states in which some of the f electrons go into d-states being lower. If it were not for the broad background, overlapping the exciton, the system ground-state would simply have each Sm in the state $f^5 d_\ell$ (d_ℓ is an essentially localized, t_{2g}, state). However, the fact that the conduction band extends below the localized state means that the energy will be lowered further by dumping some of the d_ℓ-electrons into the band-like states. The Fermi energy measured from the bottom of the band, $\mu_o = .6 - G_o \cong .5$ ev. To estimate the number of band electrons we take a free electron density of states for energy up to a cutoff e_c: $D(e) = 3e^{1/2}/e_c^{3/2}$, giving the number of conduction electrons (per Sm) of $b = 2(\mu/e_c)^{3/2}$ (e is the energy measured from the bottom of the conduction band.) [28] Thus with $e_c = 2.5$ ev. (the bandwidth used by Varma and Heine [12]) and $\mu = .5$ ev. we obtain $b = .18$. Hence a small number of electrons per Sm (much less than .7) goes into the band. To appreciate the difficulty of getting b close to .7, we consider the conventional picture shown in (1c') - i.e. there is no excitonic peak in the collapsed phase, but we assume still a linear descent of the band with volume. Then μ would be .58 ev. giving $b = .22$ - still about a factor of 3 too small. In order to get $b = .7$, μ must be 1.3 ev. (depicted in Fig. 1d'), more than a factor of 2 larger than the value in the linear extrapolation; it does not seem plausible to expect such large nonlinearity effects as a function of volume for only a 10-15% volume change in connection with the variation of the rather broad 5d band.

As just noted, the excitonic level remains in our model (and is in fact appreciably occupied), whereas in the other model it has disappeared. Presumably this level could disappear because of screening of the 4f hole by the very mobile conduction electrons; however, given that b is a small number per Sm as discussed above, and that therefore there is almost one localized d-electron per Sm, such screening could not be a large effect (no matter how small the screening length).

Another major motivation for our looking in a direction essentially different from that suggested by the conventional model is that we find it difficult to believe that it can lead to the nonmagnetic behavior observed [3]. The conventional model may be described as follows. The zeroth order ground state is one in which most (~70%) of the Sm's are in $4f^5$ states, the corresponding f-electrons having been dumped into the d-band; the remaining Sm's are in $4f^6$. The lowest $4f^5$ state in the crystal field is a Kramers doublet and is therefore magnetic. Qualitative arguments were given [3,11], based on the example considered in the Kondo effect, which led to a nonmagnetic ground state. The only attempt [14] at a calculation in the case of interest, namely where most of the Sm-sites are magnetic "impurities" (Sm, f^5 ions) was made quite recently - it was concluded that the nonmagnetic ground state still held for this case. But an approximation was made [14] that resembles very closely the one made by Hubbard in his treatment of the single-band Hubbard model. [29] This Hubbard decoupling leads [30] to nonmagnetic behavior in an example which is known to be magnetic (the half-filled narrow-single-band Hubbard model, which is antiferromagnetic). Therefore this calculation [14] does not bring us closer to the answer to the question as to the magnetism in the model studied. Furthermore, a comparison with nickel suggests strongly that the model considered [14] (a 2-band (f and d) Hubbard model with one-electron f-d mixing, large f-f intrasite Coulomb repulsion U, the bare f and d bands being narrow and broad, respectively) should lead to a magnetic ground state. As already pointed out [13], this model should be applicable to Ni with f replaced by 3d and d replaced by 4s; furthermore the requirement that the Fermi level lie in the narrow band, considered [13,14] to be crucial, is satisfied in nickel. But of course the parameters are appreciably different [13]. However, the change in parameter values in going from Ni to SmS, seems to us to be in the direction of favoring magnetism (the width of the narrow band decreases, giving a larger density of states at μ, and the density of broad-band electrons decreases somewhat, preventing any additional screening of U).

Hence we have investigated a model that is a natural consequence of Fig. 1c. One can consider a zeroth-order ground state corresponding to Fig. 1c as $(f^5d_\ell)^{N(1-b)}(f^5)^{Nb}(d_k)^{Nb}$ where N is the number of Sm ions; that is, there are N(1-b) Sm ions in a configuration with f^5 and a localized d,Nb Sm ions in f^5, and Nb conduction electrons. An important point here is that we believe f^5d_ℓ acts essentially like f^5d_k or like Sm^{3+}, insofar as the volume, Mössbauer isomer shift, and XPS are concerned (these are important measures of "valence"). This point will be discussed in some detail in later sections. But then this zeroth-order state would give all Sm behaving as Sm^{3+}, in disagreement with these experiments which suggest ~70% Sm^{3+} and ~30% Sm^{2+}. We proposed a mechanism [8-10] which mixes f^5d_ℓ and f^6 and which can lead to the right order of mixing. It is this mixing matrix element, proportional to a quantity W, that causes the change from Fig. 1c to 1d; the Fermi level relative to the bottom of the band is thus reduced to ~.4 ev. lowering b to about .13 electrons/Sm. The mean field ground state is of the form $(Af^6+Bf^5d_\ell)^{N(1-b)}(f^5)^{Nb}(d_k)^{Nb}$, with $B^2/A^2 \cong 7/3$.

In section II we describe a simplified version of our model indicating how it leads to a fairly satisfactory account of volume, amount of mixing and isomer shift vs. pressure. In section III we consider various experimental properties in the collapsed phase, showing in particular how the metallic-like plasma edge frequency observed in optical absorption, and the d.c. electrical conductivity are consistent with our model despite the localized nature of most of the d-electrons in this model. Furthermore a prediction of low-energy optical absorption (≈ tenths of a volt) via $f^5d_\ell \to f^6$ is made - this can be seen roughly in Figs. 1c or 1d. Finally, our speculation as to how nonmagnetic character can occur, and general conclusions are given in section IV.

II. VOLUME AND DEGREE OF MIXING VS. PRESSURE

We first present our model in a mean-field approximation, neglecting the small number of conduction electrons. We consider only the ground state [31] and work for simplicity within a two-level model [8] involving an f^2 spin-singlet and an fd spin-singlet at each Sm-site (in place of f^6 and f^5d, respectively). In a mean-field approximation (not Hartree-Fock) we write the crystal wave function

$$\Phi = \prod_i [A\, a^\dagger_{fi\uparrow} a^\dagger_{fi\downarrow} + \frac{B}{\sqrt{2}} (a^\dagger_{fi\uparrow} a^\dagger_{di\downarrow} - a^\dagger_{fi\downarrow} a^\dagger_{di\uparrow})]\,|0> \equiv \prod_i \phi_i \quad ; \qquad (1)$$

$a^\dagger_{\mu i \sigma}$ creates an electron in Wannier function μ at site i with spin σ, and $|A|^2 + |B|^2 = 1$. We calculate $(\Phi, H\Phi)$ where H is the usual non-relativistic Hamiltonian minus the ionic kinetic energy, assuming that (A) the ionic positions always form the NaCl lattice, (B) we can retain in the Sm-Sm interactions only terms of leading order in $(\text{Sm-Sm separation})^{-1}$, and (C) we may expand all matrix elements to second order in $V-V_2$ where V is the volume per Sm and V_2 is the value of V at p=0. We thus find the energy per Sm atom,

$$E(n,V)=(1-n)\frac{1}{2}k_2(V-V_2)^2+n[\frac{1}{2}\,k_3(V-V_3)^2+c]-W(V)\;n(1-n) \quad . \qquad (2)$$

Here $n=|B|^2$, the number of d electrons in (1), V_3 is the location of the energy minimum when n=1, k_2 and k_3 are constants and $c = E(1,V_3)-E(0,V_2)$. W is the quantity $4\,z|\xi|$, estimated to be ≈ tenths of an ev in Ref. 8: W is proportional to the 2-site Coulomb matrix element that corresponds to the transition $(f_i f_j) \to (d_i d_j)$, for neighboring sites i and j, and gives rise to the mixing in (1). Using this expression we calculate V and n* vs. p, where n* minimizes E(n,V) for fixed V. Details of the calculation and the results are given in ref. 10. Good overall agreement with the observed behavior of SmS, SmSe, SmTe was found using values of W of the expected order of magnitude.

It is important to discuss the assumption made [10] that the volume V_3 associated with f^5d_ℓ (or fd_ℓ in our simplified model) is the experimental Sm^{3+} radius. (A similar assumption was made [20] in connection with SmB_6.) Arguing empirically, it is found that the Sm^{3+} radius as determined [32] from the insulator [33] Sm_2O_3 agrees very closely (to within about 1%) with the extrapolation from [13] the series RS, R=La, Ce, ... The latter is interpreted as the size appropriate to metallic Sm^{3+} S^{2-} with ~1 conduction electron per Sm, since the neighboring members of the series (NdS and GdS) have this property. Hence the presence of Sm 5d-conduction electrons does not seem to alter the size of Sm^{3+}. But it's known that even for nearly-free electrons the conduction electron charge density is not uniform, showing appreciable peaking near the atomic sites - see e.g. the example [34] of Cs. This suggests that localized 5d electrons will not appreciably alter the size of Sm^{3+}. From a more basic point of view, we note that the outermost filled Sm shells are 5s and 5p, and they are thought to provide the repulsive force in the cases $Sm^{2+}(4f^6)$ and $Sm^{3+}(4f^5)$. It is the shrinkage in size of these shells on going from $4f^6$ to $4f^5$ (the 4f radius is less than the 5s or 5p radii) that causes Sm^{3+} to be smaller than $Sm^{2+}(4f^6)$. We expect that the shrinkage of the 5s 5p shells will not be appreciably changed on adding a 5d electron to $4f^5$ since the 5d radius is larger than the 5p radius. The basic question remaining is the effect of this 5d on the repulsive forces. As far as the interaction between neighboring Sm's, the extra 5d electron would appear to give an attractive force, tending to reduce the effective size of the Sm. On the other hand, this 5d electron would presumably increase the Sm-S repulsion. However, we feel that this will not alter the situation wherein the ionic size is essentially determined by the Sm 5p electrons because (a) the 5d shell is only 1/10 filled (the 5p being completely filled), so the 5d electron can avoid the sulphur p-electrons, and (b) the σ-overlap between $Sm5d(t_{2g})$ and S3p is zero by symmetry, leaving only the relatively small π-overlap to contribute to the repulsion resulting from the exclusion principle.

We also note here that the addition of a localized 5d to $4f^5$ should not appreciably change the isomer shift (as has already been recognized [20]) because this effect is due mainly to the 5s electrons.

III. OPTICAL PROPERTIES, XPS, SPECIFIC HEAT AND ELECTRICAL CONDUCTIVITY IN THE COLLAPSED PHASE

An obvious difficulty facing the present approach is in understanding the apparent metallic nature of the d.c. conductivity [6]

σ and the plasma frequency ω_p, in the collapsed phase in view of the small number of conduction electrons. We show here that this difficulty is only apparent.

Consider σ first. Values of the electron mobility at room temperature found [23] in semiconducting SmS range from about 5 to 15 cm^2/volt-sec. A value of 10 cm^2/volt-sec. with b=.15 electrons/Sm gives $\rho \cong 200$ $\mu\Omega$-cm., the observed value [6]. It seems reasonable to expect that the mobility will not change appreciably in going from the semiconducting to the collapsed phase since the scattering is presumably dominated by phonons at 300°K. Hence the above estimate would appear to be justified, leading to the conclusion that $b \sim .1$ is probably the correct order of magnitude.

The optical reflectivity shows a plasma-edge at an energy of roughly several volts [7]. In fact, the steep rise in reflectivity with decreasing energy follows very closely Drude-type behavior [35]. Making corrections for the high frequency dielectric constant ε_∞, Battlogg et. al. [7] found a bare value of plasma frequency ω_p that corresponds, through $\omega_p^2 = 4\pi be^2/m^*$, to $b \approx 1$ (taking $m^* \sim 1$). And in fact Battlogg et. al.[7] conclude that there must be about 1 nearly free electron per Sm in collapsed SmS. This conclusion, if correct, would clearly invalidate our approach. We now show that this conclusion does not follow from the data, and in fact our model (with $b \approx .1$), is probably consistent with the data.

The ground state in the collapsed phase, as discussed qualitatively above, consists of a small number, b of holes in Φ (Eq. (1)) plus the same number of electrons occupying the free-electron-like Bloch states near the bottom of the 5d-band. Formally this would be obtained by destroying Nb (N = number of Sm's) d-electrons in Φ and creating the same number of free-electron-like d-electrons. The destruction of a d_i in ϕ_i (eq. (1)) leaves f_i (renormalized to 1). This could have been done at any site; and symmetry demands that we multiply Φ by $a_{dk\sigma} = 1/\sqrt{N} \sum_j \exp(-ik\cdot R_j) a_{dj\sigma}$ rather than $a_{dj\sigma}$.

This gives rise to a hole band, with energies ε_{1k}. Note that these states have correlation built in. We expect this band to be very narrow - sufficiently so to give rise to the large observed [6] specific heat at low T and arguments leading to such a narrow width will be given below. Thus we have a few electrons in a broad band (band 2) and a few holes in a narrow band (band 1) - this is depicted in Fig. 2 where we take cut-off effective-mass expressions for each band energy $\varepsilon_{k\nu}$. Optical absorption will be the sum of three contributions: the two ordinary intraband terms, giving the familiar Drude expressions for the conductivity, $\sigma_\nu(\omega) = (\omega_{p\nu}^2/4\pi) \times (i\omega + 1/\tau)^{-1}$ (ν=1,2) and an interband term resulting from transitions from the occupied narrow-band states between k_B and k_A (Fig. 2) to the empty broad band states above. Since $m_1^* >> m_2^*$, the intraband terms would indeed give a plasma edge at $\omega_p = (4\pi be^2/m_2^*)^{1/2}$ and,

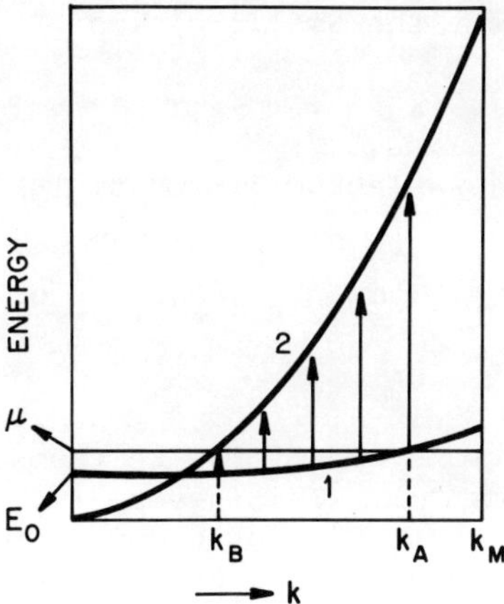

Figure 2. Energy vs k for a simplified version of the present model.

as pointed out above, would disagree seriously with experiment for $b \sim .1$ and $m_2^* \sim$ a free-electron mass. We now argue however that the interband contribution also gives rise to a Drude behavior, and can have a large coefficient.

Intuitively one might suspect an important contribution of this sort from the interband conductivity $\sigma_{12}(\omega)$, because of a possibly small excitation energy at $k \cong k_B$. To get a rough picture, we consider a two-band model of non-interacting electrons corresponding to Fig. 2. The neglected correlation effects, characteristic of our model, will be discussed below - we feel they will not alter the essential idea. At the least, this simple calculation will prove rigorously that the conclusion of Battlogg et. al.[7], namely $b \sim 1$, cannot be made unambiguously on the basis of the observed plasma frequency.

A straightforward calculation using the Kubo formula yields, in the $T \to 0$ limit,

$$\sigma_{12} = \frac{e^2}{2\pi^3 m^2} \int_{k_B<k<k_A} d^3k \; \frac{|(\psi_{k1}, p\,\psi_{k2})|^2}{\varepsilon_{k2} - \varepsilon_{k1}} \left[i\left(\omega - \frac{\varepsilon_{k2} - \varepsilon_{k1}}{\hbar}\right) + \frac{1}{\tau} \right]^{-1} \quad (2)$$

where e and m are the electron charge and mass, $\psi_{k\nu}$ is a Bloch function for band ν, $p = -i\hbar\, \partial/\partial z$, and τ is presumed to $\to \infty$. We choose $\varepsilon_{k2} = \hbar^2 k^2/2m \equiv \varepsilon_k$, $\varepsilon_{k1} = \frac{m}{m^*} \varepsilon_{k2} + E_o$ (i.e. the broad band has a free-electron mass, m^* being the effective mass for the narrow band). To get an order-of-magnitude estimate of (2) we have assumed that in the matrix element we can put the broad-band wave function $\psi_{k2} \cong \Omega^{-1/2} \exp(i\vec{k}\cdot\vec{r})$ (Ω = crystal volume). Then $(\psi_{k1}, p\psi_{k2}) \cong \hbar k_z d_k$ where $d_k \equiv \sqrt{N/\Omega} \int \exp(i\vec{k}\cdot\vec{r}) d(\vec{r})\, d^3r$ and $d(\vec{r})$ is the Wannier function for the narrow band (ψ_{k1}). We also take $d(r) = Ce^{-\alpha r}$, $\alpha^{-1} \cong 1\text{Å}$, (the maximum in r d(r) for the $Sm^{2+}(f^5d)$ ion [36]) which gives

$$d_k^{\,2} = (N/\Omega)\ 64\pi\alpha^5/(\alpha^2+k^2)^4 \ . \tag{3}$$

Then (2) can be written, replacing 1-m/m* by 1,

$$\sigma_{12} = \frac{2(ed_o)^2}{3\pi^2 m}\left(\frac{2m}{\hbar^2}\right)^{3/2}\int_{z_B}^{z_A}\frac{dz}{z}(z+E_o)^{3/2}\left[1+\frac{z+E_o}{\varepsilon_c}\right]^{-4}\left[i\left(\omega-\frac{z}{\hbar}\right)+1/\tau\right]^{-1} \tag{4}$$

To estimate the various quantities we choose b = .1, $\varepsilon_{2\ max.}$ = 3 ev., $\Omega/N = 54\ \mathring{A}^3$ (volume/Sm). This gives $\mu \cong .41$ ev., and $k_m \cong 1\ \mathring{A}^{-1}$. If we assume that the density of states of the narrow band at μ gives the observed [6] low-T specific heat, then $m^*/m \cong 89$. It follows that $E_o \cong .31$ ev., $W_1 \equiv \varepsilon_{1k_m} - E_o \cong .033$ ev. (the width of band 1), $z_A \cong 2.5$ ev. and $z_B \cong .028$ ev. Thus z_B is much smaller than the other energies appearing in the integrand: we are interested in $\hbar\omega \approx$ several ev, and for $\alpha^{-1} = 1\ \mathring{A}$, $\varepsilon_c \cong 3$ ev. Hence the asymptotic value of the integrand as $z_B \to 0$ will give an adequate approximation. Putting $\omega_o^2 = 4\pi be^2/m$, the total conductivity is

$$\sigma = (\omega_o^{\,2}/4\pi)(i\omega+1/\tau)^{-1}\ (1+\nu) \tag{5}$$

where

$$\nu = 2d_o^{\,2}\left(\frac{E_o}{\mu}\right)^{3/2}\left(\ln\frac{z_A}{z_B}\right)\left[1+E_o/\varepsilon_c\right]^{-4} \ . \tag{6}$$

The above numbers give $\nu \cong 19$. We must conclude that observation of a plasma edge at frequency ω_p (after correction for ε_∞) should be interpreted according to the formula $\omega_p^{\,2}=\omega_o^{\,2}(1+\nu)$ and therefore the number of band electrons per unit volume is $b = m\omega_p^{\,2}/4\pi e^2(1+\nu)$ rather than $m\omega_p^{\,2}/4\pi e^2$, the latter formula possibly being seriously in error. The value of ν obtained in our model is somewhat too large (by ~ a factor of 2) to give agreement with experiment, but is clearly of the right order of magnitude. In fact it is easy to see that our correlated state with one hole namely $\sqrt{2}\ a^+_{bk\sigma}\ a_{dk\sigma}\Phi$, gives a momentum matrix element which is $1/\sqrt{2}$ of the uncorrelated value. [Here $a_{dk\sigma}$ was defined as a Bloch sum of $a_{di\sigma}$ which destroys an electron in the excitonic d-state at site i i.e. in the presence of f_i^5 (rather than f_i^6); $a_{bk\sigma}$ corresponds to a free-electron-like Bloch state (b for band-like) near the bottom of the conduction band.]

We now discuss the XPS experiments [5], where the spectral intensity corresponds to ~70% Sm^{3+} character, the remainder looking like Sm^{2+}. Sm^{3+} character corresponds to energies for $f^5 \to f^4$ transitions, while Sm^{2+} involves $f^6 \to f^5$ transitions. The ground state in our model for a single site, is $Af^6 + Bf^5d$ which we simplify to $Af^2 + Bfd \equiv \phi$. If the final state is f + plane wave $(\vec{k})$ (idealization of

$f^5 + \vec{k}$), then the matrix element for the transition is, to within a sign, $M(\phi \rightarrow f,\vec{k}) = A\langle\vec{k}|p|f\rangle + (B/\sqrt{2})\langle\vec{k}|p|d\rangle$, while for a final state, d + plane wave (idealization of $f^4d + \vec{k}$), we have $M(\phi \rightarrow d,k) = (B/\sqrt{2})$ $\langle k|p|f\rangle$. For the optical experiment $|\langle\vec{k}|p|d\rangle| \gg |\langle\vec{k}|p|f\rangle|$ (since $\alpha_f \gg \alpha_d$ and $k^2 \ll \alpha_d^2$ - see eq. (3)). If this held in the XPS case, then most of the observed weight would be in the Sm^{2+} spectrum, in disagreement with experiment. However for the XPS experiment $k^2 \gg \alpha_f^{\ 2}$; this gives the opposite result, $|\langle\vec{k}|p|d\rangle| \ll |\langle\vec{k}|p|f\rangle|$; which leads to $M(\phi \rightarrow f\vec{k}) \cong A\ M(f^2 \rightarrow f\vec{k})$ and $M(\phi \rightarrow d\vec{k}) = B\ M(df \rightarrow d\vec{k})$, and this is as observed: $|M(\phi \rightarrow dk)/M(\phi \rightarrow fk)|^2 \cong B^2/A^2$ x ratio of intensities for the unmixed cases.

Finally we argue the plausibility that the width of the narrow band ε_{1k} be ~.03 ev.; As a start we consider the calculated [24] bandwidth of ~5 ev. associated with the t_{2g} states. The band calculation [24] determines 5d-states in the presence of $4f^6$ whereas our narrow band involves d-states in the presence of $4f^5$, giving an obvious tendency toward band-narrowing. A crude guess as to the amount can be obtained as follows: one-electron hopping integrals t are usually of the order of 1 Ryd. x an overlap integral Δ, and bandwidths are roughly t x the number of nearest-neighbors giving width $\approx 10^2\ \Delta$ ev., thus for a band of ~10 ev width, Δ ~.1; putting $\Delta \approx e^{-\alpha R}$ where R is the interatomic distance and $\alpha \propto$ ionic charge seen by an electron in an atomic state, α would increase by a factor of 3/2 giving, $\Delta' \approx (.1)^{3/2} \approx \Delta/3$ - a reduction factor of about 3. An important effect has to do with the form of our state Φ ; putting $\phi_{i\sigma}^{\ 3+} \equiv c_{fi\sigma}^{\ +}|0\rangle$, the hopping integral of interest is (in terms of ϕ_i of eq. (1)) $T_{ij} = \langle\phi_i\phi_{j\sigma}^{3+}|H|\phi_{i\sigma}^{3+}\ \phi_j\rangle$, which gives $T_{ij} = -(B^2/2)t'_{ij}$, where we put the effective one-electron part of $H = \sum t'_{ij}\ c_{di\sigma}^{\ +}\ c_{dj\sigma}$, t'_{ij} being the hopping integral between d-states calculated in the presence of f^5 on ions i and j. Since B ~.7 our band width at this stage is ~1/2 ev. But there is another well-known effect that we have neglected, namely the small-polaron effect; due to the term $A\ c_{fi\uparrow}^{\ +}\ c_{fi\downarrow}^{\ +}$, there will be an appreciable ionic size difference between a site with ϕ_i and one with $\phi_{i\sigma}^{\ 3+}$. In view of the large reduction factors that are known to occur from this effect [38], it does not seem unreasonable to expect a further reduction of the order of 1/10 giving a final band width $\approx$ .05 ev., which is of the order needed.

III. DISCUSSION

Understanding the magnetic behavior of SmS (and SmB_6) is very difficult on the basis of the conventional model, as discussed in Sec. I. There is considerable hope that our model will provide such understanding, as discussed earlier [10]. A most promising possibility is the following. Working within our mean-field (single-site) approximation, the basic mixing interaction W, which is an intersite interaction, provides an effective mean (electric) field at the site of interest, causing mixing of the lowest f^5d multiplet with the higher-lying f^6 J=0 state. If this mixing dominates that between other f^5d - f^6 pairs, it can be seen that the single-ion ground state must be nondegenerate [37]; and this would then have to be nonmagnetic because of time reversal invariance of the mean-field Hamiltonian. Detailed calculations on this question are in progress.

As mentioned briefly above, our model predicts low-energy optical absorption in the range of tenths of an ev. It results from the intraatomic transition between ground and excited mixed states, represented symbolically by $Af^6 + Bf^5d \rightarrow Bf^6 - Af^5d$; the corresponding momentum matrix element in our simplified 2-electron model is $\langle Af^2 + Bfd|\vec{P}|Bf^2 - Afd\rangle = \langle fd|\vec{P}|f^2\rangle$, the same as that in the semiconducting phase. Here we have assumed A and B are real, appropriate to the mean field state [8], and the single-particle matrix element of momentum between localized f and d is imaginary. Thus the intensity is expected to be appreciable. This absorption might be related to the low-energy anomaly in the dielectric function that has been found [35].

To summarize, the determination that in the collapsed phase there is a small number of conduction electrons, (by extrapolation from the semiconducting phase) plus what we see as the unlikelihood of explaining the magnetic properties in the "conventional" models has compelled us to investigate the "unconventional" model where most of the 5d-electrons in the collapsed phase are localized. We have investigated the consequences of our model, in a simplified version, for a number of experiments, and so far have not found any inconsistency. The necessary calculations for a more realistic version of our model are in progress. If our prediction of appreciable low-energy optical absorption should be found to occur experimentally, this would be strong additional evidence in favor of our approach.

Another consequence of our model is the existence of electric dipole moments on the Sm, ordered in an unknown and presumably complicated way [10].

* Work supported by NSF grant #GH-34565.

REFERENCES

1 A. Jayaraman et. al., Phys. Rev. B11, 2783 (1975).
2 J.M.D. Coey et. al., AIP Conf. Proc. no. 24, 38 (1974).
3 M.B. Maple and D. Wohlleben, Phys. Rev. Letters 27, 511 (1971).
4 T. Penney and F. Holtzberg, Phys. Rev. Letters 34, 322 (1975).
5 M. Campagna et. al., Phys. Rev. Letters 33, 165 (1974).
6 S.D. Bader et. al., Phys. Rev. B7, 4786 (1973).
7 E. Kaldis and P. Wachter, Sol. St. Comm. 11, 907 (1972).
B. Batlogg et. al., preprint.
8 T.A. Kaplan and S.D. Mahanti, Phys. Lett. 51A, 265 (1975).

9 J. Schweitzer, Phys. Rev. B13, 3506 (1976).

10 S.D. Mahanti, T.A. Kaplan, M. Barma, Phys. Lett. 58A, 43 (1976).

11 L.L. Hirst, J. Phys. Chem. Solids 35, 1285 (1974).

12 C.M. Varma and V. Heine, Phys. Rev. B11, 4763 (1975).

13 C.M. Varma, Rev. Mod. Phys. 48, 219 (1976).

14 C.M. Varma and Y. Yafet, Phys. Rev. B13, 2950 (1976).

15 C.E.T. Goncalves da Silva and L.M. Falicov, Sol. St. Comm. 17, 1521 (1975).

16 J.R. Iglesias Sicardi et al., Sol. St. Comm. 16, 499 (1975).

17 D.I. Khomskii et. al., Sol. St. Comm. 18, 985 (1976).

18 B. Alascio et. al., Sol. St. Comm. 14, 321 (1974).

19 M. Avignon et. al., Sol. St. Comm. 16, 1243 (1975).

20 A localized picture for most of the d-electrons was first discussed by Nickerson et. al., Phys. Rev. B3, 2030 (1971), in connection with SmB_6. A coherent mixing of the two configurations was not considered, in contrast to ref's. 8,9,10.

21 F. Holtzberg and J.B. Torrance, AIP Conf. Proc. no. 5, 860 (1971).

22 The interpretation of ref. 21 followed the earlier work in the analogous situation in Eu compounds, by Freiser et. al., J. Appl. Phys. 39, 900 (1968).

23 J.W. McClure, J. Phys. Chem. Sol. 24, 871 (1963). See also 39.

24 H.L. Davis, Proc. 9th Rare Earth Res. Conf., Va. Poly. Inst. and State Univ., Blacksburg, VA (1971).

25 K. Lendi, Phy. Cond. Matt. 17, 215 (1974).

26 D.L. Greenaway and G. Harbeke, Optical Properties and Band structure of Semiconductors, Pergamon Press (1968), p. 114.

27 The lowering of 0.6 ev. follows from ref. 4, with $\Delta V/V = 12\%$ across the transition.

28 Estimates were also made with a circular density of states, $D(e) \propto (1 - (\frac{2e}{e_c} - 1)^2)^{1/2}$, resulting in only minor changes.

29 J. Hubbard, Proc. Roy. Soc. (London) A276, 238 (1963).

30 D. Penn, Phys. Letters 26A, 509 (1968).

31 Our comparison of T=0 theory with experiments at 300°K seems reasonable in view of the insensitivity of p_c to temperature change found for SmS in ref. 1.

32 D.H. Templeton and C.H. Dauben, J. Am. Chem. Soc. 76, 5237 (1954).

33 N.P. Bogoroditskii, et. al. Sov. Phys. Doklady 10, 85 (1965).

34 S.D. Mahanti, Ph.D. thesis, University of California, Riverside (1968) and unpublished.

35 J.W. Allen, private communication.

36 This was found by E. Carlson (private communication) in a Hartree-Fock calculation for Sm^{2+} in the excited configuration $4f^5 5d$.

37 C.M. Varma, Bull. Am. Phys. Soc. 20, 395 (1975), discussed the removal of degeneracy for the special case of a 3x3 matrix.

38 D. Sherrington, et. al., Sol. St. Comm. 16, 1347 (1975).

39 A.V. Golubkov et. al., Sov. Phys. Sol. St. 7, 1963 (1963).

COMMENTS

NOWIK: I want to point out, as I shall discuss in my lecture this afternoon, that we have some experimental support for this model from Mossbauer studies on SmS. It turns out that the Mossbauer probe is very sensitive to the conduction electron configuration. We see the isomeric shift before the collapse and after the collapse, and there is almost no change. So the neighbor of the Sm ion doesn't feel any change. The electron that is released by the Sm must be very localized at the Sm site.

ANDERSON: I would like to make two comments. One is that I do not understand the last part of your talk where you are talking about energies of order 1-2 eV, since the binding energy of the exciton you are talking about has to be of the order of what the rest of us call the fluctuation time, this corresponding to energies much smaller than 1-2 eV. Both XPS and optical measurements probe various instantaneous averages which relate to the mean number of conduction electrons whether or not they are bound up in these excitons. So the optical measurements you are talking about cannot tell us anything about the phenomenon. The second point I want to make is that it seems to me, but I am not sure, that your mechanism is something which can be put in a different way, and has always been a part of my personal picture of these phenomena. There is no reason for the number of free electrons counted on a Fermi surface in the mixed valent phase being equal to the number of f electrons that you see by some high energy frequency probe. Your excitonic mechanism seems to be one way of doing it. The other way is to think of f electrons as a resonance placed near the bottom of the band where the density of states is changing rapidly. Haldane and I have shown that the number of actual f electrons in that resonance is not necessarily one. There is a renormalization factor which can be either much less or much greater than unity. So one can think of it as if the number of electrons that you measure physically (the actual number in the f level) is not equal to the number of electrons in that virtual level. Another way of saying it is if within the Anderson model one looks at the composition of the resonance one finds a certain number of f electrons and a certain number of free-like electrons. In the conventional Anderson model that ratio is always one to one since one uses a constant density of states. However, large renormalization effects can occur. So just within the one electron theory of resonance you can have this mixing effect.

KAPLAN: In the last part of my talk I was talking about energies of order of a few tenths eV, which might indeed be about the valence fluctuation frequency. I also am not sure about the relation of our excitonic mechanism to your picture with Haldane.

FREEMAN: I would like to make a general comment. I don't like to see people put 4f electron levels onto any energy level diagram and treat them as in the one-electron approximation. Since the f electrons are so localized in the rare earths, the Coulomb correlations are very large. Therefore, simple calculations can give misleading information about f electrons. The true excitation processes have to be determined from a many-electron point of view; e.g., one has to calculate the total energy difference between the f^6 and f^5d^1 states. This energy difference turns out to be very small, even if one ignores bonding effect corrections, resulting in the sort of picture that is evolving from the mixed valence models.

KAPLAN: In our energy level diagrams many-electron energies were shown.

PHASE DIAGRAM FOR THE KONDO LATTICE

S. Doniach*

Dept. of Applied Physics, Hansen Laboratories
Stanford University, Stanford, California 94305

The finite temperature properties of the Kondo lattice model, assumed to represent the magnetic properties of γ-Ce and Ce intermetallic compounds are discussed in terms of a Ginzburg Landau functional formalism for a 1-dimensional psuedo-spin analog model.

INTRODUCTION

The magnetic properties of the γ and β-phases of metallic Cerium and the intermetallic compounds $CeA\ell_2$ and $CeA\ell_3$ appear to be anomalous in that, although the f-level of each Ce atom has occupancy of the order of a full electron, these materials either do not exhibit magnetic order, or if they do order magnetically at low temperatures, have an anomalously low magnetic transition temperature. In a recent paper [1] the author has proposed that this kind of behavior may be understood in terms of the properties of a Kondo lattice model in which an f-electron spin on every site of the lattice is exchange coupled to a system of otherwise non-interacting conduction electrons. It is suggested that, as a result of the closeness of the f-level to the fermi level, strong "Kondo quenching" of the local f-moments occurs in which the spin of the f electron is either wholly or nearly compensated by conduction electron spin-moments, thus inhibiting the formation of a magnetic ground state.

---

*Research supported in part by Army Research Office, Durham, N.C.

FREE ENERGY OF THE 'KONDO NECKLACE' MODEL

In [1] the ground state properties of the Kondo lattice were discussed in terms of a mean field calculation of the ground state properties of a one dimensional local spin analog model, the Kondo necklace', in which the spin density excitations of the electron gas are modeled by those of a quantum x-y chain of psuedo spins, $\vec{\tau}_i$:

$$H = J\sum_i S_i\tau_i + W\sum_i(\tau_i^x\tau_{i+1}^x + \tau_i^y\tau_{i+1}^y) \quad .... \quad (1)$$

In order to discuss the properties of the model at finite temperatures it is necessary to go beyond the mean field approximation, since for a one dimensional system the fluctuations of the order parameter cannot be neglected even at low temperatures. (This is an artificial feature of the model which would not occur for a 3-dimensional system).

In the present paper the temperature dependent properties of this model are discussed using an approximate Ginzburg Landau free energy functional formulation. In the strong coupling regime $J \sim W$ the order parameter fluctuations are treated in the 'static approximation' by introducing a set of classical local statistical variables $\vec{\varphi}_i = (\varphi_i^x, \varphi_i^y)$ on each site of the lattice using the Hubbard-Stratanovitch [2] technique:

$$Z = \mathrm{Tr}\, e^{-H/T} \rightarrow \int \prod_i \left(\frac{\pi}{2}\right) d\,\varphi_i^x d\,\varphi_i^y \,\mathrm{Tr}\, e^{-\mathcal{H}[\varphi]/T} \quad .... \quad (2)$$

where

$$\mathcal{H}[\varphi] = J\sum_i \vec{S}_i\cdot\vec{\tau}_i + \sum_{ii'} w_{ii'}(\tau_i^x\varphi_{i'}^x + \tau_i^y\varphi_{i'}^y) + (\vec{\varphi}_i)^2 \quad .... \quad (3)$$

In this expression the coupling between τ-psuedo spins on adjacent sites in (1) has been replaced by a non local coupling to the order parameter fluctuations. The nature of the static approximation is that quantum fluctuations of the order parameter have been neglected.

The non local interaction in (3) is given by

$$w_{ii'} = \frac{1}{N}\sum_k e^{ik(a_i - a_i')} \, 2i\sqrt{W\cos ka} \quad .... \quad (4)$$

where a_i are the site positions on the chain. The trace in (3) may be evaluated approximately by a cumulant expansion in powers of w, leading to an approximate free energy for the model

$$F \cong - T \log\int \prod_i d^2\varphi_i \; e^{-\mathfrak{F}[\varphi]/T} \quad .... \quad (5)$$

where

$$\mathfrak{F}(\varphi) = \sum_i \left\{ (\vec{\varphi}_i)^2 + W\,\chi^o(T)\,\vec{\varphi}_i \cdot \vec{\varphi}_{i+1} + W^2 M_4(T)(\vec{\varphi}_i)^4 \right\} \ldots. \quad (6)$$

where

$$\chi^o(T) = \int_o^{1/T} d\sigma\, \langle \tau^x(\sigma)\,\tau^x(0)\rangle_J \ldots. \quad (7)$$

and the average is taken with respect to $H_J = J\,\vec{S}\cdot\vec{\tau}$ and

$$M_4(T) = \frac{1}{3!} \int_o^{1/T} d\sigma_1 d\sigma_2 d\sigma_3\, \langle T_\sigma(\tau^x(\sigma_1)\tau^x(\sigma_2)\tau^x(\sigma_3)\tau^x(0))\rangle \ldots. \quad (8)$$

In equation 6 the non local character of the fourth order term has been neglected.

It may now be seen that the second order transition for $J/W=1$ found in [1] corresponds to the $T=0$ limit of a phase boundary

$$W\chi^o(T_B) = 1 \ldots. \quad (9)$$

for an antiferromagnetic solution of the order parameter

$$\varphi_i = \varphi \cos Ka_i \ldots. \quad (10)$$

where $K = \pi/a$ is the antiferromagnetic zone boundary, at which the coefficient of the quadratic term in (6) goes negative. On evaluating (7) this gives an antiferromagnetic solution for

$$\frac{W}{J}\,\frac{1 - (1 - 4J/T_B)e^{-4J/T_B}}{1 + 3e^{-4J/T_B}} \leq 1 \ldots. \quad (11)$$

which gives the line shown in Fig. 1 for T_B. As may be seen, the boundary is essentially temperature independent for $T/W \lesssim 1$.

ORDER PARAMETER FLUCTUATIONS IN THE ANTIFERROMAGNETIC REGIME

Because of its one dimensional nature, the model cannot exhibit long range order, so that the above solution (10) will be nearly degenerate with a set of goldstone modes

$$\varphi_i^x + i\varphi_i^y = |\varphi| e^{i\theta_i} \cos Ka_i \ldots. \quad (12)$$

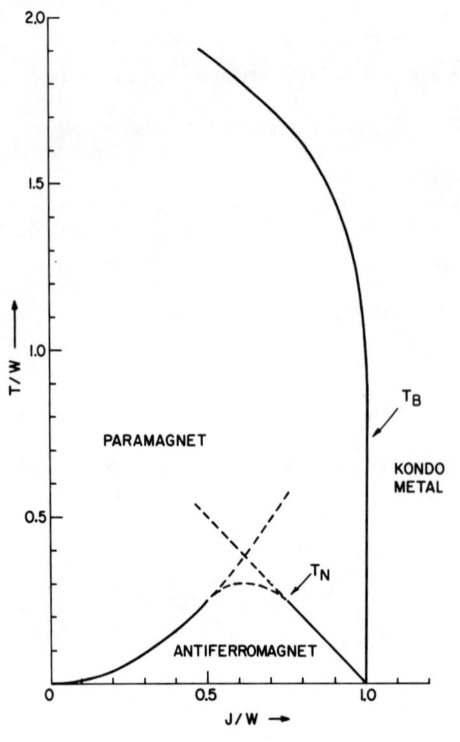

Fig. 1: Phase diagram of the Kondo lattice model (schematic). The phase boundary T_B denotes the line of second order line, phase transitions at which Kondo quenching is broken by interatomic coupling. The Neel line T_N is pieced together qualitatively from weak and strong coupling approximations.

in which the phase angle θ_i varies slowly from site to site. At low temperatures, fluctuations in the magnitude $|\varphi|$ may be neglected and $|\varphi|$ can be evaluated by minimizing $\mathfrak{F}$ using (10) to give

$$|\varphi(T)| = \sqrt{\frac{W\chi^0(T) - 1}{2\,W^2 M_4}} \quad \ldots\ldots \tag{13}$$

which corresponds to the weak antiferromagnetic moment discussed in [1].

The fluctuations of phase angle θ_i cannot be neglected at finite temperatures and lead to an order parameter correlation function at site a_i relative to the origin of

$$\langle \varphi_i^* \varphi_0 \rangle = Z^{-1} |\varphi^2| \cos Ka_i \int \prod_k d\theta_k \, e^{-k^2 a \xi \theta_k \theta_{-k}} \, e^{i\theta_i} \, e^{-i\theta_0} \quad \ldots\ldots \tag{14}$$

where the coherence length for phase fluctuations is given by

$$\xi(T) = \frac{a\,|\varphi^2(T)|}{T} \quad \ldots. \tag{15}$$

and θ_k are the Fourier components of θ_i

$$\theta_k = \sum_i e^{ika_i}\,\theta_i$$

Evaluation of Equation (14) leads to an exponentially decaying order parameter

$$\langle \varphi_i^* \varphi_0 \rangle = |\varphi^2| \cos ka_i \; e^{-a_i/\xi(T)} \quad \ldots. \tag{16}$$

As $K_B T$ increases relative to $|\varphi^2|$, ξ becomes of order a and the neglect of order parameter magnitude fluctuations in (14) is no longer valid. M_4 is of order J^{-3} at low temperatures, hence the condition $\xi/a = 1$ will define an "effective Neel temperature" given by

$$\frac{T_N}{W} \cong \left(\frac{J}{W}\right)^2 (W\chi^0 - 1) = \frac{J}{W}\,(1 - J/W) \quad \ldots. \tag{17}$$

At temperatures $T_N < T < T_B$ the system will be expected to exhibit weak local moment paramagnetism. In a 3 dimensional system a similar reduction of the Neel temperature for J/W approaching a critical value may be expected as the partial Kondo quenching of the f-spin moments lowers the magnon excitation energies (corresponding to the above 1-dimensional phase fluctuations).

WEAK COUPLING BEHAVIOR

The above results are an extrapolation of a strong coupling expansion (6). For weak coupling, $J/W \ll 1$, this will not be valid and the usual perturbation approach in powers of J/W leads to the Ruderman-Kittel-Yosida theory in which $T_N \propto J^2/W$. Note that for $J = 0$, mean field theory of the psuedo spin chain would still lead to an antiferromagnet. However this is suppressed by phase fluctuations of the order parameter leading to a behavior

$$\langle \varphi_i^* \varphi_0 \rangle_{J=0} = \frac{\cos Ka_i}{a_i}\, e^{-a_i T/aW} \quad \ldots. \tag{18}$$

corresponding to Friedel oscillations of a 1-dim Fermi gas with $\mathcal{E}_F \propto W$ (see appendix) at finite temperature.

In the intermediate coupling regime there must be some interpolation between the above antiferromagnetic phase boundary behaviors, indicated by the dotted line in Fig. 1.

REFERENCES

[1] S. Doniach, "The Kondo Lattice and Weak Antiferromagnetism," Proceedings of the Oxford Conference on Itinerant Electron Magnetism, September 1976, to be published in Physica.

[2] S. Wang, W. Evenson and J.R. Schrieffer, Phys. Rev. Letters, 23, 92 (1969).

<u>Appendix</u> by S. Doniach and L. Turkevich

THE TEMPERATURE DEPENDENT RESPONSE FUNCTION OF A 1-DIM FREE ELECTRON GAS

We start from the response function

$$\chi^0{}_q = \frac{1}{4\pi}\int_{-\infty}^{\infty} dk \, \frac{1}{k^2-(q/2)^2} \, \frac{1}{e^{(\mathcal{E}_k-\mathcal{E}_F)/T}+1} \quad .... \tag{A1}$$

At low temperatures $T \ll \mathcal{E}_F$ this may be evaluated to give after some manipulation

$$\chi^0{}_q = \frac{1}{4\pi q}\, \ell n \left(\frac{(1+q\ 2k_F)^2 + \frac{1}{2}(\pi T/\mathcal{E}_F)^2}{(1-q\ 2k_F)^2 + \frac{1}{2}(\pi T|\mathcal{E}_F)^2} \right)^{1/2} \quad .... \tag{A2}$$

The spatial transform (corresponding to the decay of order parameter correlation discussed in Equation (18) is

$$\tilde{\chi}^0(r) = \int_{-\infty}^{\infty} dq \, e^{iqr} \, \chi^0 \tag{A3}$$

For large r, the dominant contribution to the integral comes from $q \cong \pm 2k_F$ giving

$$\tilde{\chi}^0(r) = \exp(-\sqrt{2}\pi k_F rT/\mathcal{E}_F) \, \frac{\mathrm{Cos}\ 2k_F r}{4k_F r} \quad .... \tag{A4}$$

COMMENTS

HIRST: I agree that your model is appropriate for many systems including Cerium; however, the criterion you use to justify that is not, I think, the correct one. What you are saying is that your model applies because U is many times larger than the mixing width. One should really compare the mixing width not to U but to the interconfiguration excitation energy which can be as small as zero at the configuration crossover.

DONIACH: I think that you have a valid point in the sense that U in the usual Hubbard sense does not apply. But I think the fact that you cannot have two electrons with two holes in the same f orbital is extremely important in these systems.

ANDERSON: How many electrons do you have in your model? I expect that if the electron/atom ratio changes from unity to >>1 you will get ferromagnetism instead of antiferromagnetism.

DONIACH: One electron per atom. I think you are right.

VARMA: What would happen in 3D?

DONIACH: I expect the same phase boundary but the fluctuation effects will be weaker.

WOHLLEBEN: Suppose we look at $CePd_3$. There we have exact degeneracy between $4f^0$ and $4f^1$ and therefore the interconfiguration excitation energy is very small. In the case of $CeAl_2$ and $CeAl_3$ the interconfiguration excitation energy is larger than in $CePd_3$; therefore, the occupation of the $4f^1$ state is 0.99 or 1 compared to 1/2 in $CePd_3$. I think that in the case of $CeAl_2$ and $CeAl_3$ one can get things you are talking about, but $CePd_3$ appears to represent quite a different regime.

DONIACH: It seems to me that even if you have half of the Ce f-electrons carrying moments Kondo correlations will be strong.

TOURNIER: I think that your model doesn't apply to TmSe. The ground state of that system is a singlet and antiferromagnetism is of the exchange-induced type.

BUSCHOW: The UCu_5 compound seems to fit your model. There one has a large Kondo resistance and magnetic ordering is of the AF-type at low temperatures.

COQBLIN: Have you obtained phase diagrams with the following sequence when temperature decreases: antiferromagnetism and the Kondo effect (instead of the Kondo effect and then antiferromagnetism, as you presented here)? In compounds such as $CeAl_3$ there is obviously a competition between the Kondo effect and the antiferromagnetic ordering, but I think that the Kondo effect occurs at lower temperature than an eventual antiferromagnetic ordering.

DONIACH: No, I have not.

SUSCEPTIBILITY OF THE ASYMMETRIC ANDERSON MODEL: RELEVANCE TO VALENCE FLUCTUATION COMPOUNDS

H. R. Krishna-murthy, K. G. Wilson, and J. W. Wilkins

Physics Department and the Materials Science Center,

Cornell University, Ithaca, N.Y. 14853

This paper addresses the question of the relevance of the single-impurity Anderson model[1] in understanding the temperature dependence of the susceptibility of valence-fluctuation (or inter-configuration-fluctuation) compounds.[2] Our principal conclusion is that while the susceptibility of asymmetric Anderson model can mimic the observed susceptibilities, the required energy separations between various configurations of the impurity are very much smaller than present microscopic theory suggests.

I. INTRODUCTION

Valence-fluctuation compounds are metallic rare-earth compounds such as SmB_6, SmS, $YbAl_3$, TmTe, etc., in which, under certain conditions (application of pressure, for example), the rare-earth 4f shell apparently demagnetises (i.e., neither orders magnetically nor shows a strongly temperature-dependent susceptibility at low temperatures). Through measurements of magnetic susceptibility, lattice constant, resistivity, specific heat, Mössbauer isomer shift, x-ray absorption and x-ray photoemission in such compounds, evidence has been adduced[3] that the apparently non-magnetic rare-earth ions fluctuate between two ionic configurations of different valences (i.e., different integral occupations of the 4f shell). A model representing such a situation has been developed by Hirst[4] in which it is assumed that there are two distinct states of the rare earth atom and its environment, one with an integral number n of electrons in the 4f shell, and the other with (n-1) electrons in the 4f shell and one delocalised electron in the environment. The energies of these two states are denoted by E_n and E_{n-1} and the difference by $E_{ex} = E_n - E_{n-1}$.

Both of these configurations are also assumed to have well defined total angular momenta J_n and J_{n-1}, and the corresponding Hund's rule magnetic moments μ_n and μ_{n-1}. Electrons are permitted to hop between the rare earth shell and the conduction band, and this causes an admixture of the two configurations. A natural measure of the mixing strength is the width

$$\Gamma \equiv \pi\rho|V|^2 \quad (1)$$

where ρ is the conduction band density of states at the Fermi surface and V the hopping matrix element.

There has not been, as yet, any really satisfactory theoretical treatment of the Hirst model, either in the dilute limit (i.e., only a small number of rare-earth impurities in a metallic environment) or in the concentrated limit (i.e., rare-earth metals or compounds). But, basing themselves on the Hirst model, treating it in the spirit of its dilute limit, and using some very reasonable (albeit ad hoc) arguments, Wohlleben and his coworkers[2,3] have been able to account satisfactorily for the temperature dependence of the suscepbility of many valence-fluctuation compounds: Namely, at high temperatures $[T \gg \max(|E_{ex}|,\Gamma)]$, χ has a Curie-like behaviour; and as T decreases $[T \lesssim \max(|E_{ex}|, \Gamma)]$, χ exhibits a broad maximum at a temperature $T \sim -E_{ex}$ (and hence in some cases at zero temperature if $E_{ex} > 0$).

The main purpose of the present paper is to present the results of numerical calculations of the impurity susceptibility of the asymmetric Anderson model[1] and to show that they share many of the features of the susceptibility of valence-fluctuation compounds. In section II we discuss the model Hamiltonian, in section III we present and discuss the susceptibility results, and in section IV we discuss the implications of our results in the context of our understanding of valence fluctuation compounds.

II. THE ANDERSON MODEL

The Anderson model, which is a model for a single impurity immersed in a metallic environment, is of interest in understanding valence-fluctuation compounds because it is perhaps the simplest model containing the possibility of the admixture of different ionic configurations. It has three basic ingredients:

1. A conduction band, which we will take to be isotropic, and of a constant density of states ρ in a width 2D about the Fermi level, represented by the Hamiltonian

$$H_s = \sum_{k\sigma} \varepsilon_k \, c^+_{k\sigma} c_{k\sigma} . \quad (2)$$

2. A single localised, nondegenerate impurity orbital, represented by the Hamiltonian

$$H_d = \sum_\sigma \varepsilon_d c^+_{d\sigma} c_{d\sigma} + U n_{d\uparrow} n_{d\downarrow} \tag{3}$$

The orbital energy ε_d can be positive or negative (i.e., lie above or below the Fermi level of the conduction band). U, the coulomb energy, is assumed to be positive, which makes it energetically expensive to have two electrons in the impurity orbital. By choosing ε_d and U appropriately, we can make the ground state configuration of the impurity either magnetic ($n_d = 1$) or nonmagnetic ($n_d = 0$ or 2).

3. Electrons hopping between the impurity orbital and the conduction band are represented by the Hamiltonian

$$H_{sd} = V \sum_{k\sigma} (c^+_{k\sigma} c_{d\sigma} + c^+_{d\sigma} c_{k\sigma}). \tag{4}$$

As in the Hirst model, the natural measure of the hopping strength is the parameter Γ defined by equation (1). The full Anderson model is the sum of the above three pieces:

$$H_A = H_s + H_d + H_{sd}. \tag{5}$$

In this paper, our interest will be in discussing the temperature dependence of the impurity susceptibility of such a model. There is a two-fold reason why such a discussion should have any relevance at all to valence-fluctuation compounds. Firstly, even though the lowest two or three 4f configurations that are considered in the Hirst approach are modelled by the configurations of a nondegenerate orbital (where only one of the configurations, viz., $n_d = 1$, carries a magnetic moment), it seems unlikely that this approximation will cause a total loss of qualitative information. (At least, the Yb compounds of reference 2 are presumably dominated by the two configurations corresponding to a filled f shell and to one hole in it.) Secondly, 4f orbitals are so localised that the different 4f atomic orbitals in a rare-earth compound overlap very little with each other. Accordingly a single impurity model can be expected to yield some qualitatively correct information. The results for the impurity susceptibility of the Anderson model presented here have been obtained using the numerical renormalisation group techniques developed by Wilson[5] for the Kondo problem.

Fig. 1 attempts to depict the manifold of states of H_A when $\varepsilon_d < 0$ and the hopping term H_{sd} has been turned off. With the conduction electron system being in its ground state (i.e., a filled Fermi sea), the impurity orbital can be empty ($n_d = 0$), occupied by one electron ($n_d = 1$) or completely filled ($n_d = 2$). These states are represented by the thick lines in Fig. 1. For each of these three configurations of the impurity orbital, one can have the

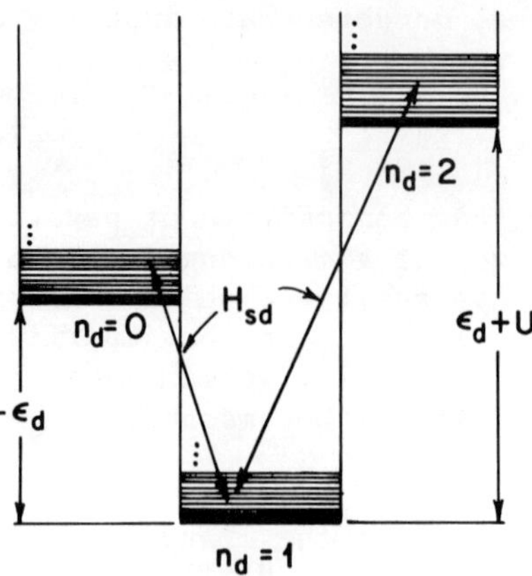

Fig. 1. Manifold of states of the Anderson Hamiltonian when local orbital-conduction electron mixing term H_{sd} is turned off. The thick lines represent states in which the conduction band is in its ground state, but the impurity orbital takes up various configurations. For each of the impurity configurations, the conduction band can also be in any of its many-body excited states, and these are indicated by the light lines lying at higher energy. The three sets of states overlap in energy. The inclusion of H_{sd} causes transitions between the states, as indicated by the diagonal arrows.

conduction electron system in any of its continuum of many-body excited states. These states are represented by the thin lines in Fig. 1. H_{sd} causes transitions between these various states as shown in Fig. 1, and therefore mixes them up. But when Γ is small compared to the separations between the various subspaces (i.e., $\Gamma \ll |\varepsilon_d|, U$), Fig. 1 still retains some meaning and helps greatly in understanding the temperature dependence of the susceptibility, which we discuss in the next section.

Finally we note that a particle-hole transformation (which corresponds to $c_k \leftrightarrow c_{-k}^+$ and $c_d \leftrightarrow -c_d^+$) takes H_A into itself with ε_d replaced by $-(\varepsilon_d + U)$. This can be seen easily by noting that under the transformation, $n_d \to (2 - n_d)$ so that the $n_d = 0$ and $n_d = 2$ subspaces in Fig. 1 are interchanged. As a consequence of this, we have the result that

$$\chi[U, \Gamma, \varepsilon_d, T] = \chi[U, \Gamma, -(\varepsilon_d + U), T]. \tag{6}$$

It is therefore sufficient for us to discuss the region of the parameter space for which $\varepsilon_d \geq -U/2$. When $\varepsilon_d = -U/2$, the Anderson model respects particle-hole symmetry and this case is called the symmetric case. In the case of Yb where the relevant configurations are f^{14} and f^{13}, the particle-hole transformation allows us to map the Yb problem on the $n_d = 0$ and $n_d = 1$ configurations which then correspond to f^{14} and f^{13}, respectively. This is convenient since our discussion for the asymmetric Anderson model is exclusively concerned with the mixing between $n_d = 0$ and $n_d = 1$ configurations.

III. SUSCEPTIBILITY RESULTS FOR THE ANDERSON MODEL

A. The Symmetric Case

The susceptibility results for the symmetric Anderson model have been discussed elsewhere.[6] Here we merely recapitulate the principal features. When $\varepsilon_d = -U/2$, the $n_d = 0$ and the $n_d = 1$ subspaces are both U/2 above the $n_d = 1$ subspace (c.f., Fig. 1). When $T >> U$, all the impurity configurations are equally thermally populated, and the splittings and the couplings between them are irrelevant. The situation is as if U and Γ are both zero, so that $T\chi$ is close to 1/8, as is illustrated by the two plots in Fig. 2, and this is referred to as the free orbital regime.[7]

Now consider the case when $\Gamma << U$. Then, for T much smaller than U, to a first approximation one can set $\Gamma = 0$ and ignore the subspaces for which $n_d = 0$ or 2 (c.f., Fig. 1). Within the subspace in which $n_d = 1$, the impurity degree of freedom looks like a spin-½ object so that we expect $T\chi \simeq ¼$. The next level of approximation is to permit virtual transitions induced by H_{sd} as shown in Fig. 1. This results in effective interactions between the impurity spin and the conduction electron degrees of freedom. It can be verified[8] that the most dominant of such interactions is a spin-spin interaction of the form

$$H_{int} = - J_{eff}\, \vec{s}(o) \cdot \vec{S}_i ; \quad \rho J_{eff} \equiv - \frac{8\Gamma}{\pi U} \tag{7}$$

where $\vec{S}_i$ is impurity spin operator and $\vec{s}(o)$ the conduction electron spin density at the impurity site. This interaction leads to the Kondo effect[6,9], as a consequence of which $T\chi$ gradually decreases and vanishes as $T \to 0$. Furthermore, χ obeys a scaling law, namely that $T\chi$ is a universal function[6] of $T/T_K(J_{eff})$, where the scaling temperature, called the Kondo temperature, is given by

$$T_K(J_{eff}) \sim U \exp\,(- 1/\rho|J_{eff}|). \tag{8}$$

The plot for $\rho J_{eff} = -.064$ in Fig. 2 illustrates this result. The zero-temperature behaviour corresponds to the situation in which the impurity degree of freedom is frozen out, leaving behind only

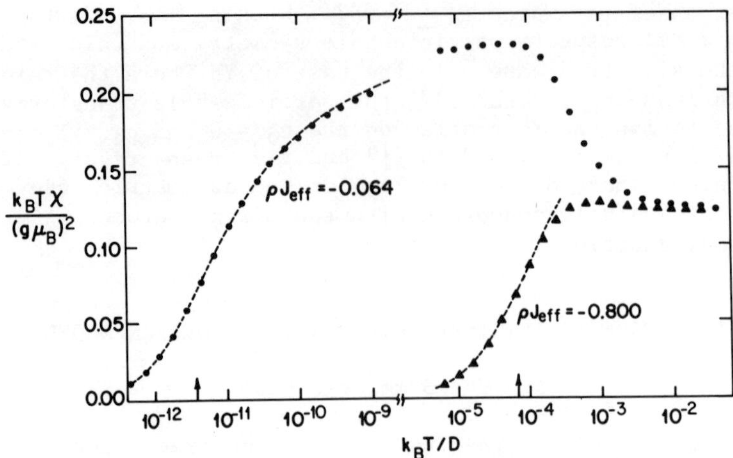

Fig. 2. Plots of $k_BT\chi/(g\mu_B)^2$ versus $\ln(k_BT/D)$, where $\chi(T)$ is the impurity susceptibility for the symmetric Anderson model, for two different values of Γ/U. The value of U/D is 10^{-3} for both plots. The values of $\rho J_{eff}(= - 8\Gamma/\pi U)$ for the two plots are marked in the figure. The dashed curves show the mapping between the above plots and the universal susceptibility curve for the Kondo Hamiltonian. The vertical arrows mark the effective Kondo temperatures for the two plots.

residual interactions between renormalised conduction electron degrees of freedom, which situation corresponds to a constant χ at zero temperature. This regime is referred to as the strong coupling regime. The regime above T_K is referred to as the local moment regime, in which χ is approximately Curie-Weiss like but with a diminished effective moment. The various regimes are illustrated in Fig. 3.

Next we consider what happens when we increase Γ keeping U fixed. The basic effect is that $|J_{eff}|$ (and hence T_K) increases, the local moment regime shrinks and the strong coupling regime grows, until, when Γ is of order U, the two transition regions merge, as illustrated by the second plot (for $\rho J_{eff} = -.8$) in Fig. 2 and as depicted in Fig. 3. When $\Gamma \gg U$, there is a direct transition from the free orbital regime for $T \gg \Gamma$ to the strong coupling regime for $T \ll \Gamma$.

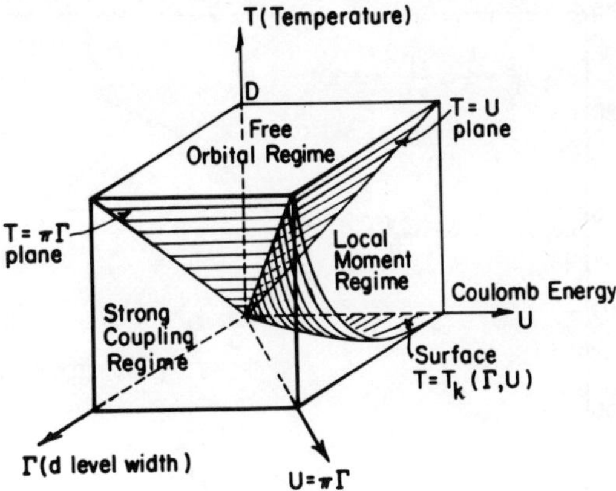

Fig. 3. Schematic sketch of the various regimes arising from the susceptibility results for the symmetric Anderson model. Note that the separations between the various regimes, in contrast to the drawing, are actually quite fuzzy.

B. Susceptibility Results for the Asymmetric Case

Consider first temperatures which are much larger than any of the energies ε_d, Γ, or U. Then all the impurity configurations are equally populated, the splittings between the various configurations are irrelevant and one again has the free orbital regime, characterised by $T\chi = 1/8$. As the temperature is lowered the separations between the various impurity configurations become important. The rest of our discussion will concentrate on the situation when $U \sim D$ and ε_d and Γ are both much smaller than U. Referring to Fig. 1, we see that for $T \ll U$ the $n_d = 2$ subspace of states can be ignored. If T is still much greater than ε_d or Γ, the $n_d = 0$ and $n_d = 1$ subspaces of states will be equally populated, and one can now neglect the splittings between these states. The situation is close to what would obtain if ε_d and Γ were both zero and U were infinite, so that $T\chi \simeq 1/6$, as confirmed by the numerical results plotted in Fig. 4. It is only when T is smaller than $|\varepsilon_d|$ or Γ, whichever is larger, that the situation becomes more complicated.

We first discuss these complications for the case when ε_d is negative and $|\varepsilon_d| \gg \Gamma$. Then the picture in Fig. 1 (with the $n_d = 2$ subspace erased) is relevant. For $T \ll |\varepsilon_d|$, as a first approximation we may ignore all but the $n_d = 1$ subspace of states. Within this subspace of states the impurity degree of freedom looks like a spin-½ object, so that $T\chi \simeq \frac{1}{4}$. As the next stage of approximation, consider virtual transitions, induced by H_{sd}, between states in the $n_d = 1$ subspace and states in the $n_d = 0$ subspace. This leads

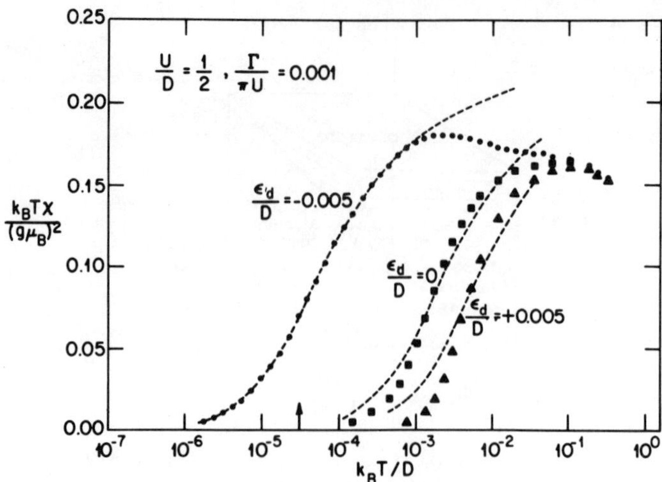

Fig. 4. Plots of $k_BT\chi/(g\mu_B)^2$ versus $\ln(k_BT/D)$ where $\chi(T)$ is the impurity susceptibility for the asymmetric Anderson model. The parameter values for the various plots are indicated in the figure. The dashed curve represents the universal susceptibility curve of the Kondo Hamiltonian. While the dotted plot can be mapped onto the universal curve with an effective Kondo temperature indicated by the vertical arrow, the other two plots are non-universal and cannot be so mapped.

to effective interactions between the impurity spin and the conduction electron degrees of freedom, and it can be shown[8] that the form of the interaction now includes potential scattering in addition to the spin-spin interaction in (7):

$$H_{int} = - J_{eff}\,\vec{s}(o)\cdot\vec{S}_i + K_{eff}\,n(o), \tag{9}$$

where n(o) is the electron number density operator at the impurity site. In terms of ε_d and Γ,[8]

$$\rho J_{eff} \cong - \frac{2\Gamma}{\pi|\varepsilon_d|} \, , \; \rho K_{eff} \cong + \frac{\Gamma}{2\pi|\varepsilon_d|} = \frac{|\rho J_{eff}|}{4} . \tag{10}$$

Potential scattering is expected to have little effect on the susceptibility, and the low temperature ($T \ll |\varepsilon_d|$) susceptibility should hence be mappable onto the universal susceptibility curve for the Kondo Hamiltonian (just as in the symmetric case). The plot in Fig. 4 for which $\varepsilon_d = -.005D$ ($\rho J_{eff} \cong .2$) illustrates this behaviour.

Now consider what happens when $|\varepsilon_d|$ is decreased keeping Γ fixed. One the one hand, the effective coupling constant $|J_{eff}|$ (and K_{eff}) increases as a result of which T_K increases and the universal region shrinks. On the other hand, there is increased admixture with the $n_d = 0$ subspace of states. When $|\varepsilon_d|$ decreases to become of order Γ, one can no longer distinguish between these two effects, and it is found that the $T\chi$ versus lnT plot can no longer be mapped onto the universal curve. This behaviour continues to be true as ε_d passes through zero and becomes positive, when the drop in $T\chi$ from the high temperature regime $[T \gg \max(\varepsilon_d, \Gamma)]$ to the low temperature regime becomes steeper. The plots in Fig. 4 for the cases when $\varepsilon_d = 0$ and $\varepsilon_d = +.005D$ illustrate this effect. The sharpening drop in $T\chi$ is easy to understand in the specific case when ε_d is positive and $|\varepsilon_d| \gg \Gamma$. In this case (see Fig. 1), it is the $n_d = 0$ subspace that lies below the $n_d = 1$ subspace. If as a first approximation we set Γ equal to zero, then

$$T\chi = \tfrac{1}{2}\, e^{-\varepsilon_d/T} \,/\, [1 + 2e^{-\varepsilon_d/T}], \tag{10}$$

which drops rapidly from 1/6 when $T \gg \varepsilon_d$ to zero when $T \ll \varepsilon_d$. When, as the next approximation, we include virtual transitions induced by H_{sd} between the $n_d = 0$ and the $n_d = 1$ subspaces, the resulting effective electron-electron interactions produce a susceptibility χ which goes to a constant at zero temperature [instead of going to zero as in (10)].

The non-universal, sharp drop in the $T\chi$ versus lnT plots in Fig. 4 corresponds to a broad maximum when we plot χ versus T. This is illustrated by the four plots in Fig. 5. In contrast to these plots, the universal susceptibility plot yields a χ that increases monotonically as T decreases.[5] Fig. 5 also illustrates the fact that the non-universality begins where ε_d is negative but $|\varepsilon_d|$ is of order Γ, and that there is nothing special about $\varepsilon_d = 0$. Note that the zero temperature regime is always the strong coupling regime, in the sense that the impurity degree of freedom is always frozen out leaving behind only residual interactions between renormalised conduction electron degrees of freedom. When the basic Hamiltonian does not obey particle-hole symmetry then neither do the residual interactions, and they will hence include potential scattering.

Within the range of parameters we have investigated, the numberical results seem to suggest the following general trends for the dependence of χ or the parameters U, Γ, and ε_d: (1) For fixed T, U and Γ, χ decreases with increasing ε_d (from $-U/2$); (2) For fixed T, ε_d, and Γ, χ increases with increasing U; (3) For fixed T, U and ε_d, χ decreases with increasing Γ; (4) For the symmetric case ($\varepsilon_d = -U/2$), for fixed T and Γ, χ increases with increasing U, a result which is a consequence of (1) and (2). These general

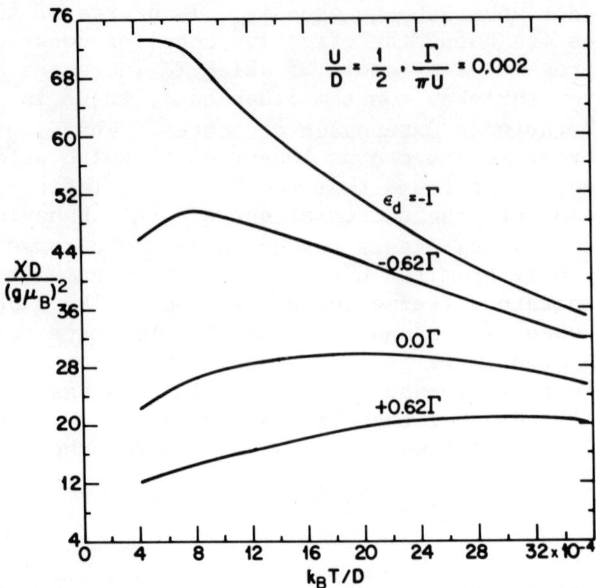

Fig. 5. More susceptibility results for the asymmetric Anderson model, plotted the conventional way. The parameter values are indicated in the figure. The plots display features similar to those seen in valence-fluctuation compounds.

trends can be used as a consistency test for any approximate solution of the Anderson model.[10]

IV. IMPLICATIONS OF THE SUSCEPTIBILITY RESULTS

It seems clear that, as a phenomenological description of susceptibility results for a wide class of valence-fluctuation compounds, Wohlleben and coworkers' ansatz for χ works fine.[2] The crucial question is: Are the underlying picture and the reasoning that led to their ansatz believable?

Now we have presented above reliable calculations of the susceptibility for a model where the underlying picture is the same as the one Wohlleben seems to have in mind. The results show the same features as the susceptibilities of the compounds considered in reference 2. The physical discussion we have presented does have some rough correspondence with the arguments of Wohlleben, if we identify $-\varepsilon_d$ as playing the role of E_{ex}. The parameter range corresponds to ε_d and Γ being of the same order, and the broad maxima occur at temperatures of the order of ε_d (or Γ). All this can be taken to mean that the answer to the question posed above is yes.

However, the numerical values of the parameter E_{ex} needed to obtain quantitative agreement with experimental results, namely E_{ex} of the order of 100 K, brings Wohlleben's picture into serious contradictions with our understanding, from a microscopic point of view, of 4f configurations in rare earth metals. There have been extensive calculations[11] of 4f excitation energies, i.e., differences between energies of various 4f configurations (and E_{ex} is presumably one such energy) in rare earth metals. The results of these calculations are in remarkable agreement with XPS measurements, which are believed to measure such energies. All the excitation energies come out to be of the order of eV's (10^4 K), and it is hard to reconcile this with the E_{ex} of the order of 100 K needed by Wohlleben. [Γ values on the order of 100 K are reasonable from the point of view of the microscopic theory.]

REFERENCES AND FOOTNOTES

Research supported in part by the National Science Foundation under grant numbers DMR74-23494 and PHY74-19018.

1. P. W. Anderson, Phys. Rev. 124, 41 (1961).

2. B. C. Sales and D. K. Wohlleben, Phys. Rev. Lett. 35, 1240 (1975) and references therein.

3. For a review, see M. B. Maple and D. K. Wohlleben in "Magnetism and Magnetic Materials - 1973", AIP Conference Proceedings No. 18, edited by C. D. Graham, Jr., and J. J. Rhyne (American Institute of Physics, New York, 1974) p. 447.

4. L. L. Hirst in "Magnetism and Magnetic Materials - 1974", AIP Conference Proceedings No. 24, edited by C. D. Graham, Jr., J. J. Rhyne and G. H. Lander (American Institute of Physics, New York, 1975) p. 11.

5. K. G. Wilson, Rev. Mod. Phys. 47, 773 (1975).

6. H. R. Krishna-murthy, K. G. Wilson and J. W. Wilkins, Phys. Rev. Lett. 35, 1101 (1975).

7. In all of our discussion we take $g\,\mu_B = 1$, although it is shown on the figures. The calculation of the factor of 1/8 and all subsequent such factors is done thus:

$$T\chi = (1/3)\,\langle S^2\rangle = \sum_{n_d} \frac{g_d}{3}\, S_d(S_d + 1)/\sum_{n_d} g_d$$

where S_d is the value of the spin in the n_d configuration, and g_d is the "spin-degeneracy" of that configuration. Hence in this case we have $2(\frac{1}{4})/4 = 1/8$.

8. J. R. Schrieffer and P. A. Wolff, Phys. Rev. 149, 491 (1966).

9. See various review articles in "Magnetism, Vol 5" edited by G. T. Rado and H. Suhl (Academic Press, New York, 1973).

10. For example, the approximations used by C. M. Varma and Y. Yafet (Phys. Rev. B13, 2950 (1975)) seem to lead to results contrary to trend (1).

11. J. F. Herbst, R. E. Watson and J. W. Wilkins, Phys. Rev. B13, 1439 (1976) and references therein, esp., J. F. Herbst, D. N. Lowy, and R. E. Watson, Phys. Rev. B6, 1913 (1972).

COMMENTS

ORBACH: I wonder if you could comment on the lack of universality. Does that mean that the Schrieffer-Wolf transformation is leading you astray in that regime?

WILKINS: There is no reason to expect universality when I start mixing the configurations. Also, the S-W transformation would not even predict universality.

A NEW RARE-EARTH IMPURITY MODEL EXHIBITING MIXED-VALENCE BEHAVIOUR

F. D. M. Haldane

Department of Physics, Princeton University

P.O. Box 708, Princeton, N.J. 08540 USA

INTRODUCTION

Presented here as an attempt to identify the simplest model showing mixed-valence (MV) behaviour, is an extension of the Anderson model of a magnetic impurity[1] to include a term representing screening of charge fluctuations. Mean-field theory (MFT) suggests that, unlike the original model, this model exhibits anomalous slow valence fluctuations in the inconfigurational fluctuation regime, which may be identified with the MV effect. A preliminary study of the model in MFT is presented, and some Kondo-like problems arising in a more exact treatment are briefly mentioned.

(A) WHY IS AN IMPURITY MODEL RELEVANT?

In the "low-density limit" of a system of rare-earth (RE) atoms embedded in a metal, interactions between RE atoms become negligible, and the system may be treated as a superposition of systems with a single RE atom in a metal. Real mixed-valence systems correspond to the high-density limit of the problem, but experimentally these materials seem to behave as "low-density" systems at intermediate temperatures.

Resistivity measurements[2,3] (Fig.1) on the MV compound $CeAl_3$ show that as the material is cooled: (a) the resistivity initially rises to a very large value[2] that at its maximum can be identified[3] with strong incoherent scattering by each RE atom, so that the l=3 scattering cross-section is at its theoretical maximum - the "unitarity limit" - corresponding to an f-level resonance at the Fermi level, and hence strongly fluctuating valence; (b) at much lower temperatures coherence develops between valence fluctuations on different sites[3] and the material condenses into a Fermi liquid regime, characterised

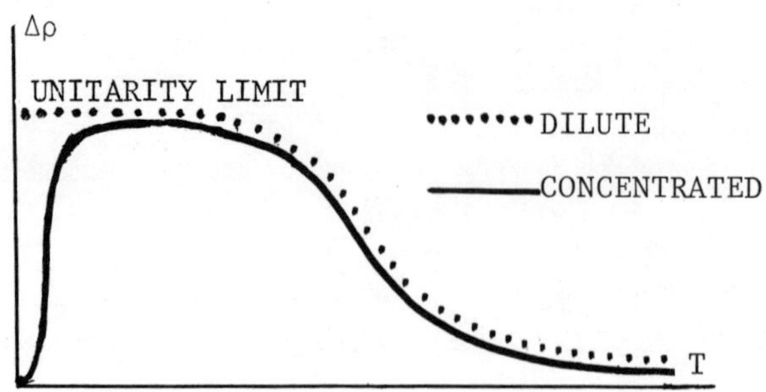

Figure 1. (Schematic). Temperature dependence of the extra resistivity per rare-earth atom (Δρ) in the dilute limit (single impurity) and concentrated limit (MV compound).

by a very large effective density of states at the Fermi level.

These results suggest that the electronic state of the RE atoms is essentially derived from the interaction of a single atom with the conduction band, leading to a fluctuating valence (and spin) state, and that the low temperature behaviour is the result of coherence between the fluctuations on different sites arising through the weak residual coupling between atoms. Our approach to the problem must thus be based on first understanding the fluctuating state using a suitable single impurity model, then treating the residual interactions between such fluctuating entities on different sites. This is in marked contrast to the treatment appropriate to a normal RE material, where valence fluctuations are energetically unfavorable, and the interaction between atoms causes a magnetic transition that stabilises the atoms against spin fluctuations.

In this paper, preliminary results of a study of an appropriate impurity model are presented.

(B) GENERALISATION OF THE ANDERSON MODEL TO INCLUDE SCREENING

To model the conduction band of the metal, which is a d-band, we use an isotropic Fermi gas; two representations, plane waves or spherical waves, are possible: for impurity problems, spherical waves are appropriate.

$$H^c = \sum_{klm\sigma} \varepsilon_{kl} c^{\dagger}_{klm\sigma} c_{klm\sigma} \tag{1}$$

The impurity is modelled by an "extra" f-orbital, orthogonal to the conduction band, plus intra-atomic interaction terms.

$$H^{imp} = E_f \sum_{m\sigma} n_{fm\sigma} + U \sum_{m\sigma \neq m'\sigma'} n_{fm\sigma} n_{fm'\sigma'} + \text{(exchange terms)} \tag{2}$$

We must also introduce a hopping term allowing for hybridisation of the impurity orbital with a combination of l=3 spherical waves of the conduction band. This will be an energetically small term, as the f-orbitals are highly localised, and, as the conduction band is a d-band, l=3 waves are derived from combinations of d-orbitals of neighbouring atoms, not the impurity site d-orbital itself, which has l=2 symmetry.

$$H^{hyb} = \sum_{\substack{km\sigma \\ l=3}} V_k^* c^\dagger_{klm\sigma} c_{fm\sigma} + h.c. \tag{3}$$

So far, these terms correspond to the usual Anderson model[1] where the l≠3 channels of the conduction band may be discarded, as they are not coupled to the l=3 channels where all the action takes place. This model is usually specialised so the impurity orbital is non-degenerate (l=0 replaces l=3 as the important channel) and its parameters chosen so that it describes the Kondo effect, which is characterised by anomalous slow <u>spin</u> fluctuations in the impurity orbital. However, in situations where <u>charge</u> fluctuations are important, this model is physically incomplete, as it does not contain the screening terms necessary to satisfy the Friedel sum rule. As E_f is varied, the phase-shift $\delta_{l=3}$ will vary, but $\delta_{l\neq 3}$ will remain zero. Clearly, if the Friedel rule is to be satisfied, a screening term causing a phase-shift in at least one other channel is required. As the material is a d-band metal, the important screening processes take place in the d-orbitals of the impurity atom itself; these are composed of l=2 spherical waves of the conduction band. The screening term used will be:

$$H^{scr} = g \sum_{m\sigma} n_{fm\sigma} \sum_{m'\sigma'} n_{dm'\sigma'} = \frac{g}{N} \sum_{m\sigma} n_{fm\sigma} \sum_{\substack{kk'm'\sigma' \\ l=2}} c^\dagger_{klm'\sigma'} c_{k'lm'\sigma'} \tag{4}$$

g is not a free parameter of the theory, but must be chosen to satisfy the Friedel rule. This is easily done in the V→0 limit, where $\sum_{m\sigma} \delta_{l=3}(\mu) = \pi n_f$, ($n_f$ is the charge state of the impurity orbital, which is a good quantum number in this limit). One easily finds that $\tan\delta_{l=2}(\omega) = -\pi g n_f \sigma_d(\omega)$, where $\sigma_d(\omega)$ is the conduction band density of states, normalised to unity. If d_o is the multiplicity (including spin) of the screening channels (d_o=10 for l=2) the Friedel rule requires $\delta_{l=2}(\mu) = -\pi n_f / d_o$. When the largest allowed value of n_f is much less than d_o, $\delta \ll \pi$, and we can linearise $\tan\delta$. The Friedel rule then becomes $d_o g \sigma_d(\mu) = 1$, and the short range screening potential (4) is thus valid when the conduction band multiplicity is large.

The simplification of the model by choosing a non-degenerate extra orbital restricts n_f to 0,1, or 2, and removes the need for any complex intra-atomic exchange terms in (2). As far as valence fluctuations are concerned, this simpler model retains the basic character of the original, and the condition for linearity of the screening response, $d_0 \gg n_f$, is easily satisfied. Discarding the irrelevant $l \neq 2,3$ channels, we simplify the notation, dropping l-labels, condensing $m\sigma$ into one label $\alpha = 1,..,d_0$, and writing $l=2$ fermion operators as $d^\dagger_{k\alpha}$ to emphasise that they act in a different space from the $l=3$ operators, $c^\dagger_{k\sigma}$. $H = H^{And} + H^{scr}$, where

$$H^{And} = \sum_{k\sigma} \varepsilon_k c^\dagger_{k\sigma} c_{k\sigma} + E_f \sum_\sigma n_{f\sigma} + U n_{f\uparrow} n_{f\downarrow} + \sum_{k\sigma} V^*_k c^\dagger_{k\sigma} c_{f\sigma} + \text{h.c.} \tag{5}$$

$$H^{scr} = \sum_{k\alpha} E_k d^\dagger_{k\alpha} d_{k\alpha} + (N d_0 \sigma_d(\mu))^{-1} \sum_{kk'\alpha\sigma} n_{f\sigma} \, d^\dagger_{k\alpha} d_{k'\alpha} \tag{6}$$

and where

$$N d_0 \sigma_d(\omega) = \sum_{k\alpha} \delta(\omega - E_k) \tag{7}$$

H^{And} is just the standard Anderson model of Ref.(1); H^{scr} is the new term.

(C) THE BOSON REPRESENTATION OF SCREENING

Within linear (Born) approximation, H^{scr} gives a good representation of the screening process, and one could go on to examine the properties of (5),(6) directly. However, I will take the possibly idiosyncratic step of replacing charge fluctuations in the screening channels by a suitable boson field, which if properly chosen exactly reproduces the linear approximation to the dynamics of H^{scr}. The cost of this procedure is the loss of some physical transparency, but the gain is a model with simpler structure, and exactly linear screening behaviour. As the original screening term was only an approximation, I see no harm in making this further simplification; those who would be happier working with a purely fermion model may rest assured that the results from the boson model presented here would be qualitatively unchanged.

The "bosons" are combinations of density fluctuations, projected on an energy variable, of the form:

$$b^\dagger_\omega \propto \sum_{kk'\alpha} \delta(E_k - E_{k'} - \omega) d^\dagger_{k\alpha} d_{k'\alpha} \tag{8}$$

these obey boson commutation relations "in the mean". The details of this "Tomonaga transformation" are explained fully in Ref.(4), and in this new representation, the term H^{scr} takes the form:

$$H^{scr} = \sum_i \omega_i b_i^\dagger b_i + \sum_\sigma n_{f\sigma} \sum_i (\alpha_i b_i + \alpha_i^* b_i^\dagger) \quad (9)$$

this is fully characterised by the spectral density $f(\omega)$:

$$f(\omega) = \sum_i |\alpha_i|^2 \delta(\omega-\omega_i) \quad . \quad (10)$$

As $\omega \to 0$, $f(\omega) \sim \omega/d_0$ (which reflects the Friedel rule); for ω greater than the conduction electron bandwidth, $f(\omega)$ is zero. In mean-field theory, $f(\omega)$ is entirely characterised by a single parameter, the positive quantity

$$C = {}_0\int^\infty d\omega \; 2f(\omega)/\omega \quad (11)$$

($-C$ is the static value of the propagator for the bosons coupled to the impurity).

In an exact treatment, the infra-red properties as $\omega \to 0$ turn out to be very important: they mimic the Fermi-surface anomaly connected with the low energy response of systems with a free Fermi-surface to a time-varying potential (this plays a crucial role in the theory of the Kondo effect and the X-ray edge). The cause of the infra-red difficulties is that the low energy boson modes take infinitely long to relax (the long-time behaviour is governed by a factor $t^{-1/2d_0}$) and that there is zero overlap between boson configurations that are fully relaxed about different values of n_f (as in MFT). This behaviour can of course also be derived for the fermion representation of the screening. (In fact, similar infra-red problems come up in the "bare" Anderson model; the screening term merely compounds them.)

(D) MEAN-FIELD THEORY RESULTS

A systematic mean-field analysis of the Anderson model with the boson screening term is given in Ref.(5). The Hartree-Fock (HF) equations take the form:

$$E_\sigma^{eff} = E_f - C\langle n_{f\sigma}\rangle + (U-C)\langle n_{f-\sigma}\rangle \quad (12)$$

$$\langle n_{f\sigma}\rangle = \lim_{\delta\to 0} \frac{\mathrm{Im}}{\pi} \int_{-\infty}^{\infty} \frac{d\omega}{e^{\beta\omega}+1} \left(\omega - E_\sigma^{eff} - \sum_k \frac{|V_k|^2}{\omega-\varepsilon_k+i\delta}\right)^{-1} \quad (13)$$

However, multiple solutions of these equations may exist, and a free-energy functional must be derived in order to determine which solution is the correct one. A random phase approximation (RPA) treatment will allow the rejection of locally unstable HF solutions (which correspond to saddle-points or maxima of the free-energy functional), but the energy function is vital in order to determine which of two locally stable solutions (minima) is <u>globally</u> stable.

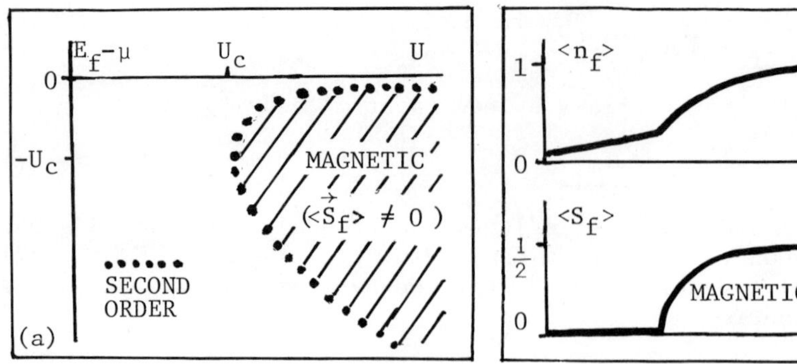

Figure 2. The Anderson model without screening; (a). mean-field theory phase diagram; (b) behaviour of $\langle n_f \rangle$, $|\langle\vec{S}_f\rangle|$ across magnetic "phase boundary".

Four parameters characterise the MFT results: $(E_f-\mu)$, U, C, and $U_c(T)$, a critical value of U below which no magnetic solutions of the HF equations exist. For T=0, $U_c \sim \pi^2 \sum_k |V_k|^2 \delta(\mu-\varepsilon_k)$ (a factor of π times the resonance width of the f-level); $U_c(T)$ rises with temperature and $U_c \sim 4T$ for $T \gg U_c(0)$, as the thermal smearing of the Fermi-level becomes greater than the width of the f-level resonance.

Figure (2a) shows the phase diagram of the MFT solutions of the standard Anderson model, C=0, in the space of the parameters $(E_f-\mu)$, U. The diagram shows the region of parameter space where the stable MFT solution breaks spin rotation symmetry and becomes magnetic, characterised by non-zero mean spin $\langle\vec{S}_f\rangle$. Such states are rotationally degenerate, and in fact RPA indicates a zero-energy "Goldstone boson" collective excitation mode associated with the rotation of the mean spin.

Figure (2b) shows how the mean charge and spin on the f-orbital change as the magnetic phase is crossed. The transition is <u>second-order</u>: in the region near the phase boundary there is essentially a single stable solution for a given value of the parameters (counting the rotationally degenerate magnetic solution as a single solution); this solution changes its character <u>continuously</u> from magnetic to non-magnetic as the phase boumdary is crossed.

If the screening term is added, the picture is basically unchanged until C beomes comparable to U_c. Then the second-order transition progressively becomes first-order, and in addition an instability between two non-magnetic states with $\langle n_f \rangle \sim 0$ and ~ 2 develops. $U_c=C$ is a tricritical point, and for $C>U_c$, phase boundaries divide the parameter space into two non-magnetic and one magnetic region, separated by <u>first</u> order transitions (Fig.(3)).

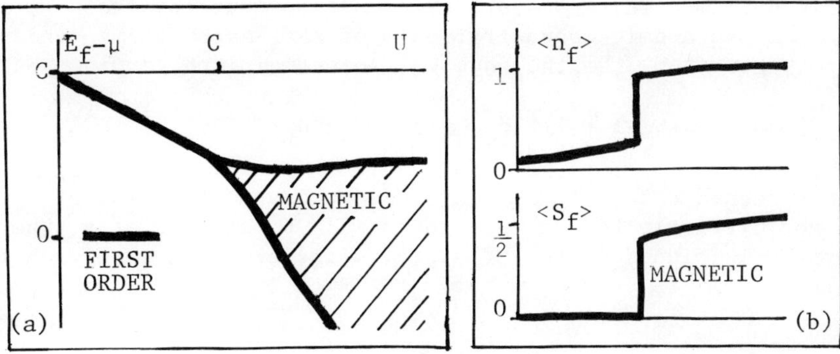

Figure 3. As figure (2), but with the screening term added ($C>U_c$).

Figure (3b) shows how charge and spin are discontinuous across the first order magnetic transition. In the neighbourhood of this transition there are two locally stable HF solutions competing for global stability, which shifts from one solution to the other across the phase boundary. This coexistence of two quasi-degenerate locally stable solutions, with very different charge and spin, in the neighbourhood of the magnetic "transition", is what makes the screened model qualitatively different from the unscreened one.

The phase transitions and the magnetic state predicted by the MFT are of course spurious. Finite systems do not have phase transitions, and it is well known that the properties of the Anderson model are smooth functions of its parameters. Fluctuations restore continuity in parameter space. Deep inside the "magnetic" region, where $<n_f> \sim 1$, and $<S_f> \sim {}^1/_2$, the spin fluctutions necessary to restore rotational invariance become very slow; this is the Kondo effect and is related to the infra-red properties of the Fermi gas mentioned earlier. From Fig.(2b), we can see that in the unscreened Anderson model, the Kondo effect will smoothly die out as the edges of the magnetic region are approached from inside; Fig.(3b) implies that it persists right up to the first-order line in the screened case. Near this line we also expect that the true state of the system resembles a mixture of the two available HF configurations. Again because of the infra-red problems the lifetime of each configuration will be greatly enhanced. We tentatively identify this interconfigurational fluctuation state with the mixed-valence effect.

(E) PINNING OF A COMPOSITE SYSTEM IN A MIXED-VALENCE STATE

In the case of a single impurity, it would be highly unlikely that the parameter ($E_f-\mu$) would take exactly the value that places it right on a phase boundary and in a mixed-valence state. In a concentrated system, however, the chemical potential μ must be

determined from a self-consistency condition that involves n_f .
If there is an atomic concentration c of rare-earth atoms, and n_o
electrons per atom, the the chemical potential is determined by

$$\int_{-\infty}^{\infty} d\omega (\exp\beta(\omega-\mu) + 1)^{-1} d_o \sigma_d(\omega) \quad + \quad c\langle n_f \rangle_{(E_f-\mu)} \quad = \quad n_o \qquad (14)$$

In the screened case, as $\langle n_f \rangle$ is a discontinuous function of μ,
it is not difficult to see that there will be a range of n_o where
the system is pinned in the MV state, with $(E_f-\mu)$ precisely on the
transition.

(F) BEYOND MEAN-FIELD THEORY ?

One goal of a more exact treatment is the study of the fluctuating regime and the determination of the characteristic timescale. Since the hybridisation V between the f-orbital and the band is by far the smallest parameter physically $(U_c \sim V^2 \sigma_d \ll C \ll U)$, one thinks of perturbation theory in V. As this is infra-red divergent, some sort of renormalisation approach may be necessary to study the non-analyticity in V. A start has been made[6] by finding an "assympt-otically exact" path integral expansion for the partition function in powers of V about the V=0 "atomic limit". This has similar (but more complex) form to the Anderson-Yuval-Hamann[7] approach to the Kondo singularity, and may too be amenable to a scaling treatment.

(G) CONCLUSION

To sum up: inclusion of the physically necessary screening term in the Anderson model seems to qualitatively alter its character in the valence fluctuation regime. Mean-field theory suggests that the new model may well mimic the mixed-valence phenomenon. If this is so, the fact that no account of phonons and coupling to the lattice was taken suggests that, contrary to much speculation, the strong coupling to the lattice may play no essential role in the <u>dynamics</u> of the problem, but merely renormalises essentially electronic parameters.

REFERENCES

1. P.W.Anderson,Phys. Rev. 124, 41 (1961)
2. K.H.J.Buschow,H.J.van Daal, Solid State Commun. 8 363 (1970)
3. K.Andres,J.E.Graebner,H.R.Ott, Phys.Rev.Lett.35 1779 (1975)
4. F.D.M.Haldane, Phys.Rev.B To be published.
5. F.D.M.Haldane, Phys.Rev.B To be published.
6. F.D.M.Haldane, unpublished.
7. P.W.Anderson,G.Yuval,P.R.Hamann,Phys.Rev.B1 4464 (1970)

COMMENTS

HIRST: I would like to say first of all that it is probably true that the relevant comparison is to the spin-only Anderson model; but, nevertheless, it is a fact that if you go to an orbitally degenerate case, then you normally get out of the ordinary Anderson model (without the screening term) a first order, instead of a second order, onset of magnetization and of charge in the local orbital. Is it not true that such an onset in an impurity problem certainly must be an artifact of the mean field approximation?

HALDANE: Yes. Mean field theory gets the energetics almost completely correct, but it gets the details and actual symmetries of the ground state wrong. What mean field theory is telling us is that in that region of the parameter space of the model where _two_ locally stable but very different Hartree-Fock solutions are energetically available, the true ground state is going to resemble a superposition of the two.

HIRST: Right. So you have to argue that perhaps the mean field theory might work better in the concentrated problem than in the impurity problem since the result is a bit unphysical.

HALDANE: I don't think mean field theory will work at all for the real properties of this model. It's going to have Kondo-type fluctuation properties associated with very unpleasant infrared behavior. I'm only using mean field theory to tell me the energetics of the model.

DONIACH: You said you had to apply the Friedel sum rule. That's coupled with the coupling constant, C.

HALDANE: The Friedel sum rule in part determines the coupling constant.

DONIACH: Yes, but you were drawing plots where C was a free parameter. It should be linked to the position of the d level.

HALDANE: C depends on the density of states and the degeneracy of the bands, d_0. C _is_ to some extent a free parameter.

DONIACH: The point is that you can crank it around a bit to change the relative screening properties.

HALDANE: Yes. On its own, the Friedel sum rule only determines the <u>infrared</u> properties of the boson field, which do not affect mean field theory.

KASUYA: You expect a discontinuous change in the Hartree-Fock approximation. However, in such a single impurity model, anyway, it should be continuous; so what is your argument: just that the transition becomes sharper?

HALDANE: I am using mean field theory to learn qualitatively what happens to the properties of the model as I vary parameters. It tells me not to expect any new kind of behavior in the unscreened Anderson model when the impurity level is at the Fermi level; with the screening term added, the first order discontinuity in mean field theory suggests new behavior may appear.

KASUYA: So you expect a continuous change but sharper?

HALDANE: Starting in the Kondo regime of the screened Anderson model, as you move the bare impurity level through the Fermi level, the properties will change in some smooth way. Mean field theory predicts a discontinuous change. We know that's not true and fluctuations will round it off slightly.

KASUYA: It will be much smoother.

HALDANE: I don't think so; the first order discontinuity in mean field approximation suggests it will be rapidly varying.

DONIACH: That depends on how you plot it.

ANDERSON: It <u>is</u> sort of important how you plot it. The chemical potential is fixed by the self-consistency condition (given by Eq. (14) of Haldane's preceding paper). A simple graphical solution shows that there will be a particular range for which pinning will occur and a region where pinning will be absent.

WILKINS: What I don't understand is on what time scale these bosons are operating. Do they provide a mechanism for seeing different things by XPS than we see from magnetic properties?

ANDERSON: I think these are fast bosons.

HALDANE: The screening processes modeled by the bosons provide just such a mechanism: the relaxation times should be much slower than XPS time scales, and the Hartree-Fock effective energy level of the resonance should be seen. Due to boson relaxation, this is either considerably above or below the Fermi level, depending on occupation, not <u>at</u> it as in the unscreened resonant case.

SNAFU IN MIXED VALENCE

C. M. Varma

Bell Laboratories

Murray Hill, New Jersey 07974

ABSTRACT

The experimentally observed properties of mixed valence compounds are those of a two component Fermi liquid, one of which has a very low degeneracy temperature. The behavior is moreover akin to that for non-interacting f resonances in the strong coupling regime. Interaction between two mixed valence f resonances, under certain conditions, is such that this behavior may be rationalized microscopically.

I. INTRODUCTION

The homogeneous mixed valence materials[1] that have attracted considerable attention recently are metals that are characterized by the fact that both the very heavy f electrons that obey atomic spectral rules to a good approximation and the much lighter s-d electrons coexist at the Fermi surface. Two of the principal problems posed by these materials are:

1) What is the ground state; how to theoretically describe it?

2) How and why does one get to this particular ground state?

A strong suggestion for the answer to the first part of (1) is provided by the experimental results. The ground state and the low-lying excitations of the mixed valence compounds are characteristic of a two-component Fermi liquid, one of which has a very low degeneracy temperature $T_0 < 0.1$ eV. This heavy liquid can only be primarily associated with the f electrons and T_0 has the order of magnitude expected of f resonance width in a s-d band. The evidence for this heavy Fermi liquid is the specific heat results[2] which are approximately T/T_0 at low temperature and the susceptibility results[3] which are approximately μ^2/T_0. The inelastic neutron scattering results[4] also strongly suggest a heavy Fermi liquid, as we discuss below. If one looks carefully at the specific heat results in SmS or the shape of the phase diagram for SmS,[5] there is clear evidence that the f electron entropy is exhausted for $T \gg T_0$. This is again characteristic of an electron liquid in the classical regime. Phenomenologically, it would seem therefore that the mixed valent compounds should be describable by a free energy functional depending on $n_{\underset{\sim}{k}\mu_1}$, $n_{\underset{\sim}{k}\mu_2}$; the Landau occupation functions which depend on the wave vector $\underset{\sim}{k}$ and the band indices, μ_i, and on the deviations $\delta n_{\underset{\sim}{k}\mu_1}$, $\delta n_{\underset{\sim}{k}\mu_2}$. The $n_{k\mu_i}$ themselves are obtained from a diagonalization of the density matrix matrices

$$\begin{pmatrix} n_{\underset{\sim}{k}\alpha\alpha} & n_{\underset{\sim}{k}\alpha\beta} \\ n_{\underset{\sim}{k}\beta\alpha} & n_{\underset{\sim}{k}\beta\beta} \end{pmatrix} , \qquad (1)$$

$n_{\underset{\sim}{k}\alpha\beta} = \langle c_{\underset{\sim}{k}\alpha} c^{\dagger}_{\underset{\sim}{k}\beta} \rangle$,,etc. where α and β are the atomic orbital indices (assuming, for example, that tight binding is valid for both orbitals, as it certainly is for f and d orbitals). The experimental observation of fractional valence, fluctuations in valence, and the fact that different experiments with time scales much slower or faster than V^{-1} observe different things are easily deducible from such a free energy functional if it is remembered that the off-diagonal elements in (1) are proportional to V, the f-d transfer integral to a first approximation.

It is one thing to prescribe such a free energy functional and quite another to derive it, or more properly to derive the Landau parameters. There are interesting difficulties in this: The problem is to hybridize correlated band orbitals, i.e., to determine $n_{\underset{\sim}{k}\alpha\beta}$. The difficulty arises because in hybridization not only wave vector but spin has to be conserved. Mean-field approximations to this problem, which consider average configurations and then hybridize them can lead to misleading results. One must consider the hybridization of different configurations, and then average over them giving them proper weight according to how frequently they occur. Some calculations with this viewpoint have been performed by W. Weber and myself using extensions of the Gutzwiller method,[6] but I do not have space to describe them here.

II. MIXED VALENCE PROBLEM AND THE SINGLE IMPURITY PROBLEM

To come back to the instructions the experimental results provide us, one finds that the behavior of mixed valence compounds is characteristic of that of a metal containing so-called magnetic impurities with their single particle level, ε_0, as described by, say, the Anderson Hamiltonian lying close to the Fermi level. The mixed valence compounds behave like a metal containing these impurities, but not in dilute concentration, but with a concentration of order 1! The big snafu (b.s.) is, what happened to the interaction between these "impurities?" How did the interaction renormalize itself away? Or is there no interaction to begin with? The answers to these questions would go a long way toward solving the problem (2) listed in the beginning, since the single impurity problem itself is well understood now.

To put the b.s. in sharper focus, consider the Anderson Hamiltonian for a magnetic impurity at a site i in a metal:

$$H_{ai} = \varepsilon_0 \sum_{\sigma} n_{i\sigma} + \sum_{\underset{\sim}{k}<k_F,\sigma} \varepsilon_{\underset{\sim}{k}} n_{\underset{\sim}{k}\sigma} + \frac{U}{2} \sum_{\sigma} n_{i\sigma} n_{i-\sigma} + \sum_{\underset{\sim}{k}\sigma} V_{\underset{\sim}{k}i} (c^{\dagger}_{\underset{\sim}{k}\sigma} c_{i\sigma} + \text{c.c.}) \qquad (2)$$

The single particle level ε_0 is determined by condition of charge neutrality; the chemical potential at the impurity site must coincide with the Fermi level. This means that if the conduction electrons have a flat density of states, ε_0 and ε_0+U must be symmetrically on either side of the Fermi level ε_F. The magnetic susceptibility for (2) over the whole range of temperature has been determined by Krishnamurthy, Wilson and Wilkins[8] using renormalization group methods. There are three regimes. For $T \gg U$, the susceptibility is Curie law, that of a classical moment; for $T_k \ll T \ll U$, Curie law, that of a local moment; and for $T \ll T_k$, in the strong coupling regime, it is given by the Pauli law.

To be a mixed valent impurity, two different charge states at the impurity site must be approximately degenerate, and close to the Fermi level. One way this can happen is by electron-phonon coupling, as was pointed out by Varma and Heine.[9] Haldane[10] has pointed out that the shift of ε_0 due to electron-electron Coulomb repulsion can also be included as a coupling to bosons. (These Coulomb interactions can also lead to the interesting possibility of coherent configurational fluctuations in mixed valence compounds.[1])

So we add to the Anderson Hamiltonian

$$\bar{H}_i = H_{ai} + H_{bi} + H_{a-bi} \tag{3}$$

$$H_b = \sum_{\underset{\sim}{g}} \omega_{\underset{\sim}{g}} \left(b^+_{\underset{\sim}{g}} b_{\underset{\sim}{g}} + \tfrac{1}{2}\right) \tag{4}$$

$$H_{a-bi} = \sum_{\sigma,\underset{\sim}{g}} C_{gi} n_{i\sigma} \left(b_g + b^+_g\right) . \tag{5}$$

We can now shift ε_0 arbitrarily, e.g., in going from one material to the next, hoping the bosons take care of charge neutrality. If now ε_0 approaches the Fermi level, the effective exchange integral (the Schrieffer-Wolf exchange integral) increases and the Kondo temperature moves up. (In the talk at the conference I gave results of a poor man's scaling for the asymmetrical Anderson model, but I do not have space to present these results here.) For ε_0 within $\sim \Delta \approx V^2/W$ (W is conduction electron bandwidth) of the Fermi level, the local moment regime in the susceptibility is eliminated and one passes directly from the classical to the strong coupling regime.[11] The characteristic temperature at which the crossover occurs is $T_k \simeq \Delta$, consistent with the variational prediction of Yafet and myself.[12]

Now, this behavior of the susceptibility for a single impurity with ε_0 close to the Fermi level is qualitatively that found for mixed valent materials. We can compare the behavior in detail with the experiments in Ce_xTh_{1-x}.[3b] In the γ phase the f level is below (but not far below) the Fermi level. As the temperature is decreased and one gets closer to the transition to the mixed valent α phase, the Curie-Weiss constant increases, in accordance with increase of the exchange integral and finally in the α phase Pauli susceptibility is obtained. In the γ phase, all inelastic spin-flip processes involve virtual scattering of f electrons to the Fermi surface and have the characteristic energy of the exchange integral. In the α phase the energy scale is simply the hybridization width -- or the degeneracy temperature of the heavy fluid -- T_0. This is in qualitative accord with the experiments.[4] This then brings us back to the b.s.: what happened to the interaction among f orbitals at different sites?

III. INTERACTION AMONG MIXED VALENT IMPURITIES

The interactions among the f orbitals at different sites will not matter if the moment at a site scales to zero at a temperature much greater than the characteristic energy of interactions that provide an effective magnetic field at a given site. This possibility was mentioned earlier.[1] Another is for the dominant interactions to be "nonmagnetic." To examine these possibilities I have considered the interaction between two mixed valent impurities and find the latter behavior possible. This problem has been looked at by a number of authors;[13] however, the interaction problem with coupling to bosons, so essential for mixed valence, has not been considered. As discussed below this makes for very important differences in the final behavior.

The Hamiltonian we consider is that of two Anderson impurities at sites i and j, coupled to bosons, and with a term giving particle transfer from site i to site j:

$$H = H_i + H_j + H_b + H_t \quad , \tag{6}$$

$$H_t = \sum_{\sigma} \tilde{V}_{ij} \left(c^{+}_{i\sigma} c_{j\sigma} + c.c. \right) \quad , \tag{7}$$

and H_i is the Anderson Hamiltonian at site i with a linear coupling to the bosons. In the Hartree-Fock approximation the coupling to bosons among other things will renormalize the transfer integral $\tilde{V}_{ij}$ to V_{ij}. V_{ij} may represent the direct transfer or transfer via the wide band as is more appropriate for the rare-earth mixed valent compounds. In the Hartree-Fock approximation the solution to (6) is given by the set of self-consistency equations

$$n_{\sigma} = \tfrac{1}{2} + \frac{1}{2\pi} \Big[\tan^{-1}(y(x-n_{-\sigma}+zn_{\sigma})-v_{-\sigma}) + \tan^{-1}(y(x-n_{-\sigma}+zn_{\sigma})+v_{-\sigma}) \Big] \quad , \tag{8}$$

$$d_{\sigma} = \frac{y((1-z)d_{\sigma}-zd_{\sigma})}{2\pi\ \Delta v_{-\sigma}} \Big[\tan^{-1}(y(x-n_{-\sigma}+zn_{\sigma})-v_{-\sigma}) - \tan^{-1}(y(x-n_{-\sigma}+zn_{\sigma})+v_{-\sigma}) \Big] \quad , \tag{9}$$

where

$$n_{\sigma} = \tfrac{1}{2}(n_{i\sigma}+n_{j\sigma}) \quad ,$$

$$d_{\sigma} = \tfrac{1}{2}(n_{i\sigma}-n_{j\sigma}) \quad ,$$

and

$x = (\varepsilon_f - \varepsilon_0)/U$, $y = U/\Delta$, $z = C/U$, (assuming q independent C in (5))

$$v_\sigma = V_\sigma/\Delta \ ,$$

$$V_\sigma = \left[\{(U-C)d_\sigma - Cd_{-\sigma}\}^2 + V_{ij}^2\right]^{\frac{1}{2}} \ .$$

With coupling to bosons missing, z=0 and this set of equations is the Alexander-Anderson[13a] solution. With $V_{ij} = 0$, $d_\sigma = 0$, and the solution is that for an isolated magnetic impurity with coupling to a boson field,[10] which for a certain range of parameters is a mixed valent impurity.

I will now briefly describe the results of the solution of (8-9). For negligible z, and ε_0 far below the Fermi surface, (8-9) lead to a RKKY coupling (or kinetic exchange coupling) between the moments at i and j. This involves polarizing the conduction electrons through virtual transitions. As ε_0 is moved toward the Fermi surface, the strength of the moment declines and so does the coupling energy. For ε_0 within Δ of the Fermi level, the coupling is ferromagnetic. There is reduction in kinetic energy due to real transfer of some charge density from one site to the other, and such charge transfers are favored for parallel alignment of the moments. This is the so-called phenomena of double exchange. It is important to note that it is favored only when the charge density in the localized orbital is less than unity.

The above discussion is for the case when there is equal charge at the two sites, i.e., $n_{i\sigma} = n_{j\pm\sigma}$ ($d_\sigma = 0$ or $d_\sigma = d_{-\sigma}$); (+) for ferro and (-) for antiferro alignment, respectively. In the mixed valent situation we must consider alignments with unequal amounts of charge on the two sites. There are two important points to be noted here. For z of order unity multiple stable solutions are obtained. For some values of z, they are degenerate and lie near the Fermi surface -- the mixed valent situation. The second point is that the term involving z acts as a feedback term, making n_σ close to integral values. This can easily be seen for the single impurity case ($d_\sigma = 0$) from (8); for large y and $z \approx 1$ the solutions are $n_\sigma \simeq 1$, $n_{-\sigma} \simeq 0$ (or vice versa) and $n_\sigma = n_{-\sigma} \simeq 1$. For the two (mixed valent) impurity case, if we consider the case of equal charge of magnitude ≈ 1, and with the resonances near the Fermi energy at both the sites, the interaction energy is close to zero, being of order V_{ij} $(1 - n_{i\sigma})$. If we consider one site doubly occupied with moment ≈ 0, and the other singly occupied; $n_{i\sigma} = n_{i-\sigma} \simeq 1$, $n_{j\sigma} = 1$, $n_{j-\sigma} = 0$; the energy is reduced due to V_{ij}, and the reduction is of order V_{ij}. This reduction is the same for $n_{j-\sigma} = 1$, $n_{j\sigma} = 0$; i.e., the system is indifferent to the sign of the net moment.

It is speculative to suggest from this two impurity case that a homogeneous mixed valent compound will have a net moment of zero for the same reasons, but not unfair to say that it is suggestive. One gains some confidence in the suggestion from the fact that the only known homogeneous mixed valence compound that magnetically orders, TmSe, is also the only one in which both the charge states (neglecting the crystal field splitting, which is small compared to the hybridization energy) have magnetic moments.

The physics of the situation is that with one site being a singlet either an up or down spin can equally well be transferred to the other site and the gain in kinetic energy is indifferent to the direction of spin at the other site. The situation comes about in two impurity problem if <u>integral</u> charge occupation of different magnitudes are nearly degenerate. One can think of other mechanisms besides the coupling to phonons (or charge fluctuations) that favor integral occupation, e.g., Hund's rule couplings.

Double exchange is the mechanism for ferromagnetic ordering in inhomogeneous mixed valent compounds[14] like $(La_{1-x}Ca_x)Mn^{3+}_{1-x}Mn^{4+}_xO_3$. It is not unnatural to expect that related physics must play an important role in homogeneous mixed valent compounds. The double exchange Hamiltonian[15] is proportional to $|\underset{\sim}{S}_i+\underset{\sim}{S}_j+\underset{\sim}{s}|$, where $\underset{\sim}{S}_i$ and $\underset{\sim}{S}_j+\underset{\sim}{s}$ are the moments at the two sites i and j, with unequal charges. If $\underset{\sim}{S}_j+\underset{\sim}{s} = 0$, there is no preference for a net moment.

It is noteworthy that the indifference free energy (entropy contribution) $k_B(T/T_0)T$, from this mechanism at T_k is of order k_BT_k. We may expect the Kondo renormalizations to take over below that temperature at the (instantaneously) singly occupied site unhampered by prejudicial interactions with the other site. Also V_{ij} would tend to stabilize the mixed valent regime

IV. SUMMARY

I have tried to point out the snafu in mixed valence compounds -- most of the properties can be understood simply from the renormalization of noninteracting f levels in a s-d conduction band, leading to a Fermi liquid behavior. The real problem is to show why the interactions between f levels do not interfere and lead to magnetic ground states. Toward this end I have considered the interaction between two mixed valence impurities. I have found that if two different, nearly integral, charge states are degenerate, as is favored in the mixed valence state, and one of them is a magnetic singlet, the interactions prejudice the behavior in favor of a nonmagnetic state.

REFERENCES

1. A recent review is C. M. Varma, Rev. Mod. Phys. 48, 219 (1976).
2. S. D. Bader, N. E. Phillips and D. B. McWhan, Phys. Rev. B7, 4686 (1973); K. Andres, J. E. Graebner and H. R. Ott, Phys. Rev. Lett. 35, 1779 (1975).

3a. M. B. Maple and D. Wohleben, Phys. Rev. Lett. 27, 511 (1971).

3b. J. M. Lawrence, M. C. Crofts and R. D. Parks, Phys. Rev. Lett. 35, 289 (1975), and Proceedings of this conference.

4. R. J. Birgeneau and S. Shapiro, Proceedings of this conference and private communication.
5. A. Jayaraman, P. D. Dernier and L. D. Longinotti, High Temperatures-High Pressures 7, 1 (1975).
6. M. C. Gutzwiller, Phys. Rev. 137, A1726 (1965).
7. P. W. Anderson and G. Yuval, in Magnetism, Vol. V, ed. by H. Suhl (Academic Press, N.Y., 1973); K. G. Wilson, Rev. Mod. Phys. 47, 773 (1974); P. Nozieres, J. Low Temp. Phys. 17, 31 (1975).
8. H. R. Krishnamurthy, K. G. Wilson and J. W. Wilkins, Phys. Rev. Lett. 35, 1101 (1975).
9. C. M. Varma and V. Heine, Phys. Rev. 11, 4763 (1975).
10. F. D. S. Haldane, Phys. Rev. (to be published) and Proceedings of this conference.
11. H. R. Krishnamurthy, J. W. Wilkins and K. G. Wilson, Proceedings of this conference.
12. C. M. Varma and Y. Yafet, Phys. Rev. 13, 2950 (1976).

13a. S. Alexander and P. W. Anderson, Phys. Rev. 133, A1594 (1961).

13b. D. J. Kim and Y. Nagaoka, Prog. Theo. Phys. (Kyoto) 30, 743 (1963); T. Moriya, ibid. 33, 157 (1965); B. Caroli, J. Phys. Chem. Sol. 28, 1427 (1967).

14. G. H. Jonker and J. H. Van Santen, Physica 16, 337 (1950).
15. See for example, P. W. Anderson in Sol. State Phys. 14, 99 (1963).

COMMENTS

ORBACH: I don't think you are right about the Sm^{2+} and Sm^{3+}. You would be right if the Sm^{3+} had no electrons on it, but it has a core, and if you think of the delocalization energy, which is what you are talking about, it would in fact favor a ferromagnetic spin array of the Sm sites. Now, you may wonder, when you transfer the electron to one of the Sm sites, where is the moment? There's a Van Vleck induced moment coming from the exchange between the two. So I would disagree with you and say that you have ferromagnetic interaction.

VARMA: The statement I have made is correct only if the splitting between the $J = 1$ and $J = 0$ state is significantly larger than the hopping integral. There is one experimental way of checking this thing about interactions between two impurities. This is the experiment of dilute Ce in La where you can move the renormalized Ce level with respect to the Fermi level by, say, applying pressure. That would be a lovely way of studying, as a function of concentration, the interaction between the two ceriums as we bring the Ce local level closer to the Fermi level. This study is probably already buried in some of the data that was shown yesterday.

KAPLAN: I can see how you can get out of the ferromagnetic coupling, but you still have a moment left. You still have an odd number of electrons. You'll have a Curie law. You've got a problem with your argument, I think.

VARMA: No, I don't.

KAPLAN: You have an odd number of electrons; you can't get out of it entirely. Six and five is eleven electrons. You have at least a spin 1/2 left over for the pair. It's just a simple point.

VARMA: Well, the point is that some of the electrons are in a very wide band so you can forget about these. You can regard their susceptibility as Pauli if their bandwidth is much larger than the local Hund's rule coupling.

KAPLAN: But that doesn't answer the question concerning the f-electrons.

ANDRES: In TmSe, where one seems to fluctuate between two different magnetic states (although one of them may be an exchange-induced singlet), one does seem to enter a ferromagnetic state at about 4°K in fields above 5000 gauss. In lower fields, one seems to enter a more complicated antiferromagnetic state. So one seems to be able to switch the exchange interactions with very moderate fields.

FONER: Do you mean that, when you turn on the field, you go into a paramagnetic or a ferromagnetic phase? In normal ordered systems, like an antiferromagnet, when you turn on the field and align all the spins the paramagnetic phase is achieved. The paramagnetic phase joins the high temperature phase with no boundary.

ANDRES: Above the critical field, one seems to cool into a ferromagnetic phase

FONER: So you mean, it's not a complete alignment from, say, a canted system.

ANDRES: No, it's not, simply a forced alignment by the external field.

WOHLLEBEN: There is some evidence from studies on TmTe under pressure that indicates that TmTe is not intermediate valent at atmospheric pressure while apparently at 19 kbars, it is. There you get non-linear magnetization curves in that state.

FONER: Well, the thing that bothers me is that then the phase boundary must have another branch. You should have to go through a line someplace which separates the paramagnetic phase at high temperatures from this ferromagnetic phase at low temperatures, and experimentally one would like to see that.

PENNEY: I want to advertise some papers about TmSe in the poster session. There is a model presented there which is a completely integral valent (non-intermediate valent) model for TmSe where you can have a change from an antiferromagnetic to a ferromagnetic interaction which is field-induced and which is pressure-dependent. Basically the model consists of Tm^{2+} impurities which mediate the interaction between the Tm^{3+} ions. I would just say that the question whether TmSe is an intermediate valence system or whether it is integral valent is very wide open at this point.

KASUYA: You say that double exchange is just through the conduction electrons, but the conduction band is very wide so you can't expect double exchange.

VARMA: The relevant quantity for the double exchange in this connection would be the matrix element squared of the f-electron for the d-electron, times the density of states of the d-electrons. This quantity would have to be much smaller than the Hund's rule coupling, and there is no reason to believe it is not.

JEFFERSON: I would like to comment that there is a scaling method for the asymmetric Anderson model which is also based on Anderson's poor man's scaling method which can handle the case where the localized level is arbitrarily close to the Fermi level. I am presenting this method in the poster session this afternoon in case anyone is interested.

MÖSSBAUER EFFECT STUDIES OF RARE EARTH MIXED VALENT MATERIALS

J. M. D. Coey and O. Massenet

Groupe des Transitions de Phase, C.N.R.S.

38042 Grenoble, France

I. INTRODUCTION

Fluctuating or intermediate electronic configurations pervade the natural sciences. They are at the heart of such varied phenomena as the working of haemoglobin, the colours of many rocks or the electrical conductivity of metals. Robin and Day [1] differentiated three classes of materials in their extensive survey of mixed valence chemistry. Another distinction between homogeneous and inhomogeneous mixed valence compounds with equivalent sites by Varma [2] and Batlogg [3] is particularly relevent to the rare earths. The former have narrow localised states hybridised with broad band states, both at the Fermi level, while the latter have only one type of state, localised or else forming a very narrow partly-filled band.

This article attempts to explain what has been learned so far about rare-earth compounds and alloys with mixed valence using Mössbauer spectroscopy. The work on iron compounds has been reviewed fairly recently by Reiff [4]. Whether the electronic configuration of an atom is seen as static, fluctuating or intermediate (homogeneous) is largely a matter of the time scale inherent in the measurement. The time scale inherent in the Mössbauer measurement τ_M turns out to be of order 10^{-8} - 10^{-10} sec., depending on the isotope and the hyperfine interactions involved, so a classification of mixed valent materials is made here on the basis of τ, the duration or relaxation time of a particular electronic configuration of the rare-earth. The data lends itself to interpretation in terms of interconfigurational fluctuations, but another possibility is discussed in section III. For materials in <u>category i</u>, τ is essentially infinite so that a rare earth in one of two stable

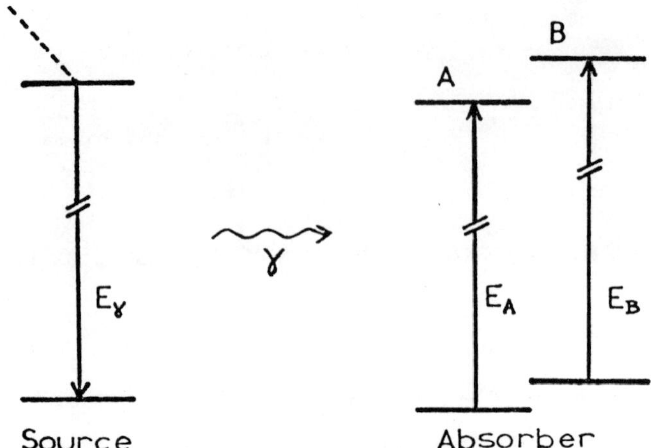

Fig. 1 Schematic representation of a Mössbauer experiment where the absorber contains nuclei with two different electronic configurations A and B. The electric monopole interaction gives different isomer shifts $\delta = E_{A,B} - E_\gamma$ for the two configurations, provided the fluctuations between them are sufficiently slow.

TABLE I

Data on some useful nuclei for Mössbauer absorbtion spectroscopy of mixed valence compounds. [6,7]

Isotope	E_γ KeV	Γ_{nat} mm/sec	$\partial\delta/\partial n_{4f}$ mm/sec	Time corresponding to 1 mm/sec
^{149}Sm	22.5	1.71	- 0.9	8.8×10^{-9} sec
^{152}Sm	121.8	1.59	- 1.64	1.6
^{151}Eu	21.6	1.31	-14.6	9.1
^{153}Eu	103.2	0.68	17.3	1.9
^{169}Tm	8.4	8.14	0.9	23.4
^{170}Yb	84.3	2.03	- 0.36	2.3

$1 eV \equiv 1.5 \times 10^{15} sec^{-1}$

electronic configurations occupies inequivalent lattice sites. For category ii, the sites are macroscopically equivalent, but $\tau \gtrsim \tau_M$ so that the two configurations are distinguished, or at least relaxation effects appear in the Mössbauer spectrum. For category iii, the sites are equivalent but $\tau \ll \tau_M$ so only the average configuration is detected in the Mössbauer measurement. This classification nearly coincides with that in reference [3]. It will be compared to a more fundamental one, where τ_M is replaced by the natural timescale in the problem, that of the phonon frequencies, in the last section.

Electronic configurations are differentiated using the Mössbauer effect by their hyperfine interactions with the nucleus.[5] These interactions are of three types, electric monopole, magnetic dipole and electric quadrupole, all of which produce different shifts or splittings of the nuclear energy levels, hence slightly different energies at which a γ-ray is resonantly abosrbed in a transition of the nucleus from its ground to excited states. The absorbtion spectrum consists of resonance lines separated by energies of order 10^{-7} eV. whereas E_γ the energy of the γ-ray itself is of order $10^4 - 10^5$ eV. As it is impossible to measure such energies with an absolute accuracy of one part in 10^{13}, they are measured relative to some standard source of the Mössbauer γ-rays which is moved with a variable Doppler velocity, v, to produce the tiny energy modulation required. The energy scale of a Mössbauer spectrum is conventionally given in mm/sec where:

$$\Delta E/E_\gamma = v/c \quad (1)$$

In what follows, we will mainly be concerned with the electric monopole interaction which arises from the electrostatic interaction of the nuclear charge distribution with the electronic charge, essentially of s character, at the nucleus. Since the nuclear radius is different in the ground and excited states, the resonance absorbtion occurs at an energy slightly displaced from the resonent emission in the source (the isomer shift, δ) if the electronic configurations of source and absorber differ. The process is illustrated schematically in Figure 1-fr an absorber with the Mössbauer isotope having two possible electronic configurations A and B differing by one 4f electron. The isomer shifts are given as:

$$\delta_{A,B} = \frac{2\pi}{3} Ze^2 \, \Delta|\psi(0)|^2_{A,B} \Delta r^2 \quad (2)$$

where $\Delta|\psi(0)|^2_{A,B}$ is the difference between the electronic charge at the absorber nucleus for configuration A or B, and that at the source nucleus. Δr^2 is the nuclear radius parameter. The valence of a rare earth ion, defined here as its charge after any 5d and 6s electrons have been removed, depends on the number of 4f electrons it possesses. Most rare earths are trivalent, but some may also be di- or quadrivalent. At most two 4f configurations are possible in a rare earth mixed valence compound because of the huge 4f-4f correlation energy, $\sim$7eV.

In the rare-earths, a change of 4f electron occupancy has greater influence on the isomer shift than either 5d or even 6s character, because of the strong indirect shielding of the nucleus from the 5s shell produced by the 4f electrons. Increasing the number of 4f electrons thus decreases $|\psi(0)|^2$. The same process gives a large increase in ionic size with decreasing valence. The ratio $\partial\delta/\partial n_{4f}:\partial\delta/\partial n_{5d}:\partial\delta/\partial n_{6s}$ is roughly -4:-1:2.

The rare earth elements are well endowed with isotopes suitable for Mössbauer spectroscopy and data on some of these are given in table I. E_γ must be $\lesssim$ 25KeV if the resonance is to be observed at room temperature, otherwise liquid helium temperatures for source and absorber are necessary because the probability of recoilless absorbtion or emission is given by:

$$f = e^{-\langle x^2\rangle E\gamma^2/k^2c^2} \qquad (3)$$

where $\langle x^2\rangle$ is the mean square displacement of the nucleus. The natural linewidth is a limitation on the resolution of the spectrum, governed by the lifetime of the nuclear excited state. Γ_{nat} is the best full width at half maximum theoretically possible for a lorentzian line in the absorbtion spectrum. The best experimental linewidths may be considerably greater. The change of isomer shift on passing from $4f^n$ to $4f^{n+1}$ configurations will depend somewhat on the compounds involved because of the influences of covalency and overlap distortion on the outer electron shells whose extent is changed by a change in 4f character. The time corresponding to 1 mm/sec is derived from (1) and the uncertainty principle.

The two configurations A and B will be distinguished in a Mössbauer experiment by their isomer shifts if their relaxation time is long compared with the product of the last two columns in Table I, always provided the experimental linewidth is not too great. If it is short, a single line will be seen in an intermediate position. To illustrate what is meant, some calculated spectra are shown in figure 2. Both suppose trivalent ($4f^n$) and divalent ($4f^{n+1}$) configurations with a ratio 2:1. Those in fig. 2a) are for the ^{151}Eu resonance and those in fig. 2b) for the ^{149}Sm resonance. Instead of the values in table I, good experimental linewidths are used for Γ (3.1 and 2.1 mm/sec, respectively) and the values taken for $\partial\delta/\partial n_{4f}$ are appropriate for sulphides (-0.75 and -12.0 mm/sec). These spectra are calculated using formulae for random fluctuations

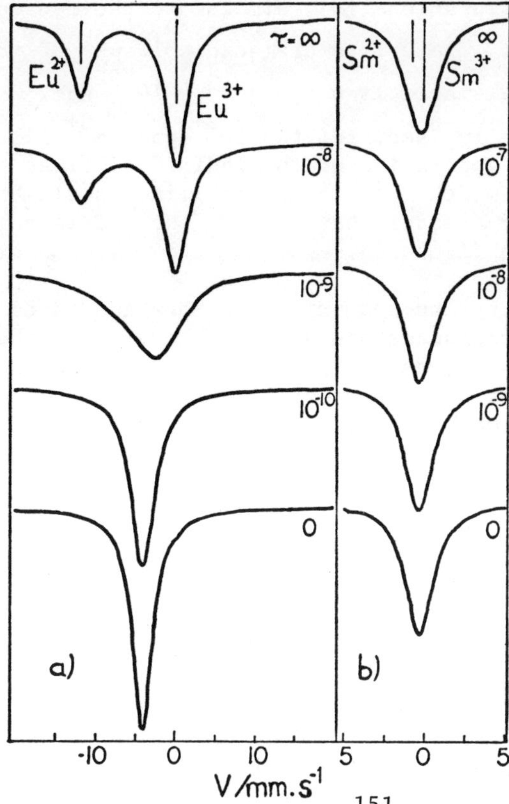

Fig. 2 Spectra calculated for a) the ^{151}Eu and b) the ^{149}Sm resonances as a function of relaxation time, supposing a relative population of 1:2 for the divalent and trivalent configurations.

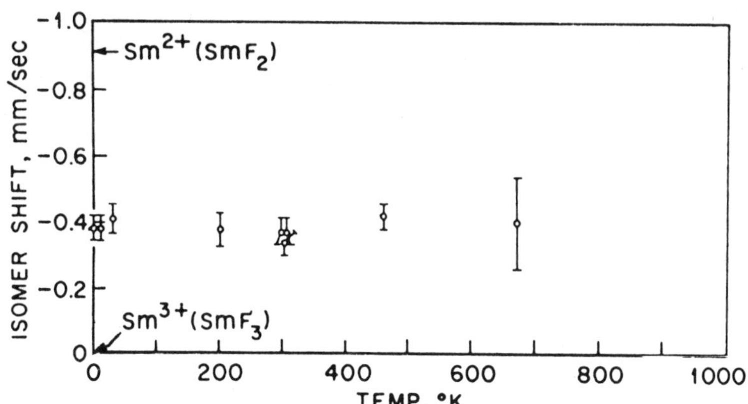

Fig. 3 ^{149}Sm isomer shift of SmB_6 as a function of temperature. Isomer shifts of typical di- and trivalent compounds are indicated. (From reference 9).

between two configurations [8], similar to those used for motional narrowing in NMR. There is a dramatic collapse of the Eu^{2+} and Eu^{3+} resonances into a single line around $\tau = 10^{-9}$ sec. The same collapse, around $\tau = 10^{-8}$ sec, for the Sm^{2+} and Sm^{3+} lines is less evident because the separation of the individual lines is just over a third of the linewidth. The composite line is slightly asymetric and significantly broader when $\tau > 10^{-8}$ sec. This point is illustrated further in figure 8c.

We will now try to summarise the Mössbauer data on mixed valence rare-earth compounds and alloys. Inevitally the picture will be rather unbalanced, as we have tended to concentrate on our own results simply for reasons of familiarity. The reader should correct his impression by referring to other articles in this book.

II. RESULTS

2.1 Samarium

The first Mössbauer work on a mixed-valence samarium compound was reported by Cohen, et al for SmB_6 [9]. Figure 3 shows the isomer shift, independent of temperature and intermediate between those typical of Sm^{2+} and Sm^{3+}. Some 30-40% of the samarium is divalent, depending exactly on how the isomer shift is interpreted, while the rest is trivalent. There is negligable line broadening, and no magnetic hyperfine splitting at 1.2K, even in an applied field of 12 kOe. This indicates that the Sm^{3+} is in some sort of nonmagnetic state (Sm^{2+}, like Eu^{3+} is a nonmagnetic ion in any case, with J=0). This could be explained if the sixth, $5d^{1-\varepsilon}6s^{\varepsilon}$ electron, which should be localised because SmB_6 is a semiconductor [10], were coupled to the $4f^5$ core in such a way as to give negligable resultant moment. However, the width of the absorbtion line at room temperature, 2.15(8) mm/sec [11], proves that there is not a static or slowly fluctuating mixture of the two Sm species in SmB_6 (which would broaden the line to more than 2.45 mm/sec), but rather the relaxation time of the distinct samarium configurations is less than 10^{-9} sec. If the Sm^{3+} ion would have possessed a moment, then the lack of any magnetic hyperfine interaction at low temperatures in applied magnetic fields shows that the relaxation time must be short ($<10^{-11}$ sec) compared with the nuclear Larmor precession time [9].

The first order phase transition in SmS [12] has been studied, also using the ^{149}Sm resonance, at room temperature as a function of pressure from 0-11 kbar. [11,13] Some typical spectra appear in

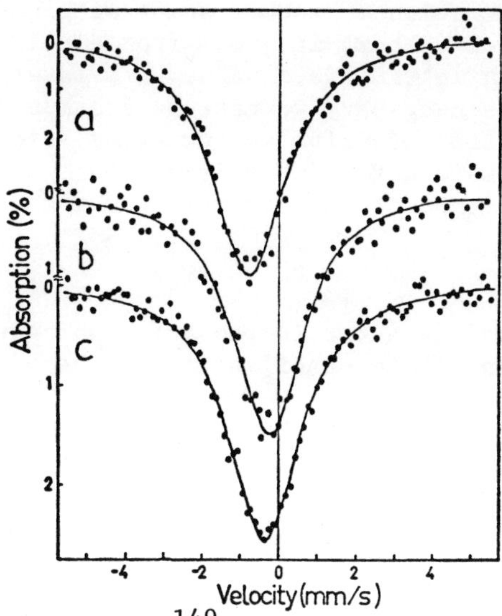

Fig. 4 Room-temperature ^{149}Sm Mössbauer spectra of a) SmS at atmospheric pressure, b) SmS at 11 kbars and c) $Sm_{0.77}Y_{0.23}S$ at atmospheric pressure. (From reference 13)

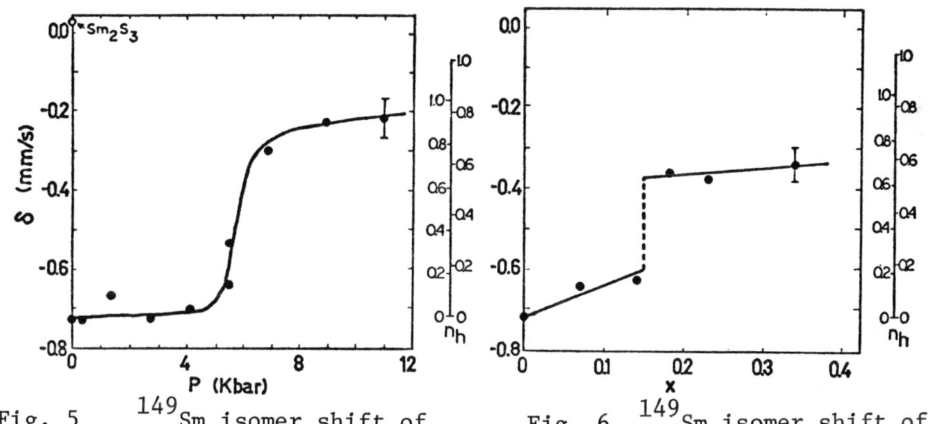

Fig. 5 ^{149}Sm isomer shift of SmS as a function of pressure.

Fig. 6 ^{149}Sm isomer shift of $Sm_{1-x}Y_xS$.

TABLE II

Variation of the ^{149}Sm isomer shift.

Type of electron n	$\partial\delta/\partial n$ mm/sec
4f (localised)	-0.75
5d (localised)	-0.20
5d (delocalised)	-0.10
6s (delocalised)	+0.45

figure 4, and the variation of isomer shift with pressure is given in figure 5. At about 6 kbar, it jumps from a value typical of divalent samarium to an intermediate value. To interpret the isomer shift in detail it is necessary to make an assumption about the electronic configuration of trivalent samarium, since one measured variable (δ) suffices to determine only one parameter, (n_h). We suppose that the intermediate configuration may be represented as

$$(1-n_h)(4f^6) + n_h(4f^5 5d^1) \tag{4}$$

The neglect of any 6s character in the Sm^{3+} configuration might be justified by ligand field stablisation of the 5d (t_{2g}) orbitals.

A comparison of the isomer shifts of samarium compounds and alloys with different electronic configurations allows us to evaluate the effects on the isomer shift of changes of electronic density of different character [11]. The results are given in Table II. $\partial\delta/\partial n_{4f}$ is the value appropriate for sulphides. Using these values and (4), the two scales of fractional trivalent character, n_h, on figures 5 and 6 are obtained; the one on the left assumes the 5d electron to be localised, the other delocalised.

Changes similar to those produced by pressure have been obtained in solid solutions of SmS and a trivalent rare earth or yttrium sulphide [14,15]. In $Sm_{1-x}Y_xS$, there is an abrupt decrease in lattice parameter and change in colour from black to gold near $x = 0.15$ [16]. The isomer shift, figure 6, clearly indicates that a change in electronic configuration occurs at this composition. Yttrium doping is not exactly equivalent to pressure, however. There are signs of some trivalent samarium appearing already in the black phase where $x < 0.15$, and the trivalent character of the gold phase is less pronounced than under pressure. These points are made more clearly on figure 7, where the data of the previous two figures is plotted as a function of unit cell volume. Curiously, the pressure data extrapolates to a point corresponding to SmS with a localised 5d electron. The fact that the points for the gold phases produced by pressure and yttrium doping do not coincide suggests that the average cation (or unit cell) volume in the mixed crystals is smaller than that actually occupied by the intermediate-valence samarium ion. It is quite plausible that the Sm dilates its surrounding cage of sulphur in the mixed crystals, and this would be a source of error in deriving the valence from the lattice parameters, systematically leading to values a little too high. Also shown on figure 7 is the normal volume dependence of the isomer shift expected if there was no change in 4f occupancy [17]. It is far smaller than either the changes observed or the errors in the measurements. Consequently it was impossible to decide whether the lattice was rigid or relaxed along with the electronic configuration fluctuations [11].

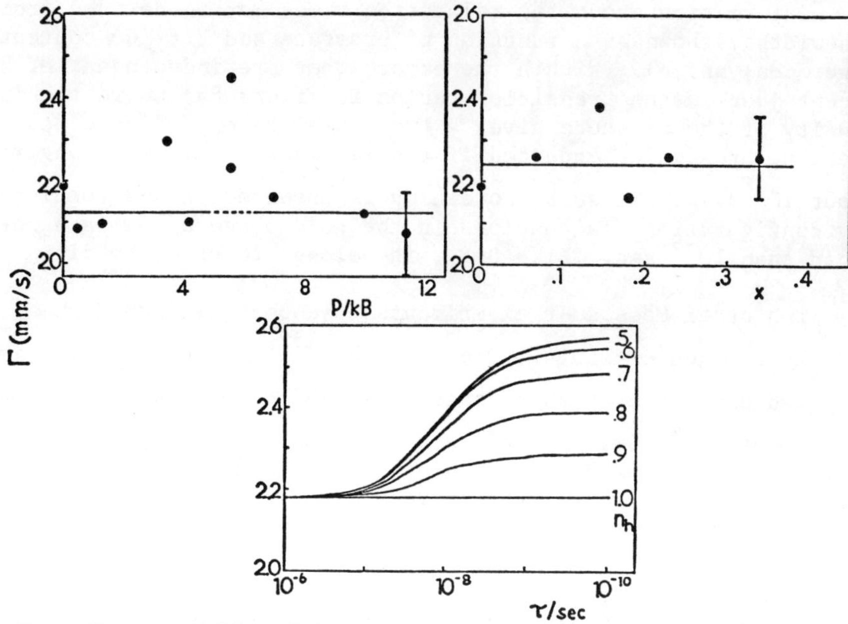

Fig. 7 Isomer shift of SmS as a function of pressure (0) and of $Sm_{1-x}Y_xS$ (0), plotted as a function of unit cell volume. The dotted line shows the volume dependence of the isomer shift expected if the samarium remained $4f^6$ throughout. X and + would be the points for $Sm^{3+}S$ if the 5d electron were localised or delocalised, respectively.

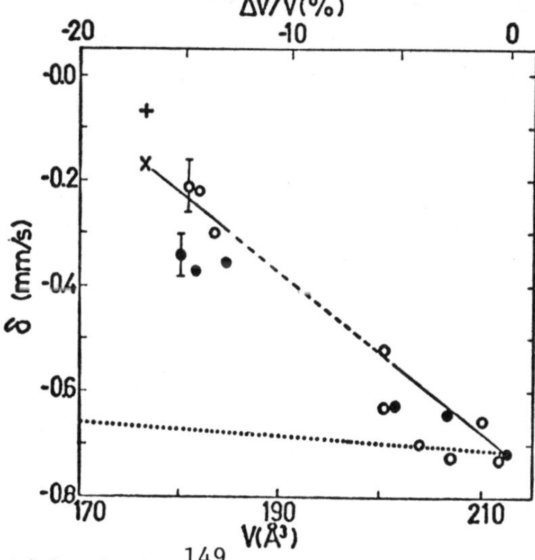

Fig. 8 Linewidth of the ^{149}Sm resonance a) measured in SmS as a function of pressure, b) measured in $(Sm_{1-x}Y_x)S$ and c) calculated as a function of relaxation time for different values of n_h, the fraction of trivalent samarium.

Information about the relaxation times may be derived from the linewidths, shown as a function of pressure and Yttrium content in figure 8a) and b). Within the errors they are independent of both, except just in the transition region in figure 8a) where the inhomogeneity of the pressure gives a two phase region. Figure 8c) shows the line broadening expected if the relaxation time was longer than about 10^{-8} sec. No such broadening is observed, so we conclude that any configuration fluctuations in the gold phase of SmS are more rapid than 10^{-9} sec. Like SmB_6, it belongs to category iii.

Two other Mössbauer experiments have been performed on $Sm_{1-x}Ln_xS$ mixed crystals. One uses the ^{152}Sm resonance, and is reported elsewhere at this conference [18]. The other is a source experiment on ^{151}Sm, the radioactive parent of ^{151}Eu. The Eu daughter impurities do not change their valence on passing from the black to the gold phase, but there are smaller changes in isomer shift due to changes in the 5d and 6s character [19].

Finally we note that an isomer shift of -0.19 (4) mm/sec has been reported for Sm_3S_4 [20]. Assuming the electronic configuration is:

$$(1-n_h)4f^6 + n_h(4f^5) \qquad (5)$$

one finds n_h = 0.71 (5), in agreement with the value 0.67 expected from the formula $Sm^{2+}Sm_2^{3+}S_4$. Since it has the Th_3P_4 structure, with equivalent samarium sites, this compound cannot belong to category i. A study of the linewidth as a function of temperature is needed to decide if it belongs in category ii or iii.

2.2 Europium

Eu_3O_4 crystallises in the $CaFe_2O_4$ structure, with inequivalent Eu^{2+} and Eu^{3+} sites. Thus it does belong to category i. Seperate Mössbauer resonance lines at -12.5 and 0.6 mm/sec were found at all temperatures above 6.2K, with an intensity ratio of approximately 1:2 [21]. Below 6.2K, magnetic hyperfine splitting of the line at -12.5 mm/sec appears as the Eu^{2+} sublattice orders magnetically. The line at 0.6 mm/sec remains unbroadened because the Eu^{3+} has no net moment [21].

Similar behaviour is found in the non-cubic phase of Eu_3S_4, stable below the phase transition at 160K [22-24]. The magnetic ordering temperature in this case is 3.1K. Above the phase transition, the compound acquires the Th_3P_4 structure, with equivalent europium sites. Nonetheless the two resonance lines remain separate

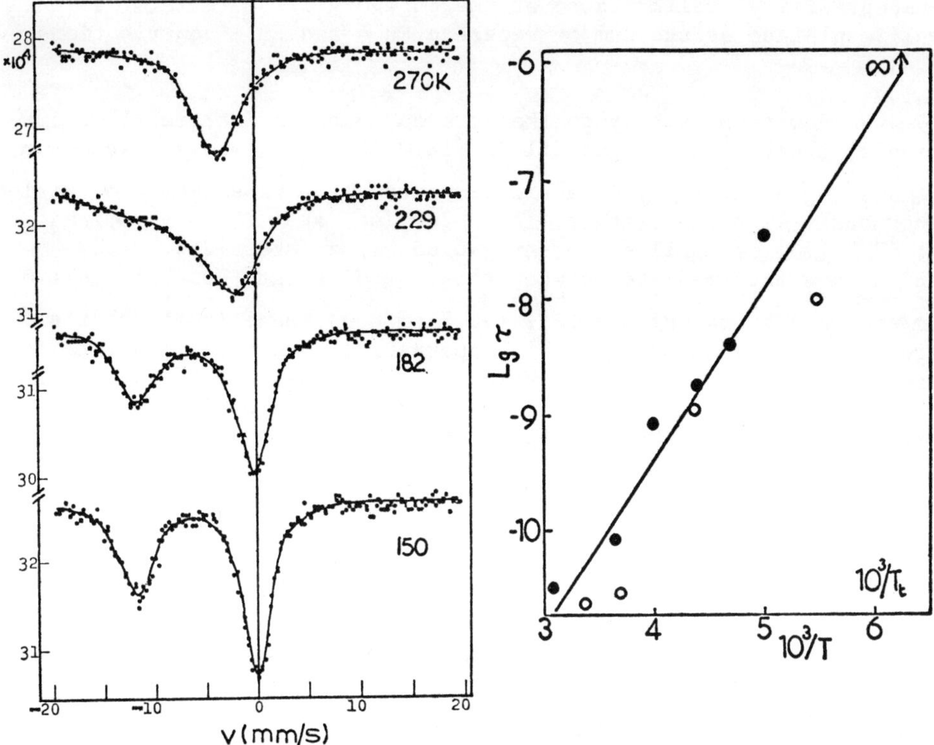

Fig. 9 (left) ^{151}Eu Mössbauer spectra of Eu_3S_4 as a function of temperature. A structural phase transition occurs at 160K.

Fig. 10 (right) Relaxation time in Eu_3S_4 as a function of 4T (O, from reference 24; O, this work).

up to about 200K, where they begin to broaden and coalesce progressively with increasing temperature. The sequence of spectra with increasing temperature, figure 9, exactly resembles that with decreasing relaxation time, figure 2b. Berkooz, et al discovered this behaviour in 1968, and interpreted it as due to a thermally activated hopping process [25]. Fitting the results to an expression:

$$\tau = \tau_o \, e^{Ep/KT} \qquad (6)$$

yielded an activation energy of 0.24 eV, in agreement with the activation energy for electrical conductivity [25], (figure 10). The pre-exponential factor τ_o is of order of the inverse phonon frequencies, 10^{-14} sec. Eu_3S_4 at temperatures above its crystallographic phase transition is at present the only really well established category ii compound among the rare earths. Its susceptibility, unlike those of the category iii materials, is well explained as the sum of seperate Eu^{2+} and Eu^{3+} contributions [24].

We now turn briefly to the work on europium intermetallic compounds, treated in more detail in [18]. $EuCu_2Si_2$ shows a spectrum consisting essentially of a single narrow line whose position varies enormously with temperature, from -2 mm/sec at 4.2K to -7 mm/sec at 670K [26] or applied pressure ∿0.05 mm/sec/kbar [27]. The shifts are intermediate between those of divalent $EuAg_2Si_2$ (-11.6 mm/sec) and trivalent $EuFe_2Si_2$ (-0.7 mm/sec) whose isomer shifts are essentially independent of temperature. From the linewidth it was deduced that the configuration fluctuations must be more rapid than 3.5×10^{-11} sec. The temperature dependence of the isomer shift was explained by Bauminger, et al [26,28] as a result of thermal population of the Eu^{2+} level by electrons at the Fermi energy, supposed to be $E_{ex} = 0.07$ eV below. Kaindl, et al [24] suggest that the unit cell volume is a more appropriate independent variable for determing the europium configuration.

$EuRh_2$ is analogous to $EuCu_2Si_2$ in that the spectrum is a single line, whose isomer shift decreases strongly with inceasing temperature [28]. The behaviour was likewise explained in terms of fast fluctuations of an electron at the Fermi level to the Eu^{2+} level, some 0.1 eV above.

Solid solutions of $EuRh_2$ have been studied with $LaRh_2$ and $PrRh_2$ which have larger unit cells, and are therefore expected to make the europium more divalent [29]. A new aspect of these spectra are the influences of different nearest-neighbour environments on the ^{151}Eu isomer shift. Overlapping lines are associated with europium atoms having 0, 1, 2, 3 or 4 europium neighbours, and each has a different temperature dependence. An inherent width of the Eu^{2+} level of ∿0.01 eV was included in the fit.

Spectra of solid solutions of europium alloys with different transition metals, nearly trivalent $EuRh_2$ or $EuIr_2$ with divalent $EuPt_2$ or $EuAl_2$, have also been reported [30]. They likewise have a complicated structure which has been interpreted in terms of fast electronic fluctuations ($\tau < 4 \times 10^{-11}$ sec) with an excitation energy which depends on composition, temperature and the number of each transition metal neighbours. Pressure studies of some of these alloys have been made recently [31].

2.3 Other Rare Earths

The isomer shift for ^{145}Nd in $NdCl_{2.37}$ is intermediate between those of $NdCl_2$ and $NdCl_3$ [32]. No time resolution is possible because of the large width of the resonance line and $NdCl_3$ might just be a two phase mixture.

A study of the ^{169}Tm resonance is TmSe, published by Triplett, et al [33] shows well-defined magnetic hyperfine splitting below the magnetic ordering temperature of 2K. Distinct, though overlapping, hyperfine patterns arise from Tm^{2+} and Tm^{3+}, and fluctuations are slow compared with the nuclear Larmor precession time.

Magnetic hyperfine splitting is also found in a source experiment on ^{170}Yb in $Tm_{1-x}Yb_xFe_2$ alloys, but the spectrum there is composed of the magnetic structure due to Yb^{3+} superimposed on an unsplit, though broad Y^{2+} line [28].

III. CONCLUSIONS

Table III summarizes the Mössbauer results on mixed valence rare earths. The variables x and y, respectively, signify solid solutions in which the rare earth or the other sublattice is substituted. 'Yes' or 'no' in the temperature column indicates if there is temperature dependence of the electronic configuration. Assignment of the materials to category i, ii, or iii is usually in agreement with the assignment that would be made on the basis of magnetic susceptibility measuements, despite the rather longer characteristic time of the Mössbauer measurement.

The picture we have used throughout is one of electronic configuration fluctuations with a characteristic relaxation time. However, the most that can be said of the category iii materials, (and almost all the mixed valence rare earths belong in this category), is that if there are any fluctuations, they must be more rapid than a time of order 10^{-9} sec. The Mössbauer data is silent about the appropriateness of a fast fluctuation picture rather than a ground state represented by a mixed wave function. However, hybridization of the 4f level with the 5d band gives it an energy width of order 10^{-2} eV, and the lifefime associated with an electron of 4f character is thus $\sim 10^{-13}$ sec. The two descriptions are not mutually exclusive.

The distinction between categories ii and iii on the basis of the Mössbauer time scale is an arbitrary one. There is, however, a natural time scale in the mixed valence problem, that of the lattice response to the change fluctuations, which is not all that far from the Mössbauer time scale. If the relaxation time is longer than the inverse phonon frequency ($\sim 10^{-14}$ sec), then a local lattice distortion will surround the rare earth ion. Conductivity by f electrons will be by polaron hopping. However, if the relaxation

TABLE III

Mössbauer data on mixed valence rare earths

Compound	Category	Experimental Variable T	P	x	y
SmS	iii		[11,13]	[13]	
SmB_6	iii	[9]no			
Eu_3O_4	i	[20]no			
Eu_3S_4	i, ii	[24]no			
$EuCu_2Si_2$	iii	[25]yes	[26]		
$EuRh_2$, $EuIr_2$	iii	[27]yes	[30]	[28]	[29]
TmSe	ii	[33]			

TABLE IV

Trivalent fraction in the electonic configuration of mixed valence compounds determined by different methods.

Compound	Mössbauer Effect	Lattice Parameter	XPS	References
SmB_6	∿0.7	0.7	0.71	[9, 34, 35]
$Sm_{0.81}Y_{0.19}S$	0.60	0.70	0.58	[11, 36]
TmSe	0.35	0.80	0.80	[33, 37]

time is less than the inverse phonon frequency, the lattice cannot follow the charge fluctuations, and more truely metallic conductivity results.

Ideally one would like a measurement on the time scale of phonon frequencies. Magnetization measurements are useful in this respect because Curie-Weiss behaviour is only seen when Boltzman population of the Zeeman levels can be established, that is if configuration fluctuations are slower than the spin-lattice relaxation time. A particularly suitable technique should be diffuse magnetic neutron scattering since the time for which a thermal neutron interacts with an atom is only $\sim 10^{-13}$ sec.

Most of the above materials have also been studied by x-ray photoemission, where distinct 4f structure is seen corresponding to the di- and trivalent parent configurations. As the time scale is $\sim 10^{-16}$ sec, any fluctuations must be less rapid than this, so a bracket $10^{-16} < \tau < 10^{-9}$ is defined for the category iii materials in table III.

A comparison of the values of n_h, the fractional trivalent character determined by different methods is given in table IV. The large measure of agreement between the results is very encouraging, and leaves little doubt about the reality of the mixed valence state.

In conclusion, Mössbauer spectroscopy is a useful technique for probing mixed valence systems. It can yield the average electronic configuration, and measure relaxation times of order 10^{-7} - 10^{-10} sec. It entails a negligible perturbation of the electronic system, and as a microscopic probe it can give information about local inhomogeneity in disordered alloys. Most of the applications to date have depended on the different electric monopole interactions (isomer shifts) of different electronic configuration, but more use may be expected to be made of the magnetic hyperfine interactions in measurements at low temperatures.

REFERENCES

[1] M. B. Robin and P. Day, Adv. Inorg. Chem. and Radiochem. 10, 247 (1967).
[2] C. M. Varma, Rev. Mod. Phys. 48, 219 (1976).
[3] B. Batlogg, E. Kaldis, A. Schlegel, G. von Schultess and P. Wachter, Solid State Comm. 19, 673 (1976).
[4] W. M. Reiff, Mössbauer Effect Methodology 8, 89 (1973).
[5] N. N. Greenwood and T. C. Gibb, Mössbauer Spectroscopy (Chapman & Hall, London) 1971.
[6] J. G. Stevens and V. E. Stevens (eds), Mössbauer Effect Data Index 1973 (Plenum, New York) 1974.
[7] G. M. Kalvius and G. K. Shenoy, Atomic Data and Nuclear Data Tables 14, 639 (1974).
[8] H. H. Wickman, M. P. Klein and D. A. Shirley, Phys. Rev. 152, 345 (1966).
[9] R. L. Cohen, M. Eibschütz and K. W. West, Phys. Rev. Lett. 24, 383 (1970).
[10] J. C. Nickerson, R. M. White, K. N. Lee, R. Backmann, T. H. Geballe, and G. W. Hull, Phys. Rev. B3, 2030 (1971).
[11] J. M. D. Coey, S. K. Ghatak, M. Avignon and F. Holtzberg, Phys. Rev. B14, December 1976.
[12] A. Jayaraman, V. Narayanamurti, E. Bucher, and R. G. Maines, Phys. Rev. Lett. 25, 1430 (1970).

[13] J. M. D. Coey, S. K. Ghatak and F. Holtzberg, AIP Conf. Proc. 24, 38 (1975).
[14] L. J. Tao and F. Holtzberg, Phys. Rev. B11, 3842 (1975).
[15] A. Jayaraman, P. Dernier and L. D. Longinotti, Phys. Rev. B11, 2783 (1975).
[16] T. Penney and F. Holtzberg, Phys. Rev. Lett. 34, 322 (1975).
[17] G. M. Kalvius, U. F. Klein and G. Wortmann, J. Physique 35, C6-139 (1974).
[18] I. Nowik, These proceedings.
[19] R. L. Cohen, Abstracts of the International Conference on Magnetism, Amsterdam 1976.
[20] M. Eibschütz, R. L. Cohen, E. Buehler and J. H. Wernick, Phys. Rev. B6, 18 (1972).
[21] H. H. Wickman and E. Catalano, J. Appl. Phys. 39, 1248 (1968).
[22] H. H. Davis, I. Bransky and N. M. Tallan, J. Less Common Metals 22, 193 (1970).
[23] E. Görlich, H. U. Hrynkiewicz, R. Kniec, K. Latka and K. Tomala, Phys. Stat. Sol. b64, K147 (1974).
[24] O. Massenet, J. M. D. Coey and F. Holtzberg, J. Physique 37, C6- (1976).
[25] O. Berkooz, M. Malamud and S. Shtrikman, Solid State Comm. 6, 185 (1968).
[26] E. R. Bauminger, D. Froindlich, I. Nowik, S. Ofer, I. Felner and I. Mayer, Phys. Rev. Lett. 21, 1053 (1973).
[27] G. Kaindl, et al, J. Physique 37, C8- (1976).
[28] E. R. Bauminger, I. Felner, D. Froindlich, D. Levron, I. Nowik, S. Ofer and R. Yanovsky, J. Physique 35, C6-61 (1974).
[29] E. R. Bauminger, I. Felner, D. Levron, I. Nowik and S. Ofer, Phys. Rev. Lett. 33, 890 (1974).
[30] E. R. Bauminger, I. Felner, D. Levron, I. Nowik and S. Ofer, Solid State Comm. 18, 1073 (1976).
[31] G. Wortmann, et al, J. Physique 37, C8- (1976).
[32] G. Kaindl, Z. Physik 240, 100 (1970).
[33] B. B. Triplett, N. S. Dixon, P. Boolchand, S. S. Hanna and E. Bucher, J. Physique 35, C6-653 (1974), and to be published.
[34] J. P. Mercurio, Thesis, University of Bordeaux, (1975).
[35] J. N. Chazabriel, M. Campagna, G. K. Wertheim and P. H. Schmidt, Solid State Comm. 19, 725 (1975).
[36] R. A. Pollak, F. Holtzberg, J. F. Freeouf and D. E. Eastman, Phys. Rev. Lett. 33, 820 (1974).
[37] A. Campagna, G. K. Wertheim, E. Bucher, AIP Conf. Proc. 24, 22 (1974).

COMMENTS

BIRGENEAU: The phonon frequencies in these materials is of the order of 10^{-12} sec. On the other hand, e.g., in both the γ and α phases of Cerium, the spin fluctuation time is always much faster than the phonons so the lattice has no time to respond. Similar is the situation in Eu_3S_4 where the fluctuation frequency is $\sim 10^{-14}$ sec which is again two orders of magnitude faster than the harmonic vibrations of Eu and an order of magnitude faster than the S vibrations. So it appears again as if the lattice has no time to respond.

CAMPAGNA: Is charge ordering in Eu_3S_4 a correct interpretation in view of the fact that the same transition has been observed in La_3S_4 which obviously has no f electrons?

COEY: At present experiments do not permit the assertion of charge ordering in Eu_3S_4. All we can say is that we have very slow fluctuations.

HIRST: Could I say a word about the question of determining the time scale of charge fluctuations. This is an example of where this terminology can lead people into a pitfall. The fact of the matter is at very low temperatures, say in the limit $T \to 0$, the charge fluctuations represent (in the type of mixed valent state which we hope is occurring in the simplest types of theories) zero point motion. There is absolutely no motion in the sample. The charge at the nucleus is constant and so consequently the broadening must always go to zero in the T limit $T \to 0$ according to such a theory. In the high T limit which means T high with respect to the nominal fluctuation rate then it would indeed be correct to use the simple type of charge fluctuation estimate which you are referring to. When you refer to the charge fluctuation, time as determined from these spectra the picture is that there is a net of one 4f-electron flipping in and out of the ion in a random gaussian fashion, which is characterized with this time.

COEY: That is right. Viewed from the nucleus it looks like a stochastic process.

UNKNOWN: Has anybody done Mossbauer experiments on TmSe?

COEY: Tm is not a good nucleus in which to look for isomer shifts but one can measure magnetic hyperfine interactions at very low temperatures on the Tm nucleus and distinguish the magnetic hyperfine fields in the ordered phase of TmSe.

ORBACH: I would like to follow Birgeneau's remark. If in fact the phonon frequency is so slow how do you explain the physics of the hopping rate?

COEY: When I mentioned 10^{-14} sec, I could have been off an order of magnitude because of the $1/T \to 0$ extrapolation.

GÜNTHERODT: From the Raman scattering experiment we find the phonon frequency to $\sim 10^{-14}$ sec.

WACHTER: Can you say from your measurements on SmS whether the high pressure phase has a different mixing than the SmS doped with Yttrium?

COEY: In the Y-doped phase there is less of 3^+ than in the high pressure phase of SmS. An increase in pressure on SmS of 11 kbar gives you already 0.9 for the fraction of 3^+; another 5-10 kbar would probably drive it completely 3^+.

EIBSCHUTZ: In your experiment do you see a phase transition? Do you have charge fluctuations below the phase transition?

COEY: One can't say since the spectrum above the transition already appears static. There is a small change in the isomer shift at the transition.

A THEORETICAL MODEL OF MIXED VALENCE

ELECTRONIC AND ELASTIC PROPERTIES

S.K. Ghatak* and M. Avignon

Groupe des Transitions de Phases, C.N.R.S.

B.P. 166, 38042 Grenoble Cedex, FRANCE

I. INTRODUCTION

The rare-earth elements are characterized by strongly localized and highly correlated f-states which retain their atomic-like character in compounds. In general, the rare-earth is in a well-defined configuration with an integral number of electrons in the 4f-shell. In the great majority of compounds and alloys, this corresponds to trivalent ions with core configuration $4f^{n-1}$. Accordingly, most of the rare-earth monochalcogenides are metallic with one 5d electron in the conduction band. Exceptions are Sm, Eu, Yb which appear as divalent with a $4f^n$ core configuration in sulphides, selenides and tellurides /1/. The final occupied f-level then lies below the bottom of the empty conduction band which is formed from 5d-6s orbitals and these compounds are semiconductors. The case of Tm is rather critical since its valence changes from 3^+ to 2^+ in going from the sulphide to the telluride /2/. By application of pressure, essentially because of the increasing strength of the cubic crystalline field acting on 5d-orbitals /3/, the separation between f-level and conduction band decreases resulting the possibility of valence instability. This instability is obviously involved with the transfer of some electrons from the f-level to the much wider d-band and a partial transformation of divalent towards trivalent ions. This is clearly indicated by a systematic study of the lattice constant throughout the whole series since trivalent ions are ≃ 20 % smaller than divalent ions. In SmS the f-electrons can be thermally excited into the conduction band /4/. The transition is isostructural (NaCl type).

* Now at : Institut für Theoretische Physik, Freie Universität Berlin, 1 Berlin 33, Arnimallee 3, West Germany.

Near the pressure where such transitions occur, many physical quantities exhibit anomalous behaviour, e.g. change in resistivity /5/, in Mössbauer isomer shift /6/, in X-ray and ultra-violet photoemission spectra /7/, in P-V relationship /5,8/, anomalous increase in compressibility /9/. A large electronic specific heat is also observed in the metallic phase of SmS /10/. This transition can also be induced by doping with trivalent rare-earths La, Gd .., or with Yttrium which form an isostructural compound whose lattice constant is smaller . For example $Sm_{1-x} Y_x S$ undergoes a transition at room temperature from a black semiconducting phase to a gold metallic phase at $x \simeq 0.15$ /11,12/. In this case the decrease of the lattice constant is not the only point. For example, no transition is observed when doping with Ca /13/ even though the lattice constant of CaS is much smaller than that of SmS. The position of the conduction band of the dopant relative to the SmS conduction band is probably important. In addition, each Yttrium donates an extra electron to the conduction band. Most of these properties have been largely reviewed recently /14,15/.

An important fact concerns the type (first order or second order) of the pressure-induced transition. Taking only the Sm monochalcogenides at room temperature, it is discontinuous in SmS while it is continuous in SmSe and SmTe. It is not yet known whether or not it remains continuous down to T = 0. Nevertheless, this seems a possibility that has to be considered. If it were so, it would be difficult to explain it in the framework of Mott's theory /16/. On the basis of band structure calculations, Davis /17/ has invoked two different mechanisms for the transition, i.e. a $f \to d$ delocalization in SmS and a simple closing of the Se or Te (p) - Sm(5d) band gap. In this case, the 4f-states are found in the p-band in SmSe and SmTe. However, the XPS data show similar electronic structures with the f-states in the gap between the Sm(5d) conduction band and the p-valence band /7/. This suggests that the same type of mechanism is probably responsible for the transition in the three compounds.

To investigate this transition, we consider a model which is essentially based on the Coulomb interaction between localized and itinerant states. When a f-electron is promoted from the f-shell to the conduction band, the conduction electrons feel a crystal potential which is more attractive. This results in a lowering of the d band and a possible configuration crossover /18/. The mixing interaction between these two kinds of states is necessary to obtain intermediate valence at T = 0 and high value of the electronic specific heat in the metallic phase. Although small, the hybridization plays an important role in the transition. It has been incorporated in the model and treated in the framework of the virtual bound state approach. As we will see later, depending on the relative magnitude of the energy gain and the width of the f-level, the transition can be abrupt or smooth. This simple model has been successfully applied to interpret the Mössbauer data /19,12/. The elastic properties will be derived from the electronic configuration since the volume depends on the valence. Part of the calculations on elastic properties have been reported earlier /20/.

II. THE MODEL

As stated in the introduction, we are dealing with two kinds of states : highly correlated atomic-like f-states and itinerant d-states. The total hamiltonian can be splitted in two parts : H_f which represents the interaction inside the ions and H_c describing the conduction electrons. The description of the correlations inside the atomic electron shell requires the introduction of the so-called atomic operators /21/. The energy of the configuration with n f-electrons can be written as :

$$E_n = T_o n + \frac{U}{2} n (n - 1)$$

T_o is the one electron energy. U the intra-atomic Coulomb interaction being of the order of $\frac{1}{2}$ Ry $\simeq$ 7 eV in rare-earth metals /22/ so that two different configurations have energies differing by such a quantity. At most, two configurations $4f^n$ and $4f^{n-1}$ can be close in energy.

We treat the transition in the frameword of a one electron problem and consider the following hamiltonian :

$$H = \sum_k \varepsilon (k) a_k^+ a_k + E_o \sum_i b_{im}^+ b_{im} - G \sum_i b_{im}^+ b_{im} a_k^+ a_k$$

$$+ \sum_{i,k} V_{ki} (a_k^+ b_i^+ + a_k b_i) \qquad (1)$$

where a_k^+ (a_k) creates (destroys) an electron in the conduction band ; b_{im}^+ (b_{im}) creates (destroys) a hole in the mth orbital of atom i and changes the configuration from f^n to f^{n-1} (or from f^{n-1} to f^n) - E_o represents the energy required to excite an electron from the f^n shell into the conduction band at Fermi level taking into account many-body effects. E_o is directly accessible to experiment /7/. Since at most one electron can be taken out of f^n shell, the localized quasi-particle level will be assumed to contain one state per atom. The third term represents the Coulomb interaction between the conduction electrons and holes in the f-shell and provides a driving force for the transition. Hybridization between f- and band states is represented by the last term. It mixes the two configurations f^n and $f^{n-1}d$. V_{ki} is formally different from a single-electron f-d mixing

$$V_{ki} = \langle n | H | n-1, k \rangle$$

This mixing interaction is difficult to calculate and will be considered in a phenomenological way in the present model /23/. Several authors have developped models along similar ideas /24,25/. Iglesias Sicardi et al /26/ did not use the ionic model and considered the system only from the resonant scattering point of view without introducing any intrinsic driving mechanism for the transition.

We consider the rare-earth monochalcogenides as a collection of independent rare earths impurities. This approach has proven to be useful in the past /27/. The hamiltonian given by (1) can be written in terms of creation (destruction) $C_i^+(C_i)$ operators for electrons in localized states :

$$H = \sum_k (\varepsilon(k)-G) a_k^+ a_k - E_o \sum_i C_i^+ C_i + G \sum_i C_i^+ C_i a_k^+ a_k$$

$$+ \sum_{k,i} V_{ki} \; (a_k^+ C_i + a_k C_i^+) \tag{2}$$

The energy E_o can be taken as reference energy and the bottom of the empty conduction band as the gap Δ. We employ Zubarev's formalism /28/ to write the equations of motion for the different Green's functions :

$$\omega \ll C_i | C_i^+ \gg = 1 + \sum_k V_{dk} \ll a_k | C_i^+ \gg + G \sum_k \ll a_k^+ a_k C_i | C_i^+ \gg \tag{3a}$$

$$\omega \ll a_k | C_i^+ \gg = \varepsilon_{(k)} \ll a_k | C_i^+ \gg + \quad V_{ki} \ll C_i | C_i^+ \gg$$

$$+ G \ll C_i^+ C_i a_k | C_i^+ \gg \tag{3b}$$

$$\text{and } \omega \ll a_k | a_k^+ \gg = 1 + \varepsilon(k) \ll a_k | a_k^+ \gg + V_{ki} \ll C_i | a_k^+ \gg$$

$$+ G \ll C_i^+ C_i a_k | a_k^+ \gg \tag{3c}$$

In Hartree-Fock approximation, higher-order Green's functions are decoupled in terms of the thermal average of the occupation number in the conduction band n_c and in localized states n_f :

$$n_c = \frac{1}{N} \sum_k < a_k^+ a_k >$$

$$n_f = < C_i^+ C_i >$$

$$\text{Then } G_{ff}(E) = \ll C_i | C_i^+ \gg = \frac{1}{E - Gn_c - \Sigma(E)} \tag{4}$$

$$\text{with } \Sigma(E) = \sum_k \frac{V_{ki}^{\;2}}{E - \varepsilon(k) - Gn_f + i\eta} \tag{4a}$$

and $$G_{kk}(E) = \ll a_k | a_k^+ \gg = \frac{1}{E - \varepsilon(k) - Gn_f} + \frac{V_{ki}^2 \ll c_i | c_i^+ \gg}{(E - \varepsilon(k) - Gn_f)^2} \qquad (5)$$

Then the density of localized states is obtained :

$$D_f(E) = \frac{1}{\pi} \frac{\delta(E)}{\left[E - Gn_c - R(E) \right]^2 + \delta(E)^2} \qquad (6)$$

where R(E) and $\delta(E)$ are the real and the imaginary parts of $\Sigma(E)$

$$R(E) = P \sum \frac{V_{ki}^{\ 2}}{E - \varepsilon(k) - Gn_f}$$

and $$\delta(E) = \pi \sum_k V_{ki}^{\ 2} \, \delta\left(E - \varepsilon(k)\right) \qquad (7)$$

In a metallic phase, when the localized level lies inside the conduction band, taking a constant mixing parameter V_{ki}

$$\delta(E) = \pi V^2 D_c(E) \qquad (8)$$

then considering $\delta(E)$ as a constant seems a reasonable approximation. This can be applied to the $\alpha \to \gamma$ transition in cerium since both phases are metallic /29/. When the self-consistent level E_o' (E_o' is a solution of $E - Gn_c - R(E) = 0$) is outside the spectrum $\varepsilon(k)$, the level remains sharp ($\delta(E) = 0$) but the strength of the δ-function is renormalized. The total number of f-states it contains N_f becomes less than one /30/ :

$$N_f = \left. \frac{1}{1 - \frac{d\Sigma}{dE}} \right|_{E = E_o'} \qquad (9)$$

The way N_f varies as E_o' approaches the bottom of the conduction band depends very much on the shape of the conduction band /31/. In order to take into account hybridization effects in the semi-conducting phase in a very simple way, we choose to take $\delta(E)$ as a constant δ throughout the whole transition. In this way, no discontinuous character is introduced by the hybridization itself contrary to the treatment of Iglesias Sicardi et al /26/. We believe that this does not alter the physics of the problem. The average number of holes $n_h = 1 - n_f$ is thus given by

$$n_h = \frac{1}{2} + \frac{1}{\pi} \operatorname{Im} \psi\left(\frac{1}{2} + \frac{\delta}{2\pi kT} - i \frac{E_F - Gn_c}{2\pi kT} \right) \qquad (10)$$

where Im means the imaginary part and $\psi(x - iy)$ is the di-gamma function /32/.

At T = 0,

$$n_h = \frac{1}{2} - \frac{1}{\pi} \arctan \frac{E_F - Gn_c}{\delta} \qquad (11)$$

The Fermi level E_F is determined in terms of n_c :

$$n_c = \int \frac{1}{e^{\frac{E - E_F}{kT}} + 1} D_c(E)dE \qquad (12)$$

$D_c(E)$ being the actual density of states of the conduction band. If one neglects the effect of hybridization on the itinerant states then, the conduction band is just shifted downwards by Gn_h

$$D_c(E) = D_c^\circ (E + Gn_h) \qquad (13)$$

For the sake of simplicity, we take a rectangular band of width W :

$$D_c^\circ(\omega) = D \quad \text{for} \quad \Delta < \omega < \Delta + W$$
$$= 0 \quad \text{otherwise}$$

Then

$$E_F = \Delta + \frac{n_c}{D} - Gn_h + kT \ln \left(1 - e^{-\frac{n_c}{DkT}}\right) \qquad (14)$$

where we have assumed $\frac{W}{kT} >> 1$ which is valid for bandwidth $\simeq$ 2-3 eV. The self-consistent equation for n_h (or n_c) is obtained from the conservation of the number of electrons

$$n_h = n_c \qquad (15)$$

Before going to the general solution, let us consider the situation at T = 0. Then the self-consistent equation reduces to :

$$\phi(n_h) = \cot \pi n_h - Yn_h = \frac{\Delta}{\delta} \qquad (16)$$

where $Y = (\frac{1}{D} - 2G)/\delta$

The parameter Y is a measure of the energy gain relative to the width of the level in delocalizing an electron and hence its magnitude governs the nature of the transition. The fonction $\phi(n)$ is plotted in Fig. 1 for different values of Y. For large values of $\frac{\Delta}{\delta}$ there is only one solution ; n_h is very small and increases as the gap decreases with increasing pressure. For $Y \geqslant -\pi$ (curves (c) - (d) of Fig. 1), the change in n_h is continuous from small to large values of n_h. On the contrary, when $Y < -\pi$ (curves (a) and (b) of Fig. 1) as $\frac{\Delta}{\delta}$ decreases we reach a point where two solutions appear ;

then, for smaller $\frac{\Delta}{\delta}$ three solutions are present. The state corresponding to the highest n_h is first 'metastable' as can be seen from the energy of the system

$$E = \Delta n_h + \frac{Y\delta}{2} n_h^2 - \frac{\delta}{\pi} \mathrm{Ln}\,(\sin \pi n_h) \tag{17}$$

This state becomes stable for $\Delta/\delta = -\frac{Y}{2}$; this defines the transition. Thus in the model, the discontinuous transition does not take place when the f-level becomes degenerate with the conduction band /26/ but occurs at a finite gap. This is consistent with the value of the gap in SmS $\simeq$ 0.15 - 0.2 eV /4,33/ and a rate of decrease with pressure $\frac{d\Delta}{dp} \simeq -$ 10 meV/Kb /14/.

The parameter $(2G - \frac{1}{D}) = -Y\delta$ plays also a key role in determining the density of states at the Fermi level in the metallic side of the transition :

$$D_f(E_F) = \frac{1}{\pi\delta} \frac{1}{1 + Y^2 (n_h - \frac{1}{2})^2} \tag{18}$$

This goes to the maximum value $\frac{1}{\pi\delta}$ as $n_h \to \frac{1}{2}$. For Y = - 5, δ = 4 meV then $D_f(E_F) \simeq$ 30 st/eV. The very large value in SmS /10/ makes us believe that the transition in SmS is not far from being critical.

At finite temperature, the self-consistent equation can be written as follows :

$$\phi(n) = G(n) - Yn - \frac{kT}{\delta} \ln\,(1 - e^{-\frac{n}{DkT}}) = \frac{\Delta}{\delta} \tag{19}$$

where G(n) is the inverse function of (10) defined as follows :

$$n = \frac{1}{\pi}\left[\frac{\pi}{2} + \mathrm{Im}\,\psi(x - iy)\right] = G^{-1}\left(\frac{y}{x - \frac{1}{2}}\right)$$

This is plotted in Fig. 2 for Y = - 10, D = 2 st/eV and $\frac{\delta}{2\pi kT} = \infty$, 0.2, 0.1. It is seen that the critical value $(\frac{\Delta}{\delta})$ for transition does not change very much with temperature. In Eq. (19), the temperature dependent term gives a very small contribution. When temperature increases, the transition occurs for $(\frac{\Delta}{\delta})_c$ larger thus at a lower pressure and the phase boundary terminates at a critical point. Contrary to experiment /14/, in the T-p diagram, the slope is negative (Fig. 3). Qualitatively, the same result is obtained by da Silva and Falicov /25/. As they mentioned this can be attributed to the fact that the entropy associated with the localized moments has been neglected. The phase boundary is almost a vertical line since $(\frac{\Delta}{\delta})_c$ varies by less than 0.1 between T = 0 and $\frac{\delta}{2\pi kT} = 0.2$. For $\delta \simeq 20$ meV, this gives an increase in Δ of 1 meV. Then the sign of $\frac{dTc}{dp}$ can be changed due to the thermal expansion of the lattice. For instance, for GdS, the thermal expansion coefficient is found to be $\simeq 10^{-5}$°K^{-1} /34/ while the above reduction of Δ corresponds to an increase of volume $\frac{\Delta V}{V} \simeq 2.10^{-4}$ for a deformation potential $\Sigma \simeq 5$ eV /14/.

In Fig. 4 for comparison, we have plotted the number of conduction electrons vs. pressure for two different values of δ together with the values derived from Mössbauer experiments /19/. Fig. 5 represents $\frac{1}{n_h}$ vs. pressure for SmS, SmSe, SmTe. The curves look very similar to the resistivity vs. pressure (Fig. 5 of reference 14). We cannot demand more than qualitative agreement since our model is very crude in many respects.

III. PRESSURE - VOLUME RELATIONSHIP

We assume that the lattice constant depends linearly on the 'average valence' $v = 2 + n_h$ according to Vegard's law :

$$a(P,n_h) = n_h a_3(P) + (1 - n_h)\, a_2(P) \tag{20}$$

where $a_3(P)$ and $a_2(P)$ are the lattice parameters for pure tri- and divalent SmS respectively. These parameters depend on pressure. At room temperature and normal pressure, we can take for SmS /35/

$$a_3(o) = 5.62 \text{ Å} \quad \text{and} \quad a_2(o) = 6.00\text{Å}$$

$a_3 = 5.75$ Å, $a_2 = 6.22$ Å for SmSe, and $a_3 = 6.15$ Å, $a_2 = 6.6$ Å for SmTe.

First of all, the decrease of the lattice constant when electrons are delocalized helps the transition. This effect can be easily added to the electronic contribution considered in Part II. This is essentially introduced through the decrease of the gap Δ as

$$\Delta(P) = \Delta_o(P) - \beta(P)\,P \tag{21}$$

Then the parameter Y becomes $\frac{1}{D} - 2G - \beta$; β can be estimated $\simeq 1$ eV knowing $\frac{d\Delta}{dp}$ /14/, the compressibility at low pressure ($n_h \simeq o$) $\chi = 2.10^{-3}$/Kbar /36/ and $\frac{da}{dn_h}$ from (20) ; it is pressure dependent

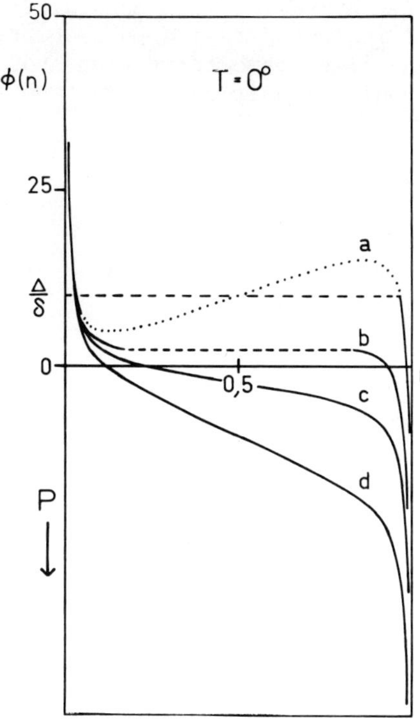

Fig. 1 : Plot of n vs. $\frac{\Delta}{\delta}$ for Y = −20 (a), − 5 (b), + 5 (c), + 20 (d) at T = 0°K.

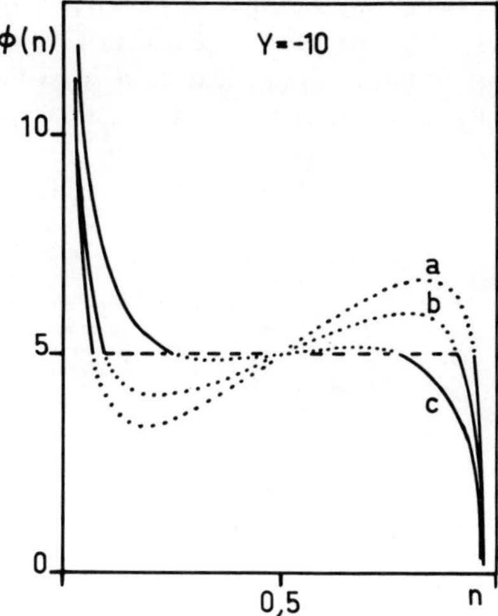

Fig. 2 : $\phi(n)$ vs n (n vs $\frac{\Delta}{\delta}$) for Y = − 10 for different temperatures a) T = 0,

(b) $\frac{\delta}{2\pi kT} = 0.2$

(c) $\frac{\delta}{2\pi kT} = 0.1$

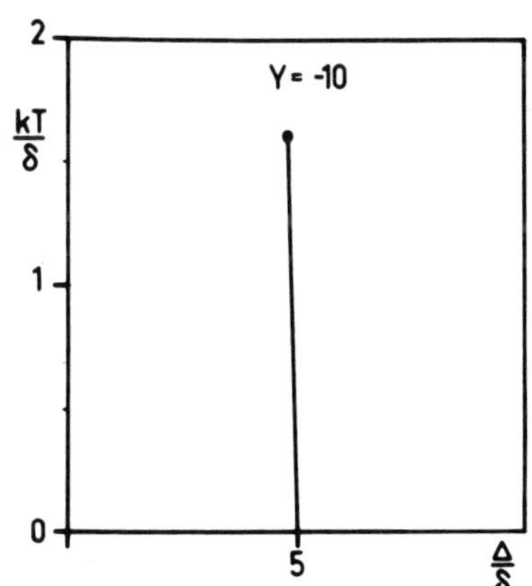

Fig. 3 : Phase diagram T − Δ for Y = − 10

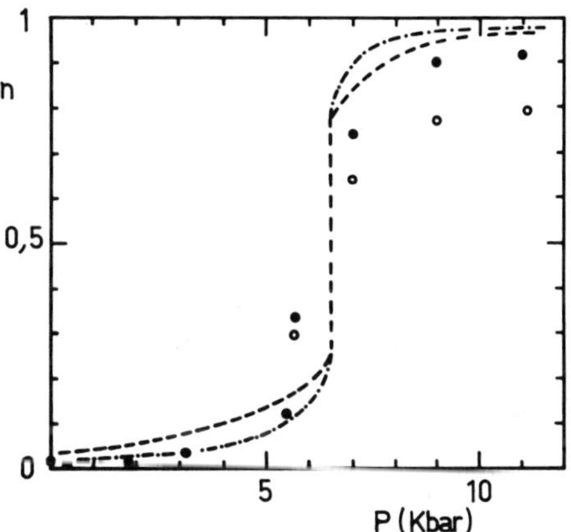

Fig. 4 : The fraction n of Sm^{3+} in SmS as a function of pressure. The theoretical curves have been calculated for Y = − 10, $\frac{\delta}{2\pi kT} = 0.1$ δ = 5 meV (−·−·−) and δ = 10 meV (− − − −). The experimental points are from Ref. 19 assuming delocalized 5d electrons (0) localized 5d-electrons (•).

and this makes the parameterization of the problem much more difficult. The strain $e(P)$ can be separated in two parts : a pure compression effect $e_1(P)$ in the absence of any electronic transition and $e_2(P)$ the strain related to the electronic transition :

$$\frac{\Delta V}{V} = e(P) = e_1(P) - e_2(P)\, n_h \tag{22}$$

with $e_1(P) = 3\,\dfrac{a_2(P) - a_2(o)}{a_2(o)}$

$$e_2(P) = 3\,\frac{a_2(P) - a_3(P)}{a_2(o)}$$

To estimate these terms and get their pressure variation one can use the Birch equation for $\xi = o$ /37/. This equation is known to be followed by compounds which have mainly ionic character up to a volume strain of the order of 30 %. Accordingly :

$e(P) = -\,aP + bP^2$ where $a = \chi_o$ the compressibility at normal pressure and $b = \dfrac{5\chi_o{}^2}{2}$

To a first approximation

$$e_2(P) \simeq 3\,\frac{a_2(o) - a_3(o)}{a_2(o)} + (\Delta\chi_o)P \tag{23}$$

where $\Delta\chi_o$ is the compressibility difference between tri- and divalent compounds. The compressibility at normal pressure for hypothetical trivalent compounds can be estimated from an empirical relationship between bulk modulus B_o and specific volume V_o /8/. The P-V relationships obtained from equation (22) using the results derived in Part II are plotted in Fig. 6. We can see that the theoretical curves are in good agreement with experimental results. The compressibility is given by :

$$\chi(P) = \chi_1(P) + e_2(P)\,\frac{\partial n_h}{\partial P} + \frac{\partial e_2}{\partial P}\, n_h \tag{24}$$

The first term χ_1 obtained from Birch equation decreases with pressure while the second and third therms depend very much on the nature of the transition. The second term is the most important . For very sharp transitions, this contribution becomes large. After the transition this is very small since the number of conduction electrons remains almost constant. The last contribution is negative. So in the metallic phase, the compressibility will be less than that resulting from pure compression. Therefore, we see that the compres--sibility will be very large around the transition ; this is shown in a semi-quantitative way in Fig. 7. This qualitative behaviour is observed in both SmS /8,9/ and $Sm_{1-c}Y_cS$ /38/.

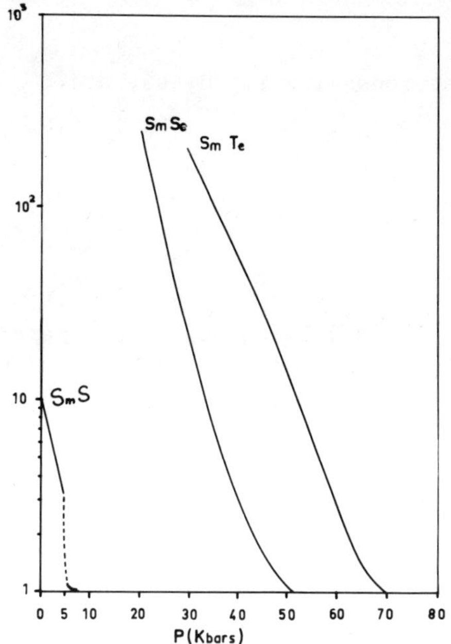

Fig. 5 : Plot of $\frac{1}{n}$ vs. pressure for $\frac{\delta}{2\pi kT} = 0.1$ and Y = - 10 (SmS) - 6 (SmSe) - 5 (SmTe)

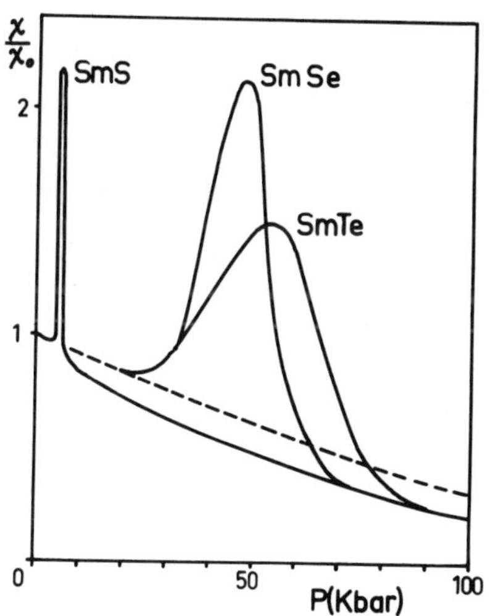

Fig. 7 : Semi-quantitative variation of the compressibility vs. pressure for SmS, SmSe, SmTe

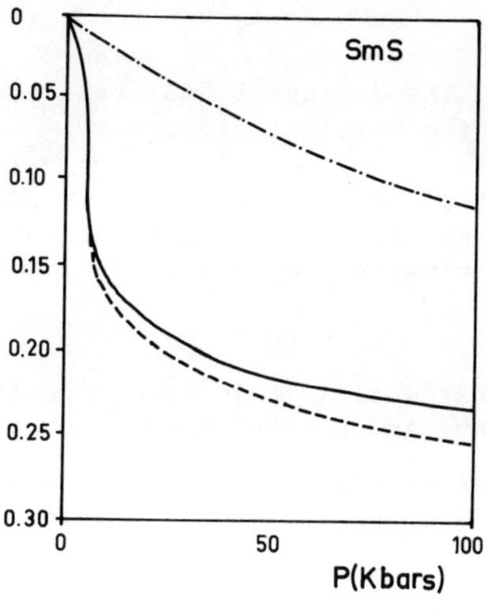

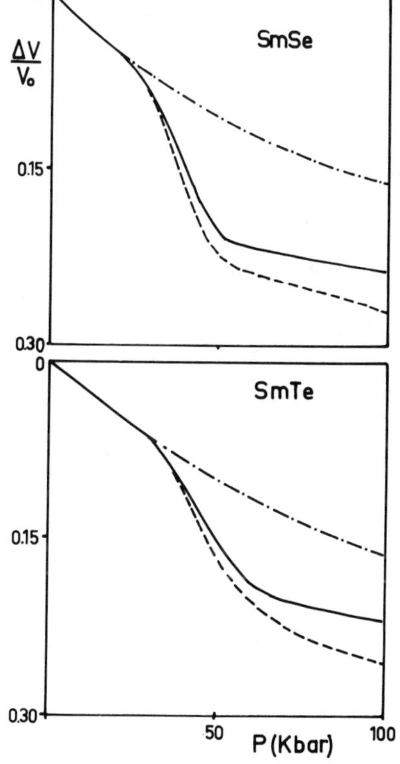

Fig. 6 : The variation of volume $\frac{\Delta V}{Vo}$ as a function of pressure. The full line is the experimental curve, the theoretical curves are the pure pressure effect (-·-·-) and the variation due to the electronic transition (- - - -).

IV. MIXED COMPOUNDS

The doping of SmS with trivalent ions like Y, Gd, La, etc.. has two distinct effects :

a) the ligand distances decrease and the crystal field splitting increases. In this case the gap Δ decreases with the concentration of trivalent impurities c :

$$\Delta = \Delta_o - \alpha c$$

To estimate α is quite difficult and it is like an adjustable parameter. An order of magnitude estimate can be made by combining experimental results $(\frac{\partial a}{\partial c})$ for small concentration and $\frac{\partial \Delta}{\partial P}$. Then $\alpha \simeq .4$ eV for Y-doped samples.

b) the trivalent rare-earth have one extra conduction d-electron and the total number of conduction electrons is

$$n_c = c + (1 - c)\, n_h$$

In these mixed compounds the conduction band density of state will be different from pure SmS. For example YS is a good metal with a fairly broad d-band and the d-band is expected to be wider in $Sm_{1-c}Y_c$ S. We assume that the d-electrons from YS and from the trivalent SmS form a common d-band with an average density of states

$$D = (1 - c)\, D(SmS) + c\, D(YS)$$

with D(SmS) = 4 st/eV and D(YS) = 2 st/eV. Taking Y = - 6 and δ = 5 meV, we calculate n_h (Fig. 8). This result fits the Mössbauer data from ref. 19 in a reasonable manner. We also calculate the lattice constant assuming :

$$a = c\, a(YS) + (1 - c) \left[n_h a_3 + (1 - n_h) a_2 \right]$$

The result is given in Fig. 9 together with experimental data from ref. 11. In this figure, we also reported the variation of the bulk modulus. The value for pure YS is 1000 Kbar.

V. DISCUSSION

The equation of state obtained for SmS, SmSe and SmTe is in semi-quantitative agreement with experimental results, particularly in the transition region. This is also true for the number of conduction electrons. In the metallic phase, there is deviation from experimental data. One reason for this is the neglect of the screening of the Coulomb interaction. The interaction parameter G being large in the semiconducting phase is expected to be reduced as the number of

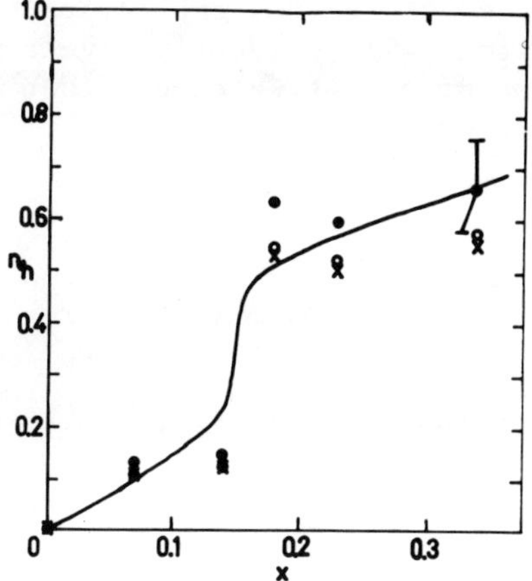

Fig. 8 : Plot of n_h vs. concentration for $Sm_{1-c}Y_c$ S. The experimental points are from Ref. 19 as in Fig. 4.

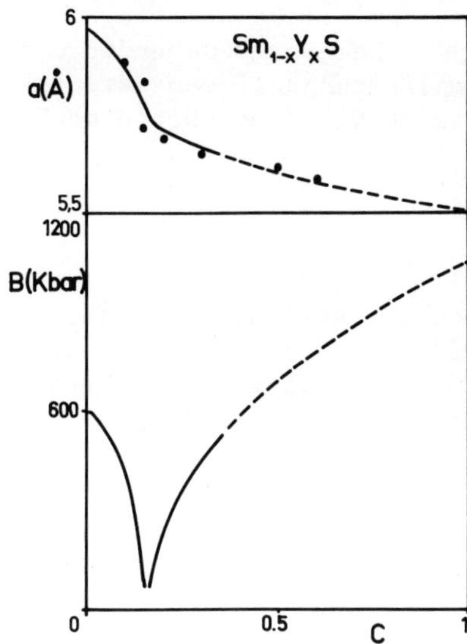

Fig. 9 : The lattice constant and bulk modulus vs. c for $Sm_{1-c}Y_c$ S.

carriers increases. This will slow down the transition and broaden the intermediate valence region. The structure of the conduction band is also important in this respect. If the band structure is such that the Fermi level moves up rapidly, this will help to stabilize the intermediate valence state. This point is in agreement with the fact that the valence in the metallic phase is smaller in $Sm_{1-c}Y_c$ S than in SmS /11,19/. In $Sm_{1-c}Y_c$ S the Fermi energy increases more rapidly due to the extra d-electrons from Y and to the larger d-band width in YS. In this sense, the intermediate valence is related to the position of Sm impurities f -'virtual bound state' relative to the Fermi level in YS.

Another reason for stability of intermediate phase may be the increase in lattice energy /39/. We have considered here a linear relation between the variation of volume and the average valence. We have not included any non-linear or anharmonic effect /39-42/. At present, it is not easy to discern the correct driving mechanism and probably both electronic and lattice contributions intervene to some extent. For instance, in the lattice contribution βn^2 we have estimated $\beta \simeq 1$ eV and a Coulomb term $G \simeq 1.5$ eV has been recently calculated in pure cerium /43/.

In this paper, the width of the f-level is almost considered in a phenomenological way. The magnitude of the mixing depends on many factors like the location of the f-level with respect to the band, the electron-phonon interaction /44/. Attempt has also been made to include an effective mixing due to the Coulomb interaction G /45/. The energy dependence of the width will evidently give a more complicated self-consistent equation for n_h but we expect that the physics will not change much.

REFERENCES

/ 1/ A. Iandelli, in Rare Earth Research, ed. by E.V. Kleber (Mc Millan, New York, 1961).
/ 2/ H.R. Ott and P.S. Wang, this conference.
/ 3/ B. Batlogg, J. Schoenes and P. Wachter, Phys. Lett. 49A, 13 (1974).
/ 4/ V.P. Zhuze et al, Sov. Phys. Solid State 6, 205, 213 (1964). R. Didchenko and F.P. Gortsema, J. Phys. Chem. Solids, 24, 863 (1963).
/ 5/ A. Jayaraman, V. Narayanamurti, E. Bucher and R.G. Maines, Phys. Rev. Lett. 25, 1430 (1970).
/ 6/ J.M.D. Coey, S.K. Ghatak and F. Holtzberg, A.I.P. Conf. Proc. 24, 38 (1975).
/ 7/ M. Campagna, G.K. Wertheim and E. Bucher, Structure and Bonding 30, 125 (1976).
/ 8/ A. Jayaraman, A.K. Singh, A. Chatterjee and Usha Devi, Phys. Rev. B9, 2513 (1974).
/ 9/ A. Chatterjee, A.K. Singh and A. Jayaraman, Phys. Rev. B6, 2285 (1972).
/10/ S.D. Bader, N.E. Phillips and D.B. McWhan, Phys. Rev. B7, 4686 (1973).
/11/ T. Penney and F. Holtzberg, Phys. Rev. Lett. 34, 322 (1975).
/12/ M. Avignon, S.K. Ghatak and J.M.D. Coey, J. of Magn. and Magn. Materials, 3, 88 (1976).
/13/ F. Holtzberg (unpublished).
/14/ A. Jayaraman, P.D. Dernier and L.D. Longinotti, High Temperatures - High Pressures 7, 1 (1975).
/15/ C.M. Varma, Rev. Modern Phys. 48, 219 (1976).
/16/ N.F. Mott, Phil. Mag. 30, 403 (1974).
/17/ H.L. Davis, Proc. 9th R.E. Conference p. 3, Virginia Polytechnic Institute, Blacksbury, Va. October 1971.
/18/ L.M. Falicov and J.C. Kimball, Phys. Rev. Lett. 22, 997 (1969). L.M. Falicov in New Developments in Semiconductors, ed. by P.R. Wallace, R. Harris and M.J. Zuckermann, Noordhoff. Int. Publ. Leyden (1971).
/19/ J.M.D. Coey, S.K. Ghatak, M. Avignon and F. Holtzberg, Phys. Rev. (in press).
/20/ M. Avignon and S.K. Ghatak, Conf. on Electronic Properties of Solids under High Pressure, Leuven, Sept. 1-5, 1975. Europhysics Conf. Abstract 1A, 46 (1975). S.K. Ghatak in Phonon Scattering in Solids, ed. by L.J. Challis, V.W. Rampton and A.F.G. Wyatt, Plenum Publishing Corp.N.Y.(1976)
/21/ J. Hubbard, Proc. Roy. Soc. A 277 , 237 (1964) ; A 285, 542

(1965).
/22/ J.F. Herbst, D.N. Lowy and R.E. Watson, Phys. Rev. B6, 1913 (1972).
/23/ M. Avignon and S.K. Ghatak, Solid State Comm. 16, 1243 (1975).
/24/ A.N. Kocharjan and D.I. Komskii, Sov. Phys. - Lebedev Inst. Reports N 8, 3 (1974).
/25/ C.E.T. Conçalves da Silva and L.M. Falicov, Solid State Comm. 17, 1521 (1975).
/26/ J.R. Iglesias Sicardi et al, Solid State Comm. 16, 499 (1975).
/27/ B. Coqblin and A. Blandin, Adv. Phys. 17, 281 (1967).
/28/ D.N. Zubarev, Sov. Phys. Uspekhi 3, 320 (1960).
/29/ B. Alascio, A. Lopez and C.F.E. Olmedo, J. Phys. F : Metal Physics 3, 1324 (1973).
/30/ P. Nozières, Theory of interacting Fermi systems, p. 89 (W.A. Benjamin Inc. 1964).
/31/ M. Avignon (to be published).
/32/ M. Abramowitz and I.A. Stegun, eds Handbook of Mathematical functions (Natl. Bur. Std. Washington 1964).
/33/ B. Batlogg, E. Kaldis, A. Schlegel and P. Wachter, Phys. Rev. (to be published).
/34/ Yu. M. Goryachev and T.G. Kutsenok, High Temp. High Pressures 4, 663 (1972).
/35/ R.A. Pollak et al, Phys. Rev. Lett. 33, 820 (1974).
/36/ E. Kaldis and P. Wachter, Solid State Comm. 11, 907 (1972).
/37/ F.J. Birch, J. Geophys. Res. 57, 227 (1952).
/38/ T. Penney et al, A.I.P. Conf. Proc. 29, 392 (1975).
/39/ T. Penney and R.L. Melcher, Proc. of the International Conf. on Metal-Nonmetal Transitions, Autrans June 28 - July 1, 1976 J. de Physique 37, Suppl. C4 (1976) ; T. Penney (this conference).
/40/ P.W. Anderson and S.T. Chui, Phys. Rev. B9, 3229 (1974).
/41/ L.L. Hirst, J. Phys. Chem. Solids, 35, 1285 (1974).
/42/ C.M. Varma and V. Heine, Phys. Rev. B11, 4763 (1975).
/43/ B.N. Cox and P. Lloyd, J. Phys. F : Metal Phys. 6, 797 (1976).
/44/ D. Sherrington and S. von Molnar, Solid State Comm. 16, 1347 (1975).
/45/ D.I. Khomskii and A.N. Kocharjan, Solid State Comm. 18, 985 (1976).

COMMENTS

KAPLAN: Sherrington and von Molnar find polaron narrowing in the density of states of the order of 10 to 100. Woudn't this effect make your density of states too large?

AVIGNON: In the model I described the f- level width is a free parameter of the theory, we may consider that polaron narrowing is already included.

KAPLAN: My point is that you can start with a certain f- level width and the narrowing will occur as a result of polaron effects.

AVIGNON: That's correct and this may explain why the f-level width δ is so small. It is generally assumed to be $\sim 10^{-2}$eV. In the case of pure Ce, Hill and Kmetko [J. Phys. F. 5, 1119 (1975)] proposed that sd-f hybridization is at least one order of magnitude greater giving $\delta \sim 1$eV. Then polaron narrowing would make it $\sim 10^{-2}$eV. However, in the metallic phase there effects will be reduced because of screening.

HIRST: You showed an increase in the number of electrons with the amount of Sm^{3+} as a function of pressure, this increase being strong below the transition. This is, I presume, just a function of how much hybridization you assume. You can change the slope (in the pretransition region) by changing the mixing parameter without having intermediate valence after the transition or without having a first order transition.

AVIGNON: Yes, it can be as shown in Fig. 4. However, the behavior in the transition region is not dependent solely on the mixing parameter.

PENNEY: Do you go at some pressure completely to Sm^{3+}?

AVIGNON: Yes. In this model you go completely to the Sm^{3+} state at sufficiently high pressure. It is difficult to quote the exact number since it depends on the parametrization of the gap with respect to the pressure. If you believe it is always changing at the rate of 10 meV/kbar, which is difficult to believe in the metallic phase, you get something like $2 \sim 5$ times the critical pressure.

UNKNOWN: Do you believe for the transition to take place, e.g., in SmS, there should be a jump in the number of carriers before the localized level touches the bottom of the band?

AVIGNON: From this model I get that the transition occurs before the gap is completely closed. Recent measurements by Wachter have shown the gap to be 0.015 eV. In this case I think the transition occurs before the gap is closed.

UNKNOWN: Matsubara and coworkers have shown theoretically that it is possible to have this kind of discontinuous change in n before the localized level touches the bottom of the conduction band.

AVIGNON: Matsubara and coworkers have adopted essentially the same model as us but they consider the energy dependence of the f-level width. Then in addition to the renormalization effects on the f-level you have to take into account the mixing effect on the conducting electrons particularly around the resonance.

BIRGENEAU: I wonder if you have actually tried to do a real wave function calculation of, say, hybridization as a function of lattice constant. Part of what we seem to be missing here is the chemical bond sort of approach to this problem which obviously in the end will have to be folded in. One knows what a Sm 4f-electron looks like and there is some idea of what the 5d-electron looks like on a neighboring site. So I would have thought a molecular orbital type of calculation could be done to estimate the various parameters.

AVIGNON: No, we haven't tried.

LINEAR RESPONSE OF INTERMEDIATE VALENCE COMPOUNDS: MÖSSBAUER EFFECT AND PHOTOELECTRON EMISSION

B. Alascio

Comisión Nacional de Energía Atómica-Centro Atómico Bariloche

8400 San Carlos de Bariloche-Argentina

We wish to discuss here two examples of the application of linear response theory (L.R.) to intermediate valence compounds: Mössbauer Effect (M.E.) and photoelectron emission (X.P.S.). The reason for this choice of examples is that while the Mössbauer spectra of intermediate valence compounds show isomer shifts that fit an average of the values corresponding to the two possible valence states, the photoemission spectra yield a weighted superposition of the spectra of each of the valence states involved.

We start in Section 1 by defining the model Hamiltonian for intermediate valence. This section includes also a short discussion of the statistical properties in terms of Green's functions. In Section 2 we calculate the response function for Mössbauer effect. Section 3 resumes the results of the response function for X.P.S. Finally, in Section 4 we briefly comment on our results and compare them with existing calculations.

1. THE HAMILTONIAN

The Hamiltonian we shall use to describe rare earth intermediate valence compounds includes two types of states:

a) Highly correlated "localized" states representing the 4f shells of the rare earths (r.e.).

b) Extended, single electron states representing the itinerant states in the compounds.

Most experimental data have been obtained in concentrated systems. However, the interaction between different sites is usually small and does not affect essentially the properties we wish to discuss here. In order to avoid the complications arising from the interactions between different r.e. ions and since we are interested here in the L.R. to external fields we will restrict ourselves to the treatment of a single r.e. atom immersed in a crystalline solid. For that reason we shall take a model Hamiltonian describing a single r.e. impurity connected to an infinite set of extended states. If, furthermore, we neglect Coulomb interactions between localized and band electrons, the Hamiltonian H_e can be separated into three parts: H_L and H_B describing the 4f and itinerant electrons respectively, and H_{hyb} mixing these states.

The localized states will be represented by the symbol $|n,M>$, where n indicates the number of electrons in the localized shell and M corresponds to the different eigenstates of these electrons including all interactions between them.

This type of description which is suitable for highly correlated states has been developed by several authors (1),(2),(3),(4).

In terms of these states the Hamiltonian describing the 4f electrons is then:

$$H_L = \sum_{nM} E_{n,M} \, |n,N><n,N| \tag{1}$$

where $E_{n,M}$ are the energies of the localized states. Each set of states corresponding to a given number of electrons (n) is well separated from the others by energies of the order of the intraatomic Coulomb repulsion, while within each set the energy separation is determined by LS coupling.

The Hamiltonian for the band states is described in terms of creation and annihilation operators $C^+_{k\sigma}$, $C_{k\sigma}$, and the energies of these states are ε_k, so that

$$H_B = \sum_{k\sigma} \varepsilon_k \, C^+_{k\sigma} C_{k\sigma} \tag{2}$$

Finally the hybridization Hamiltonian is such that it makes possible the transfer of an electron from a localized to an extended state

$$H_{hyb} = \sum_{\substack{n,N,M,\\ k\sigma}} V^{(n)}_{NM}(k\sigma) \, |n,N><n+1,M| \, C^+_{k\sigma} + \text{h.c.} \tag{3}$$

The total electronic Hamiltonian is then

$$H_e = H_L + H_B + H_{hyb} \tag{4}$$

We will now define "creation" and "annihilation" operators for the localized states by

$$F_{NM}^{(n)+} = |n+1, M><n,N| \quad ; \quad F_{NM}^{(n)} = |n,N><n+1, M| \tag{5,6}$$

Note that these operators change the number of electrons by one, and thus they anticommute with $C_{k\sigma}$ and $C_{k\sigma}^{+}$, while

$$\left| F_{NM}^{(n)}, F_{N'M'}^{(n)+} \right|_{+} = \delta_{nn'}\delta_{NN'}\delta_{MM'}(|n,N><n,N|+|n+1,M><n+1,M|) \tag{7}$$

Using these operators we can define Green's functions (in Zubarev's notation (5)), which will allow us to calculate the statistical mechanics of the system. We will study particularly:

$$G_{NM}^{(n)}(f) = <<F_{NM}^{(n)}(t); \; F_{NM}^{(n)+}(0)>> \tag{8}$$

The equation of motion for the Fourier transform of $G_{NM}^{(n)}(t)$ is obtained from Eq. 4 and reads:

$$\begin{aligned}
\omega G_{NM}^{(n)}(\omega) = {} & \frac{Pn,N+Pm,M}{2\Pi} + (E_{n+1,M}-E_{n,N})\; G_{NM}^{(n)}(\omega) + \\
& + \sum_{k\sigma N'} V_{N'M}^{(n)}(k\sigma)<<|n,N><n,N'|C_{k\sigma} \; ; F_{NM}^{(n)+}>> \\
& + \sum_{k\sigma M'} V_{NM'}^{(n)}(k\sigma)<<|n+1,M'><n+1,M|C_{k\sigma}; \; F_{NM}^{(n)+}>> \\
& - \sum_{k\sigma M'} V_{MM'}^{(n-1)*}(k\sigma)<<C_{k\sigma}^{+}|n,N><n+2,M'| \; ; F_{NM}^{(n)+}>> \\
& - \sum_{k\sigma N'} V_{N'M}^{(n-1)*}(k\sigma)<<C_{k\sigma}^{+}|n-1, N'><n+1,M|; \; F_{NM}^{(n)+}>>
\end{aligned} \tag{9}$$

where $P_{n,N}=<|n,N><n,N|>$ is the thermal average of the projection operator into the $|n,N>$ state.

To simplify the algebra as much as possible we will now decouple the Green's functions appearing in Eq. (9) in a way similar to that proposed by Hewson (6):

$$<<|n,N><n,N'|C_{\vec{k}\sigma} \; ; \; F_{NM}^{(n)+}>> \simeq \delta_{NN'} P_{n,N} <<C_{\vec{k}\sigma} \; ; \; F_{NM}^{(n)+}>> \quad (10)$$

$$<<|n+1,M'><n+1,M| C_{\vec{k}\sigma} \; ; \; F_{NM}^{(n)+}>> \simeq \delta_{MM'} P_{n+1,M} <<C_{\vec{k}\sigma} \; ; \; F_{NM}^{(n)+}>> \quad (11)$$

The other two Green's functions appearing in Eq. (9) are taken equal to zero.

The equation of motion for $<<C_{\vec{k}\sigma} \; ; \; F_{NN}^{(n)+}>>$ is approximated by:

$$(\omega-\varepsilon_{\vec{k}\sigma}) <<C_{\vec{k}\sigma} \; ; \; F_{NM}^{(n)+}>> = V_{NM}^{(n)*}(k\sigma) \; G_{NM}^{(n)}(\omega) \quad (12)$$

This decoupling procedure has already been used in connection with intermediate valence (7), (8), (9).

Using now Eqs. (9) to (12) it is possible to find $G_{NM}^{(n)}(\omega)$. The expression we obtain is:

$$G_{NM}^{(n)}(\omega) = \frac{1}{2\Pi} \; \frac{P_{nM} + P_{n+1,M}}{\omega-(E_{n+1,M}-E_{n,M})-(P_{n,N}+P_{n+1,M}) \; \Delta_{NM}^{(n)}(\omega)} \quad (13)$$

where

$$\Delta_{NM}^{(n)}(\omega) = \sum_{\vec{k}\sigma} \frac{|V_{NM}^{(n)}(k\sigma)|^2}{\omega-\varepsilon_{\vec{k}\sigma}} \quad (14)$$

More elaborate treatments have been devised to deal with the Friedel-Anderson model in the approximation of large correlation. That given by Hewson and Zuckermann (10) can also be adapted to the present problem, the only difference being that the factor $(P_{n,N}+P_{n+1,M})$ in the denominator of $G_{NM}^{(n)}(\omega)$ is modified. In what follows we will ignore it for the sake of simplicity

$$G_{NM}^{(n)}(\omega) = \frac{1}{2\Pi} \frac{P_{n,N} + P_{n+1,M}}{\omega-(E_{n+1,M}-E_{n,N}) - \Delta_{NM}^{(n)}(\omega)} \tag{15}$$

using now Zubarev's formulae we can calculate the thermal mean value of the product $F_{NM}^{(n)+} F_{NM}^{(n)}$. Since this product is a projection operator on the $|n+1,M>$ state, Eq. 15 allows us to calculate

$$P_{n+1,M} = (P_{n,N}+P_{n+1,N}) \int_{-\infty}^{+\infty} d\omega f(\omega) \frac{1}{\Pi} \frac{\Delta_{NM}^{(n)}(\omega)}{|\omega-(E_{n+1,M}-E_{n,N})|^2+|\Delta_{NM}^{(n)}(\omega)|^2} \tag{16}$$

where $f(\omega)$ is the Fermi function, (we take the Fermi energy = 0).

Following the arguments given in (11) it can be seen that if the quantities $\Delta_{NM}^{(n)}(\omega)+T,(k_B = 1)$ are small compared to the intraionic splittings, all P_nM are very small except at most two, corresponding to the ground state configurations of two consecutive values of n, and from Eq. (16) this situation is found when

$$(\Delta_{NM}^{n}(\omega) + T) \simeq |E_{n+1,M} - E_{n,N}| \tag{17}$$

N and M belong to the ground state configurations.

We now give an example calculation of the probability of occupancy for each valence state under the following simplifying assumptions:

a) All $\Delta_{NM}^{(n)}(\omega)$ are equal and independent of ω.(Δ).

b) Both ground states n^{th} and $(n+1)^{th}$ configurations correspond to eigenstates of the total angular momentum of magnitude J and J'. We can now write:

$$\Phi = \frac{1}{2} - \frac{1}{\Pi} \mathrm{Im}\ \Psi\left(\frac{1}{2} + \frac{\Delta}{2\Pi T} + i\ \frac{E_{n+1,M} - E_{n,N}}{2\Pi T}\right) \tag{18}$$

where Ψ is the Digamma function (12) and with this definition, we have:

$$P_{n+1,M} = (P_{n+1,M} + P_{n,N})\Phi \qquad (19)$$

In the absence of magnetic field $E_{n+1,M}$ is independent of M and $E_{n,N}$ is independent of N. We will denote them by E_{n+1} and E_n. As a consequence $P_{n,N} = P_n$ and $P_{n+1,M} = P_{n+1}$. These are the probabilities of finding the ion in a given state of each configuration. P(n), the probability of finding the ion in any state corresponding to n electrons is then $P(n) = (2J+1)P_n$; similarly $P(n+1) = (2J'+1)\,P_{n+1}$.

Since the system under consideration has to be in either of these states we also have:

$$(2J+1)\,P_n + (2J'+1)\,P_{n+1} = P(n) + P(n+1) = 1 \qquad (20)$$

using now Eqs. (19) and (20) we find:

$$P(n+1) = \frac{(2J'+1)\Phi}{(2J+1)-(2J-2J')\Phi} \qquad (21)$$

It is worth noticing that qualitatively, Φ can be approximated by a modified Fermi function (11)

$$\Phi(E_{n+1,M}-E_{n,N}, \Delta, T) = \left[\exp\left(\frac{E_{n+1,M}-E_{n,N}}{T+\Delta'}\right) + 1\right]^{-1} \qquad (22)$$

(Δ' is a parameter of the order of Δ).

Using Eq. (20) in (21) we find:

$$P(n+1) = \left[\frac{2J+1}{2J'+1}\exp\left(\frac{E_{n+1,M} - E_{n,N}}{T + \Delta}\right) + 1\right]^{-1} \qquad (23)$$

Similarly we can use $G^{(n)}_{NM}$ to calculate $P_{n,N}$ in the presence of an external magnetic field, and from them the susceptibility. When the approximation of Eq. (22) is used for Φ in the expression for the magnetic susceptibilities we obtain the formulae used by Sales and Wohlleben (17) to fit their measurements in Yb compounds.

Analizing Eq. (21) it follows from this treatment of Hamiltonian H_e that mixing between localized and band states produce the non-integer occupancy of the n electron states. Saturation of the magnetic susceptibility at low temperatures also follows in a natural way.

2. MÖSSBAUER EFFECT RESPONSE FUNCTION

Mössbauer experimental results on compounds containing r.e. atoms with non integral occupation numbers are interpreted according to the Wickman , Klein and Shirley (13) (Wi.K.S.) equations, since they provide a satisfactory description of the experimentally determined isomer shifts, line widths, and their temperature dependence. (14), (15), (16).

To study the R.F. for the system described by the Hamiltonian H_e we must consider its coupling to the external perturbation. In the case of Mössbauer effect, it is necessary to include the nuclear levels in our Hamiltonian and the states corresponding to the localized levels will contain a new subscript ν, indicating the excitation state of the nucleus. Thus $|n,N,\nu>$ indicates the Nth state of the n-electron configurations in the νth nuclear state.

Since, as shown in Section 2, only two valence states can coexist, we will reduce our Hamiltonian by including only two values of n. To simplify the algebra further, we will consider non degenerate configurations for each valence, so that we drop the indexes N and σ from H_e, and will assume that we can have only two states for the nucleus (ν = 0,1).

With these assumptions, H_e is reduced to:

$$H_e = E_n \sum_\nu |n,\nu><n,\nu| + E_{n+1}|n+1,\nu><n+1,\nu| + \sum_k \varepsilon_k C_k^+ C_k$$
$$+ \sum_{k\nu} \{V\, |n,\nu><n+1,\nu| C_k^+ + V^* C_k |n+1,\nu><n,\nu| \qquad (24)$$

The populations of the n and n+1 electron states are then obtained from Eq. (19) and (20) taking J = J' = 0.

To have the complete Hamiltonian we must add to Eq. (24) the energies corresponding to the nuclear levels, taking into account the fact that, because of the isomer shift, they will be in general different for the nth and $(n+1)^{th}$ electron states. We will denote the nuclear excitation energies by ω_n and ω_{n+1}. Thus we add to Eq. (24) the nuclear-electronic part

$$H_{ne} = \omega_n |n,1><n,1| + \omega_{n+1}| n+1,1><n+1,1| \qquad (25)$$

the total Hamiltonian being

$$H = H_e + H_{ne} \tag{26}$$

We must now consider the interaction of the system with the external perturbation. If we represent the electromgnetic field associated to the γ rays by a scalar ϕ, the interaction Hamiltonian can be written:

$$\begin{aligned} H_{int} &= A\,\phi(|n,0\rangle\langle n,1| + |n+1,0\rangle\langle n+1,1|) \\ &\quad + A^*\,\phi^*\,(|n,1\rangle\langle n,0| + |n+1,1\rangle\langle n+1,0|) \\ &= A\,\phi a^+ + A^*\,\phi^*\,a \end{aligned} \tag{27}$$

Here A is a matrix element which we take constant, and ϕ, ϕ^* corres pond in quantum theory to creation and annihilation of a photon.

Hint, thus describes the absorption or emission of a photon with the excitation or decay of the nuclear levels into any of the valence states.

It can be shown, using the formulae given in ref. (5) that the L.R. is proportional to:

$$x(\omega) = \frac{1}{\Pi}\int_{-\infty}^{+\infty} d\omega' \quad \tanh\frac{\beta\omega'}{2} \quad \frac{\mathrm{Im}\left|\langle\langle a,a^+\rangle\rangle_{\omega'+i0}\right|}{\omega-\omega' + i0} \tag{28}$$

here $\langle\langle a; a^+\rangle\rangle_\omega$ is the Fourier transform of the retarded Green's function defined in such a way that for $t\to 0_+$ it reduces to the anticonmutation of a, a^+, β is the inverse temperature ($k_B = 1$) and Im $|\ |$ indicates the imaginary part of $|\ |$. The absorptive part of the R.F. is then:

$$x''(\omega) = -\tanh\frac{\beta\omega}{2}\,\mathrm{Im}\left|\langle\langle a,\ a^+\rangle\rangle_{\omega+i0}\right| \tag{29}$$

In order to compare with the Wi.K.S. equations it is convenient to separate the Green's function into two parts:

$$<<a,a^+>> = <<\sum_\nu |n,\nu><n,\nu|a;\ a^+>> + <<\sum_\nu |n+1,\nu><n+1,\nu|a;\ a^+>>$$
$$= G_n + G_{n+1} \qquad (30)$$

Using the equations of motion we obtain from $H = H_e + H_{ne}$

$$(\omega-\omega_n)\ G_n = I_n + V^2\sum_k\left[(\omega-\varepsilon_k+E-\omega_n)^{-1}+(\omega+\varepsilon_k-E-\omega_{n+1})^{-1}\right] G_n$$
$$-V^2\sum_{kk'}\left[(\omega-\varepsilon_k+E-\omega_n)^{-1}+(\omega+\varepsilon_{k'}-E-\omega_{n+1})^{-1}\right]<<(1-C^+_{k'}C_k)a,a^+>> \qquad (31)$$

and

$$(\omega-\omega_{n+1})G_{n+1} = I_{n+1}+V^2\sum_k\left[(\omega-\varepsilon_k+E-\omega_n)^{-1}+(\omega+\varepsilon_k-E-\omega_{n+1})^{-1}\right]G_{n+1}$$
$$-V^2\sum_{kk'}\left[(\omega-\varepsilon_k+E-\omega_n)^{-1}+(\omega+\varepsilon_{k'}-E-\omega_{n+1})^{-1}\right]<<C^+_{k'}C_k a,a^+>> \qquad (32)$$

where

$$I_n = \frac{1}{2\Pi}\ P_{(n)} + V\sum_k\left(\frac{-<\sum_\nu|n+1,\nu><n\nu|C_k>}{\omega-\varepsilon_k+E-\omega_n} + \frac{<C^+_k\sum_\nu|n,\nu><n+1,\nu>}{\omega+\varepsilon_k-\omega_{n+1}}\right) \qquad (33)$$

$$I_{n+1} = \frac{1}{2}\ P_{(n+1)} + V\sum_k\left(\frac{<\sum_\nu|n+1,\nu><n\nu|C_k>}{\omega-\varepsilon_k+E-\omega_n} - \frac{<C^+_k\sum_\nu|n,\nu><n+1,\nu>}{a+\varepsilon_k-E-\omega_{n+1}}\right) \qquad (34)$$

and $E = E_{n+1} - E_n$.

To obtain a closed system of equations the simplest decoupling procedure consists of neglecting correlations between band states and nuclear excitations, so we replace

$<<c^+_k\ c_{k'}\ a;\ a^+>>$ by $<c^+_k\ c_{k'}><<a;\ a^+>> \simeq \delta_{kk'}<n_k>(G_n + G_{n+1})$

Equations (31) and (32) reduce then to:

$$(\omega-\omega_n)\, G_n = I_n + \sum_n G_n - \sum_{n+1} G_{n+1} \tag{35}$$

$$(\omega-\sum_{n+1})\, G_{n+1} + I_{n+1} + \sum_{n+1} G_{n+1} - \sum_n G_n \tag{36}$$

where

$$\sum_n = V^2 \sum_k \langle n_k \rangle \left[(\omega-\varepsilon_k+E-\omega_n)^{-1} + (\omega+\varepsilon_k-E-\omega_{n+1})^{-1} \right] \tag{37}$$

$$\sum_{n+1} = V^2 \sum_k (1-\langle n_k \rangle) \left[(\omega-\varepsilon_k+E-\omega_n)^{-1} + (\omega+\varepsilon_k-E-\omega_{n+1})^{-1} \right] \tag{38}$$

At this point it is necessary to distinguish between metal or semiconducting hosts.

We will restrict ourselves to the first type and characterize them by defining the zero of energy as the Fermi level in a band of constant density of states ρ, which extends from $-\infty$ to $+\infty$. Corrections arising from a finite band extending from -W to +W in energy are of the order Γ/W, where Γ is the width of the resonant level $\Pi v^2 \rho$. For all cases of interest involving rare earths, Γ/W will be very small.

When studying the R.F. for ω near ω_n or ω_{n+1}, we can neglect $|\omega-\omega_{n+1}|$ or $|\omega-\omega_n|$ in (37) and (38). Then, we obtain

$$\sum_n = -2i\Gamma\left[1-f(E)\right] \quad ; \quad \sum_{n+1} = -2i\Gamma f(E) \tag{39, 40}$$

In this case Eqs. (31) and (32) are identical to those proposed by Wi.K.S. with

$$\frac{1}{\tau_n} = 2\Gamma f(E), \text{ and } \frac{1}{\tau_{1+n}} = 2\Gamma\left[1-f(E)\right] \tag{41, 42}$$

The behaviour of the R.F. is then the same as that obtained in Ref. (13). P_A in their equations is to be identified with P_n. Notice that the relation

$$P_{(n+1)} = \frac{\tau_n}{\tau_n + \tau_{n+1}}$$

is verified with the thermal population of the localized orbital calculated to zeroth order in V. This is a consequence of our approximation which involved calculating the population of the band states to zeroth order in (39). The approximation is satisfactory for temperatures larger than the "fluctuation frequency" Γ.

We have shown in this section, that the use of the Wi.K.S.formulae is correct for interpreting Mössbauer experimental results in metals. The parameters entering in theWi.K.S. equations are given in terms of the microscopic parameters of our Hamiltonian. Even though this indicates that the interpretation in terms of temporal fluctuations is correct, one must keep in mind that these fluctuations are originated by the mixing between a localized state whose energy falls within a continuum of energies corresponding to extended states.

3. PHOTOEMISSION SPECTRA

Following a procedure similar to that given in Section 2, we can calculate the R.F. for XPS. Assuming that the situation corresponds to coexistence of the two valence states $|n\rangle$ and $|n+1\rangle$ ($E_{n+1} - E_n \sim \Gamma + kT$), it is found that the emission resonances are at energies $\varepsilon_{n+1} = \omega + E_{n+1} - E_N$ and $\varepsilon_n = \omega + E_n - E_{n-1}$ where ω is the energy of the incident X-Rays. The effect of hybridization is to shift and broaden these peaks by energies of the order of Γ. Since Γ is below experimental resolution and is small compared with any of the energies involved, these results are not significant. The integrated intensities of the peaks are roughly given by P_{n+1} and P_n respectively. Here it is important to calculate P_n an P_{n+1} including hybridization as done in Section 1. Notice that, because of the assumption of intermediate valence between the $|n+1\rangle$ and $|n\rangle$ states, the peak at ε_{n+1} coincides, within an energy of order Γ with the threshold for ejection from the Fermi surface, while the peak at ε_n is separated from it by an energy of the order of the intraatomic Coulomb repulsion.

4. DISCUSSION

We have studied the L.R. for M.E. and X.P.S. In the case of M.E. we have shown that hybridization leads to the Wi.K.S. equations, and thus it has been possible to explain the merging of the lines corresponding to the two valences and the effects of "motional narrowing".

In the case of X.P.S. the superposition of the spectra results from the cohexistence of the two valences (P_n and P_{n+1} of the same order of magnitude).

Hybridization, consequently, can explain experimental results in both cases.

The inclusion of mixing in the Hamiltonian describing X.P.S. leads again to a pair of coupled equations with the structure of

Eqs. (35) and (36). However the widths and shifts are not given by the same expressions. For that reason it is necessary to treat M.E. and X.P.S. separately.

Both, for M.E. and X.P.S. our results can be compared with the model of Gonçalves da Silva and Falicov (18) by taking an infinitely narrow band in H_e. The approximate solutions obtained by the decouplings are smilar to the exact ones. The decouplings used in the L.R. Green's functions, are consistent with those used in Section 1 , (Hewson, Ref. 10). This procedure leads to a very reasonable description of intermediate valence, except perhaps in those properties related to magnetic behaviour.

The results presented here have been obtained in colaboration with T. Huber, C. Balseiro, M. Foglio and H. Wio. They will be published elsewhere "in extenso".

The author is grateful to C.E.T. Gonçalves da Silva for clarifying discussions. to the staff o the Instituto de Física of the University of Sao Paulo (UNICAMP) for its hospitality during the writing of this paper, and to the OAS for financial assistance.

REFERENCES

1) M.E. Foglio, Rev. Mexicana Fis. 17, 137 (1969)
2) J.H. Jefferson and K.W.H. Stevens, J. Phys. C 9, 2151 (1976)
3) T.Huber and B. Alascio (to be published)
4) M.E. Foglio (to be published)
5) D.N. Zubarev, Soviet Phys. (Uspekhi) 3, 320 (1960)
6) A.C. Hewson, Phys. Rev. 144 , 420 (1966)
7) B. Alascio, A. Lopez and C. Almedo, J. Phys. F 3, 1324 (1973)
8) C.M.Varma and Y. Yafet, Phys. Rev. B 13, 2950 (1976)
9) C.E.T. Gonçalves da Silva and L.M. Falicov, Solid State Comm. 17. 15221 (1975)
10) A.C. Hewson and M. Zuckermann, Phys. Lett. 20, 219 (1966)
11) B. Alascio, Solid State Comm. 16 717 (1975)
12) B. Coqblin and A. Blandin, Adv. Phys. 17, 281 (1968)
13) H.H. Wickman, M.P. Klein and D.A. Shirley, Phys. Rev. 152, 345 (1966)
14) E.R. Bauminger, D. Froindlich, I. Nowik, S. Ofer, I. Felner and I. Mayer, Phys. Rev. Lett. 30, 1053 (1973)
15) O. Berkooz, M. Malamud and S. Strikman, Solid State Comm. 6, 185 (1968)
16) J.M. Coey, S.K. Ghatak, M. Avignon and F. Hltzberg Phys. Rev. (to be published)
17) B. Sales and D. Wohlleben, Phys. Rev. Lett. 35, 1240 (1975)
18) C.E.T. Gonçalves da Silva and L.M. Falicov, Phys .Rev. B 13 3948 (1976)

COMMENTS

MAPLE: Have you made any fits to experimental data using the expression you derived for the susceptibility?

ALASCIO: No, not here. It's relatively easy to fit the susceptibility in the case of the Yb compounds. However, you have to take values for the excitation energies and widths which differ from the values used by Wohlleben and Sales. Consequently, the occupation probabilities vary much more slowly with temperature using our formula. However, there were no measurements of occupation or lattice parameter.

HIRST: I made a comment earlier concerning the business of identifying the fluctuation time with what you see in the Mössbauer spectra, and I think this is correct at high temperatures, but incorrect at low temperatures. Basically, what's happening at low temperatures is that the fluctuation rate is staying constant but the amplitude of the fluctuations is going to zero. This leads to the prediction that the line width should increase linearly with temperature; however, it may be too small to see.

ALASCIO: I haven't been able to derive a fluctuating equation for the charge occupation. There is a similar problem if you take a simple harmonic oscillator coupled linearly to an infinite set of harmonic oscillators. If you study the evolution of the amplitude of the first oscillator as a function of time, assuming that at the initial time the infinite set of states is at a certain temperature, you can show that this amplitude can be described by a Langevin equation, since the problem is that for an oscillator with damping and with a stochastic force acting on it. If one could derive a similar equation for the case of intermediate valence, this would give a more precise meaning to the word fluctuation in connection with this problem. In this case, also, the fluctuating force is proportional to the temperature and has a mean frequency which is independent of force.

HIRST: The approach I would propose is that one adopts the idea that one has a density of quasiparticle states which is overlapping the Fermi surface. Then, you can derive an expression valid at low temperatures for the fluctuation of charge onto the nuclear site simply from the Fermi function for these quasiparticles. The answer you get is that the fluctuation frequency is, roughly, kT (isomer shift)^2 (density of 4f states)^2, up to a numerical factor.

COEY: Why should the amplitude of the charge fluctuations go to zero as T goes to zero in a homogeneous mixed valence material?

HIRST: The problem is partly the problem of language. People in quantum chemistry use the word resonance and people use in various contexts the term zero point motion. But in all these problems the ground state is actually static. The fluctuation frequency characterizes not a motion which is actually occurring in the ground state, but rather a motion which occurs if you disturb the ground state and watch how it restores itself. In order to see such a disturbance, you have to either use finite temperature or you have to make a measurement at finite frequency. The second case would apply to XPS. There you're pumping in a lot of energy and then, of course, you see big fluctuations. At zero temperatures there should be no real spontaneous fluctuation in the charge density or anything else which you can really measure.

MOSSBAUER AND XPS STUDIES OF EUROPIUM AND SAMARIUM MIXED VALENT SYSTEMS

Israel Nowik*

Henry Krumb School of Mines
Columbia University
New York, New York 10027

INTRODUCTION

Among the various techniques employed in the study of the phenomenon of mixed valences, measurements of lattice dimension, conductivity, susceptibility, Mossbauer and X-ray photoelectron spectroscopy (XPS), only the last two are of microscopic nature with resolution in time and space. All other methods measure a bulk and time average of the phenomenon. The Mossbauer technique is capable of detecting dynamic phenomena if the change rate is not too fast ($< 10^{11}$ sec^{-1}). It turns out that charge fluctuations in metallic systems are faster than $10^{11} sec^{-1}$ so that only an average valency is observed. On the other hand, the XPS technique measures the phenomenon within a time of $\sim 10^{-17}$ sec, which turns out to be short compared to the inverse charge fluctuation rate. Thus the XPS spectra display the "instantanous" picture, two subspectra corresponding to the two different valences. Due to the spectroscopic nature of both the Mossbauer and XPS techniques, both can identify inequivalent sites in inhomogenous mixed valent systems. In the Mossbauer case, different sites have different hyperfine interactions (most convenient to study is the isomer shift, expressing the electronic charge density in the nucleus). In the XPS case, inequivalent sites display different binding energies of the various electrons, in particular, those in the valence band.

In the present paper, we review the experimental results and analysis of Mossbauer studies of Eu^{151} in $EuSi_2Cu_2$, $EuRh_2$, Eu_{1-x}

*On leave from the Hebrew University, Jerusalem.

$R_xA_{2-y}B_y$ (R=rare earth, A=Rh, Ir, Ni, B=Pt, Al), Eu^{153} in $Sm_{1-x}R_xS$, Sm^{152} in $Sm_{1-x}R_xS$, and XPS studies of $EuRh_{2-x}Pt_x$ and $Sm_{1-x}R_xS$. It is shown how the Mossbauer and XPS data can be analyzed to yield interconfiguration excitation energies, charge fluctuation rates and local environment and temperature dependence of the excitation energies. The XPS studies yield the local environment dependence of the electronic binding energies. The study of a dilute Mossbauer probe in a mixed valent system (Eu^{153} in $Sm_{1-x}R_xS$) yields additional information on the nature of the mixed valent state of the host ions.

MOSSBAUER SPECTRA OF MIXED VALENT SYSTEMS

We interpret observed Mossbauer spectra within the following model for the mixed valent state. A virtual bound 4f state of half width at half height Δ is localized at energy E_{exc} (interconfiguration excitation energy) above the Fermi energy (Fig.1). The probability to find an ion in the divalent state is given by[1]

$$P_2(T) = \int_{-\infty}^{\infty} \frac{\Delta/\pi}{\Delta^2+(E-E_{exc})^2} \times \frac{dE}{1+(Z_3/Z_2)\exp(E/kT)} \qquad (1)$$

where for europium Z_2=8 and $Z_3=1 + 3\exp(-480/T) + 5\exp(-1330/T) + 7\exp(-2600/T)$ and for samarium Z_2 is equal to Z_3 of europium and $Z_2(Sm)=6+8\exp(-1500/T)$. When the ion is in the divalent state

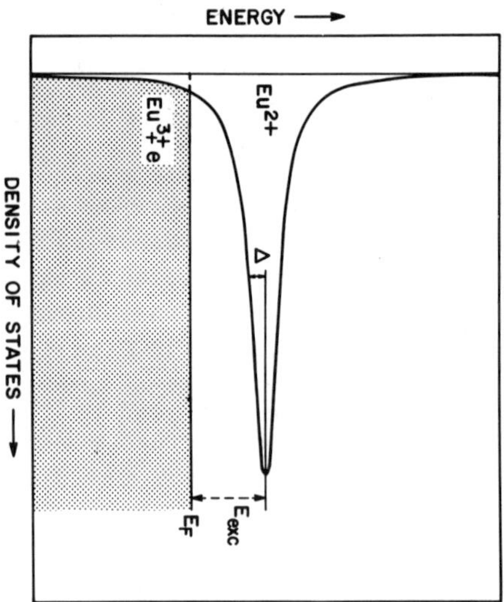

Figure 1. Schematic representation of the localized 4f level with width at half height 2Δ, at an energy E_{exc} above the Fermi level.

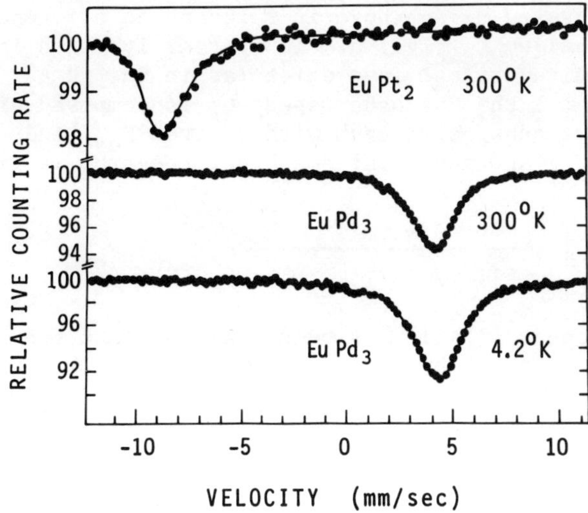

Figure 2. Mossbauer spectra of stable Eu^{2+} in $EuPt_2$ and Eu^{3+} in $EuPd_3$.

the electronic charge density in the nucleus leads to an observed isomer shift S_2, in the trivalent state the isomer shift is S_3 (Fig.2). Since in the mixed valent state the Mossbauer frequency fluctuates between S_2 and S_3 at a rate $1/\tau_f = 2\Delta/\hbar$, the observed Mossbauer spectrum is given by the well known Wickman formula[2]. In the case of $\tau_f(S_2-S_3) \gg 1$, two Mossbauer absorption lines of positions S_2 and S_3 and intensities p_2 and $1-p_2$ are observed. In the charge fluctuating metallic systems, it turns out that $\tau_f(S_2-S_3) \ll 1$ and only an average Mossbauer absorption line is observed. The position and half width of the absorption line are given by[2,3]

$$S(T) = (S_2-S_3)p_2(T) + S_3 \tag{2}$$

and

$$\Gamma_{exp} = \Gamma + \pi(S_2-S_3)^2 \tau_f \,(1-(2p_2-1)^2) \tag{3}$$

It is obvious that a Mossbauer isotope will be most sensitive to changes in τ_f if $\tau_f(S_2-S_3)^2/\Gamma$ is large. This makes the Sm^{152} a favourable isotope for samarium charge fluctuation studies. In the metallic mixed valent systems of $EuRh_2$[1], $EuCu_2Si_2$[3], collapsed SmS[4] and SmB_6[4,5] the limit $\tau_f\,(S_2-S_3) \ll 1$ has been realized. In the charge hopping case of Eu_3S_4[6] the limit $\tau_f\,(S_2-S_3) \gtrsim 1$ is also observed.

The discussion given above is relevent to homogenous systems which have a unique E_{exc} value. In systems in which inequivalent sites of the mixed valent ions exist (as in $Eu_{1-x}R_xA_{2-y}B_y$ and $Sm_{1-m}R_xS$ systems) the Mossbauer spectrum is composed of a superposition of various subspectra each with a given E_{exc} and relative probability $q(E_{exc})$ and the total Mossbauer spectrum is given by

$$I(\omega) = \int_{-\infty}^{\infty} \frac{\Gamma/\pi}{\Gamma^2 + (\omega - S(T,E_{exc}))^2} q(E_{exc})dE_{exc} \tag{4}$$

In some cases[7,8] it has been realized that the average excitation energy is given by; $\bar{E}_{exc} = E_o + E_1 n + E_2 x + E_3 T$, where n is the number of first nearest neighbours of Eu (or Sm) ions replaced by R ions, or A ions replaced by B ions. Since this formula for $\bar{E}_{exc}$ is only of approximate nature, we replace $q(E_{exc})$ by

$$q(E_{exc}) = \frac{\Delta_2/\pi}{\Delta_2^2 + (E_{exc} - \bar{E}_{exc})^2} Q_n^N(x) \tag{5}$$

where $Q_n^N(x)$ is a binomial expression[8] giving the relative probability of having n out of N first rearest neighbours of type R (or B). Δ_2 is an inhomogeneous extra width to E_{exc}.

We shall see that this description is not always applicable. In some systems, as in $EuNi_{2-x}Pt_x$, E_{exc} (n, x, T) is not at all a linear function of the variables (it turns out that also the binding energy measurable by XPS in $Sm_{1-x}R_xS$ is not a linear function in these variables). For any combination of (x, n) the Eu ion is either divalent or trivalent almost independent of temperature ($0 < T < 400K$). We shall now demonstrate various experimental results to which the above formulas can be applied and conclusions concerning $E_{exc}(T)$ and Δ can be derived.

MOSSBAUER AND XPS STUDIES OF $EuRh_{2-x}Pt_x$

This system has been studied by the Mossbauer effect[7,8] (fig.3) and by X ray photoelectron spectroscopy[1] (fig.4). The Mossbauer spectra show a sharp temperature dependence (fig.3), the measured isomer shift is displayed in fig.5. The XPS studies yield the absolute values for the populations in the two valence states. Thus the experimental value for $p_2(300K)$ in $EuRh_2$ together with the Mossbauer value for S(300K) yield through eq.2 the value S_3= 3.8 mm/sec. where S_2-S_3 is assumed to be as that in insulators, -13 mm/sec. Once S_3 is known the value of S(4.2K) = 1.2 mm/sec yields p_2(0 K) which according to eq.1 should equal $\frac{1}{\pi}\arctan(\Delta/E_{exc})$. At this stage, the temperature dependence of the isomer shift in $EuRh_2$ could be calculated as a function of the single free parameter

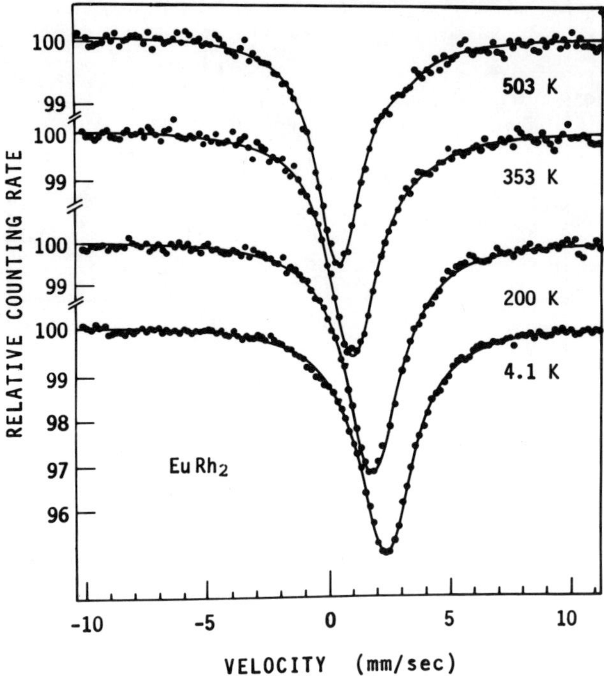

Figure 3. Mossbauer spectra of Eu^{151} in $EuRh_2$ at various temperatures.

(E_{exc}) and compared to the experimental observations. Very good agreement with the experimental results (fig.5) is obtained with E_{exc} = 1550 K, independent of temperature ($0 < T < 500$ K). This leads to a value of 2Δ = 1250K for the total width of the 4f level (.1eV). This is the first determination of the width of the virtual bound 4f level and the charge fluctuation time $\tau_f = .6 * 10^{-14}$sec.

THE TEMPERATURE DEPENDENCE OF THE ISOMER SHIFT IN $EuCu_2Si_2$

Mossbauer studies of $EuCu_2Si_2$[3] show a very strong temperature dependence of the Eu^{151} isomer shift (fig.6). Assuming S_2-S_3 = -13 mm/sec and S_2= -12.5 mm/sec (as that of the Eu^{2+} stable component in $EuCu_2Si_2$) one can try and analyze S(T) to yield E_{exc} and Δ. It turns out that if one assumes that E_{exc} is temperature independent eq.2 and eq.1 cannot reproduce the experimental temperature dependence of the isomer shift. Assuming a linear temperature dependence $E_{exc} = E_o + E_T T$, one can obtain with E_o= 1400K and E_T= - 1.76 the solid curve shown in fig.6. These also yield 2Δ = 1650 K and hence $\tau_f = .45*10^{-14}$ sec.

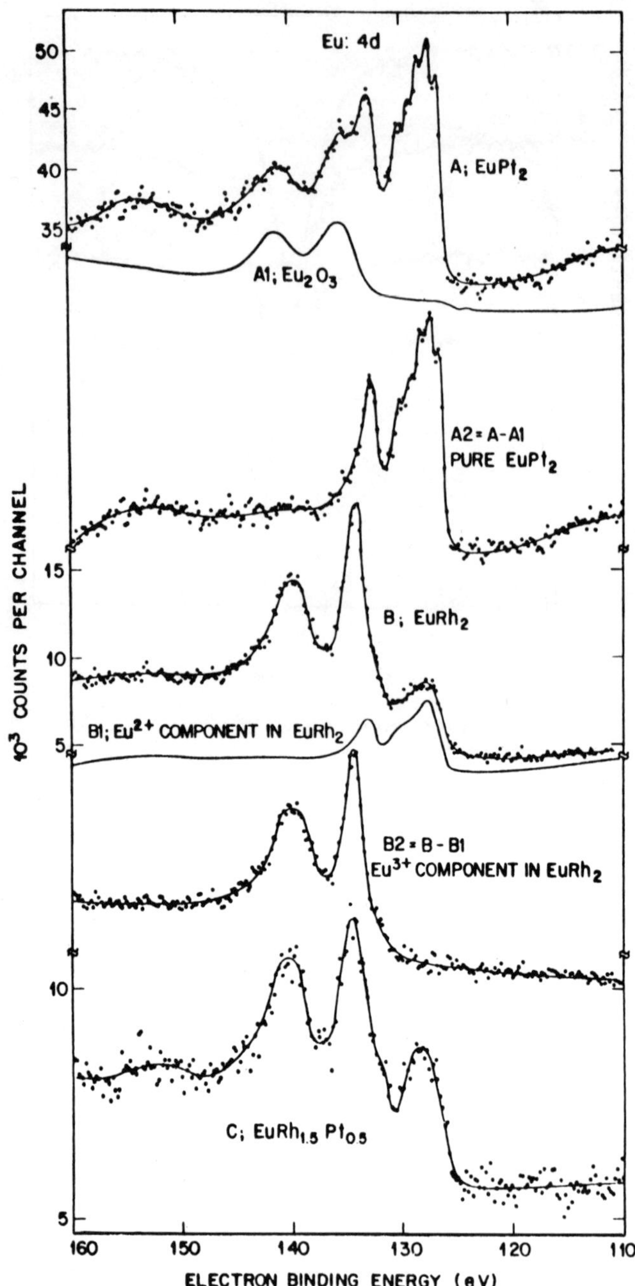

Figure 4. Eu 4d photoelectron spectra at 300K from $EuPt_2$(A), $EuRh_2$(B) $EuRh_{1.5}Pt_{.5}$(C) and Eu_2O_3(A1).

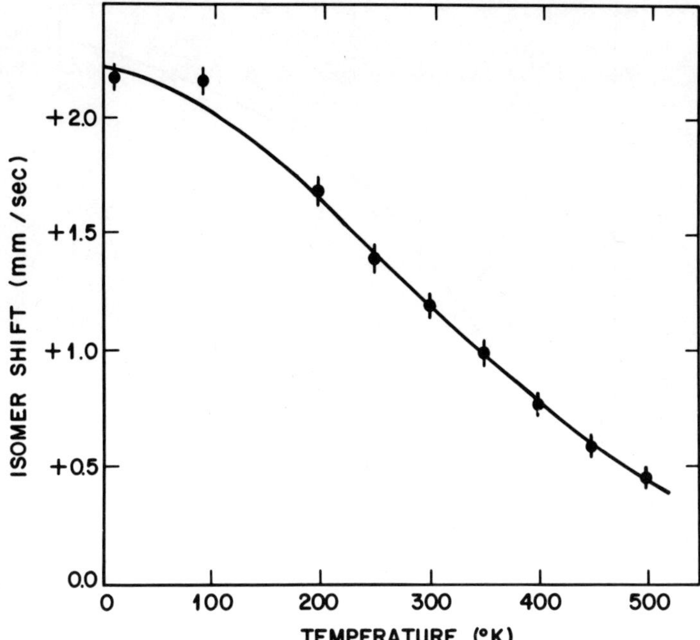

Figure 5. The temperature dependence of the isomer shift in $EuRh_2$. Solid curve is obtained from theory (eq.2) with $E_{exc} = 1550^oK$.

The observation that E_{exc} is strongly temperature dependent is consistent with recent Mossbauer studies of $EuCu_2Si_2$ under pressure[9]. It has been shown[9] that the observed changes in the isomer shift as a function of temperature are in great part due to the changes in volume. The sharp volume change is driven by the $Eu^{3+} \rightarrow Eu^{2+}$ transition caused by the decrease of E_{exc}. It is clear that all the parameters of the system are interconnected and also Δ must be temperature dependent. When the temperature is raised, not only electron relative populations are changed but also interatomic distances (due to ordinary expansion and due to the electron redistribution). The change in interatomic distances affects the interactions between localized and conduction electrons thus changing Δ. The change in interatomic distances also causes to a redistribution of electrons in the conduction band which in turn affect E_{exc}. Hence, we realize that the picture adopted to analyze the $EuCu_2Si_2$ Mossbauer observations is only a convenient simple phenomenological description of the much more complicated reality.

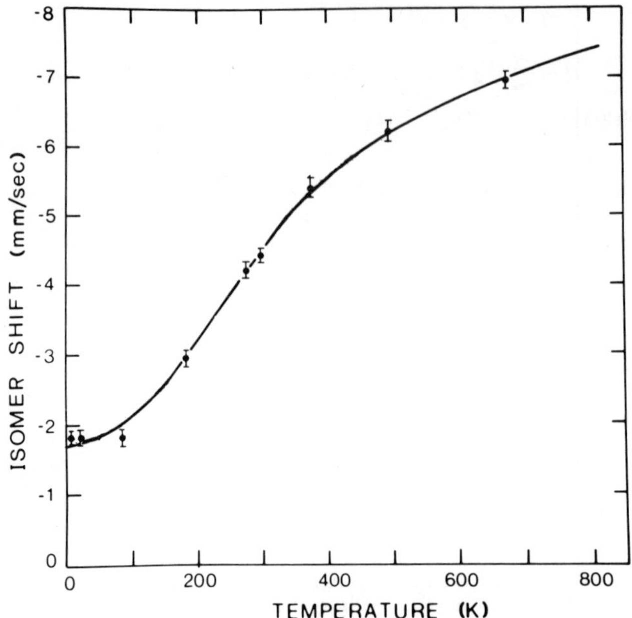

Figure 6. The temperature dependence of the isomer shift in $EuCu_2Si_2$. Solid curve is obtained from theory (eq.2) with $E_{exc} = (1400 - 1.76\ T)^oK$.

MOSSBAUER STUDIES OF $Eu_{1-x}R_xA_{2-y}B_y$ (7,8)

The experimental results of the $Eu_{1-x}La_xRh_2$ (7), $EuRh_{2-x}Pt_x$ and $EuIr_{2-x}Pt_x$ (8) studies were originally analyzed in terms of eq.5. In the original analysis (7,8), S_2-S_3 was assumed to be much less than the value we believe it should be (-13±1 mm/sec) and S_3 was chosen to be 2 mm/sec instead of the value of 4 mm/sec we obtain in pure $EuRh_2$ (1). For these reasons, one should treat the derived values for E_{exc} and Δ_2 reported in Ref.7,8 with some reservations. However, the experimental observations lead to some quite general conclusions:

a) In these systems the interconfiguration excitation energy depends strongly on local environment and in particular on the nature of the first nearest neighbors. b) The average excitation energy for a given first nearest neighbour shell composition (E_{exc} (T,n)) is temperature dependent. c) The analysis of the Mossbauer spectra can yield the detailed dependence of the excitation energies on composition (x,y), environment (n) and temperature. d) The necessity for a finite value of Δ_2 indicates that the

excitation energy depends also on second and farther neighbours.
e) The extreme limits of the observed isomer shifts indicate that Δ is not extremely small relative to E_{exc} and the ionic states are not pure 4f states.

Recent Mossbauer studies[10] of $EuNi_{2-x}Pt_x$ and $EuIr_{2-x}Al_x$ show spectra which are composed of narrow Eu^{2+} and Eu^{3+} absorption lines. The spectra are temperature independent ($T \leq 300K$). The ratio of the relative intensity of the two absorption lines changes with composition (x), however, it does not correspond to a subdivision according to the statistics of first nearest neighbours. These results tend to indicate that E_{exc} can be very sensitive to local environment and change drastically from $E_{exc} \gg kT$ to $-E_{exc} \gg kT$ as the local environment is changed due to changes in x. The small experimental value for the difference in isomer shift between the two observed lines ($\sim$ -10 mm/sec) indicates that though $|E_{exc}| \gg kT$ ($T \leq 300K$), Δ is not extremely small relative to E_{exc}, leading to nonpure Eu^{2+} and Eu^{3+} ionic configurations. This is consistent with conclusion e) for the other $Eu_{1-x}R_xA_{2-y}B_y$ systems.

MOSSBAUER STUDIES OF AN Eu^{153} PROBE IN $Sm_{1-x}R_xS$

The high sensitivity of the Eu isotopes' isomer shift to changes in conduction electron density tempted Cohen et al.[11] to study $Sm_{1-x}R_xS$ (R=Ca, Y, La, Tm,Gd) through the Mossbauer emission spectra of Eu^{153}. Neutron irradiation of $Sm_{1-x}R_xS$ leads to $Sm^{152}(n,\gamma)Sm^{153}(\beta^-)Eu^{153}$. In this procedure the Eu^{153} Mossbauer nucleus is extremely dilute in the host lattice. The observed dependence of the isomer shift on R and x is shown in fig.7. If we consider that the isomer shift is strongly volume dependent[12], it turns out that almost the whole change in observed isomer shift is due to lattice changes in the $Sm_{1-x}R_xS$ system and very little due to the extra electron released to the conduction band by the $Sm^{2+} \rightarrow Sm^{3+}$ transition. This result proves that this extra electron is of a predominant d character and possibly localized at the Sm site.

Sm^{152} MOSSBAUER STUDIES AND XPS STUDIES OF $Sm_{1-x}R_xS$ AND SmB_6

The mixed valent state of Sm in SmS under pressure and in $Sm_{1-x}R_xS$ has been most widely studied by a variety of techniques. Mossbauer studies utilizing the Sm^{149} isotope have been reported by Coey et al.[4] with the conclusion from the absorption line width that $\tau_f < 10^{-9}$ sec. XPS studies of the same systems yield well resolved Sm^{2+} and Sm^{3+} components indicating that $\tau_f \gg 10^{-17}$sec. Mossbauer studies of SmB_6[4,5] have shown that Sm in this compound is in the mixed valent state. This conclusion was confirmed by XPS studies[13]. The samarium charge fluctuating state can be studied also by the Sm^{152} isotope which has several advantages over the Sm^{149} isotope (easily prepared strong source and large isomer shift). Recent studies of Sm^{152} in SmS, $Sm_{.3}Pr_{.7}S$, SmB_6 and $SmBe_{13}$[14]

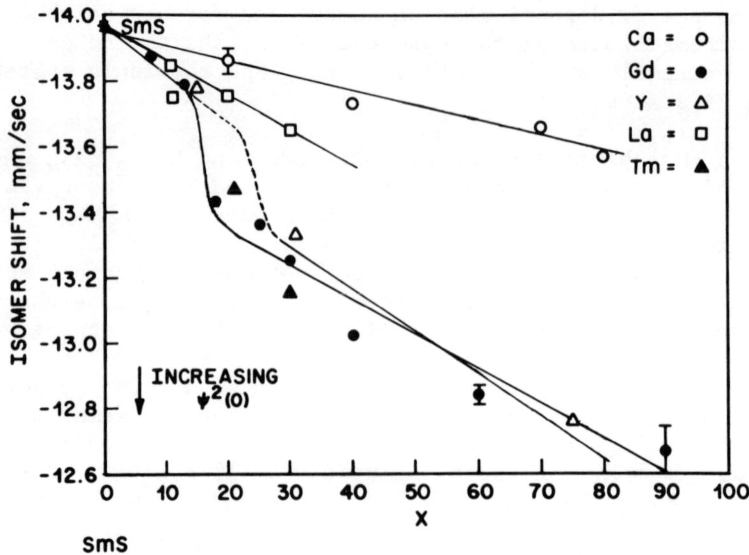

Figure 7. Isomer shift of Eu in $Sm_{1-x}R_xS$ as a function of x for R=Ca, Y, La, Tm and Gd. The observed discontinuity in the isomer shift in the transition region can be accounted for mainly by the large volume change.

(fig.8) lead to the following conclusions: a) Sm^{152} in comparison to Sm^{149} is more sensitive to charge fluctuations because it has a higher $\tau_f(S_2-S_3)^2/\Gamma$ value. b) The observed width and position of the absorption line in $Sm_{.3}Pr_{.7}S$ and SmB_6 prove that the mixed valent state even at 4.2K is a fluctuating charge state with $\tau_f<10^{-10}$ sec. In $SmBe_{13}$ at 12K (above T_c, fig.8) the line is wide due to Sm^{3+} paramagnetic spin relaxation phenomena. The absence of broadening in SmB_6 and $Sm_{.3}Pr_{.7}S$ proves that the Sm ions are by no means pure Sm^{3+} ions. c) The inequivalent sites existing in $Sm_{.3}Pr_{.7}S$ have very similar isomer shifts proving that unlike in the case of Eu in $Eu_{1-x}R_xA_{2-y}B_y$, the local environment has little effect on the Sm state. The Sm state is determined by long range cooperative phenomena.

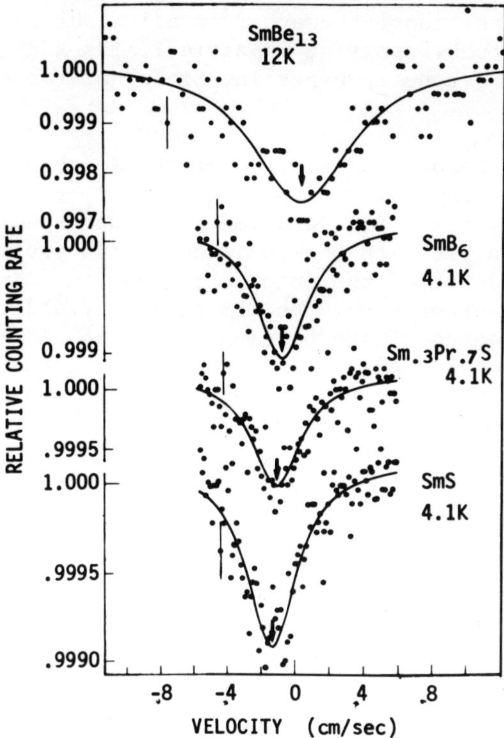

Figure 8. Sm^{152} Mossbauer spectra in SmS, $Sm_{.3}Pr_{.7}S$, SmB_6 and $SmBe_{13}$.

Recent XPS studies[15,16] of $Sm_{1-x}R_xS$ (R=Y, Ca, Gd, Th) in the x region before the crystal goes over to the collapsed phase, show spectra in the valent band region which indicate that the binding energy of the extra 4f electron depends strongly on local environment. In the collapsed phase region all sites seem equivalent in close correspondence to the Mossbauer observations[4,14]. It seems that in the metallic phase conduction electrons screen the local environment effects.

CONCLUSIONS

We can summarize our findings as follows: Eu isotopes are most suitable to investigate charge fluctuation phenomena by the Mossbauer effect due to the large value of $(S_2-S_3)/\Gamma$, and due to the wide range of temperatures one can study Eu^{151}. Sm, Tm and Yb which also show charge fluctuation phenomena and have Mossbauer isotopes are of limited excess due to small isomer shifts and/or high energy γ-transitions. However, in all cases one can study the mixed valence state by applying magnetic fields and observing the magnetic response (magnetic hyperline field) of the charge fluctuating state.

The study of homogenous systems by the Mossbauer effect and XPS are especially rewarding. Both E_{exc} and τ_f can be derived as shown by the $EuRh_2$ case[1]. Detailed Mossbauer studies of inhomogenous systems can yield the composition, local environment and temperature dependence of the interconfiguration excitation energies. On the other hand, the XPS studies yield the local environment dependence of the binding energies. Additional information about the mixed valent state can be obtained by studies of dilute Mossbauer probes in the mixed valent systems as demonstrated by the Eu^{153} in $Sm_{1-x}R_xS$ study[11].

REFERENCES

1. I. Nowik, M. Campagna and G.K. Wertheim, Phys.Rev. Letters (1976).
2. I. Nowik and H.H. Wickman, Phys. Rev. Letters 17, 1949 (1966).
3. E.R. Bauminger, D. Froindlich, I. Nowik, S. Ofer, I. Felner and I. Mayer, Phys. Rev. Letters 30, 1053 (1973).
4. J.M. Coey, S.K. Ghatak and F. Holtzberg, AIP Conference Proceedings No.24, 38 (1975), J.M. Coey, S.K. Ghatak, M. Avignon and F. Holtzberg, Phys. Rev. (1976).
5. R.L. Cohen, M. Eibschutz and K.W. West, Phys. Rev. Letters 24, 382 (1970).
6. O. Berkooz, M. Malamud and S. Shtrikman, Solid State Comm. 6, 185 (1968).
7. E.R. Bauminger, I. Felner, D. Levron, I. Nowik and S. Ofer, Phys. Rev. Letters 23, 890 (1974).
8. E.R. Bauminger, I. Felner, D. Froindlich, D. Levron, I. Nowik, S. Ofer and R. Yanovsky, J. de Physique 35, C6-61 (1974), E.R. Bauminger, I. Felner, D. Levron, I. Nowik and S. Ofer, Solid State Comm. 18, 1073 (1976).
9. J. Rohler, H. Balster and G. Kaindle, Proceedings of 1976 Mossbauer Conference, J. de Physique, Jan. 1977.
10. E.R. Bauminger, I. Felner, I. Nowik and S. Ofer (unpublished).
11. R.L. Cohen, I. Nowik, K.W. West and E. Bucher, presented to Phys. Rev. (1976).
12. G.M. Kalvius, O.F. Klein and G. Wortmann, J. de Physique 35, C6-139 (1974).
13. J.N. Chazalviel, M. Campagna, G.K. Wertheim and P.H. Schmidt Solid State Comm. 19, 725 (1976).
14. R. Yanovsky, I. Nowik and I. Felner (unpublished).
15. M. Campagna, S.T. Chui, G.K. Wertheim and E. Tosatti, Phys. Rev. B14, 653 (1976).
16. M. Campagna, I. Nowik, G.K. Wertheim and D.N.E. Longinotti (presented to Phys. Rev.).

COMMENTS

CAMPAGNA: If you extrapolate the Mössbauer peak to T = 0°K for $EuCu_2Si_2$, do you obtain 100% trivalency?

NOWIK: No. It's about 16-17% divalent even for T→0; for $EuRh_2$, it's about 12% divalent for T→0.

ORBACH: I have two questions: (1) do you still believe that at the dilute Eu end in $LaRh_2$ that the excitation energy changes sign and (2) do you have any evidence for 5d electrons in the trivalent Eu phase?

NOWIK: As for the first question, there is no doubt that dilute Eu in $LaRh_2$ is purely divalent. As for the second question, I don't know the answer.

COEY: Would you say that your decomposition of the spectra, e.g. in Eu-La alloys, in terms of nearest neighbor effects and fast fluctuations is unique? Isn't it possible that some of the line broadening is coming from slower fluctuations?

NOWIK: We don't believe it's strictly a nearest neighbor effect, and that is why the theory which we used to analyze these spectra is not as simple as the model I presented. We assumed that the average excitation energy depends on the number of first nearest neighbors, but this excitation energy has an inhomogeneous width because of contributions from second and third nearest neighbors. I assume that the fluctuations are very fast, since e.g., the experiments give, for $EuRh_2$, $2\Delta \sim 100$ meV.

EIBSCHUTZ: Are your data on $EuRh_2$ corrected for second order Doppler shifts or is this a small contribution?

NOWIK: The second order Doppler shift is less than approximately 3% of the shift.

MAGNETIZATION ALONG PRINCIPAL DIRECTIONS OF TmSe AT HIGH HYDROSTATIC PRESSURES

F.P. Missell,* † S.Foner* and R.P. Guertin∓ §

*Francis Bitter National Magnet Laboratory, ** Massachusetts Institute of Technology, Cambridge, Massachusetts 02139.
∓Physics Dept., Tufts University, Medford, Massachusetts 02155.

ABSTRACT

The hydrostatic pressure dependence of the magnetization, σ, of mixed-valence TmSe is reported. Measurements of σ(H,T,P) were made on single crystals with applied fields, H, along the [100], [110] and [111] directions for $1.4 \leq T \leq 20$ K, $0 \leq H \leq 60$ kG and $0 \leq P \leq 10$ kbar. The magnetic properties of TmSe are anisotropic at low temperatures. Above $T \sim 3$ K, S-shaped σ vs H transitions are observed for H ∥ [100] and [110], but not for H ∥ [111]. For $P > 0$, the fields, H_T, for the higher field transitions, increase rapidly as pressure is increased: at low P, $dH_T/dP \simeq +1.7, +2.0$ and $+2.1$ kG/kbar for H ∥ [100], [110], and [111], respectively. For the sample which shows a well-defined susceptibility anomaly at $T_0 \sim 3$ K, $dT_0/dP \simeq (+0.1 \pm 0.03)$ K/kbar at low P. The low-field susceptibility, χ_0, decreases rapidly with pressure [e.g., for H∥[100], $(1/\chi_0)(\Delta\chi_0/\Delta P) \simeq 0.034$/kbar for $P \simeq 8$ kbar] and approaches a constant value for $P \geqslant 7$ kbar. Two independent arguments, one based on the measured lattice constant plus the compressibility of TmSe, and one based on the σ(H,T,P) data, show that for $P \geqslant 20$ kbar TmSe should become trivalent. The high field-high pressure characteristics are found to be similar for two TmSe samples which have quite different H-T phase diagrams. The data are discussed in terms of the mixed-valence nature of TmSe and compared with results for more nearly trivalent Tm in TmS.

† On leave from and supported in part by Universidade de São Paulo, Brazil. Supported in part by Fundação de Amparo à Pesquisa do Estado de São Paulo.
** Supported by the National Science Foundation.
§ Supported by National Science Foundation Grant #DMR 75-09494. Visiting Scientist at the Francis Bitter National Magnet Laboratory, MIT.

INTRODUCTION

In a previous paper[1] we reported the hydrostatic pressure dependence of the multiboundary field-temperature phase diagram of a single crystal sample of TmSe. The determination of the phase diagram and its pressure dependence for an applied field H ∥ [100] was made by measurements of the magnetization, σ, as a function of field, temperature and pressure. Here we report more extensive observations of $\sigma(H,T,P)$ for H ∥ [100] and, for the first time, measurements with H parallel to the other two principal crystallographic directions, [110] and [111]. The $\sigma(H,T,P)$ data, obtained for $0 \leqslant H \leqslant 60$ kG, $1.4 \leqslant T \leqslant 20$ K, and $0 \leqslant P \leqslant 10$ kbar, show that the phase diagrams for TmSe are anisotropic. The results of these measurements will be discussed in terms of the mixed-valence characteristics of TmSe.[2]

PREVIOUS STUDIES OF TmSe

Evidence for the mixed-valence (Tm^{2+}-Tm^{3+}) of TmSe comes from several sources. The effective Curie-Weiss moment[3,1] lies between that expected for divalent Tm (4.54 μ_B) and trivalent Tm (7.56 μ_B). Lattice constants between 5.71 Å and 5.64 Å have been reported.[3-6] These values lie between those expected[3] for Tm^{2+}Se (5.96 Å) and Tm^{3+}Se (5.63 Å). Campagna et al[4] report evidence for the coexistence of Tm^{2+} and Tm^{3+} in TmSe based on x-ray photoemission spectroscopy (XPS). Triplett et al[7] report evidence for mixed-valence in TmSe based on Mössbauer measurements. However, Batlogg et al[6] find no evidence for valence mixing in TmSe from optical reflectivity studies and conclude that the Tm ion is trivalent in their sample. The references cited above[3-7] show that there is a large disparity of estimates of Tm^{2+}- Tm^{3+} concentration in TmSe.

Several of the above studies yield microscopic information related to the valence of a particular Tm site. In contrast, static magnetization measurements yield macroscopic data averaged over many atoms and for long times. A feature of TmSe, which so far appears to be unique among mixed-valence systems, is the presence of complex phase boundaries in the H-T plane at low temperatures in selected TmSe samples. These phase boundaries were first reported by Ott et al,[8] who used magnetostriction, thermal expansion and ac susceptibility techniques for H ∥ [100] at atmospheric pressure (P=0). Although some reported data[8,9] suggested long-range magnetic order at low temperature for TmSe, neutron scattering measurements[10] to date have not shown evidence for crystal-field levels or long-range magnetic order in TmSe.

EXPERIMENTAL DETAILS

Several single crystal samples of TmSe were investigated. Most of the data were taken on a single crystal sample (A) obtained from M. Campagna and was part of the same sample[11] used in the XPS measurements,[4] in the studies by Ott et al.,[8] and in our reported results.[1] The sample was oriented along [100], [110] or [111] and then placed in a small high-purity beryllium-copper clamp device capable of producing hydrostatic pressures to $\simeq$10 kbar. The entire clamp-plus-sample assembly was attached to a vibrating sample magnetometer adapted to a 60 kG high-homogeneity superconducting solenoid.[12] The clamp was surrounded by a small volume Dewar inserted into the detection coils. Measurements for $4.2 \leqslant T \leqslant 20$ K were made in He^4 exchange gas, and measurements for $1.4 \leqslant T \leqslant 4.2$ K were made with the clamp assembly vibrated in liquid He^4 condensed directly in the small Dewar. For data at $P=0$, taken without the pressure clamp, the crystal orientation was well within 2°. For $P>0$ the pressure clamp was employed and the crystallographic orientation with respect to $\vec{H}$ was somewhat less precise.

RESULTS

Selected magnetic properties of TmSe for $P=0$ at low temperatures and for H along the principal directions are illustrated in Fig. 1 for sample A2. (Numerals indicate different portions of A.) The low field susceptibility, χ_0, for $1.4 \leqslant T \leqslant 4.2$ K is shown in Fig. 1a for samples A1 and A2. (A1 was discussed in Ref. 1.) The temperature dependence of χ_0 demonstrates a clear anomaly at $T \simeq 3$ K. Above $T \simeq 3$ K, χ_0 is nearly isotropic and below 3 K there are variations which depend on the particular sample and its orientation. The temperature dependence of χ_0 shows a Curie-Weiss behavior for $30 \leqslant T \leqslant 280$ K and begins to deviate for $T < 30$ K. Least squares fits yield $\theta = -29.5 \pm 0.5$ K and an effective moment $\mu_{eff} = 6.14\ \mu_B$. An interesting feature at relatively low fields is illustrated in Fig. 1b where σ vs H is shown for H ‖ [100] for an initial increase of H after cooling the crystal in zero field to 1.4 K. A small step is observed on the first field-upsweep; for subsequent field sweeps the step is very difficult to detect.[13] To reproduce the step in σ vs H the sample must first be warmed up above $\simeq$ 3 K. This type of hysteresis is also observed along the [110] direction, but the step occurs at higher H. For H closely aligned along the [111] direction, no evidence for a step in σ vs H is detected. In Fig. 1c we show σ vs H at $T = 1.4$ K for higher H. A rapid rise in σ vs H is observed for all directions. The small step of σ in Fig. 1b is barely visible on this scale. As T is increased, the sharp rise

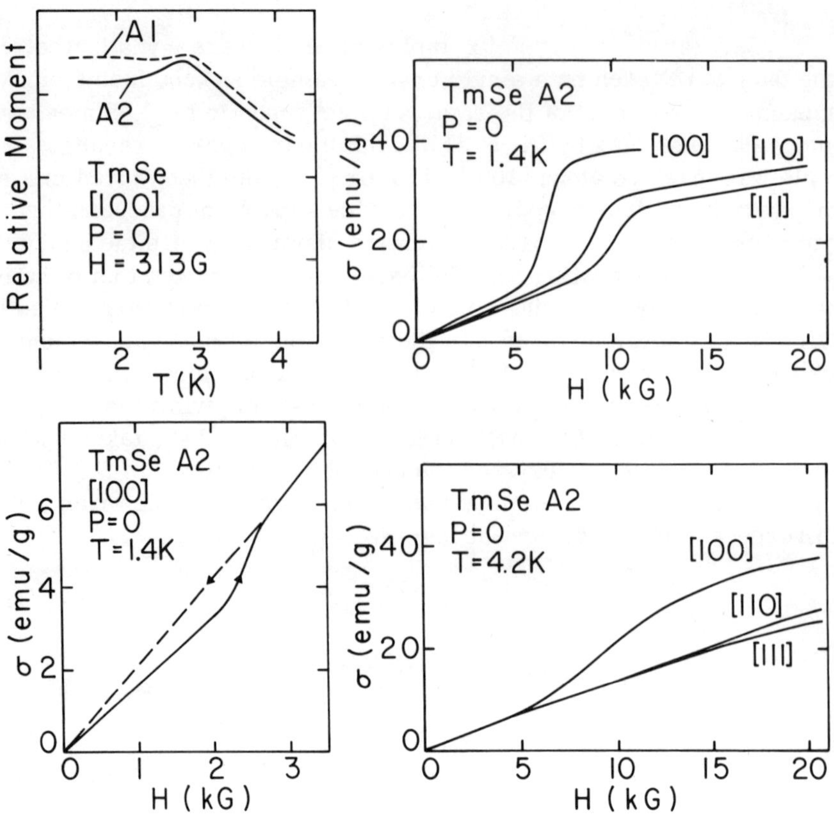

Fig. 1 Magnetic properties of TmSe at P=0. No demagnetization corrections are made for figures in the text. The demagnetization factors varied between $\sim 4\pi/2$ and $4\pi/4$ for the irregularly shaped samples.

(a) Relative moment vs temperature for two pieces of TmSe, Sample A. The anomaly at T $\simeq$ 3 K defines a point on boundary I of Fig. 2.

(b) Moment σ vs applied field H at T = 1.4 K for the first cycle of H after cooling the crystal in zero field. The step, which is difficult to detect for subsequent field sweeps, defines a point on boundary II of Fig. 2.

(c) Moment at 1.4 K vs field along the [100], [110] and [111] directions. The inflection point of σ vs H corresponds to a point on boundary III of Fig. 2.

(d) Moment vs field at 4.2 K for three principal directions. The inflection point of the broad S-shaped feature, which is present for H ∥ [100] and [110] (but not for H ∥ [111]), corresponds to a point on boundary IV.

becomes less pronounced, and its position moves to somewhat lower fields up to T ≃ 3 K. The characteristics of σ vs H at T = 4.2 K are illustrated in Fig. 1d. Broad S-shaped curves of σ vs H are seen for H ∥ [100] and [110], but the S-shaped feature is absent for H ∥ [111]. For H ∥ [100] the transition becomes broader at higher T; at T = 21 K, σ vs H is linear up to H ≃ 30 kG. The magnetic anomalies illustrated in Fig. 1 correspond to points on the phase boundaries of TmSe in the H-T plane; the anomalies in Figs. 1a, 1b, 1c and 1d correspond to the boundaries labelled I, II, III and IV, respectively, by Ott et al[8] for H ∥ [100]. We define the transition fields, H_T, as the fields for which dσ/dH is a maximum determined by numerical differentiation of the σ vs H data.

The phase diagrams obtained from analysis of the P = 0 data for various orientations are shown in Fig. 2. Our data for H ∥ [100] are in good agreement with those of Ref. 8. Although it is difficult to determine the order of the various transitions from the σ vs H data, independent magnetostriction measurements[14] combined with the data of Ref. 8 show that transitions III and II appear to be first-order and transition IV appears to be higher than first-order. An interesting feature of Fig. 2 is the absence of boundaries II and IV for H ∥ [111]. The transitions associated with boundary I are defined by the susceptibility anomalies obtained from data such as those in Fig. 1a.

The effect of pressure on transitions III is illustrated in Figs. 3a, 3b and 3c for the different crystallographic orientations. As the pressure is increased, the following features are observed: 1) the transition fields, H_T, increase at a rate dH_T/dP = +1.7 kG/kbar, +2.0 kG/kbar and +2.1 kG/kbar for H ∥ [100], [110] and [111], respectively; 2) the step in σ vs H decreases for all orientations; 3) the low field slope of σ vs H (the susceptibility) decreases and appears to approach a constant value; and 4) the high field ($H > H_T$) magnetization increases slightly with field, but

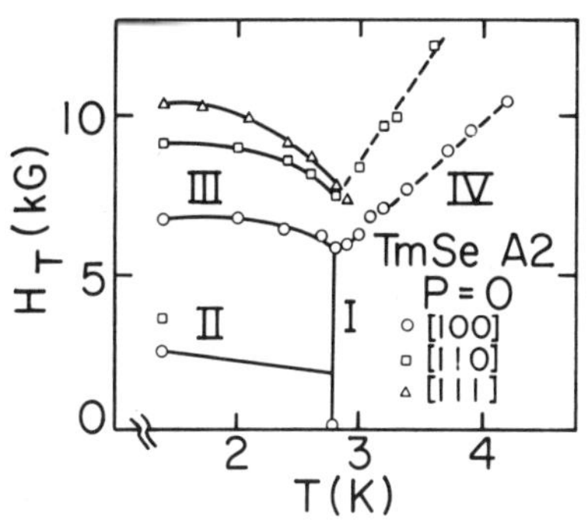

Fig. 2 Transition field, H_T, vs T for three principal directions. H_T is defined as the field at which dσ/dH is a maximum. Roman numerals I to IV label phase boundaries at P = 0 as in Ref. 8.

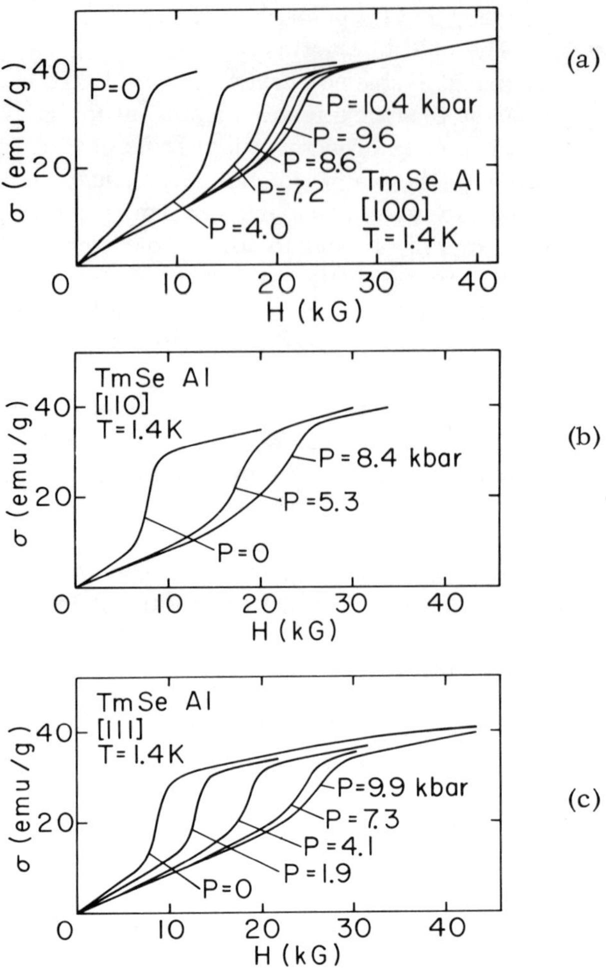

Fig. 3. σ vs H for TmSe at 1.4 K for several pressures.

the rate of increase depends on orientation. Extrapolation of the high pressure-low field slope of σ vs H to high fields yields an estimate of the field, H_{TC}, at which the step in σ would vanish. This procedure was used earlier for data along the [100] direction (see Ref. 1, Fig. 2). Although H_{TC} differs for the three orientations, the extrapolated pressure at which the step in σ disappears is about 20 kbar for all three orientations. We conclude that for $P \geqslant 20$ kbar, transitions III and IV would vanish, independent of the marked anisotropy of the magnetic characteristics of TmSe.

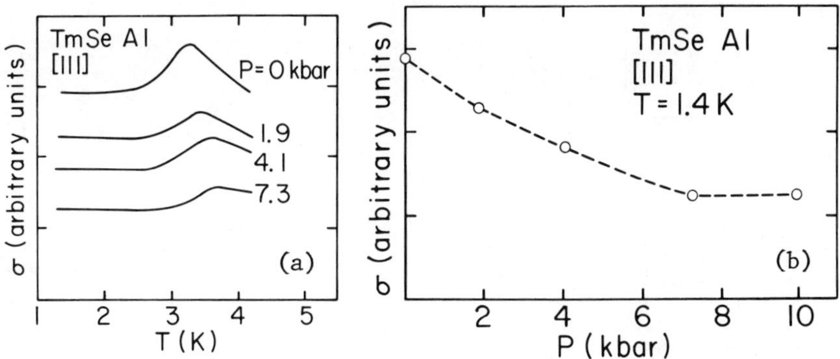

Fig. 4 Pressure-temperature dependence of moment of TmSe for H=3.86 kG parallel to [111].

(a) Temperature for maximum of σ vs T increases with increasing P.

(b) σ (H=313 G) vs P at 1.4 K becomes independent of P above P~ 7 kbar.

The pressure dependence of the low-field magnetization for H ∥ [111] is illustrated in Fig. 4a. For each orientation the maximum (which defines transitions I in Fig. 2) is less well defined at higher pressures. There appears to be an increase in the temperature of the maxima, T_0, with increasing pressure ($dT_0/dP \simeq 0.1$ K/kbar) in Fig. 4a. Although boundary II is not observed for H ∥ [111], for H nearly parallel to [111], $dH(II)_T/dP \simeq +0.6 \pm 0.2$ kG/kbar. It is interesting to note that the initial pressure dependence of the susceptibility at T=4.2 K is about an order of magnitude larger than previously observed for (trivalent) Pr- and Tm-monopnictides and Pr-monochalcogenides.[15] Fig. 4b illustrates the pressure dependence of χ_0 at T = 1.4 K for H ∥ [111]; χ_0 decreases rapidly at first and approaches a constant value for P ⩾ 7 kbar, similar to that for H ∥ [100] (see Ref. 1). In general, for a given TmSe sample at fixed field, the magnetic moment is decreased as pressure is increased.

Figs. 5a, b and c show three dimensional plots of the H-T-P phase diagrams for sample A1 for the three crystallographic orientations. These diagrams were derived from σ (H, T, P) measurements. For H ∥ [100] (Fig. 5a) data for P = 8.6 and 9.4 kbar were omitted for visual clarity. Boundaries II are also omitted from Figs. 5a and b, and extrapolations of boundaries III to T=0 were made for clarity. The positions of boundaries III and IV move rapidly to higher fields with increasing pressure.

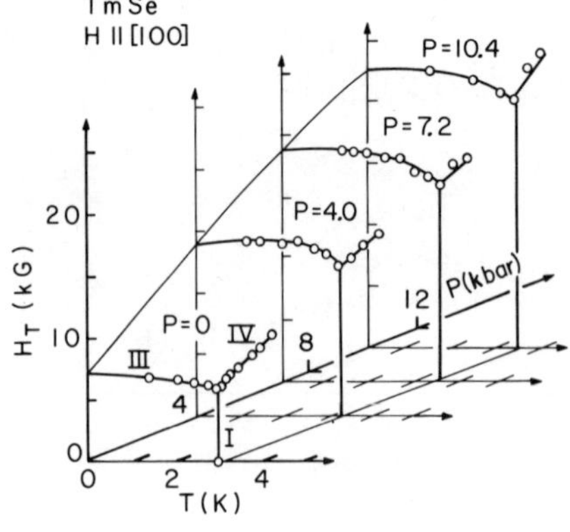

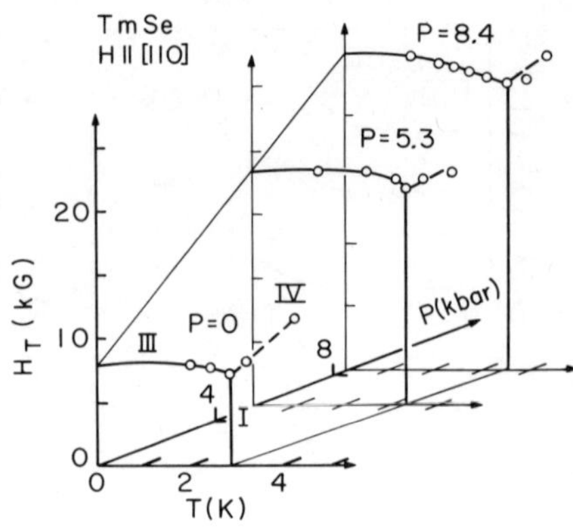

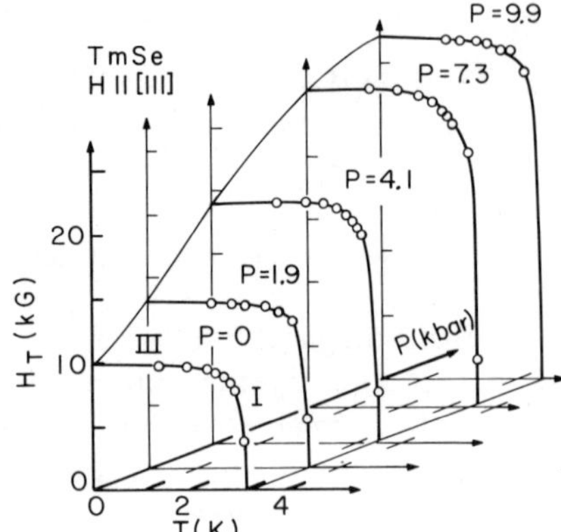

Fig. 5 H-T-P phase diagrams for TmSe. Boundary II of Fig. 2 is not shown.

(a) H ∥ [100] (b) H ∥ [110]

(c) H ∥ [111]

Boundary I terminates at the intersection of boundaries III and IV in Figs. 5a and 5b, and the confluence of these three boundaries is presumed to be a multicritical point. (The transitions constituting III are apparently first-order, whereas those constituting IV are higher order[14,8].) The locus of these multicritical points forms a critical line, L_1, on the H-T-P surface. The locus of points derived from extrapolation of boundaries III to P $\gtrsim$ 20 kbar forms another critical line, L_2, on the H-T-P surface. It is interesting to speculate on the extrapolation of L_1 to the region of L_2: if L_1 and L_2 meet tangentially, they are part of a single critical line; if they meet nontangentially, their intersection defines yet another critical point.

DISCUSSION

A. Effect of Pressure on TmSe

In our previous work[1] we presented two independent arguments which showed that pressure caused TmSe to become more trivalent, and that complete conversion to $Tm^{3+}Se$ was expected for $P \gtrsim 20$ kbar. The first argument, which did not involve any magnetic data, was based on the compressibility of TmSe, the lattice constant, a_o, of our sample, and the pressure needed to reduce a_o to the value[3] required for $Tm^{3+}Se$. Recent measurements of the elastic constants[16] show that the compressibility of TmSe agrees with our previous estimate[1] and thus confirm our earlier conclusions. The second argument was based on the extrapolations of the pressure dependence of boundary III for H ∥ [100]. These results were also consistent with complete conversion to $Tm^{3+}Se$ for $P \gtrsim 20$ kbar. The data presented here for H ∥ [110] and [111] are in agreement with Ref. 1.

The (trivalent) Tm-monopnictides[15] have a crystal-field singlet ground state, Γ_1, and a first excited triplet state, Γ_4, separated by a splitting Δ. The low field susceptibility, χ_o, of these materials is van Vleck-like and independent of temperature for $T << \Delta \simeq 30$ K. Assuming that χ_o at high pressures for TmSe is close to that for $Tm^{3+}Se$, we calculate $\Delta \simeq 60$ K for a Γ_1-Γ_4 crystal field level sequence for $Tm^{3+}Se$.

B. Comparison of TmSe Materials

All of the data described above were obtained with pieces of sample A, whose origin was the same as those used by Ott et al.[8] and Campagna et al.[4] These samples showed the complex phase boundaries at low temperatures. Recently we have reexamined another[17] TmSe sample (B1) in order to determine which features of TmSe are sensitive to details of sample preparation and which features are more general. For sample B1, fits to the Curie-Weiss law were similar to that of sample A, with $\theta = -26.2 \pm 0.5$ K and with $\mu_{eff} = 6.24\ \mu_B$. The lattice constants were $a_o = 5.73$ Å and 5.72 Å for samples A and B1, respectively.

The low field susceptibility of sample B1 at P=0 for $1.4 \leq T \leq 4.2$ K is shown in Fig. 6a. In contrast to Fig. 1a (sample A), there is no clear cut anomaly near T = 3 K. The field dependence of σ for sample B1 at T=1.4 K is shown in Fig. 6b. Similar data for sample A2 are shown for comparison. For sample B1 the σ vs H transitions were slightly broader than for sample A2, phase boundary III for B1 was slight-

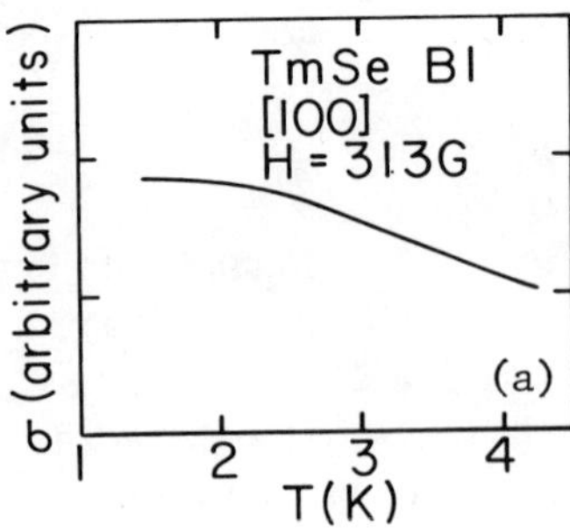

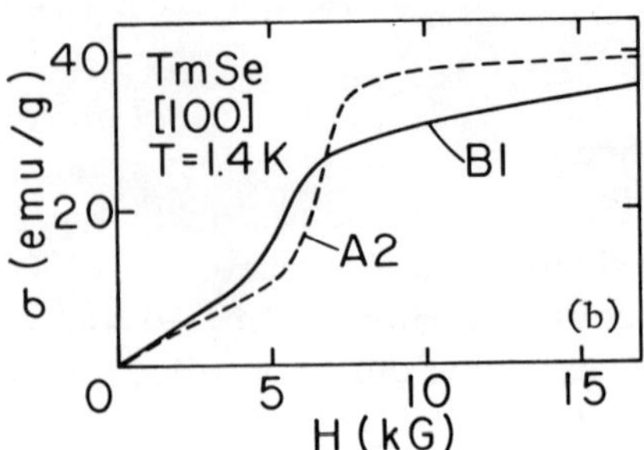

Fig. 6 Magnetic properties of TmSe, Sample B1.

(a) Moment vs temperature does not show anomaly near 3 K. [Compare with Fig. 1(a).]

(b) Comparison of σ vs H at 1.4 K for samples B1 and A2.

ly lower than for A2, and boundary IV increased with increasing temperature at a somewhat higher rate than for A2. The cusp between III and IV was still present. However, I and II were not apparent for sample B1. The absence of boundaries I and II for B1 is reminiscent of the behavior of A1 at high pressure, where these transitions become more difficult to observe. Furthermore, the pressure dependence of χ_0 at 4.2 K for samples A1 and B1 was the same $[(1/\chi_0)(\Delta\chi_0/\Delta P) = -(0.034 \pm 0.001)/\text{kbar}$ for $P \simeq 8$ kbar]. We conclude that the high field and high pressure properties are essentially the same for samples A and B. Thus, even TmSe which does not show all of the complex phase boundaries shows the expected high field-high pressure features.

C. Models for Long-Range Magnetic Order in TmSe

In Ref. 1 we noted that the phase boundaries of sample A could not be described by simple antiferromagnetic or metamagnetic long-range magnetic order. Boundary III is consistent with a simple metamagnet, but boundary IV is not observed in a simple metamagnet (not to mention boundary II). Considering only features common to samples A and B (i.e., boundaries III and IV), there is one ordered magnetic structure which might be appropriate; this is the HoP structure which is apparently observed[18] for DySb. In DySb this structure transforms to a ferrimagnetic structure at 4.2 K for $H \simeq 20$ kG along [100], has a second transition to saturation at $H \simeq 40$ kG, and shows complex boundaries[19] at temperatures near T_N. A boundary such as IV would thus correspond to that for the ferrimagnetic phase which should saturate at higher fields. Previously[1] we reported no transitions above boundary III for H ∥ [100] up to 150 kG. We have extended our measurements to 230 kG at P = 0 and T = 1.5 K in dc fields, and have not observed a transition to saturation for H ∥ [100] or for H ∥ [111] in sample A. If TmSe has an ordered HoP structure, these high field data point to a very large anisotropy of TmSe. As emphasized earlier, neutron data on well characterized TmSe would be extremely valuable. Preliminary early results[10] showed no evidence for long-range magnetic order or for crystal-field splitting in TmSe. If TmSe is generally inhomogeneous[20], and if the order is sensitive to inhomogeneities, then we might expect no long-range order, but possibly short-range order. Our magnetic data do not permit us to determine the microscopic behavior, but can be used to test consistency of a given ordered model.

D. Boundaries II and IV

The magnetic properties of sample A for H ∥ [111] are unusual in that boundaries II and IV are not observed. If we assume that boundary II involves reorientation of the tetragonal crystallographic domains, boundary II should not be observed because, due to symmetry, there would be no driving force for a random distribution of domains. For H ∥ [110] the transitions for boundary IV become more difficult to detect than for H ∥ [100]. Furthermore, dH_T/dT for H ∥ [110] is larger than dH_T/dT for H ∥ [100]. Whether (dH_T/dT) becomes very large for H ∥ [111] and thus is not detected by an H-sweep, or whether it is not present is not clear.

E. Effect of Pressure on TmS

An alternate means for reducing the effective lattice parameter of TmSe is to substitute an S atom for Se. Since pressure increases the trivalency of Tm in TmSe, we expect Tm in TmS also should be more trivalent. This is found to be true; Bucher[3] found $\mu_{eff} = 7.19\,\mu_B/Tm$, close to the Tm^{3+} value ($7.56\,\mu_B/Tm$). Our measurements on similar (polycrystalline) material at P = 0 show a susceptibility maximum at $T_0 = 5.2$ K and $dT_0/dP = +0.15$ K/kbar. S-shaped σ vs H transitions are observed below T_0. The P-dependence of χ_0 at 4.2 K is $(1/\chi_0)(\Delta\chi_0/\Delta P) = (-0.007 \pm 0.001)$/kbar, much smaller than for TmSe. The large μ_{eff} and small pressure dependence of the magnetic properties for TmS is in agreement with our conclusions based on more extensive data for TmSe.

ACKNOWLEDGEMENTS

We are grateful to M. Campagna and L.D. Longinotti of Bell Laboratories for furnishing the TmSe samples. We wish to thank E.J. McNiff, Jr., for measuring the magnetic moments to 230 kG, Y. Shapira for measuring the magnetostriction at high fields and discussions, and B. Lüthi for providing a value for the compressibility of TmSe. We also wish to thank W.M. Walsh, Jr., for informative discussions

REFERENCES

1. R.P. Guertin, S. Foner, and F.P. Missell, Phys. Rev. Lett. 37, 529 (1976).
2. C.M. Varma, Rev. Mod. Phys. 48, 219 (1976).
3. E. Bucher, K. Andres, F.J. diSalvo, J.P. Maita, A.C. Gossard, A.S. Cooper, and G.W. Hull, Jr., Phys. Rev. B 11, 500 (1975).
4. M. Campagna, E. Bucher, G.K. Wertheim, D.N.E. Buchanan, and L.D. Longinotti, Phys. Rev. Lett. 32, 885 (1974).
5. R.W. Ward, B.P. Clayman, and T.M. Rice, Solid State Commun. 17, 1297 (1975).
6. B. Batlogg, K. Kaldis, A. Schlegel, and P. Wachter, Phys. Lett. 56A, 122 (1976).
7. B.B. Triplett, N.S. Dixon, P. Boolchand, S.S. Hanna, and E. Bucher, J. Phys. (Paris), Colloq. 35, C6-653 (1974).
8. H.R. Ott, K. Andres, and E. Bucher, in Magnetism and Magnetic Materials – 1974, AIP Conference Proceedings No. 24, edited by C.D. Graham, Jr., G.H. Lander, and J.J. Rhyne (American Institute of Physics, New York, 1975), p. 40.
9. See Ref. 7 quoted in Ref. 8 above.
10. See Ref. 8 quoted in Ref. 8 above.
11. M. Campagna, private communication.
12. R.P. Guertin and S. Foner, Rev. Sci. Instrum. 45, 863 (1974).
13. The data of Fig. 1b show that the susceptibility below the step is smaller than that above, contrary to what was stated in Ref. 1.
14. Y. Shapira, private communication.
15. R.P. Guertin, J.E. Crow, L.D. Longinotti, E. Bucher, L. Kupferberg, and S. Foner, Phys. Rev. B 12, 1005 (1975).
16. The compressibility of TmSe at $T = 4.2$ K is 2.0×10^{-12} cm^2/dyne. (B. Lüthi, private communication).
17. This material was kindly furnished by L. D. Longinotti of Bell Laboratories.
18. G. Busch and O. Vogt, J. Appl. Phys. 39, 1334 (1968). T.O. Brun, G.H. Lander, F.W. Korty and J.S. Kouvel, in Magnetism and Magnetic Materials – 1974, op. cit., p. 244.
19. P. Streit, G.E. Everett and A.W. Lawson, Phys. Lett. 50A, 199 (1974).
20. Apparently various TmSe crystals show variations from batch to batch and variations in homogeneity between different portions of the same crystal with resultant changes in electrical and magnetic anomalies. See paper by W.M. Walsh, Jr., K. Andres, L.W. Rupp, Jr., and L.D. Longinotti in these proceedings.

COMMENTS

LUTHI: Do you see a transition pressure of 20 kbar [in sample B]?

GUERTIN: The only thing we've looked at is the pressure dependence of the susceptibility at 4°K, to $\sim$10 kbar, but I'm fairly confident that it would be something like that.

FONER: Wouldn't the pressure dependence of the lattice parameter provide an independent estimate of that transition?

GUERTIN: Yes, the lattice constant argument based on your [Luthi's] compressibility results would indicate that it would be about 20 kbar, because it's starting off with only a slightly smaller lattice constant.

HOLTZBERG: What is the ordering temperature for B compared to A?

GUERTIN: We just don't see any peak in the susceptibility. The cusp between transitions III and IV occurs at 3°K, and yet if we look at low fields, we just don't see this susceptibility peak like we do in sample A.

CAMPAGNA: Permit me to say that one does not understand what is going on at 3°K even in sample A, since, as Professor Orbach pointed out, it is a very shallow cusp and, furthermore, it shifts towards higher temperatures, which is in the opposite direction to what I would expect in the presence of tri-valency.

ANDERSON: There are a lot of impossible things in the data of course, but it seems to me the most impossible is the disappearance of transition IV. Higher order transitions don't disappear.

GUERTIN: Yes, I agree. We've looked at the magnetization up to 20°K and that S-shaped character just eventually melts away. We feel we just don't have the sensitivity to see it.

LOW TEMPERATURE THERMAL PROPERTIES OF Tm-MONOCHALCOGENIDES

H.R. Ott[+], B. Lüthi[*], P.S. Wang[*]

[+] Laboratorium für Festkörperphysik, ETH-Hönggerberg, 8093 Zürich, Switzerland.

[*] Physics Department, Rutgers University, New Brunswick, New Jersey 08903, USA.

ABSTRACT

In order to search for manifestations of crystal electric field (CEF)- effects in Tm-monochalcogenides we have studied several thermal properties (elastic constants, thermal expansion, magnetostriction) in these materials at low temperatures. We find evidence for CEF-effects in TmTe and, to a lesser extent, in TmS. A striking absence of CEF-effects in TmSe indicates the instability of the ion-configuration in this compound. In addition we discuss the magnetic phase diagram of TmSe, presenting experimental data on the magnetic field dependence of elastic constants and the electrical resistivity.

INTRODUCTION

Interconfigurational fluctuations (ICF) were attributed to Tm-monochalcogenides on the basis of measurements of lattice constants,[1] magnetization, magnetic[2] susceptibility χ, electrical resistivity ρ[1], pressure effects on χ[2] and ρ[1], and X-ray photoemission spectra[3]. Another major argument in favour of valence fluctuations of the Tm-ions in these compounds was the negative result of neutron scattering experiments[4] concerning the detection of distinct crystal electric field (CEF) energy levels of the 4f-configuration ground state multiplet of the Tm-ions. It seemed therefore of interest to use other techniques for investigating the possible existence of distinct CEF energy levels in these materials which should be present in the normal case. As has been shown recently[5], the elastic constants, the thermal expansion and the magnetostriction show CEF-effects and measurements of such properties are therefore providing information on the structure of the CEF energy level scheme. If the effects can be detected at all, one can distinguish between different valence states (in our case Tm^{2+} and Tm^{3+}) because different configurations also lead to different splittings of the 4f ground state. One of the purposes of this work was to look for such effects in these compounds by investigating the elastic constants in TmSe and TmTe, the thermal expansion and the magnetostriction in TmS, TmSe and TmTe.

In addition we also present results for the electrical resistivity as a function of temperature and magnetic field in TmSe. Apart from CEF-effects we were also interested in magnetic properties and we have studied the magnetic phase transition of TmS and TmSe. Especially the recently found magnetic phase diagram of TmSe[6] was investigated in some more details. In the next section we present and discuss zero magnetic field properties followed by a section on magnetic field dependent properties. The theory involved in these experiments has recently been outlined in ref. 5 and 8.

ZERO MAGNETIC FIELD PROPERTIES

We present and discuss experimental results on elastic constants and thermal expansion. Due to the difficulty in growing large enough single crystals of TmS the results on this substance are not sufficient for a quantitative interpretation mainly due to the lack of elastic constants data. The most detailed quantitative CEF-investigation we can present for TmTe. In TmSe we observe a marked absence of CEF-effects. Material constants deduced from this study are summarized in table I.

Elastic Constants

In fig. 1a and 1b we present the temperature dependences of several elastic constants for TmSe and TmTe respectively. As mentioned above we failed to investigate TmS due to a lack of suitable single crystals of this compound. Comparing figs. 1a and 1b it may be seen that CEF-effects are absent in TmSe bur rather pronounced in the case of TmTe. In TmSe one notices a phase transition at $T_o = 3$ K which will be discussed below. For $T > T_o$ we observe normal elastic behaviour for all modes. Note that also the bulk modulus $c_B = (c_{11} + 2c_{12})/3$ has a normal temperature dependence above T_o and and does not show any softening. A distinct anomaly for $T \lesssim T_o$ appears in the $c_{11} - c_{12}$ mode and is shown in more detail in fig.1c. The strong velocity change for $T < T_o$ can be attributed to domain wall

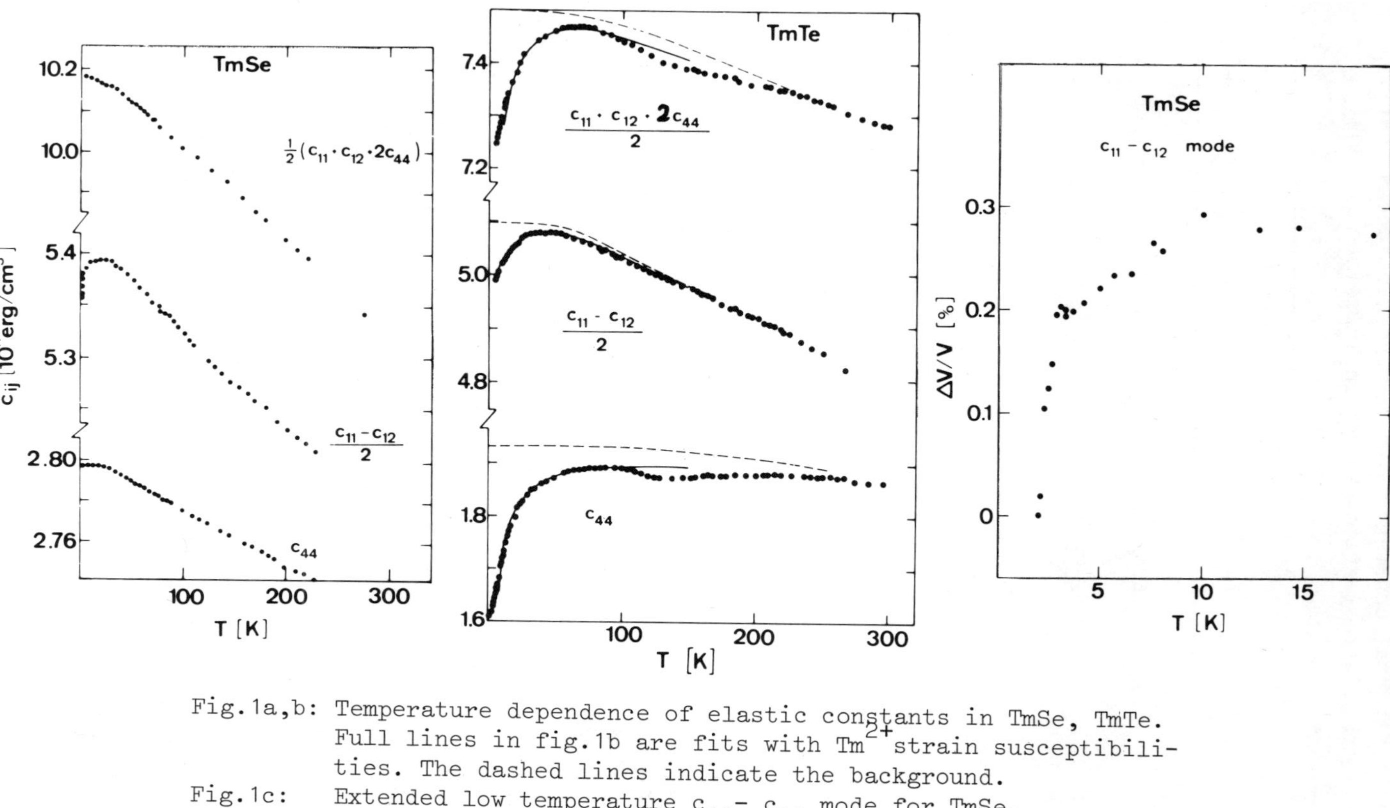

Fig. 1a,b: Temperature dependence of elastic constants in TmSe, TmTe. Full lines in fig. 1b are fits with Tm^{2+} strain susceptibilities. The dashed lines indicate the background.

Fig. 1c: Extended low temperature $c_{11} - c_{12}$ mode for TmSe.

Table I: Material constants for Tm-monopnictides as derived from this work (except for T_o in TmTe (see ref.1)).

	T_o [K]	a [Å]	c_{44}	$\frac{c_{11} - c_{12}}{2}$	$\frac{c_{11} + 2c_{12}}{3}$
			at 300 K	$[10^{11}$ erg/cm$^3]$	
TmS	5.5	5.42			
TmSe	2.9	5.71	2.7	5.2	5.1
TmTe	0.21[1]	6.346	1.86	4.8	3.8

stress effects. We investigated three different TmSe single crystals from the same batch with the same results. The same material was used before in photoemission studies[3] and previous experiments on thermal properties[6,12]. The significance of the absence of any CEF-effects in the paramagnetic region of TmSe will be discussed after having presented the other experimental results.

In TmTe we observe very distinct CEF-effects (fig.1b) as have been seen in many other rare earth compounds before[5,8,10]. We note a remarkable softening of 14% of the c_{44} mode from 80 K down to 2 K. The full lines are theoretical fits by calculating the strain susceptibilities[8] for the level scheme $\Gamma_8 - \Gamma_7(10\ K) - \Gamma_6(16\ K)$. The proposed total level splitting of the multiplet of 16 K is an upper limit for the following reason: If the CEF energies would exceed 17 K the theoretical strain susceptibilities would exhibit minima for $T > 4$ K and this is clearly not observed in the experiment. Our proposed level scheme gives good agreement between theory and experiment for all three elastic constants as shown in fig.1b, except for the strong softening of the c_{44}- and the $c_{11} + 2c_{12} + c_{44}$ modes predicted by our calculations due to the Γ_8 ground state, which is not observed at such strength. From the fits we deduce the following magnetoelastic coupling constants[8,10]: $g_2^2 = 0.81$ mK and $g_3^2 = 24.5$ mK. As discussed previously[8] we have no explanation for the slight anomaly around 120 K yet. Recently the CEF-effects on the elastic constants of TmTe were interpreted on the basis of a level scheme $\Gamma_6 - \Gamma_8(10\ K) - \Gamma_7(16\ K)$ with equally nice fits. This level scheme, however, cannot explain other experimental results. A CEF level scheme based on the Tm^{3+} configuration cannot explain the low temperature anomalies of TmTe. This may be seen by comparing the results in fig.1b with corresponding results for TmSb[10]. One expects a singlet ground state followed by a triplet as the first excited state for a Tm^{3+} monochalcogenide[11] (the same as in the case of TmSb) and there-

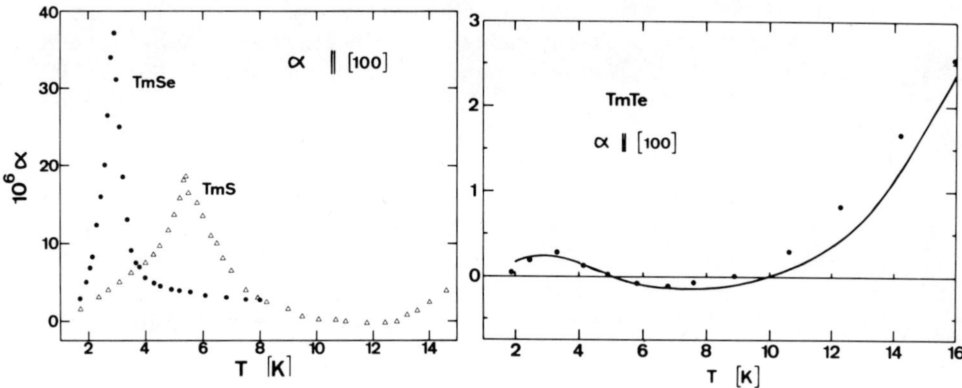

Fig.2: Linear thermal expansion coefficients along the (100)-direction for TmS, TmSe and TmTe as a function of temperature. The solid line for TmTe in fig.2b is a theoretical fit assuming the same CEF level scheme as for the elastic constants.

fore similar strain susceptibilities should be observed in both cases. This is, however, not seen, especially for the $c_{11}-c_{12}$ mode.

The absolute values of the symmetry elastic constants of the Tm monochalcogenides (table I and fig.1) are somewhat higher than the corresponding values for the Eu-monochalcogenides[13] (c_{44}, $c_{11}-c_{12}$, $c_{11}+2c_{12}$). A similar behaviour was found in the Tm- and Eu-monopnictides (see discussion of this point in ref. 10).

Thermal Expansion

In figs.2a and 2b we present results for the low temperature thermal expansion coefficients α along the (100)-axes of TmS, TmSe (fig.2a) and TmTe (fig.2b) respectively. Strong anomalies are clearly discernible for TmS and TmSe in the vicinity of the phase transitions at T = 5.5 K (TmS) and at T = 2.9 K (TmSe). The high temperature part of the thermal expansion of TmS indicates that CEF-effects similar to those observed in TmSb[7] might be present. In TmSb we have a CEF-splitting of the lowest two levels $\Gamma_1-\Gamma_4$ of 25 K. The strong magnetic anomaly in TmS of $20\times10^{-6}\,K^{-1}$ at T_N masks the expected (in analogy with TmSb), much smaller Schottky type anomaly due to CEF-effects of the order of $10^{-6}\,K^{-1}$. In analogy to the results of fig.1a for the elastic constants we do not expect any CEF-effects in the thermal expansion of TmSe. The results of fig.2a are compatible with this expectation.

In fig.2b we show the linear thermal expansion coefficient of TmTe as a function of temperature. Its behaviour can be interpreted as a CEF-effect similar to that reported previously for TmSb[7]. The solid line is a fit based on the CEF level scheme used for the interpretation of the elastic constant measurements (fig.1b). The low temperature part shows good agreement between experiment and calculation with the following Grüneisen parameters[7] (defined as $\gamma_i = -(d\ln E_i/d\ln V)$): $\gamma(\Gamma_7)=1.26$ and $\gamma(\Gamma_6)=-1.52$. Above 8 K additional thermal expansion due to lattice contributions becomes important. The resulting T^5-term is of the order of $2\times10^{-12}K^{-6}$. From the γ-values we can calculate magnetoelastic coupling constants associated with the volume strain ε_V[5] and entering the Hamiltonian

$$H_{me} = -\tilde{g}_{11}\,\varepsilon_V\,O_4 - \tilde{g}_{12}\,\varepsilon_V\,O_6 \tag{1}$$

where $O_4 = O_4^o + 5\,O_4^4$ and $O_6 = O_6^o - 21\,O_6^4$. We obtain $\tilde{g}_{11} = -18.6$ mK and $\tilde{g}_{12} = 0.28$ mK.

Specific Heat

The specific heat of the Tm-monochalcogenides was measured previously[1]. Anomalies were found at the phase transitions of TmS and TmSe similar to the anomalies of the thermal expansion shown in fig.2. In the case of TmTe the behaviour was tentatively ascribed to a Schottky anomaly without a quantitative interpretation. With our CEF level scheme deduced from elastic constant and thermal expansion measurements we get the maximum of the anomaly at the right temperature(6 K) and also the predicted maximum value of the anomaly is found to be in good agreement with experiment.

In summary we have shown in this section that for TmTe the Tm-ions are in the 2+ configuration and that the overall CEF-splitting of the J=7/2 ground state is about 16 K. Compared with other rare earth compounds this total splitting is anomalously small but similar to splittings observed in Ce-pnictides. We interpret our results with a level sequence $\Gamma_8 - \Gamma_7 - \Gamma_6$ which explains adequately all the various experimental results. This sequence is compatible with a rather large B_6-term in the CEF-potential (see ref.11). For TmSe we found no direct evidence of CEF-effects in all physical properties studied. This feature may be taken as another indication for the non stable ionic configuration of the Tm-ions in this compound and in that case one would be led to the conclusion that one has to deal with temporal fluctuations. Assuming a stable ionic configuration the absence of CEF-effects can also be due to anomalous broadening of the CEF energy levels. For TmS our results are compatible with the accepted Tm^{3+}-configuration.

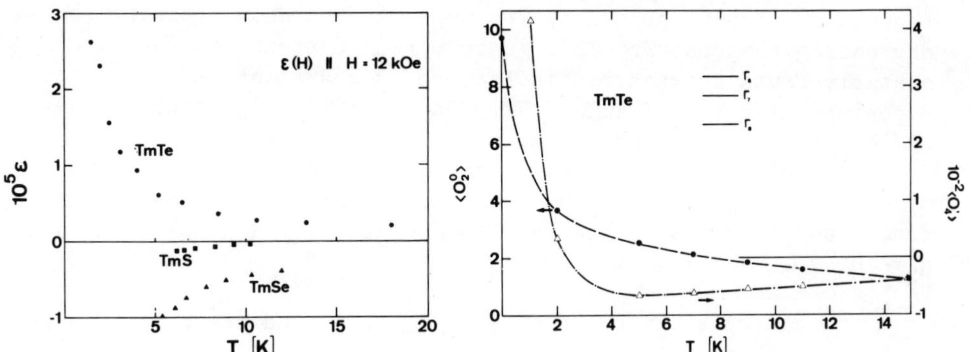

Fig.3a: Temperature dependence of the magnetostriction $\varepsilon(H)$ along the (100)-direction in a parallel magnetic field H=12 kOe for TmS, TmSe and TmTe.

Fig.3b: $\langle O_2^0 \rangle$ and $\langle O_4' \rangle$ for the TmTe level scheme as a function of temperature.

MAGNETIC FIELD DEPENDENT PROPERTIES

In this section we present and discuss magnetostriction- and magnetoresistance experiments, the latter especially in view of a discussion on the magnetic phase diagram of TmSe.

Magnetostriction

The magnetostriction was measured along the (100)-direction in a parallel magnetic field up to 15 kOe. In fig.3a we show $\varepsilon(H) = (\ell(H) - \ell(0))/\ell(0)$ for H = 12 kOe in the paramagnetic region of all three substances. The sign of $\varepsilon(H)$ in TmTe is different to that in both TmS and TmSe. The magnetostriction due to CEF-effects is a measure of the thermal average of multipole operators. Including quadrupole- and octupole terms, $\varepsilon(H)$ can be written[5]

$$\varepsilon(H) = -\frac{\sqrt{6}}{3}\sqrt{N}\left[\frac{g_2}{\sqrt{c_{11}-c_{12}}}\langle O_2^0(H)\rangle + \frac{g_4}{\sqrt{c_{11}-c_{12}}}\langle O_4'(H)\rangle\right] \qquad (2)$$

where $O_4' = O_4^0 - 7\, O_4^4$.

In fig.3b we show a plot of $\langle O_2^0 \rangle$ and $\langle O_4' \rangle$, calculated using the CEF level scheme deduced for TmTe above. Both terms show a strong temperature dependence, similar to that observed experimentally for TmTe as shown in fig.3a. For the 3+ configuration with the expected singlet ground state for Tm^{3+} both these quantities are only weakly temperature dependent and at low temperatures show a saturation with decreasing temperature. This is in marked contrast to the experimentally found characteristic feature. A quantitative fit to the experimental curve of TmTe in fig.3a is difficult because we have neglected exchange- and volume contributions. Taking those into account would lead to a many parameter fit. Considering the terms of eq.(2) only, g_2^2 is of the order of 10^{-1} mK, as has been determined from elastic constants (see section on zero field properties) and $g_4 \sim 10^{-2} \cdot g_2$.

In analogy to TmSb[5] (another singlet ground state system) we expect a rather small magnetostriction with weak temperature dependence for TmS. This is indeed what one observes (fig.3a). The sign of ε(H) in TmS as well as in TmSe is opposite to that in TmTe. This is not to be expected from CEF-considerations alone. The sign of ε(H) usually follows the sign of the appropriate Stevens factor[5] and in our case the Stevens factor α_J is positive for both Tm^{2+} and Tm^{3+}. Strong exchange striction effects may account for the behaviour in TmS as has been observed previously in other rare earth compounds[5].

For TmSe such a comparison may not be valid because no other CEF-effects have been detected and therefore an analysis of the magnetostriction data should not be based on CEF-effects. However, large exchange striction effects may account for ε(H) in TmSe as well. This would also explain the strong variation of T_o (see next section) and the lattice constant with slight variation of the stoichiometry.

Magnetic Phase Diagram of TmSe

In fig.4 we show the magnetic phase diagram of TmSe taken from ref.6. It was established by means of magnetostriction- and magnetic susceptibility measurements. Our new measurements on elastic constants and electrical resistivity confirm it. As an example we show some magnetoresistance results in fig.5, taken above and below T_o. At all temperatures ρ increases slightly in low fields and then decreases appreciably with still increasing field. The field dependence of ρ is found to be the same for increasing and decreasing fields. The characteristic features are connected with the phase boundaries shown in fig.4. Likewise the elastic constants, especially $c_{11} - c_{12}$ show noticeable kinks at phase boundary IV. The same phase diagram has recently been confirmed and extended with magnetisation measurements under pressure[12].

The phase boundaries are not compatible with normal antiferromagnetic behaviour as originally suggested[6] (spin flop transition below T_o). This may be seen from the temperature dependence of all the phase boundaries[14]. The behaviour must be connected with the

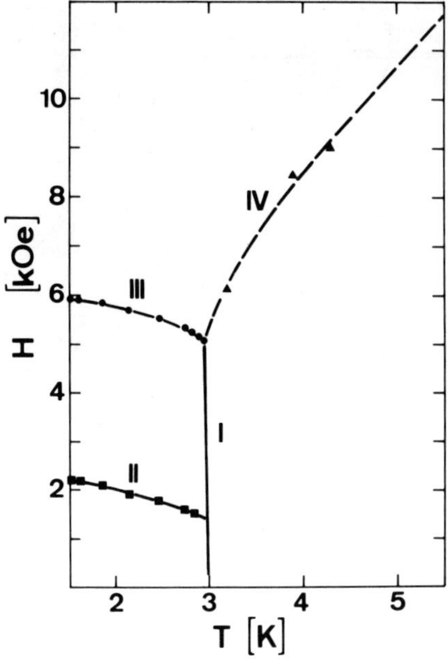

Fig.4: Magnetic phase diagram of TmSe
(taken from ref.6)

anomalous physical properties of TmSe in the paramagnetic phase as discussed in the section on zero field properties. As has been suggested before[1,3,9] all these anomalous properties of TmSe might be due to valence fluctuations of the Tm-ions (Tm^{2+}, J=7/2 and Tm^{3+}, J=6).

The resistivity data (fig.5) could qualitatively be explained by assuming a magnetic field induced suppression of such valence fluctuations. Together with the decrease of the resistivity one observes a positive relative volume change $\Delta V/V$ of the order of $5x10^{-4}$ (in agreement with the pressure dependence of phase boundary III[12]) and a tetragonal distortion. The volume change has, however, the wrong sign and is far to small in order to account for any appreciable configuration changes from 2+ to 3+. The sign of the volume change would even imply a slight localization of electrons and thus an increase of resistance. Thus external magnetic fields and external pressure obviously induce competing effects, as is also seen in the shift of phase boundary III to higher fields with increasing pressure[12]

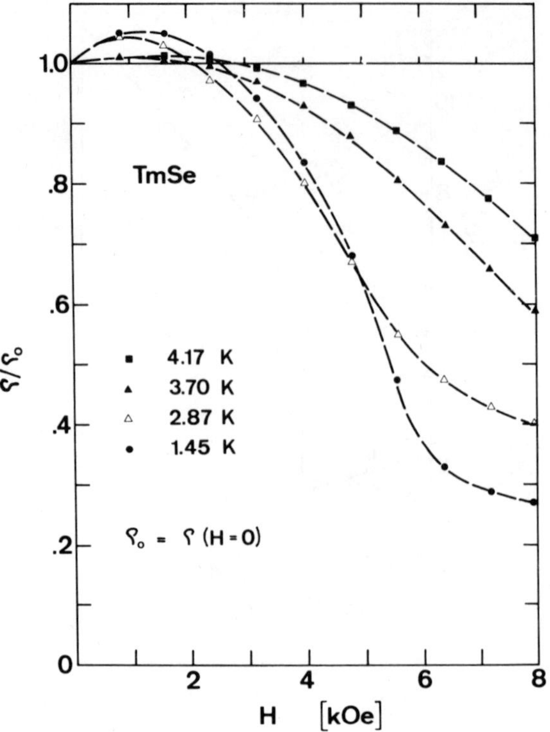

Fig.5: Low temperature magnetic field dependence of the electrical resistance of TmSe (T_o = 2.9 K)

All the experiments presented suggest a long range order phenomenon at T_o. Such an ordering is difficult to explain in a system with alleged valence fluctuations. A microscopic theory explaining all these effects consistently has not yet been given.

From numerous investigations of the magnetic phase diagram of TmSe it has been found that slight deviations from stoichiometry drastically change T_o and the phase boundaries[15]. From earlier work of Bucher and co-workers[1] it is known that a slight chalcogen surplus leads to a contraction of the crystal lattice thus favouring the 3+ configuration. We note that a 1% surplus of Tm in TmSe lowers the critical temperature T_o to ~2 K and one also observes smaller critical fields. In this case it is also found that the anomalies of the thermal expansion and the magnetostriction are weakend (by a factor of 4 in the case of $\alpha(T)$ and by a factor of 2 for $\varepsilon(H)$). On the other hand the decrease of the electrical resistance with increasing magnetic field is more pronounced in the material with rare earth surplus. Although the overall features are essentially the same in samples with varying stoichiometry it seems quite clear that one should be careful in interpreting physical properties of TmSe. This strong dependence of the low temperature properties on the stoichiometry of the compound, however, also indicates that one has delicately balanced interactions between the localized 4f-electrons and the conduction electrons which are easily disturbed by only slight variations of external influences.

For reasons of comparison we also have investigated the influence of magnetic fields on the lattice parameter of TmS below T_N. Between 0 and 15 kOe we observe one field induced phase transition leading to a phase diagram similar to that of TmSe but with less phase boundaries. In contrast to TmSe the phase transition is accompanied by a negative relative volume change. Of interest is also a comparison of the magnetostriction in the high field phases of TmS and TmSe. The most striking feature is the absolute linearity of $\varepsilon(H)$ with respect to H in both materials. The sign of $\partial\varepsilon/\partial H$ is, however, different in both cases. Linear magnetostriction has been found before in ferromagnetic metals[16] and invar type alloys[17] above their saturation field. Wohlfarth later showed[18] that this can be understood by applying the itinerant electron model of ferromagnetism. In our case it means that the high field phases are regions of saturation magnetisation (for TmSe see refs. 1 and 12) and that the pressure dependence of this magnetisation is of different sign in both materials.

In addition to the zero field properties also the magnetic field dependent features of the Tm-monochalcogenides are of some interest. Especially the phase diagram of TmSe is still not thoroughly understood and it would be of great help if another attempt to clarify the situation could be made by means of neutron diffraction experiments either under external pressure or with external magnetic fields.

ACKNOWLEDGEMENTS

We thank Prof. E. Bucher and Dr. F. Hulliger for providing the samples used in these experiments. We also acknowledge informative discussions with Dr. K. Andres and Dr. W.M. Walsh.

REFERENCES

1. a compilation of all this work is given in E. Bucher, K. Andres, F.J. di Salvo, J.P. Maita, A.C. Gossard, A.S. Cooper and G.W. Hull, Jr., Phys. Rev. B11, 500 (1975).
2. D. Wohlleben, J.G. Huber and M.B. Maple, AIP Conf. Proc. 5, 1487 (1972).
3. M. Campagna, E. Bucher, G.K. Wertheim, D.N.E. Buchanan and L.D. Longinotti, Phys. Rev. Letters 32, 885 (1974).
4. D.E. Cox, L. Passell, R.J. Birgeneau and E. Bucher, unpublished.
5. a rewiew of CEF-effects on thermal properties in intermetallic compounds is given by B. Lüthi, P.S. Wang, Y.H. Wong, H.R. Ott and E. Bucher, Proc. 2nd Int. Conf. on CEF-Effects in Metals and Alloys (Plenum Press, New York), in press.
6. H.R. Ott, K. Andres and E. Bucher, AIP Conf. Proc. 24, 40 (1974).
7. H.R. Ott and B. Lüthi, Phys. Rev. Letters 36, 600 (1976).
8. B. Lüthi, AIP Conf. Proc., Pittsburgh 1976, in press.
9. M. Campagna, G.K. Wertheim and E. Bucher, AIP Conf. Proc. 24, 22 (1974).
10. M.E. Mullen, B. Lüthi, P.S. Wang, E. Bucher, L.D. Longinotti, J.P. Maita and H.R. Ott, Phys. Rev. B10, 186 (1974).
11. K.R. Lea, M.J.M. Leask and W.P. Wolf, J. Phys. Chem. Solids 23, 1381 (1962).
12. R.P. Guertin, S. Foner and F.P. Missell, Phys. Rev. Letters 37, 529 (1976).
13. Y. Shapira and T.B. Reed, AIP Conf. Proc. 5, 837 (1971).
14. see e.g. F. Keffer, Handbuch der Physik 18/2 (Springer, Berlin 1966).
15. K. Andres, W.M. Walsh, private communications.
16. E. Fawcett and G.K. White, J. Appl. Phys. 38, 1320 (1967).
17. M. Matsumoto, T. Kaneko and H. Fujimori, J. Phys. Soc. Japan 26, 1083 (1969).
18. E.P. Wohlfarth, Proc. Moscow Magnetism Conf. 1973, 2, 28 (1974).

COMMENTS

TOURNIER: Have you studied Cerium Kondo compounds or observed crystalline field effects in TmSe?

OTT: No, but I saw yesterday in the poster session a similar phase diagram for $CeAl_2$ with fields approximately ten times larger than for TmSe, but I have not made any measurements.

LUTHI: (to Tournier) Are you asking about $CeAl_2$?

TOURNIER: My question is, do you see the crystalline splitting in a Kondo system?

LUTHI: In $CeAl_2$, if you consider that a Kondo system, the elastic constant and thermal expansion measurements have clearly shown crystalline electric field effects, the same as for TmSe. In the elastic constant (of $CeAl_2$) you can see crystal field splittings up to several hundred degrees.

OTT: The splitting you would expect in TmSe from the ground state to the next excited state, if it were to show the 3^+ behavior, is certainly of the order of 15-20 degrees. That should definitely show up if it's there.

JAYARAMAN: It would be very interesting to study the substitution of Yb in systems like TmSe. You'll expand the lattice and probably increase the Tm^{2+} content. One could study the effect of this additional Tm^{2+} on various phase transitions one is observing. Has this been done?

OTT: No.

FONER: I would say it has been done, at least indirectly, in the high field studies of TmSe, wherein the field induces a lattice expansion. If you turn on a (low) magnetic field, the sample contracts, and if you squeeze on it, you still get small effects like that which you might argue does change the Tm^{2+} concentration.

OTT: Well, I would say that the volume change at phase boundary III in TmSe is opposite to what you would expect if you argue that the high field phase is more 3^+ than the low field phase. There the volume effect is opposite in sign and much too small.

HAEN: I would like to return to the crystalline field. In the resistivity of TmS the presence of Kondo-like behavior masks any contribution from crystal field effects. Could this be a problem in the thermal expansion results?

OTT: Well, I would not think so. If you have a crystal field effect in the resistance in TmS, it is certainly not apparent in TmSe because the contribution always goes up, it never comes down, and because the splitting should be similar to TmS.

HAEN: But this effect is masked by the increase at low temperature. If one looks at the resistance in magnetic field, the curves are very similar for TmS and TmSe in the zero field limit.

OTT: Yes, but it should come down at 10°K or so.

HAEN: But that depends on the logarithmic slope of the resistivity at higher temperatures and on the crystalline field splittings, so your maximum could be displaced by competition between these two contributions.

OTT: Yes, but as I said they're strikingly absent in the elastic constants. For these effects to be there and not show up in elastic constants would require cancelling effects which would be almost impossible since we have three different strain symmetries. That they would all cancel simultaneously I would very much doubt.

ORBACH: Why are you saying the crystal field should be known for TmSe?

OTT: Because if the ordering is an induced moment ordering of a ground state singlet and one applies a very simple mean field theory, the splitting should be about the same since the transition temperatures in the selenide and sulphide are very similar. If you calculate, you get a splitting of about 15-20°K for the two lowest levels.

ORBACH: So you're arguing entirely on the basis of the transition temperatures then?

OTT: That's one thing, and the other thing is that if the high temperature part of our thermal expansion data is due to crystal field effects similar to what one sees in TmSb (which is also a singlet system), then you would expect approximately the same splitting between 15 and 20 degrees.

OBSERVATION OF LOCALIZED MOMENTS, KONDO EFFECT, SPIN GLASS REGIME, MAGNETIC ORDERING IN RARE EARTH SYSTEM WITH UNSTABLE VALENCE

R. Tournier and F. Holtzberg*

Centre de Recherches sur les Très Basses Températures,

C.N.R.S., BP 166 X, 38042 Grenoble-Cedex, France

Certain rare earth elements have an unstable valence and with compound formation or under pressure exhibit a non-magnetic ground state at low temperature. Their electronic properties are characterized by a very large γT term in their specific heat and a very high Pauli like susceptibility.[1] For example SmB_6[2], SmS in its collapsed phase [3], αCe [4], $CeSn_3$[5], and $CeAl_3$[6] are in this non-magnetic state. This behavior is generally observed on rare earth compounds with an intermediate valence near or far from an integral number which is a clear indication that the 4f level is near the Fermi level. The larger the Pauli-like susceptibility, the nearer the system is to magnetic ordering. Because of the mixing of the 4f level with the conduction electrons the orbital and spin moment[7] of the rare earth can disappear at low temperature or exchange interactions can compete and dominate the Kondo condensation to induce magnetic ordering. As examples we will discuss the case of $CeAl_2$ and $CeAl_3$.

We want to argue that such a non-magnetic compound or a compound near a magnetic instability always contains imperfections which induce localized moments. Low temperature electronic properties will be strongly affected by these accidental defects which depend on the sample history. The Kondo effect is often observed in metallic phases. This is the case for SmS, $Sm_{1-x}La_xS$, TmS, TmSe, and $CeAl_3$. A spin glass regime may also be observed due to the presence of these localized moments which magnetically interact at long distances. This has been observed for the gold $Sm_{1-x}Y_xS$ phase and for the black $SmS_{1-x}As_x$ phase.

As for 3d transition impurities, two types of local environment can play a role in the creation of localized moment on an anomalous rare earth. In the first type the impurity is surrounded by a critical number of the same impurity i.e. a cluster. This phenomenon has been studied in many dilute alloys in the last several years, e.g. the group of three cobalt atoms in gold[8] and in copper[9] and pairs of iron in gold[10] and copper[11]. In the second type when the impurity is diluted in a non magnetic binary alloy the magnetism appears discontinuously when a critical number of neighbors of a certain kind surrounds the impurity[12]. We will discuss for example the case of Ce diluted in a $La_{1-x}Th_x$ matrix.

$La_{1-x}Th_x$: Ce

Ce impurities at concentrations up to 3 % have been introduced in La-Th alloys[13]. Ce is known to be magnetic in La[14] and non-magnetic in Th.[15]

The disappearance of the log T slope of the Kondo resistivity and the study of the depression of the superconducting transition temperature versus Th concentration confirm the existence of a continuous transition from magnetism to non magnetism by increasing the concentration of Th in La at a constant Ce impurity level.

A similar transition has been observed in LaCe as a function of pressure[16] by analysis of the same physical measurements. These progressive transitions have been explained by the movement of the 4f level from below to a position above the Fermi level[17]. This model is satisfactory for Ce diluted in La because the impurity always sees the same environment but it cannot be applied to Ce diluted in $La_{1-x}Th_x$ for increasing x because the local environment changes with concentration.

The saturation magnetization, σ_s, has been measured[18] and compared to the Curie constant, C. The ratio C/σ_s appears to be independent of the impurity concentration when both are decreasing with x. This implies that the magnetic moment is constant, in fact equal to the ground state doublet Γ_7 moment of Ce in a cubic crystalline field. The number of magnetic Ce decreases rapidly with increasing Th concentration. These results establish that Ce magnetic moment does not decrease continuously but discontinuously for a minimum of Th nearest neighbors.[18]

$Sm\, S_{1-x}\, As_x$

The introduction of As^{3-} in SmS produces the appearance[19] of a moment on each of the six Sm nearest neighbors, i.e. they become trivalent. These clusters interact at long distance and give rise to a spin glass. An ordering temperature of 0.5 K is observed for 1 % of As. If its ordering is via a RKKY interaction it must depend on a very small number of electrons. The same long distance interaction has been observed[20] for V diluted in Ti_2O_3 which also has a very low electron concentration.

SmS and $Sm_{1-x}La_xS$

The presence of accidental Sm^{3+} impurities has been previously recognized by EPR measurements[21] on "pure" SmS at temperature between 4 and 20 K. The samples have concentrations as high as 5 %. The fundamental ground state of Sm^{3+} is the doublet Γ_7.

These Sm^{3+} impurities give a Kondo effect in their resistivity behaviour[22] when the SmS conduction electron concentration is increased by the addition of LaS . The measured Hall constant is independent of the temperature down to very low concentration of Lanthanum . If we substract the log T Kondo term from the total resistivity we obtain a lattice resistivity which takes the Bloch-Gruneisen form. If we superpose the T term of the lattice resistivity above 50 K and the LaS resistivity, the scale factor is approximately the Lanthanum concentration.

It is possible to saturate the Sm^{3+} magnetization in a very high field below 2 K. In fact, the saturation of other accidental rare earths impurities is included in this saturation magnetization. But very low temperature measurements, down to 0.05 K, permit the separation from the Sm^{3+} contribution, because they are easily saturated in very low field compared to the field which is necessary to saturate the small moment of Sm^{3+}. The latter probably disappears at very low temperature. Its Kondo temperature appears to be of the order of 0.05 K.

The study of the transport porperties at high concentration of Lanthanum indicates that the resistivity is nearly independent of the temperature. Although the magnetization measurements have not been made we interpret this phenomenon as due to a large increase of the Kondo temperature of Sm^{3+} when the lattice constant decreases with La addition. Samarium can, therefore, become a Kondo rare earth impurity.

$Sm_{1-x}Y_xS$[27]

The magnetization curve obtained for two concentrations in fields up to 150 kOe, x = 0.05 and 0.09 indicates the absence of a detectable saturable magnetization. Such an absence could indicate, as in the case of $Sm_{1-x}La_xS$ alloys, for $x > 0.1$, that the Kondo temperature is very high. But the fact that the residual resistivity is on an order of magnitude smaller in $Sm_{1-x}Y_xS$ alloys than in $Sm_{1-x}La_xS$ of the same concentration, indicates that some Y alloys contain an order of magnitude less Sm^{3+} than equivalent La alloys. Specific heat measurements on one of the $Sm_{1-x}Y_xS$ alloys confirms these results[23].

In the gold collapsed phase, localized moments appear again and may be the consequence of repeated expansion and dilatation with the temperature. A hyperfine specific heat is measured at very low temperature. If we attribute it entirely to Sm^{3+} we calculate a concentration of 15 % using the ionic value of the hyperfine constant. Such an impurity concentration seems too high and other rare impurities can contribute to this hyperfine specific heat. In addition a remanent magnetization is observed below 1 K by cooling under a 1.6 kOe field. The Sm^{3+} spin glass specific heat is superimposed, at very low temperature, on the linear term determined at higher temperature by Von Molnar[23]. The high temperature linear term is due to the great majority of Sm in an intermediate valence state.

Such an observation of Sm^{3+} localized moments indicates that the local appearance of magnetism is very sensitive to small changes of the local density of states introduced by environment effects due to defects.

SmS under hydrostatic pressure[24]

Properties of SmS, in its black phase, such as the lattice constant, the Van Vleck susceptibility, the transport properties, the concentration of accidental Sm^{3+} impurities[22], vary from sample to sample (see for example ref. 23). In particular, the presence of Sm^{3+} is related to crystalline imperfections[21] which create local collapses.

In order to look at the influence of the defects induced by the phase transition under pressure the SmS magnetization has been measured[24] under hydrostatic pressures up to 9 kbar in fields up to 70 kOe.

The apparent width of the first order transition is due to the inhomogeneity of the critical pressure in each part of a single crystal. It is for this reason that the crystal explodes[25] during the cycle under hydrostatic pressure. The different pressure cycles change the critical pressure of each grain. In the Vth cycle at T = 77 K (fig. II, ref. 24) the transition begins at 7 kbar and is not completed at 8 kbar. In the IXth cycle it begins at 9 kbar. Thus the SmS critical pressure is strongly dependent on the history of the sample and particularly on the residual stress in different parts of the sample after it breaks up. The defects created after several transitions from the black to the gold phase under pressure have increased the critical pressure of each part of the sample.

SmS metallic thin films[26]

The presence of Sm^{3+} detected by E.P.R. measurements[21] in the black SmS has been studied in the gold SmS by measuring the resistivity versus temperature. SmS thin films have been prepared in the black phase and converted to gold collapsed SmS phase by rubbing the surface of a film with a 2 mm diameter glass sphere. The resistivity, between 1 and 300 K, is qualitatively and quantitatively like the resistivity measured by Bader et al.[27] on bulk samples under pressure. The resistivity goes through a minimum and increases when the temperature decreases. The upper limit is attained by a simple T^2 variation. This behaviour demonstrates the presence of the Kondo effect[28][29] with Sm^{3+} impurities which have a higher Kondo temperature than in the case of dilute $Sm_{1-x}LaS$ alloys. The typical Kondo logarithmic decrease of the resistivity with increasing temperature would be obtained if it was possible to substract the lattice resistivity. From these results it can be included that the decrease of the resistivity[27] of the gold SmS increaseswith temperature and that it is not an intrinsic effect but due to Sm^{3+} impurities.

TmS, TmSe

The experiments presented[30][31][32][34] here are interpreted within a model which assumes that TmSe and TmS are essentially the same with the thulium in the trivalent state. This is in agreement with the earlier conclusions based on reflectivity measurements on these monochalcogenides[35]. The anomalies in the low temperature transport[30] and magnetic properties are interpreted here as due to accidental Tm^{2+} impurities. Generally, Tm^{3+} is non magnetic[36] at low temperatures in its singlet ground state. The exchange interactions between Tm^{3+} can induce magnetic moments and magnetic ordering by competition with the crystalline field splitting[36]. Here TmS and TmSe are exchange induced antiferromagnets. We shall argue that there is experimental evidence that these trivalent compounds give rise to a Kondo effect when their excited levels are occcupied as for example in the case of Pr impurities in $LaSn_3$[37]. In fact it is shown for the first time that both Tm^{2+} and Tm^{3+} can give the Kondo effect.

Influence of Tm^{2+} impurities on the lattice constant and the residual resistivity

The experimental lattice constant of TmS fits a smooth curve for the trivalent rare earth monosulfides, so that we can assume that thulium is in the 3^+ state, with a possible very low Tm^{2+} impurity concentration. According to Bucher[38] the lattice constant of TmSe can vary between 5.67 and 5.71 Å. The lattice constants of the samples referred to as 75.II.92 and 75.I.32 are respectively, 5.680 and 5.708 Å. The interpolated value taken from a plot of the rare earth monoselenides vs atomic number is 5.64 Å (the lattice constant reported by Iandelli[39] are obtained on powder simples). Assuming this to be the lattice constant of purely trivalent Tm in TmSe and the increase in lattice constant is due to the presence of the larger divalent ion we calculate a concentration of $\sim$ 20 % Tm^{2+} for the sample having the largest lattice constant. The greater the deviation from the pure divalent lattice constant the greater the value of the residual resistivity[30]. For example TmS has the lowest resistivity at 4 K compared to values of 3x as large for sample 75.II.92 and 6x as large for sample 75.I.32.

The plateau of the residual resistivity at low temperature is approached smoothly as in Kondo dilute alloys. The T^2 law is followed within the precision of measurement. The resistivity decrease permits an evaluation of a Kondo temperature of several K in TmSe and a 10 times larger value in TmS. We observe (ref. 30) that sample H 75.I.32 has a Kondo temperature two times smaller then B. 999.7.4.74, or in general the higher the residual resistivity the lower the Kondo temperature.

This interpretation in terms of the disappearance of the Tm^{2+} magnetic moment due to the Kondo effect, implies that it is not coupled to the Tm^{3+} magnetic moment. This point will be discussed later.

The arrival of magnetic ordering is always accompanied by an increase in resistivity, the amplitude of which depends strongly on the sample[(30)]. In each case the magnetic ordering produces an increase in the Kondo temperature of Tm^{2+} impurities, depending probably on the amplitude of the lattice contraction[(40)] below the ordering temperature equivalent to a pressure effect on the Kondo temperature.

A maximum[(30)] in the TmS resistivity is observed at 10 K, whereas Bucher[(38)] has observed the anomaly at 14 K. We attribute the maximum to the beginning of the disappearance of the Kondo resistivity of Tm^{3+} when Tm^{3+} are dropping into the singlet crystal field ground state before magnetic ordering occurs.

This phenomenon is analogous to that observed by Le Thuillier and Haen[(37)] on $La_{1-x}Pr_xSn_3$ system which shows a Kondo effect and has a singlet crystal-field ground state. The temperature of the maximum is correlated to the crystalline field splitting of the rare earth[(41)]. For TmSe the maximum should also be observed as in TmS because of the trivalent character of Thulium. For the TmSe samples B 999.7.4.74 and H 75.I.32 the maxima[(30)] are at 2.2 and 3.4 K respectively : the Tm^{2+} contribution to the resistivity probably modifies the apparent position of this maximum and for TmSe and TmS the crystal field splitting is sensitive to sample preparation.

TmS and TmSe : Exchange induced antiferromagnets

Bucher et al have observed[(38)] a peak in the susceptibility at T = 5.18 K for TmS and between 1.85 and 2.8 K for TmSe. We find[(32)] 3.1 and 1.7 K for 75.I.32 and 75.II.92 respectively. A remanent magnetization is observed below the ordering temperature, related to monodomain effects which have an uncompensated moment due to vacancies or to the presence of non magnetic Tm^{2+} on the Tm^{3+} sublattices of the antiferromagnetic. The larger the remanent magnetization, the more numerous the imperfection, the greater is the residual resistivity.

As in the case of an exchange induced antiferromagnet or ferromagnet[(36)], the specific heat anomaly[(34)(40)] at the Néel temperature is superimposed on the Schottky anomaly due to the progressive thermal occupation of the first excited levels. The specific heat peak is here also less prominent than that usually found[(36)(42)] for magnetic ordering because much of the entropy of the system has already been removed by ions dropping into the singlet crystal field ground state before magnetic ordering occurs.

The magnetic ordering is also observed by means of the hyperfine specific heat(34) (measured down to 50 mK) which reflects the hypperfine field created on the nuclear moment proportional to the electronic moment. The hyperfine field we obtain from these measurements are 3.55 MOe for TmS and 2.63 MOe for TmSe (75.I.32) whereas a sample from Bucher has a hyperfine field(43) equal to 1.93 MOe and Néel temperature equal to 2.8 K. Thus the moment depends on sample preparation. The distribution of the ordering temperature, magnetic moment, and crystalline field from sample to sample characterizes the high sensitivity of the low temperature properties to imperfections and to the lattice constant and are in agreement with the interpretation based on exchange induced ordering.

Properties of TmS and TmSe in a variable magnetic field

The magnetization, magnetoresistivity and Hall effect have been studied(31)(32) in very high fields. As it has previously been shown by Bucher(40) and Guertin(44) the magnetization increases with a jump as a function of an external field. The greater the amplitude of the jump, the greater the residual resistivity measured. Several examples are given in Fig. 1, ref. 32. Sample 75.II.92 which has no jump at 1.4 K has a resistivity 6 times smaller(30) than sample 75.I.32 which shows a large jump in the magnetization. TmS which has a residual resistivity 20 times smaller than that of sample 75.I.32 shows a very small upturn characterizing a jump of an order of magnitude smaller. The slope of the magnetization reaches a value not far from the inverse of the demagnetizing factor, indicating a possible transition towards ferromagnetism.

The high residual resistivity of TmSe is completely suppressed(31) by an external field depending on the demagnetizing factor in the direction of the external field. Under a field, resistivities smaller than the room temperature value are readily observed and are of the order of 50 μΩcm. A small hysteresis is observed on the TmS resistivity versus the field (Fig. 2 ref. 31).

We attribute these effects to the appearance of a magnetic moment on the Tm^{2+} site. When this moment is saturated the total magnetic resistivity is suppressed. The rapid decrease we observe on a TmSe needle when the field is parallel to the needle axis indicates in some samples the presence of a field induced transition (fig. II, ref. 31).

If we plot the negative magnetoresistivity at 4.2 K, in the paramagnetic region of TmSe, versus the square of the magnetization(45)(46)(47) we obtain straight lines (fig. 3, ref. 31). This result demonstrates that the resistivity is mainly of magnetic origin as in the Kondo effect. The coefficient of M^2 is 10 times smaller for 75.II.92 than for 75.I.32 whereas the magnetization weakly varies from sample to sample.

The resistivity at 4.2 K contains Tm^{2+} and Tm^{3+} contributions whereas the magnetization mainly comes from Tm^{3+}. The fact that

$-\Delta\rho \sim H^2$ continues to be respected, independent of the Tm^{2+} concentration, indicates that this law is respected not only for Tm^{2+} but also for Tm^{3+} which at 4.2 K always gives a contribution to the resistivity. It also implies that the Tm^{2+} contribution to the magnetization is proportional in low fields to the Tm^{3+} magnetization. This proportionality is possible because the molecular field λM^{3+} created on Tm^{2+} is probably larger by an order of magnitude than the external field. Thus Tm^{2+} and Tm^{3+} together contribute to a negative magnetoresistance proportional to the square of the Tm^{3+} magnetization.

The maximum Tm^{3+} contribution to the resistivity is observed on fig. I ref. 31 when a 5 kOe field is applied. In this field the resistivity is lower than the room temperature resistivity at 1.5 K and attains a value which is an order of magnitude larger at 4.2 K whereas the magnetization decreases only by 20 %. This variation is due to the scattering of the conduction electrons on the Tm^{3+} excited levels and will lead to a maximum and after a decrease in the resistivity versus temperature down to the room temperature value. All these results demonstrate the large contribution of Tm^{2+} and Tm^{3+} Kondo effect to the resistivity in each field and at each temperature.

The field necessary to suppress the Tm^{2+} Kondo resistivity of TmS is an order of magnitude larger than for TmSe. It is an indication that the Kondo temperature of Tm^{2+} is an order of magnitude higher in TmS than in TmSe.

Field induced transition[32][38][40][44]

The jump in the Tm^{2+} magnetization is not only due to the Tm^{2+} magnetization which would correspond to an excessive Tm^{2+} concentration. It is a transition from an exchange induced antiferromagnet to an exchange induced ferromagnet as also indicated by the large slope in the magnetization at the critical field, and the rapid fall of the resistivity. This phenomenon is possible if all the Tm^{3+} nearest neighbours of each Tm^{2+} have a positive interaction with Tm^{2+} moment. At a critical exchange field the Tm^{2+} moment is increasing. When the percolation limit between these clusters is attained for a critical concentration of Tm^{2+} ferromagnetism can occur. A jump or an upturn also exists at low Tm^{2+} concentration. The local exchange field produces at a critical value a local ferromagnetic cluster.

This interpretation assumes that Tm^{2+} is non magnetic in zero field in a Tm^{3+} antiferromagnetic lattice as shown by the large increase of the resistivity below the Néel temperature. In the presence of a field there is a moment on Tm^{2+} induced by a positive large molecular field proportional to the Tm^{3+} magnetization. It has been suggested that this phenomenon is possible in a fcc structure[32]. The 12 Tm^{3+} first neighbours can be antiparallel and cannot give any contribution to the molecular field on Tm^{2+}.

It is therefore, the next neighbour interaction which creates the antiferromagnetic coupling. The antiferromagnetic structure of MnO can be described in such a way.

The pressure effects on the magnetization jump have been studied by Guertin[44]. The critical field increases by a factor 4 when the pressure varies from 0 to 10 kbar. This phenomenon can be interpreted in terms of a large increase of the Tm^{2+} Kondo temperature versus pressure as it is generally observed for Cerium impurities. The formation of a ferromagnetic cluster around each Tm^{2+} needs a larger molecular field with a larger Tm^{2+} Kondo temperature.

Exchange induced ordering in Kondo compounds

A Kondo compound contains 100 % of impurities. If the Kondo temperature of the rare earth ground state is larger than the exchange interactions, no magnetic ordering and a high linear term in the specific heat is expected. Then the specific heat goes through a maximum and progressively decreases. The entropy of the ground state is found again at higher temperatures and added to the entropy from the progressive occupation of the crystal field excited levels. Some compounds such as $CeAl_3$[49] $CeIn_3$[50], contain a superimposed peak in their specific heat anomaly due to Kondo effect or due to crystal field levels. A very high linear term[51][52] continues to be observed at low temperatures in the specific heat as a function of the temperature.

We want to argue, by analogy with the observation of a magnetic ordering in singlet ground systems, that the exchange interactions can compete and dominate the Kondo condensation to induce magnetic ordering. The susceptibility per ion in the absence of exchange interaction can be approximated by a Curie-Weiss law[11] $\chi_k = \frac{C}{T+T_k}$ (for $T \gtrsim T_k/4$). If one includes the exchange interaction in the molecular field model $M = \chi_k(H+\lambda M)$

$$\frac{1}{\chi} = \frac{1}{\chi_k} - C\lambda \ ; \ \chi = \frac{C}{T+T_k-C\lambda} \ .$$

The effect of the exchange is to reduce the paramagnetic Curie temperature for a ferromagnetic exchange and to make it more negative for an antiferromagnetic exchange. Thus the effective paramagnetic temperature is $T_k - C\lambda$ depending on the sign of λ. The value of $C\lambda$ for which the susceptibility diverges at $T = 0$ gives the exchange threshold value for the observation of ferromagnetism $\lambda C\chi_{o_k} = 1$, χ_{ok} being the one impurity susceptibility at $T = 0$ K. There will also be an exchange threshold value for the antiferromagnetism depending on the type of magnetic structure we shall have. The simplest case will be $T_k = - C\lambda$.

The specific heat due to magnetic ordering would have the same superimposed character as in the current cases of exchange induced systems.[36]

$CeAl_2$

$CeAl_2$ is well known[41] as a Kondo compound which has a peak in its susceptibility[56][58], its resistivity[57], and its specific heat[54][56][59] around 4 K. Above 4 K, the susceptibility follows a Curie law.[60] Thus, this system is not far from being ferromagnetic ($T_k \gtrsim C$). Below 4 K, when we apply an external field[60] of 40 kOe a ferromagnetic transition is induced by the field. We suggest that in zero field at zero K, $CeAl_2$ is not ordered. Cornut[33] has measured no remanent magnetization down to 0.05 K. A linear term is observed below 1 K and equal to 200 mJ/mole. The initial susceptibility measured by Cornut[33] at very low temperatures is 3.9×10^{-2} uem/mole and independent of the temperature up to 1 K. This value is 20 times larger than the Pauli susceptibility deduced from the density of states derived from the linear term of the specific heat. Such a large enhancement reinforces the interpretation that this compound is nearly ferromagnetic. No interpretation about the apparent λ transition seen in the specific heat around T = 4 K is given.[55][56][59]

Pressure effects have been studied on the magnetization[60][61] and on the specific heat[59]. The critical field for the transition to the ferromagnetism is increased under pressure. The temperature of the maximum of the specific heat is increased in the same ratio. We interpret the origin of these effects is the increase of the cerium Kondo temperature under pressure.

$CeAl_3$

A small peak[59][6] in the specific heat has been observed at T = 6 K and is superimposed on the specific heat anomaly due to the beginning of the occupation of the excited crystal field levels. At very low temperatures a linear term[52] in the specific heat equal to 1620 mJ/mole.K^2 has been observed. The ratio $\frac{\gamma}{\chi}$ is about 0.5 confirming the moment disappears by Kondo effect.

The slow variation ($\chi = \frac{C}{T+23}$) of the susceptibility[62] with temperature eliminates the possibility for $CeAl_3$ to be a weak ferromagnet. The observation of the arrival of a spontaneous magnetization at T = 6 K seems to justify the presence of a magnetic ordering. This saturation depends on the sample. We believe that it is related to the imperfections[59] in an antiferromagnetic lattice and is due to domain effects.

The influence of pressure is always to increase the Kondo temperature[17] and thus to reduce the susceptibility and the linear term in the specific heat as observed[59] in the low temperature region. The ordering temperature is unchanged under pressure probably indicating that the magnetic interactions and the Kondo temperature increase together under pressure.

The peak observed at 2.6 K in the specific heat has nearly disappeared under pressure. So we believe that this compound contains imperfections and therefore many cerium ions have different

Kondo temperatures.

Conclusion

In principle, compounds of rare earth elements with an unstable valence may contain either integral valence states, a uniform intermediate valence state or some combination. In fact, when two configurations are nearly degenerate impurities or defects may have a very large effect, stabilizing one valence locally even in the presence of the others.

In these systems a variety of behavior is possible depending on the energy difference between the configurations, the concentration and type of defects and the coupling between the rare earth ions. For example localized moments which can give magnetic ordering or Kondo effect or singlet crystal field ground state can be observed on the rare earth depending on the relative importance of its interaction with the conduction band, the crystal field and the exchange interaction.

We have discussed many new results which illustrate this point of view. The presence of Sm^{3+} has been found in many SmS compounds leading to Kondo effect or to spin glass ordering. Tm^{2+} is also present as Kondo impurities in $Tm^{3+}Se$ and $Tm^{3+}S$. These last compounds are Kondo compounds when Tm^{3+} excited levels are populated. They are new examples of Van Vleck compounds giving Kondo effect similar to the effects observed in $LaPrSn_3$.

In some Kondo compounds such as $CeAl_3$, magnetic interactions compete with the Kondo condensation so that weak antiferromagnetism is superimposed on the usual properties of a system in a singlet Kondo ground state.

Acknowledgments : The authors are grateful to B. Coqblin, J. Chaussy, G. Chouteau, B. Cornut, P. Haen, T. Penney, S. Von Molnar, for many fruitful discussions.

*Permanent address : IBM Research Center, Yorktown Hts N.Y. 10598.

1 - B. Coqblin, A.K. Bhattacharjee, B. Cornut, F. Gonzalez-Gimenez, F.R. Iglesias-Sicardi, R. Jullien, Journal of Magn. and Magn. Mat. 3, 67 (1976).

2 - J.C. Nickerson, R.M. White, K.N. Lee, R. Bachmann, T.H. Geballe, G.W. Hull, Phys. Rev. B3, 2030 (1971).

3 - S.D. Bader, N.E. Philips, D.B. Mc Whan, Phys. Rev. B7, 4686 (1973).

4 - N.E. Philipps, J.C. Ho, T.F. Smith, Phys. Lett. 27A, 49 (1968). " Ce".
O.K. Lounasmaa, Phys. Rev. 133, A502 (1964).

5 - "$CeSn_3$", J.R. Cooper, C. Rizzuto, G. Olcese, J. Phys. 32, C1-1136 (1971).

6 - "$CeAl_3$"
M.H. Van Maaren, K.H.J. Buschow, H.J. Van Daal, Solid State

Comm. 9, 1981 (1971).
7 - B. Coqblin, J.R. Schrieffer, Phys. Rev. 185, 847 (1969).
8 - "AuCo"
J. Le Guillerm, Thèse, Université de Grenoble 1963 (unpublished).
P. Costa-Ribeiro, J. Souletie, and D. Thoulouze, Phys. Rev. Lett. 24, 900 (1970).
E. Boucaï, B. Lecoanet, J. Pilon, J.L. Tholence, and R. Tournier, Phys. Rev. B3, 3834 (1971).
9 - "CuCo"
R. Tournier, Thèse Université de Grenoble 1965
R. Tournier and A. Blandin, Phys. Rev. Lett. 24, 397 (1970).
10 - G. Frossati, J.L. Tholence, D. Thoulouze, and R. Tournier, Physica 84B, 33 (1976).
11 - J.L. Tholence and R. Tournier, Phys. Rev. Lett. 25, 867 (1970).
12 - V. Jaccarino, L.R. Walker, Phys. Rev. Lett. 15, 258 (1965).
13 - O. Peña, F. Meunier, Solid State Comm. 14, 1087 (1974).
14 - "LaCe", A. Benoit, R. Delaplace, J. Flouquet, Phys. Rev. Lett. 32, 222 (1974).
15 - S. Ortega, M. Roth, C. Rizzuto, and M.B. Maple, Sol. State Comm. 13, 5 (1973).
16 - M.B. Maple, J. Wittig, K.S. Kim, Phys. Rev. Lett. 23, 1375 (1969).
17 - B. Coqblin, J. Phys. 32, C1-599 (1971).
B. Coqblin, M.B. Maple, G. Toulouse, Inter. J. Mag. 1, 333 (1971).
18 - O. Peña, R. Tournier, A. Benoit, J. Flouquet (to be published).
19 - F. Holtzberg, O. Peña, T. Penney, R. Tournier, This Conference.
20 - J. Dumas, C. Schlenker, J.L. Tholence, R. Tournier, AIP Conf. Proc. 29, 431 (1975).
21 - W.M. Walsh, Jr., E. Bucher, L.W. Rupp, L.D. Longinotti, AIP Conf. Proc. 24, 34 (1974).
22 - G. Chouteau, F. Holtzberg, O. Peña, T. Penney, R. Tournier, S. Von Molnar, J. Phys. Coll. 37, C4-(1976).
23 - S. Von Molnar, T. Penney, F. Holtzberg, J. Phys. Coll. 37, C4-(1976).
24 - B. Cornut, F. Holtzberg, O. Peña, J. Voiron, This Conference.
25 - F. Holtzberg, AIP Conf. Proc. 18, 478 (1974).
26 - A. Chenevas-Paule, P. Haen, F. Lapierre, R. Tournier and L. Vieux-Rochas, This Conference.
27 - F. Holtzberg, J.C. Lasjaunias, J.M. Mignot, O. Peña, A. Ravex, and M. Vandorpe (unpublished).
28 - P. Nozières, Low Temperature Physics LT 14, edited by Matti Krusius and Matti Vuorio, North-Holland/American Elsevier, Vol. 5, 339 (1975).
29 - W.M. Star, F.B. Basters, G.M. Nap, E. de Vroede, and C. Van Baarle, Physica 58, 585 (1972).
30 - A. Berger, E. Bucher, P. Haen, F. Holtzberg, F. Lapierre, T. Penney, R. Tournier, This Conference.

31 - P. Haen, F. Holtzberg, F. Lapierre, T. Penney, and R. Tournier, This Conference.
32 - G. Chouteau, F. Holtzberg, J.M. Mignot, O. Peña, and R. Tournier, This Conference.
33 - B. Cornut, Thèse Université de Grenoble 1976, N° d'ordre CNRS A012281.
34 - F. Holtzberg, J.C. Lasjaunias, J.M. Mignot, A. Ravex, M. Vandorpe, This Conference.
35 - B. Batlogg, E. Kaldis, A. Schlegel, P. Wachter, Phys. Lett. 56A, 122 (1976).
36 - B.R. Cooper, Critical Review in Sol. State Sc. 3, 83 (1972).
37 - P. Lethuillier, P. Haen, Phys. Rev. Lett. 35, 1391 (1975).
38 - E. Bucher, K. Andres, F.J. Di Salvo, J.M. Maita, A.C. Gossard, A.S. Cooper, and G.W. Hull Jr., Phys. Rev. B11, 500 (1975).
39 - A. Iandelli, Rare Earth Research, edited by E. Kleber The Mac Millan Company, New York 135 (1961).
40 - H.R. Ott, K. Andres, E. Bucher, AIP Conf. Proc. 24, 40 (1974).
41 - B. Cornut, B. Coqblin, Phys. Rev. B5, 4541 (1972).
42 - T.E. Katila, N.E. Philipps, M.C. Veuro, and B.B. Triplett, Phys. Rev. B6, 1827 (1972).
43 - B.B. Triplett, N.J. Dixon, P. Boolchand, and S.S. Hanna, J. Physique Coll. 35, C6-653 (1974).
44 - R.P. Guertin, S. Foner, F.P. Missel, Phys. Rev. Lett. 37, 529 (1976).
45 - J.L. Genicon, F. Lapierre, J. Souletie, Phys. Rev. B10, 3976 (1974).
46 - J. Souletie, AIP Conf. Proc. 29, 294 (1975).
47 - P. Monod, Phys. Rev. Lett. 19, 1113 (1967).
48 - J. Kondo, Solid State Physics 23, 184 (1969). F. Seitz, D. Turnbull, and H. Ehrenreich, editor Academic Press, New York and London.
49 - J.V. Mahoney, V.U.S. Rao, W.E. Wallace, R.S. Craig, and N.G. Nereson, Phys. Rev. B9, 154 (1974).
50 - A.M. Van Diepen, R.S. Craig, W.E. Wallace, J. Phys. Chem. Solids 32, 1867 (1971).
51 - S. Nasu, A.M. Van Diepen, H.H. Neumann and R.S. Craig, J. Phys. Chem. Solids 32, 2773 (1971).
52 - K. Andres, J.E. Graebner, and H.R. Ott, Phys. Rev. Lett. 35, 1779 (1975).
53 - K. Andres, E. Bucher, S. Darack, and J.P. Maita, Phys. Rev. B6, 2716 (1972).
54 - R.W. Hill, J.M. Machado da Silva, Phys. Lett. 30A, 13 (1969). C. Deenedas, A.W. Thompson, R.S. Craig, and W.E. Wallace, J. Phys. Chem. Sol. 32, 1853 (1971).
55 - F. Steglich and M. Loewenhaupt, This Conference.
56 - M.C. Croft, I. Zorić and R.D. Parks, This Conference. J. Markovics, L. Kupferberg, M. Croft, I. Zorić, and R.D. Parks, This Conference.
57 - M. Nicolas-Francillon, A. Percheron, J.C. Achard, O. Gorochov, B. Cornut, D. Jerome, B. Coqblin, Solid State Comm. 11, 845

(1972).

58 - W.M. Swift and W.E. Wallace, J. Phys. Chem. Solids 29, 2053 (1968).
J.A. White, H.J. Williams, J.H. Wernick, and R.C. Sherwood, Phys. Rev. 131, 1039 (1963).
V. Nicolescu, I. Pop, M. Rosenberg, Phys. Stat. Sol.(b) 53, 701 (1972).

59 - A. Berton, J. Chaussy, G. Chouteau, B. Cornut, J. Peyrard, and R. Tournier, This Conference.

60 - B. Barbara, M.F. Rossignol, H.G. Purwins, and E. Walker, Sol. State Comm. 17, 1525 (1975).

61 - B. Barbara, H. Bartholin, D. Florence, M.F. Rossignol, and E. Walker, to appear in the ICM 76 Proceedings.

62 - A.S. Edelstein, R.E. Majewski and T.H. Blewitt, This Conf.

COMMENTS

SHAPIRO: One thing that has become clear about these Tm compounds is that their behavior is very sample dependent. If an experimentalist wanted to do some work on these what is a good sample and what is a bad sample?

TOURNIER: It depends on what you want. If you have no Tm^{2+} impurities you have no ferromagnetic transition. You can imagine that you form a ferromagnetic cluster around each Tm^{2+} for some critical exchange field. With a sufficient number of Tm^{2+} ions, you have percolation and ferromagnetism. The fact that the appearance of this magnetism does not depend on concentration proves that it's a local effect. It exists also in TmS where an upturn in the magnetism is seen. The amplitude of the magnetization is so large that it cannot be due to the Tm^{2+} ions alone; you must also consider the twelve surrounding Tm^{3+} ions.

WOHLLEBEN: If there are sufficiently fast fluctuations in TmSe, you would expect that the hyperfine Moessbauer multiplet collapses due to motional narrowing. Thus you shouldn't see a hyperfine specific heat unless the hyperfine field is very large or the fluctuation frequency very small. Have you estimated the maximum fluctuation frequency consistent with your hyperfine specific heat?

TOURNIER: No. We have a remnant magnetization combined with a peak in the susceptibility. This, I think, really indicates an ordering.

OTT: In TmS you can explain the specific heat rather nicely if you assume a Γ_1, Γ_4 splitting of 15-20°K. In the selinide, it is just about impossible. The specific heat has to be explained by assuming a dominant Tm^{3+} configuration. Thus to explain the specific heat the splitting would have to be above 40-50°K, where one should certainly see crystal field effects in the elastic constants, but doesn't.

ORBACH: You mentioned that there was a T^2 temperature dependence of the resistivity in SmS films. If you expand that as $1-(T/T_k)^2$, do you get the same characteristic temperature as from the susceptibility measurements?

TOURNIER: That's a problem. You ask a very good question. The Kondo temperature we deduced was about 30 or 50 degrees. It is approximately the same as deduced from the electronic specific heat if you attribute the electronic specific heat to a Kondo or spin fluctuation effect. The purpose of the resistivity study was to see if perhaps SmS was a Kondo metal, but the resistivity attains a plateau. The first question is why does it not go down?

The flat plateau we obtain is in fact due to accidential Sm^{3+} impurities. We have Sm^{3+} ions which are not in the same situation as the regular lattice. Walsh has done some very good work with EPR which has shown the presence of that. We have shown the manifestation of magnetization. Here we have a Kondo temperature which is larger. The Sm^{3+} which is present here has a valency different from the intermediate phase. It is more near 3^+ and has a Kondo behavior.

EDELSTEIN: I would like to make a comment which may bear on the question of what is happening at 6K in $CeAl_3$. We observe structure in the 2nd derivative at 6K of the spline fit to the resistivity both before and after it was disordered by neutron irradiation; however, this structure is absent in the difference. Hence, disorder does not effect whatever is going on at 6K. This suggests to me that it is unlikely that the anomaly is connected with the occurrence of bulk magnetic ordering.

PHOTOEMISSION, X-RAY EMISSION AND ELECTRON ENERGY LOSS SPECTROSCOPY OF MIXED-VALENT COMPOUNDS

M. Campagna

Bell Laboratories, Murray Hill, New Jersey 07974

The recent progress made in understanding the multiplet structure, lineshape and intensities in 4f-photoemission is briefly reviewed. The application of high-resolution X-ray and ultraviolet photoemission spectroscopy (XPS and UPS resp.) to the study of configurationally mixed electron states near the Fermi energy in metallic systems is discussed. Effects of configurational mixing on core levels is described, as seen by XPS and X-ray emission spectroscopy. Sm and Ce metal will be considered in this special case. An example of a detailed analysis of 4f-photoemission lineshapes in substituted $Sm_{1-x}R_xS$ alloys, R = Ca, Y and Th is presented and shown to provide new information on the role of electronic effects in the valence transition of Sm. The use of electron energy loss spectroscopy (ELS) to study intraionic f → d and charge transfer transitions will also be discussed. These results are compared with XPS and UPS data with the aim of demonstrating the possibilities of these techniques to cast new light on the electronic structure of metallic mixed-valent materials.

This work was done in collaboration with G. K. Wertheim.

COMMENTS

WOHLLEBEN: If you do XPS on valence d-electrons in transition metals, can you see the type of spectrum that you very clearly see in the rare earths?

CAMPAGNA: The answer is no. The 3d electrons in metallic systems are really kind of delocalized. For example, if you look up the XPS spectrum of Fe, you will find that you resolve neither the band character of the d-electrons nor the multiplet structure that you might expect, if, e.g., the Fe were in a d^8 or d^7 configuration. On the other hand, you might remember that we have spin-polarized photoemission experiments that cannot be easily explained on the basis of band theory. Hence the question of the 3d's is very open. Equally interesting is the question of what we will see in photoemission spectroscopy on the actinides. There the 5f electrons are not as localized as the 4f; hence, will we see the multiplet structure or not? This can be a first step before we can understand what's going on in the 3d's.

POLLAK: Eastman and Freeouf did use that scheme with the transition metal oxides and it worked pretty well for the non-conducting materials. There is a recent Physical Review Letter on that: Phys. Rev. Letters 34, 395 (1975).

DONIACH: I would like to make a comment about the relaxation shifts. When you pull an electron out of a core level, we know there are very large shifts associated with the relaxing of the atom and screening. If you pull them out of the f-level right near the Fermi surface, it worries me a little, because when the conduction electrons exchange with the f-level they're doing it very close to the Fermi level. In XPS, you're pulling them out to a kilovolt and the relaxation shifts may be quite different.

PHASE DIAGRAM, OPTICAL AND PHONONIC PROPERTIES OF THE VALENCE INSTABILITY OF SmS

G. Güntherodt, R. Keller, P. Grünberg[+], A. Frey, W. Kress, R. Merlin, W. B. Holzapfel and F. Holtzberg*

Max-Planck-Institut für Festkörperforschung

Stuttgart, Federal Republic of Germany

ABSTRACT

The lattice constant of SmS single crystals has been measured at 300K under hydrostatic pressure up to 70 kbar. Linear extrapolation to zero pressure and comparison with the lattice constant of trivalent rare earth chalcogenides at atmospheric pressure yields a Sm valence of 2.8 just after the phase transition. A new phase appears above 20 kbar with a Sm valence of 2.9. The valence state of Sm in SmS and $Sm_{1-x}Y_xS$ is identified by the final hole state multiplet structure. The configurational crossover in $Sm_{1-x}Y_xS$ ($x \gtrsim 0.15$) results in strong resonant scattering of conduction electrons at the Fermi level, as deduced from an analysis of optical data. Upon cooling metallic SmS under hydrostatic pressure from 300K to 4.2K we observe a 4% shift in the plasma reflection edge towards lower energy. Hence we can attribute the six fold increase in resistivity of SmS in the collapsed phase at low temperature to an increased scattering of conduction electrons, not necessarily associated with the 4f level of the matrix. The electron-phonon coupling has been studied for optical excitations $4f^n \rightarrow 4f^{n-1}\,5d^1$ in EuS and SmS by means of Raman Scattering. The associated volume change around the rare earth ion is seen to couple strongly to LO phonons, particularly to LO(L). The latter are about 20% "softer" in semiconducting SmS than in EuS. This softening is understood in terms of a lattice dynamic model as originating from the higher compressibility of SmS, presumably due to f-d hybridization. The extension of the model towards configuration crossover is discussed.

1. INTRODUCTION

Valence instabilities of rare earth (RE) elements and ions are found at the beginning (Ce), middle (Sm,Eu) and end (Tm,Yb) of the lanthanide series. For example in RE monochalcogenides RE-X (X=O, S, Se, Te) the half filled and filled 4f shell strongly favor the divalent state of the RE ion. For these essentially ionic compounds there is a strong correlation between the volume and the valence of the RE ion which is directly related to the 4f occupation number. Since the lattice constants of the $RE^{2+}X$ and $RE^{3+}X$ form smoothly varying curves as a function of atomic number, an intermediate valence can be defined by interpolation between the two extremes. The isostructural semiconductor-metal transition of the $RE^{2+}X$ is known to lead to a crossover of $RE^{2+}(4f^n)$ and $RE^{3+}(4f^{n-1}5d^1)$ configurations. In the special case of SmS, the first order electronic phase transition at pressures near 6.5 kbar or the discontinuous collapse of the lattice constant at a critical concentration of solid solutions formed with trivalent ions result in a nonmagnetic, intermediate-valence state. These effects have been the central theme of much research in recent years [1].

In this paper we focus on three aspects of valence instabilities in SmS: a) the volume change under hydrostatic pressure with relation to the phase diagram, compressibility and Sm valence; b) optical properties of SmS under hydrostatic pressure and of $Sm_{1-x}Y_xS$ with respect to the phase diagram and transport properties; and c) phononic properties of semiconducting SmS, interpreted on the basis of a lattice dynamic model and its possible extension towards configuration crossover.

In order to test the above interpolation scheme for the Sm valence, we have carried out accurate measurements of the lattice constant of SmS single crystals under hydrostatic pressure up to 70 kbar. The accuracy of the measurements permits an extrapolation of the lattice constant to zero pressure. This makes comparisons with the lattice constants of $RE^{3+}S$ at atmospheric pressure much more meaningful. Furthermore, accurate data on the compressibility up to 70 kbar have been obtained. This is relevant to our recent studies of the phononic properties of SmS.

The dramatic changes in color of SmS as a function of pressure, alloying or temperature naturally suggest the quantitative measurement and analysis of the optical reflectivity. The latter provides information about plasmons, changes in concentration and/or scattering time of the conduction electrons. This information in turn can be complementary to results obtained from transport measurements. As with X-ray photoemission (XPS), optical absorption is valuable in identifying the valence state of the RE ion from its final hole state (FHS) multiplet structure.

Another key question concerns the kind of phonons associated with valence instabilities. Ionic size and polaronic effects seem to play an important role in intermediate-valence materials [2-6]. We present here results of Raman scattering (RS) measurements, interpreted in the framework of a lattice dynamic model. The initial optical excitation in RS from the localized 4f levels and the electron-phonon coupling in the intermediate state provide insight into phononic properties associated with configuration crossover.

2. PHASE TRANSITION, COMPRESSIBILITY AND VALENCE OF SmS

The pressure dependence of the lattice constant of SmS single crystals has been measured at 300K under purely hydrostatic pressure up to 70 kbar. The experimental technique and the results will be described in more detail elsewhere [7]. Visual observation of the semiconductor-metal transition under a microscope indicated a transition delayed to about 10 kbar. These results do not agree with previous measurements of the pressure dependent volume change of SmS powder for which one observes a smearing of the phase transition. [8] The volume change at the first order phase transition amounts to about 14%. The derived compressibility of the semiconducting phase of SmS is $(2.28 \pm 0.05) \cdot 10^{-3}$ $kbar^{-1}$. This is in good agreement with other measurements [9,10]. Between 10 and 20 kbar a linear pressure dependence of the lattice constant is observed. The hysteretic back transformation upon releasing the pressure follows the same straight line down to about 2 kbar. This leads to a compressibility of $2.01 \cdot 10^{-3}$ $kbar^{-1}$ for the metallic collapsed phase of SmS between 10 and 20 kbar. Unfortunately, we were not able to obtain any information on the compressibility between 6.5 and 10 kbar. In this pressure range we would actually expect an enhancement of the compressibility, similar to that observed in $Sm_{1-x}Y_xS$ for $0.15 < x < 0.2$. [10] For pressures above 20 kbar, our measurements suggest the presence of a new phase. We find another fairly linear pressure dependence of the lattice constant between 20 and 70 kbar. The slope is smaller compared to that below 20 kbar. The compressibility of this high pressure phase is $1.11 \cdot 10^{-3}$ $kbar^{-1}$. The question is whether this new phase can be attributed to $Sm^{3+}S$. From a linear extrapolation of the lattice constant to zero pressure and comparison with that of $Sm^{3+}S$ (5.62Å) we deduce a Sm valence of 2.81 for the collapsed phase below 20 kbar, and 2.91 for pressures between 20 and 70 kbar. The value of 2.81 is close to that of 2.77 previously derived from the lattice constant of a powder at 10 kbar [8,11]. Although the agreement appears to be rather fortuitous, it clearly rules out any speculations inferred from questionable data [12]. The smaller Sm valence of 2.7 in the collapsed phase of $Sm_{1-x}Y_xS$ [13] may be associated with the increased conduction electron concentration introduced by the Y^{3+} ions.

A comparison of the compressibility for p > 20 kbar with that of $Gd^{3+}S$ ($0.83 \cdot 10^{-3}$ $kbar^{-1}$) [14] shows a 25% difference. A similar difference of 21% exists also for the compressibilities of rombohedral Sm and hcp Gd metal, which are generally assumed to be trivalent [15]. Despite the different crystal structures, both metals have a close packed structure. The anomalously enhanced compressibility of Sm metal is similar to that of γ-Ce and La [15]. This suggests that f-d hybridization or mixed-valent behavior in the case of Sm metal [16] are possibly responsible for this anomalous behavior. Consequently, there may still be strong f-d hybridization associated with the appearance of a new phase of SmS above 20 kbar.

The pure Sm^{3+} state is probably not stabilized up to very high pressures (p > 20 kbar) due to the presence of the $4f^5 5d^1$ configuration in the vicinity of the Fermi level E_F. This can be contrasted to the situation in $RE^{3+}X$ compounds, where the 4f states are several eV below E_F, without f-d hybridization and a large energy separation between $4f^n$ and $4f^{n-1}5d^1$ configurations.

3. OPTICAL PROPERTIES

Some of the optical properties of SmS have been discussed recently [12,17]. Optical studies of $Sm_{1-x}Y_xS$ [18] have been reviewed elsewhere [19]. Here we give a very brief summary and present some new data on SmS under hydrostatic pressure.

The pure valence state of Sm^{2+} and Sm^{3+} can be identified in optical absorption by the FHS multiplet structure as shown in Fig. 1. Energy separations of multiplet levels are taken from atomic spectroscopy and intensities have been calculated by means of fractional parentage coefficients [20]. Due to the crystal field splitting of the final 5d states we obtain two sets of multiplets (solid and dashed lines in Fig. 1). The intensities have been multiplied by the degeneracies of the 5d states. It has been noted recently in the case of GdS, GdP [19] and Gd metal [21] that binding energies of 4f states are about 2eV smaller in optical absorption compared to XPS. This might be due to different final states [19]. Thus we have placed the $Sm^{3+}(^5I)$ multiplet at about 4eV in Fig. 1, compared to the assignment at 6eV in XPS of SmAs [22]. The agreement obtained appears to be superior to our previous assignment [19].

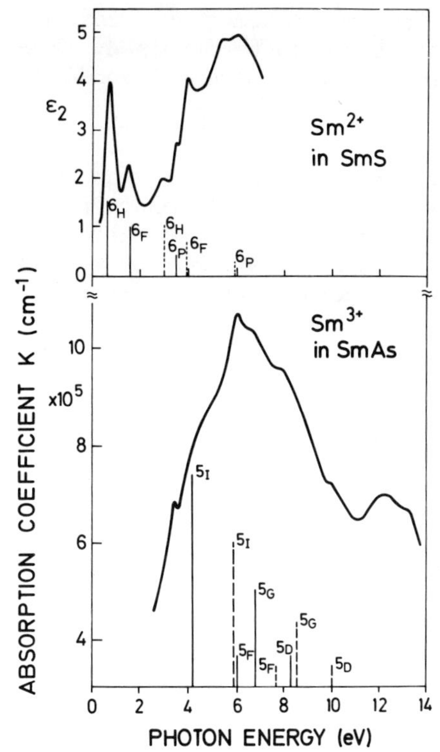

Fig. 1 Final hole state multiplet structures of $Sm^{2+}(4f^6)$ and Sm^{3+} $(4f^5)$ in optical absorption; (——) $5dt_{2g}$ and (---) $5de_g$ final states.

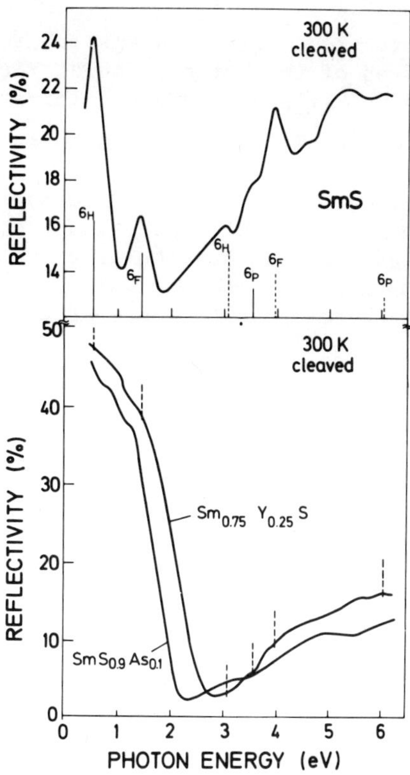

Fig. 2 Reflectivity of semiconducting SmS and metallic, collapsed $Sm_{0.75}Y_{0.25}S$ and $SmS_{0.9}As_{0.1}$.

Semiconducting SmS is black in color due to its low reflectivity all over the visible region as seen in Fig. 2. Alloying SmS with YS or SmAs results in a metallic, collapsed phase above a certain critical concentration. The associated metal-like plasma reflection edge in Fig. 2 is responsible for the golden and purple color, respectively, of $Sm_{0.75}Y_{0.25}S$ and $SmS_{0.9}As_{0.1}$. The two dopants differ in that each Y^{3+} provides one conduction electron, whereas only 1/20 of that is introduced by alloying with semimetallic SmAs. Interestingly enough, for $Sm_{0.75}Y_{0.25}S$ we can still identify the Sm^{2+} FHS multiplet structure in the collapsed phase, as indicated in Fig. 2 by the dashed lines. This also indicates clearly that there is an overlap of intra- and interband transitions. Obviously there are severe problems in decomposing them. Hence no accurate values of the optical dielectric constant ε_∞ can be obtained. The plasma reflection edges originate primarily from screened conduction electron plasma oscillations at $\omega_p^2 = \frac{4\pi ne^2}{m^*\varepsilon_\infty}$, as derived from $\varepsilon_1 = 0$ [18,19], where m* and n, respectively, are the effective mass and concentration of conduction electrons. Because of the above complications and particularly due to the ambiguities in m* one cannot determine n or the Sm valence from ω_p. Assuming values for ε_∞ we have determined the dc scattering time of conduction electrons $\tau_{dc} \sim \frac{\sigma_{dc}}{\omega_p^2}$, accounting for scattering within $E_F \pm kT$ [18,19]. The result is shown in Fig. 3. The collapsed gold phase ($x \gtrsim 0.15$) is characterized by additional, strong resonant scattering compared to the black phase ($x < 0.15$). Metallic, collapsed SmS shows at 300K (Fig. 3) a similar order of magnitude scattering as $Sm_{1-x}Y_xS$ for $x \gtrsim 0.15$. Reduction in the number of 4f states for $x \to 1.0$ results in reduced scattering. This is consistent with a model of a pinning of the Fermi level within the 4f level [23].

By cooling a $Sm_{0.72}Y_{0.28}S$ sample from 300K to 5K we see in Fig. 4 an unusually large (15%) shift of the plasma edge. This is primarily due to a localization of conduction electrons in the 4f level. Further support for this localization is given by the change in Sm valence from 2.7 (300K) to 2.4 (5K) and the concurrent increase in lattice constant [23]. Neglecting changes in ε_∞ and m* upon cooling from 300K to 5K, the 15% change of ω_p in Fig. 4 gives an upper limit of a 30% reduction in conduction electron concentration. The reduction in n is in the opposite sense expected from the measured decrease in resistivity $\rho = \frac{m^*}{e^2 n\tau}$ from 300K to 5K (by a factor of 3 for $Sm_{0.8}Y_{0.2}S$) [23]. Consequently, the decreasing resistivity has to be associated with a strong increase in τ. Indeed, this is observed for a $Sm_{0.75}Y_{0.25}S$ sample in Fig. 3. The reduced scattering is consistent with a model placing the Fermi level above the 4f level in the low temperature black phase for $x \gtrsim 0.15$ [23].

On the other hand, there is an increase in resistivity of metallic SmS ($p > 6.5$ kbar) upon cooling to 4.2K. In order to determine whether such an increase in ρ is associated with changes in n or τ, we have measured the reflectivity of SmS for $p > 6.5$ kbar from 300K to 4.2K. Simulations of this experiment on mechanically polished single crystals [12] or thin films [17] are not conclusive because of possible substrate effects. [17, 19] Our re-

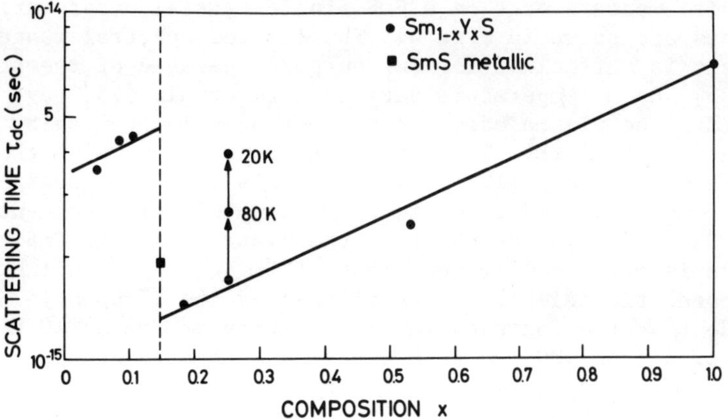

Fig. 3 DC scattering time of conduction electrons at 300K in $Sm_{1-x}Y_xS$ and metallic SmS.

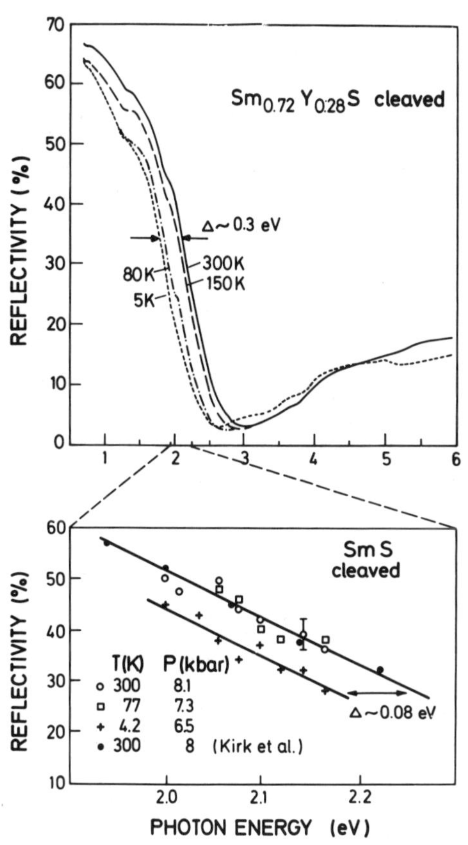

Fig. 4 Temperature dependence of the reflectivity of $Sm_{0.72}Y_{0.28}S$ and of a SmS single crystal under hydrostatic pressure.

flectivity measurements on a SmS single crystal under various gas pressures are shown in Fig. 4. The limited spectral range of a dye laser is sufficient for our purposes because of the steep plasma edge.[25] Room temperature data by Kirk et al. [25] have been included. The plasma edge shifts continuously from below 77K upon cooling down to 4.2K. These results do not agree with those made on polished surfaces [12, 17]. From this shift we obtain an upper limit of about 8% reduction in conduction electron concentration from 300K to 4.2K. The shift of the plasma edge for SmS under pressure is much smaller than that of $Sm_{0.72}Y_{0.28}S$, which has extra conduction electrons contributed by the Y dopant. Hence we conclude that the increase in resistivity of SmS at 10 kbar by a factor of 6 from 300K to 4.2K is not due to a change in n, but rather a decrease in τ. In a first approximation we associate the small shift in plasma edge of SmS upon cooling with a localization of conduction electrons in the 4f level. Consequently the increase in scattering at low temperatures is most likely not associated with the 4f level of the matrix, but rather with Sm^{3+} impurities. [26] Some evidence for the presence of these impurities in our samples has been obtained from photoluminescence measurements. [27]

From the temperature dependence of the reflectivity of metallic SmS in Fig. 4 we have no direct indication of an f-d hybridization gap appearing below 20K. [1, 28] The continuous shift of the plasma edge goes in parallel with the continuous increase in ρ. Hence we have no direct evidence of a first order gold to black transition upon cooling, as observed in $Sm_{1-x}Y_xS$ ($x \gtrsim 0.15$).

4. PHONONIC PROPERTIES

The electron-phonon coupling is of central interest in the case of valence instabilities. In this respect the intermediate-state electron-phonon interaction in RS can provide some insight. The initial state in RS of divalent RE chalcogenides $RE^{2+}X$ is the localized 4f level. The initial optical excitation $4f^n \rightarrow 4f^{n-1}5d^1$ implies a valence change from RE^{2+} to RE^{3+}. The associated volume change around the RE ion due to the much smaller RE^{3+} ionic radius should couple strongly to vibrations of the surrounding nearest neighbor (nn) anions. Such a vibration corresponds to an optical phonon mode. We have performed resonant RS measurements on EuS at 300K, with the initial optical excitation $4f^7 \rightarrow 4f^6 5d^1$. The momentum conservation of RS has been found to be relaxed in paramagnetic EuS due to spin disorder and the associated loss of translational symmetry [29]. Hence the RS is no more restricted to $\vec{k} \sim 0$, but rather density of states (DOS) from the entire Brillouin zone (BZ) will be obtained. The result for EuS is given in Fig. 5. It has been shown recently that the fundamental mode of EuS at $\omega_o = 240$ cm^{-1}, with $\omega_{TO} < \omega_o < \omega_{LO}$, originates mainly from the high density of states of the LO branch near the L point of the (BZ) [29]. The

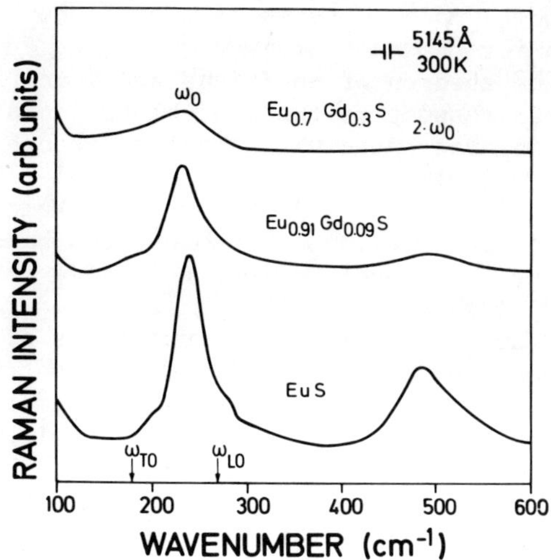

Fig. 5 Raman spectra of $Eu_{1-x}Gd_xS$ at 300K.

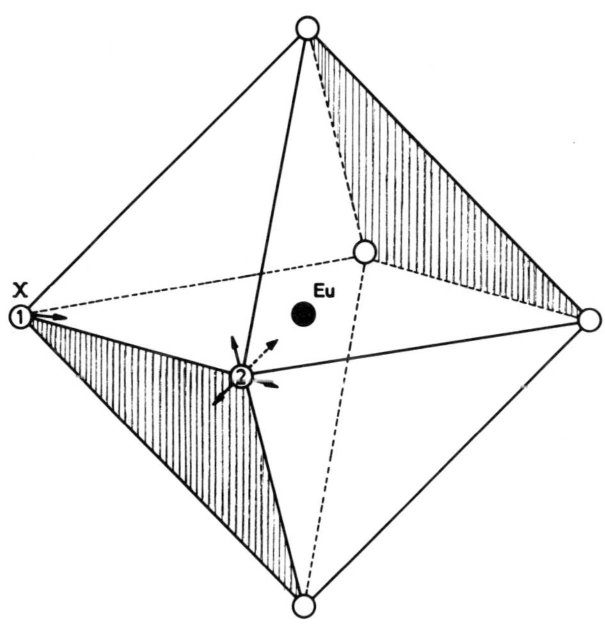

Fig. 6 Nearest neighbor S coordination of Eu, indicating the LO(L) mode symmetry.

symmetry of the LO(L) mode in terms of normal coordinates is illustrated in Fig. 6 for the nn coordination of the Eu ion. Hatched triangles of the octahedron of nn S ions are found to vibrate against each other and with respect to the central Eu ion at rest. This mode, obviously, is associated with a volume change around the Eu ion upon exciting one electron from the 4f level. This is similar to the assumption made in a model by Varma and Heine [4] that the promotion of an 4f electron into the conduction band leads to a lattice contraction. In order to test this idea further we have doped EuS with Gd^{3+}, which has a 9% smaller ionic radius. The LO(L) mode at ω_o in EuS and its overtones at multiples of ω_o are seen to be strongly quenched with increasing Gd concentration as shown in Fig. 5. The quenching of the LO(L) mode by smaller size Gd^{3+} points to the importance of ionic size - and environmental effects when exciting an electron from the 4f level. On the other hand, no quenching of the LO(L) mode has been observed in $Eu_{1-x}Sr_xS$ when substituting Eu^{2+} by isoelectronic Sr^{2+} of very similar ionic radius [30]. A quenching of the LO(L) mode in $Eu_{1-x}Gd_xS$ by free carriers due to Gd^{3+} doping can be ruled out since the Thomas-Fermi screening wave vector amounts only to at most half the Brillouin zone. Thus zone boundary phonons will not be affected by the free carriers. The ω_o mode of EuS shows a broadening and slight shift towards lower frequencies with increasing Gd concentration. This may be due to anharmonic strain effects associated with the Gd substitution. Our experimental observations in $Eu_{1-x}Gd_xS$ bear some resemblance to a model by Anderson and Chui [2], where the substitution of Sm^{3+} by Sm^{2+} in a $Sm^{3+}B_6$ matrix is found to create large anharmonic strain effects.

Semiconducting SmS is paramagnetic with a J=0 ground state. Consequently spin disorder scattering is absent. The same refers to the "forbidden" LO(Γ) Fröhlich scattering as observed in ferromagnetically saturated EuX or diamagnetic YbX [29,30]. The reason for the latter is presumably the screening of the LO(Γ) phonons by free carriers ($\sim 10^{19}$ cm^{-3}) in nominally semiconducting SmS. However, we observe defect-induced first order RS in semiconducting SmS as shown in Fig. 7. Besides the J=0 → J=1 electronic RS [31] we observe at 300K a double peak structure around 100 cm^{-1}. This coincides quite well with the calculated one-phonon DOS of EuS [29,30]. The peak around 193 cm^{-1} at 300K is identified by its temperature dependence as due to first order RS. A second order RS process at this frequency would be quenched by a factor of about 5 upon cooling from 300K to 80K. The peak at 193 cm^{-1} (300K) is positioned between the two maxima of the one-phonon DOS of TO and LO branches of EuS [29,30]. The missing double peak structure around 193 cm^{-1} in Fig. 7 may be due to either strong coupling to the LO phonons only when initially exciting from the Sm^{2+} ($4f^6$) level or due to an overlap of DOS from the TO and LO branches. In any case the LO phonons appear to be about 20% "softer" compared to those of EuS (Fig. 5). The reason for this "softening" can be

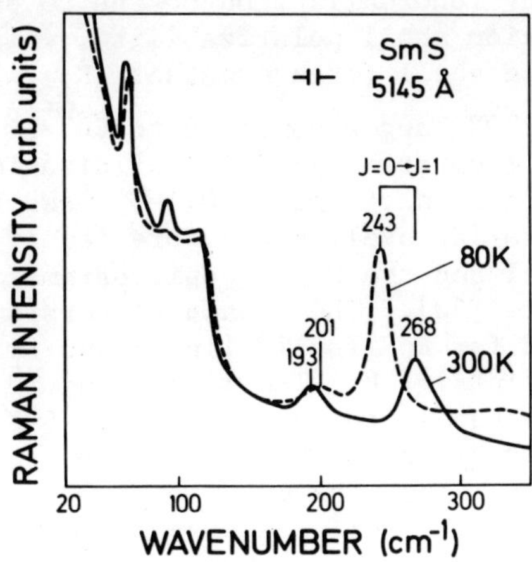

Fig. 7 Raman spectra of semiconducting SmS.

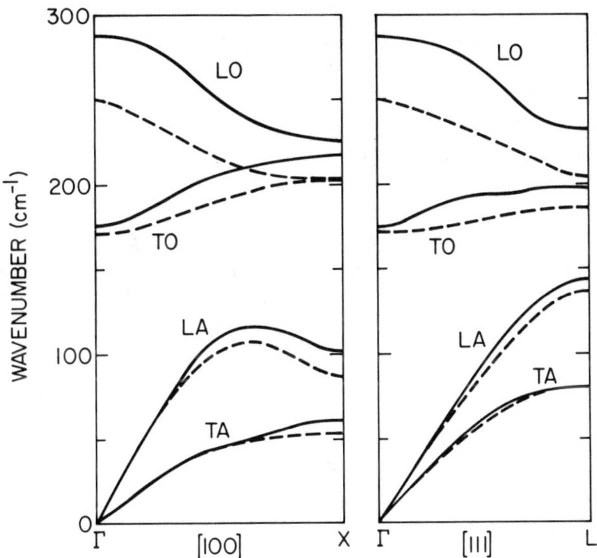

Fig. 8 Overlap shell model calculations of the phonon dispersion of semiconducting EuS(——) and SmS (---) at 300K.

understood quite generally in terms of lattice dynamic shell models. These show a strong renormalization of the LO phonon branches upon increasing the cation shell polarizability, which is inversely proportional to the shell force constant (K_1). The compressibility of $Sm^{2+}S$ is about 20% larger compared to $Eu^{2+}S$ (section 2). This presumably arises from stronger f-d hybridization in SmS ($\sim V_{mix}/Eg$) due to its very small energy gap ($\sim$0.1eV). We have applied to SmS at 300K the overlap shell model, previously used for EuS [29,30]. The model and the macroscopic parameters used will be described elsewhere [30]. The phonon dispersions obtained from the model calculations for EuS (solid lines) and SmS (dashed lines) at 300K are compared in Fig. 8. The TA, LA and TO branches are not very strongly affected in going from EuS to SmS. The strongest effect is on the LO branches which are lowered by about 15% in SmS with respect to EuS. Thus the one-phonon DOS of SmS around 193 cm^{-1}, at 300K, in Fig. 7, originates mainly from the high DOS of the LO and possibly TO branches near the zone boundary in Fig. 8.

The phonon dispersion of the metallic, intermediate-valence phase of SmS is of particular interest. Measurements of the elastic constants of $Sm_{1-x}Y_xS$ at 300K show a soft bulk modulus for $0.15 < x < 0.2$ [10]. The divergence of the mean square displacement of the S ions $\langle u_S^2 \rangle$ near the gold-black phase transition upon cooling $Sm_{0.7}Y_{0.3}S$ has been interpreted in terms of a diverging compressibility [32]. Unfortunately, as mentioned in section 2, there is no information on the bulk modulus of metallic SmS near the semiconductor-metal transition. However, we would expect in general that the configurational crossover between $4f^6$ and $4f^5 5d^1$ configurations in SmS enhances the volume compressibility of the Sm ion. The associated isotropic deformation of the S octahedron (Fig. 6) can probably best be accounted for in the "breathing" shell model [33, 34] by introducing a Γ_1 mode symmetry and assuming $K_1 = G_1 \sim 1/\kappa$. G_1 is the breathing shell force constant, which is inverse proportional to the compressibility κ. The decrease in G_1 near configuration crossover should lead to a further softening of the LO branches compared to semiconducting SmS, most pronounced at the L point of the BZ [32]. Indeed, this has been observed in preliminary neutron scattering experiments on $Sm_{0.75}Y_{0.25}S$ by H. Mook [35]. The LO branch is seen to be lowered below the TO branch.

ACKNOWLEDGEMENTS - The authors are grateful to M. Cardona and H. Bilz for valuable discussions and to W. Zinn for his cooperation. We also thank P.G. Lockwood for crystal growth, S.O. Ellmann for microprobe analysis, K. Fischer for EuS single crystals and H. Hackfort for $Eu_{1-x}Gd_xS$ thin films. We are indebted to H. Böhringer for his expert technical assistance.

REFERENCES

+ Institut für Festkörperforschung, KFA, Jülich, Federal Republic of Germany.
* IBM T. J. Watson Research Center, Yorktown Heights, NY, USA.

[1] C.M. Varma, Rev. Mod. Phys. 48, 219 (1976).
[2] P.W. Anderson and S.T. Chui, Phys. Rev. B9,3229 (1974).
[3] D. Sherrington and S. von Molnar, Solid State Commun. 16, 1347 (1975).
[4] C.M. Varma and V. Heine, Phys. Rev. B11, 4763 (1975).
[5] D. Sherrington and P. Riseborough, C.N.R.S. Colloqu. on Semi-cond.-Metal Trans., Autrans, 1976, to be publ. in J.de Phys.
[6] P.D. Dernier, W. Weber and L.D. Longinotti, to be publ.
[7] R. Keller, L.L. Hirst, G. Güntherodt, W.B. Holzapfel and F. Holtzberg, to be publ.
[8] A. Jayaraman, V. Narayanamurti, E. Bucher and R.G. Maines, Phys. Rev. Lett. 25, 368, 1430 (1970).
[9] E. Kaldis and P. Wachter, Solid State Commun. 11,907 (1972).
[10] R.L. Melcher, G. Güntherodt, T. Penney and F. Holtzberg, 1975 Ultrasonics Symp. Proc., IEEE Cat.#75, CHO 994-4SU, p. 16; T. Penney, R.L. Melcher, F. Holtzberg and G. Güntherodt, AIP Conf. Proc. 29, 392 (1975).
[11] M.B. Maple and D. Wohlleben, Phys. Rev. Lett. 27, 511 (1971).
[12] B. Batlogg, E. Kaldis, A. Schlegel and P. Wachter, Phys. Rev. B15, Nov. 15 (1976).
[13] T. Penney and F. Holtzberg, Phys. Rev. Lett. 34, 322 (1975).
[14] A. Jayaraman, P.D. Dernier and L.D. Longinotti, High Temp.-High Pressure 7, 1 (1975).
[15] B. Johansson and A. Rosengran, Phys. Rev. B11, 2836 (1975).
[16] M. Campagna, this conference.
[17] D.W. Pohl, R. Jaggi, K. Gisler and H. Weibel, Solid State Commun. 17, 705 (1975).
[18] G. Güntherodt and F. Holtzberg, Solid State Commun. 18,181 (1976).
[19] G. Güntherodt, Festkörperprobleme (Adv. in Solid State Phys.), Vol.XVI, p. 95, Vieweg, Braunschweig (1976).
[20] R. Suryanarayanan, G. Güntherodt, J.L. Freeouf and F. Holtzberg, Phys. Rev. B12, 4215 (1975).
[21] J.L. Erskine and C.P. Flynn, Bull. Am. Phys. Soc. 21, 351 (1976).
[22] R.A. Pollak, F. Holtzberg, J.L. Freeouf and D.E. Eastman, Phys. Rev. Lett. 33, 820 (1974).
[23] T. Penney and F. Holtzberg, Phys. Rev. Lett. 34, 322 (1975).
[24] S.D. Bader, N.E. Phillips and D.B. McWhan, Phys. Rev. B7, 4686 (1973).
[25] J.L. Kirk, K. Vedam, V. Narayanamurti, A. Jayaraman and E. Bucher, Phys. Rev. B6, 3023 (1972).
[26] A. Chenevas-Paule, P. Haen, F. Lapierre, R. Tournier and L. Vieux-Rochas, this conference.

[27] R. Merlin, R. Tsu, G. Güntherodt, G. Abstreiter and M.W. Shafer, to be publ.
[28] N.F. Mott, Phil. Mag. 3o, 403 (1974).
[29] P. Grünberg, G. Güntherodt, A. Frey and W. Kress, Proc. Int. Conf. on Magneto-Optics, Zurich, 1976, to be publ. in Physica B.
[30] P. Grünberg, G. Güntherodt, R. Merlin, A. Frey, W. Kress, H. Bilz, M. Cardona, K. Fischer and W. Zinn, to be publ.
[31] M.I. Nathan, F. Holtzberg, J.E. Smith, J.B. Torrance and J.C. Tsang, Phys. Rev. Lett. 34, 467 (1975).
[32] P.D. Dernier, W. Weber and L.D. Longinotti, to be publ. in Phys. Rev.
[33] U. Schröder, Solid State Commun. 4, 347 (1966).
[34] W. Kress, Phys. Stat. Sol. (b) 62, 403 (1974).
[35] H. Mook, this conference.

COMMENTS

SCHIRBER: In the single crystal work on SmS under pressure, how do you generate these pressures?

GUNTHERODT: We use a diamond anvil cell with a pressure transmitting fluid held between the anvil faces by a gasket. The low temperature measurements were done in a gas pressure cell which can reach 10 kbar.

SCHIRBER: In your x-ray work, when you go through the SmS transition, can you come back and retain the single crystal?

GUNTHERODT: Well, you can come back and your x-ray reflections will still show the same crystal structure. However, the reflections show a slight streaking, and if, after releasing the pressure, you take the sample out of the cell, you have a mosaic of little crystallites. If you then touch the sample, it just falls apart.

ORBACH: I really have trouble understanding how the spin disorder gives rise to the L-phonon in the Raman spectra. Isn't it possible that you have impurity disorder, so that in some sense the L-point is impurity-activated? For you to be right, you'd have to have an enormous coupling between the spins of the Eu and the lattice. I suppose it's possible, but I've never heard of that as being a dominating mechanism.

GUNTHERODT: This is just the exciting feature of the europium chalcogenides. You see this strong coupling in the most recent work on Raman scattering. Of course, there is the question of defects being present, but the scattering you see at high temperature is quenched when you go through the ordering temperature, or apply a magnetic field, so you have direct evidence that the spin disorder is the dominating effect. However, at low temperature the defect-induced scattering will remain. I think we understand it quite well, although some people have argued against it.

ALLEN: One thing which should be reiterated about these metallic looking plasma edges is that you really can't tell if there is a small gap in the system or not by looking at the metallic plasma edges. The response at this high an energy for a system with a small gap would still be essentially plasma-like; it would look very metallic. The only evidence concerning the presence of a gap must be either direct spectroscopic evidence based on very low energy spectroscopy, or from some kind of transport measurement. It was pointed out yesterday that the best evidence against a gap was the specific heat. You just can't distinguish that from looking at these optical properties.

WALSH: This very strong coupling of an optic mode to valence fluctuations seems to be extremely sensible on physical grounds. In fact, it seems to me that if they're going to fluctuate, then they ought to be pinned at the optical mode frequency instead of having a stochastic spectrum.

GUNTHERODT: At the present time we have no firm evidence that they are very much higher than the optical frequencies, but rather they are of the same order of magnitude. They are definately not below those frequencies. So I would say that being pinned at the optical mode frequency is a most reasonable assumption. If one lowers the fluctuation frequency below that of the LO phonon frequency, as was shown by the work of Dernier and Weber in applying a lattice dynamical model, the softening of the mean square displacement of the sulfur ions in $Sm_{0.7}Y_{0.3}S$ can no more be accounted for. I felt that this was a very interesting test.

VARMA: I think it is necessary to distinguish between two kinds of fluctuations, one of which almost certainly occurs. This is the fluctuation of an incoherent kind; that is we have charge fluctuations with a correlation which decays exponentially. This kind of fluctuation is not going to be pinned in any sense by these optical phonons. If on the other hand there are any coherent configuration fluctuations which are of a propagating nature (which may be damped slightly, of course), then it is these fluctuations which might be coupled very strongly to the phonons. The reflectivity data on SmS down to low frequency (which reveals the real part of the dielectric constant going up) seems to indicate that there are such coherent fluctuations which couple to light, that is, acquire dipole moments by coupling to phonons.

VALENCE FLUCTUATIONS IN MIXED VALENCE COMPOUNDS WITH Th_3P_4 STRUCTURE

P. Wachter

Laboratorium für Festkörperphysik, ETHZ Hönggerberg
8093 Zürich, Switzerland

ABSTRACT

Eu_3S_4 and Sm_3S_4 exhibit a temperature activated valence fluctuation. For these compounds the electronic structure has been determined by optical reflection measurements between 0.03 eV and 12 eV. The activation energy to be overcome in the fluctuation process is evaluated from the electrical conductivity and the intrinsic phonon frequency driving the fluctuation is determined by inelastic light scattering. Photoluminescence measurements exhibit a drastic change when the valence fluctuation rate is driven across the reference time scale corresponding to the life time of a photo excited electron. The magnetic susceptibility of Sm_3S_4 behaves unusual, inasmuch as the magnetic moment of Sm^{3+} is quenched above 100 K. As well Eu_3S_4 as Sm_3S_4 exhibit magnetic order.

1.INTRODUCTION

The Th_3P_4 structure is a well known crystal structure in which many rare earth (RE) compounds crystallize (1). When the RE ion is strictly trivalent, these compounds are metals with one free electron per formula unit. When the RE ion is stable in more than one valence, mixed valence compounds can be formed. This is the case in e.g. Eu_3S_4 and Sm_3S_4, where the cations are di-and trivalent. A simple valence argument then shows that the chemical formula can be written as e.g. $Eu^{2+}Eu_2^{3+}S_4^{2-}$ where di-and trivalent cations are present in the ratio 1:2. We expect such a compound to be a semiconductor. This is most evident by the observation of an energy gap of 1.7 eV by optical transmission experiments on Eu_3S_4 (2).

However, the most important aspect of the above compounds is the fact that the anion surrounding of each cation in the bcc Th_3P_4 structure ($I\bar{4}3d-T_d^6$, z=4) is identical for di-and trivalent cations and that consequently all cation sites are crystallographically equivalent. Therefore, the valence of an individual cation can be changed by transfering an electron from a divalent cation to a neighboring trivalent cation. Since the ionic radii of divalent and trivalent RE ions differ by as much as 10% such a valence change will be strongly coupled to lattice phonons. The valence fluctuations thus generated, at least at high temperatures, are a random process in which all cations participate. The process of valence change occurs via a temperature activated process and thus offers the unique possibility to scan the fluctuation rate over any reference time scale of a physical experiment by simply changing the temperature. Such a reference time scale may be given by the isomer shift of the Mössbauer effect, having a nuclear life time of $\tau_M \sim 10^{-9}$ sec for the Eu or Sm ion. When the fluctuation rate is larger than τ_M^{-1} an average valence 2.66 is observed for the cation and only one Mössbauer line for Eu_3S_4 (3) or Sm_3S_4 (4). When at low temperatures the fluctuation rate is less than τ_M^{-1} two Mössbauer lines are encountered, corresponding to di-and trivalent Eu ions. The intensity ratio of these lines is 1:2 as given by the chemical formula (3).

2. ELECTRONIC STRUCTURE

The electronic structure of any material is best obtained by optical methods yielding directly the energy separation of occupied and empty electronic states. In a mixed valence compound incorporating valence fluctuations, the interpretation of the spectra rests upon the fact that the measuring frequency is larger than the fluctuation rate. The latter is at 300 K about 10^{11} cps (3) so that photon energies down to the far IR can be used. The incident photon thus "sees" a static distribution of charges and indeed the spectra do not change when cooling the sample down to 4.2 K, as verified in the case of Sm_3S_4 (apart from the usual sharpening of absorption peaks with decreasing temperature). This perception has also another important consequence, namely that one can just as well measure a related, non fluctuating mixed valence compound and thus obtain a similar information. Such a compound is e.g. Eu_3O_4 (5) which crystallizes in the $CaFe_2O_4$ spinel structure, where Eu^{2+} and Eu^{3+} lattice positions ar inequivalent. Thus apart from minor effects due to different crystal field splitting the electronic structure of this compound should be equivalent to Eu_3S_4. This is important because the latter compound exists only in polycrystalline form which hampers optical measurements and Sm_3S_4, though available in single crystals, has a very complex multiplet structure.

The optical reflectivity of single crystals of Sm_3S_4 (6) and Eu_3O_4 (5) has been measured for photon energies between 0.03 eV and 12 eV and for pressed powder pellets of Eu_3S_4 for photon energies between 0.03 eV and 6 eV (2). These data have been analyzed in terms of the dielectric functions by means of the Kramers-Kronig relation. For Sm_3S_4 the computed optical constants are shown in Fig.1. Since these compounds contain well localized 4f states we may ask about the relative arrangement of $4f^n$ and $4f^{n-1}$ states. The latter correspond to a higher degree of ionization than the former and thus their energy will be lower by the Coulomb correlation energy U. U corrsponds to the difference between third and fourth ionization energy (5) and it is found to be between 5 and 6 eV.

Thus the highest occupied level is a $4f^n$ of a divalent cation and optical transitions are possible into crystal field split 5d bands and 6s bands. This situation is similar as in EuS and SmS which contain only divalent cations, and just as in the series of the Eu- and Sm monochalcogenides (7) the energy separation $4f^n$-5d baricenter is constant for the same cation. This implies that the larger the 5d crystal field splitting, the smaller the energy gap.

For Eu_3S_4 (1.7 eV) and Sm_3S_4 (0.1-0.2 eV) the energy gap has been found to be practically the same as in EuS (1.65 eV) and SmS (0.1-0.2 eV), respectively (2,6,7),and the crystal field splitting Δ of the 5d bands is with about 2.3 eV also the same as in the monochalcogenides in agreement with the above reasoning. Of course a distinction in t_{2g} and e_g subbands is not obvious since the local point symmetry is S_4, but in any perturbation theory the cubic term will be largest. In performing optical transitions from a localized $4f^n$ into a conduction band one has to realize that the remaining n-1 electrons can be in excited states before relaxation with the lattice. In the case of Sm_3S_4 these are the multiplets $4f^5(^6H_J, ^6F_J, ^6P_J)$ for the $4f^6 \rightarrow 4f^5$ 5d and $4f^4(^5I_J, ^5F_J, ^5G_J)$ for the $4f^5 \rightarrow 4f^4$ 5d transition. The probabilities to find the 4f shell in

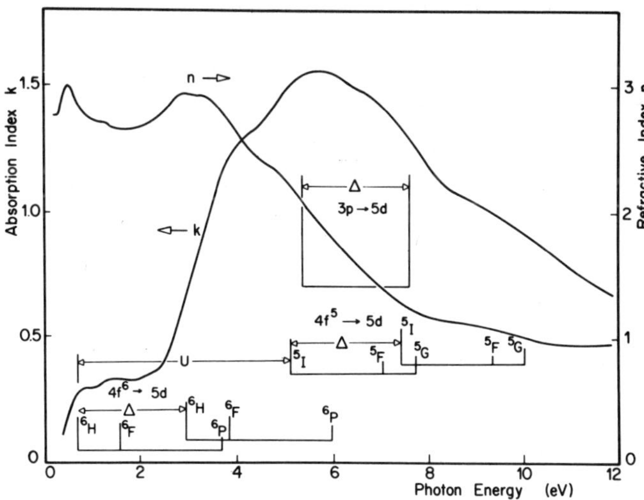

Fig.1 n and k of Sm_3S_4. U:Coulomb correlation energy, Δ:5d crystal field splitting. Baricenters of final state multiplets indicated.

one of these multiplets can be computed with the "coefficients of fractional parentage"(8). The situation is even more complex, since all optical transitions into 5d bands appear twice due to the crystal field splitting Δ of these bands. In addition to 4f-5d transitions one observes also $3p^6 \rightarrow 5d,6s$ transitions. As a whole, the complex situation in Sm_3S_4 could only be made consistent with the experiment by knowing the less complicated case of Eu_3O_4 (5) and the optical structure of EuS and SmS (7).

In Fig.2 we show the energy level diagram for Sm_3S_4 and for comparison also of SmS. To the right of each diagram the density of states for one electron bands is shown and to the left initial and final multiplets of multi-electron states.

The reflectivity measurements at 300 K just as well gave the residual ray spectrum in the far IR region. With $\nu_{LO} \simeq 300\ cm^{-1}$ and $\nu_{TO} \simeq 250\ cm^{-1}$ for Eu_3S_4 and Sm_3S_4 these phonon frequencies are also well comparable with the ones in EuS and SmS, $\nu_{LO} = 266\ cm^{-1}$ and $\nu_{TO} = 178\ cm^{-1}$ (9). The shift towards higher energies in the 3-4 compounds relative to the monochalcogenides can readily be explained by a reduced Eu-S interatomic distance in the former (1).

3.VALENCE FLUCTUATIONS AND INELASTIC LIGHT SCATTERING

Regarding Fig.2 one may wonder whether the Coulomb correlation energy U is involved in the fluctuation process. However, considering that the valence change is between neighboring cations we realize that the total energy of the pair of cations remains the same, only the energy levels between di-and trivalent cations have interchanged. But other energies are involved connected with the change of radii of the fluctuating cations. These energies have to do with the rearrangement of anions in the course of the fluctuations and with lattice modes. Therefore, inelastic light scattering is an important tool for its investigation.

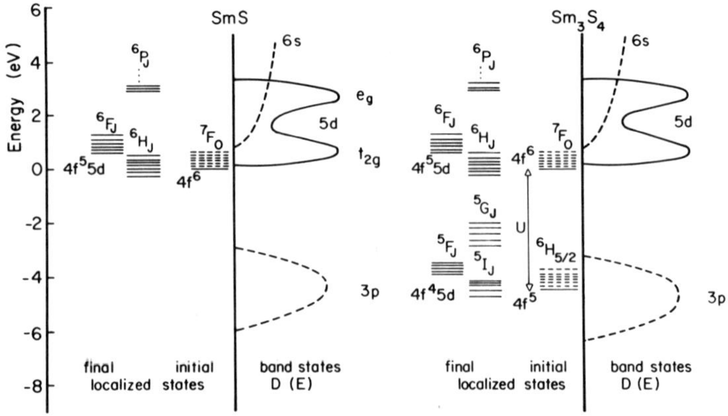

Fig.2 Energy level diagram of SmS and Sm_3S_4

In Fig.3a we show for 2 K and an excitation energy of 2.41 eV the scattering spectrum of Eu_3S_4. The lowest energy line at about 300 cm^{-1} is also observed in the IR spectrum and corresponds to the polar vibration ν_{LO} of Eu and sulfur ions. In addition, very strong phonon lines are observed at 425 cm^{-1}, followed by higher harmonics such as 850 cm^{-1} etc. (10). These modes cannot be explained by polar Eu-S vibrations, expecially since they are only slightly IR active. Instead, these phonon lines can be ascribed to a vibration involving only sulfur ions. The Th_3P_4 structure permits such an oscillation of the structure parameter u along (111) directions (1) which is indicated by arrows in the insert of Fig.3a. Here the anion polyhedron is shown and it is evident that the sulfur ions describe a combined breathing and twisting motion (10).

Valence fluctuations are proposed to be driven by these vibrations which relate to a volume change of the coordination polyhedron. To simplify the model we show in Fig.3b in linear dimension a pair of Eu cations with different valence and a sulfur ion in between. Due to the different charges and radii of the cations the sulfur ion is displaced closer to the trivalent cation. After an exchange of valence a new equilibrium position is taken and, naturally, an energy barrier E_p exists between the double potential wells. Such a model has first been proposed by Mulak and Stevens (11) and then been elaborated by Vitins (12). The valence change then goes concomitant with the positional change of the sulfur ions and the change in ionic radii of the cations. We now can write the fluctuation rate ν as

$$\nu = \nu_o \exp.(-E_p/kT) \quad (1)$$

with $\nu_o = 425\ cm^{-1} = 1.3 \cdot 10^{13}\ sec^{-1}$.

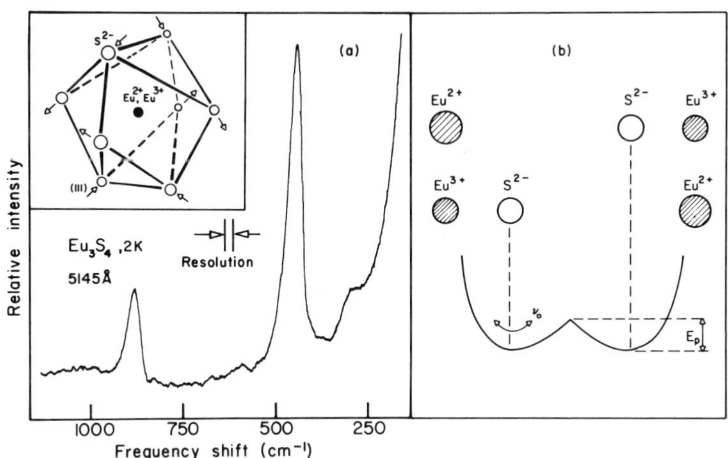

Fig.3 Inelastic light scattering (a) and hopping model of Eu_3S_4 (b). The insert in (a) shows the anion polyhedron.

The harmonics in the inelastic light scattering can be interpreted as excitations of the harmonic oscillator given by the double potential well structure. The same inelastic light scattering experiments as in Eu_3S_4 have also been performed in Sm_3S_4, but there the spectra are dominated by the electronic Raman effect, yielding intra-4f excitations. However, there can be no doubt that ν_o has the same value also in Sm_3S_4, since lattice constant, crystal structure and mass of the ions practically are identical with Eu_3S_4. In the latter compound also Mössbauer experiments have been performed successfully to determine ν_o which has been extrapolated to be (3) $4 \cdot 10^{13}$ sec^{-1}, in good agreement with the direct measurement above.

4. ELECTRICAL TRANSPORT PROPERTIES

The valence change driven by the movement of the sulfur lattice is a temperature activated process. The electron transfer from divalent to trivalent cation, therefore, is temperature activated with the activation energy E_p. Thus the mobility of the elctrons has to be of the hopping type.

By measurements of thermopower and electrical conductivity it could be established indeed that the mobility is temperature activated (3,13,14). Our conductivity measurements on Eu_3S_4 yield $E_p = 0.18$ eV in good agreement with the literature (2). The results for Sm_3S_4 single crystals yield at high temperatures $E_p = 0.142$ eV and below about 100 K $E_p = 0.132$ eV (2,6). The results on our well characterized Sm_3S_4 single crystals (6) are in disagreement with results on polycrystalline samples (15,16), also in other respects as will be shown below. It shall be mentioned only shortly that the handling of polycrystalline material ought to be done with extreme care to avoid contaminations which completely change the properties of the material.

On Eu_3S_4 and on Sm_3S_4 anomalies are observed in the electrical conductivity. In Eu_3S_4 at 165 K a discontinuous change occurs and in Sm_3S_4 a kink is found near 125 K in the Arrhenius plot (6). These anomalies have been brought in connection with a charge ordering at these temperatures as proposed by Carter (17). Direct evidence of a concomitant tetragonal distortion of the lattice is so far missing. But in any case the persistence of an activation energy in the mobility also below the critical temperatures (see above) shows that the valence fluctuations persist. This will also be demonstrated further down with other experiments.

The energy gap between $4f^6$ and 5d states in Sm_3S_4 has been found to be the same as in SmS (6) (see section 2), namely 0.1-0.2eV. One wonders whether an insulator metal can be induced by pressure, just as in SmS. The pressure coefficient of the resistivity of Sm_3S_4 is negative and one can derive $(dE_p/dp) = - 3 \pm 0.3$ meV/kbar (6) which differs by nearly a factor 2 with the same measurements on polycrystalline samples of ref.(16).

However, here we see a problem: since the $4f^6$-5d energy gap and the activation energy for hopping are practically the same (this not the case in Eu_3S_4) it is difficult to decide what really happens. With an estimated pressure of about 30 kbar the activation energy should be reduced continuously to zero. This could indicate that one has come from the double well potential of sulfur to a single well potential and then the valence change is no longer by hopping but hybridization may occur. Here we have to wait for further experiments.

5. PHOTOLUMINESCENCE

When exciting a crystal with photon energies larger than the energy gap the excited electron becomes thermalized at the bottom of the conduction band in about 10^{11} sec. During a life time τ it stays in the conduction band and then recombines with a hole under emission of light. If the excitation is from a $4f^n$ level the recombination is with the 4f hole (18). Normally the emission energy is strongly Stokes shifted with respect to the excitation energy because a lattice relaxation around the hole is encountered according to the Franck-Condon principle.

In Eu_3S_4 at 300 K one observes a broad 5d-4f emission band centered at about 1.25 eV, with a high energy shoulder extending right up to the energy gap as depicted in Fig.4a (19). Thus, practically no Stokes shift is observed. However, cooling this compound to 4.2 K makes the shoulder disappear completely and now one observes the expected Stokes shift with a magnitude as typical for a

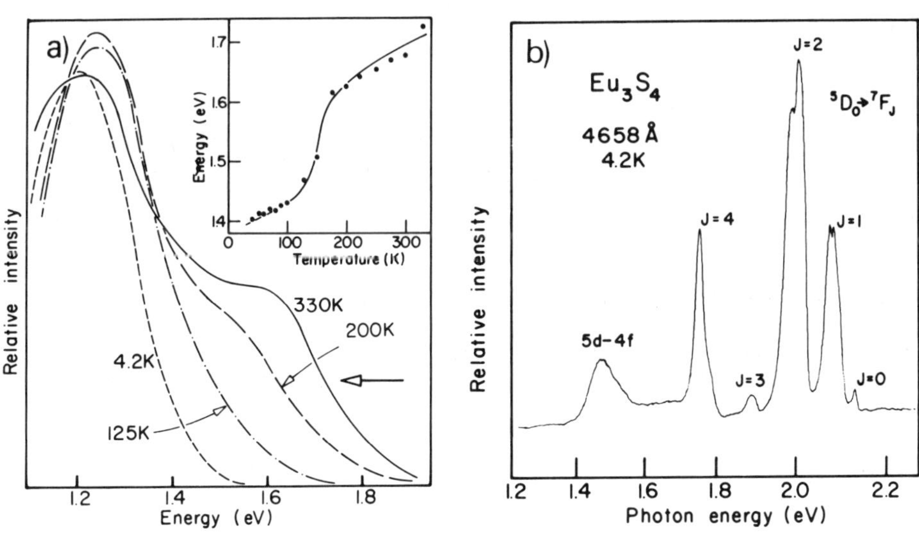

Fig.4 5d-4f (a) and intra-4f (b) photoluminescence of Eu_3S_4. The insert in (a) shows the energy of the luminescence edge

5d-4f recombination (18). Plotting an arbitrarily chosen emission intensity (indicated by the arrow) versus the temperature we obtain the insert of Fig.4a. The strongest alteration in the spectrum is near 155 K.

As explanation we propose that the Stokes shift can only be observed when the life time of a cation configuration at the site of recombination is larger than the life time τ of the excited electron. The largest change in the Stokes shift is expected when τ equals the inverse of the fluctuation rate ν. With E_p from conductivity measurements of Eu_3S_4 and ν_o from inelastic light scattering we compute for τ at 155 K about 10^{-6} sec, which is in good agreement with the life times encountered in the Eu monochalcogenides (18). A similar situation is found in the Mössbauer measurements, where strongest alteration in the isomer shift is observed near 240 K, corresponding to the nuclear life time τ_M of about 10^{-9} sec.

The Stokes shift of the photoluminescence emission is not only influenced by the lattice relaxation but also by magnetic order (18). Since it has been shown that Eu_3S_4 exhibits ferromagnetism (20) at 3.8 K (21) we expect additional effects near T_c. One has learned from the luminescence of the Eu monochalcogenides (18) that ferromagnetic order is indicated by a red shift of the emission band combined with a quenching of the emission intensity. (The inverse is true for type II antiferromagnets) This is a consequence of the large f-d exchange coupling of the excited electron which leads also to a magnetic relaxation of the excited electron in the ferromagnetic state. In Eu_3S_4 we observe (19) below about 4 K and down to 1.8 K a red shift and quenching of the luminescence peak shown in Fig.4a. Both effects can be enhanced by a magnetic field, giving clear evidence that the magnetic order is of ferromagnetic nature.

Besides the above described 5d-4f luminescence one has also observed strong intra-4f luminescence in Eu_3S_4 and in Sm_3S_4 (12,19). This luminescence results from optical transitions between spin orbit split multiplet levels of the Eu^{3+} or Sm^{2+} ion. In Fig.4b this spectrum is shown for Eu_3S_4 at 4.2 K using a laser excitation energy of 4658 Å. (Fig.4b is not corrected for the spectral sensitivity of the PM tube). The intra-4f luminescence peaks can be assigned to optical transitions from the ^5D_o to the ^7F_J multiplet levels of Eu^{3+} and the spectral positions of the lines fit very well with the baricenters as observed in EuF_3 (22). The S_4 point symmetry splits the J=1 level into a doublet and splittings in the other J levels are indicated. Although the excitation energy in Fig.4b corresponds to the transition $^7F_o \rightarrow {}^5D_2$ no luminescence was observed originating from the $^5D_2$ multiplet level.

The intra-4f luminescence of Eu_3S_4 and Sm_3S_4 is strongly quenched with increasing temperature (12,19). This is quite unusual for an intra-4f luminescence and this effect is absent in the intra-4f luminescence of Eu_3O_4, a non fluctuating compound (12). Therefore again, it is most probable that this quenching in Eu_3S_4 and Sm_3S_4 is related to valence fluctuations. In a similar reasoning as with the 5d-4f luminescence it is conceivable that the intra-4f luminescence is quenched if the life time of the excited state ($\tau > 10^{-8}$ sec) is longer than the average fluctuation rate ν^{-1}.

6. MAGNETIC SUSCEPTIBILITY

When interpreting the magnetic susceptibility of a mixed valence compound it is best to start with a non fluctuating mixed valence system such as Eu_3O_4. The molar susceptibility of this compound has been measured (23) and one can quantitatively interpret it as being the sum of the divalent and trivalent cation susceptibilities. In the case of Eu_3S_4 at low temperatures we can try the same:

$$\chi_{Eu_3S_4} = \chi_{Eu^{2+}} + 2\chi_{Eu^{3+}} \qquad (2)$$

The ground state of Eu^{2+} is $^8S_{7/2}$ and of Eu^{3+} ^7F_o. Thus crystal field effects come only into play via the higher multiplet states of Eu^{3+}, being a Van Vleck ion. The susceptibility can be writen (neglecting the crystal field effects)

$$\chi_{Eu_3S_4} = \frac{L\mu_B^2}{3k}\left\{\frac{g^2J(J+1)}{T-\Theta}\right\} + \frac{2L\mu_B^2}{k}\left\{\frac{\frac{8}{E_1}+\left[\frac{4.5}{T}+\left(\frac{15}{E_2-E_1}-\frac{8}{E_1}\right)\right]\exp.(-E_1/T)}{1+3\exp.(-E_1/T)}\right\} \qquad (3)$$

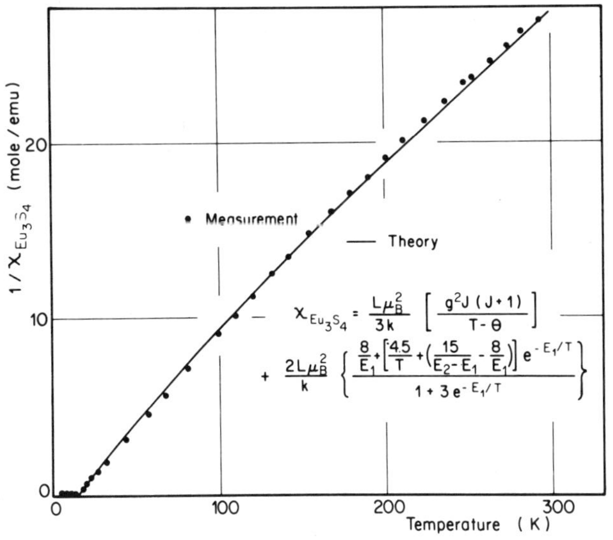

Fig.5 Inverse molar susceptibility of Eu_3S_4

with g=2, J=7/2, L = Avogadro's number. E_1 and E_2 are the energy differences $^7F_0$-$^7F_1$ and $^7F_0$-$^7F_2$, respectively, which can be taken from the intra-4f luminescence of Eu_3S_4: E_1= 510 K, E_2= 1496 K (baricenter).

In Fig.5 the results of the measurements are shown between 1.8 K and 300 K. The $1/\chi$ plot directly yields the paramagnetic Curie temperature Θ = 15 K, in good agreement with (20). In Fig. 5 is also plotted the theoretical susceptibility using eq. (3) with all parameters determined by experiment. The agreement between theory and measurement is excellent, justifying the simple model that the susceptibility of di-and trivalent cations is additive. The measurements are also in perfect agreement with data of Davis et al.(24) above 140 K, who come, however, to the wrong conclusion that the experiment should be interpreted with a Curie-Weiss law alone, arriving at a negative Θ. However, it has been mentioned already that Eu_3S_4 is a ferromagnet with T_c= 3.8 K (20,21).

In the case of Sm_3S_4 the situation is more complex (25). Here $Sm^{2+}(^7F_0)$ and $Sm^{3+}(^6H_{5/2})$ are both Van Vleck ions, the latter also with a magnetic ground state. Now crystal field effects can no longer be neglected, at least regarding Sm^{3+}. The susceptibility is taken again as the sum of the one of Sm^{2+} and 2 Sm^{3+} ions where the ground state of the latter, being a Kramers ion, is split by the crystal field of S_4 local symmetry into three doublets. In eq. (4) E_1 and E_2 are the energies of the excited $^7F_1$ and $^7F_2$ states of Sm^{2+}. $F_{i,j}$ are the off-diagonal matrix elements of the susceptibi-

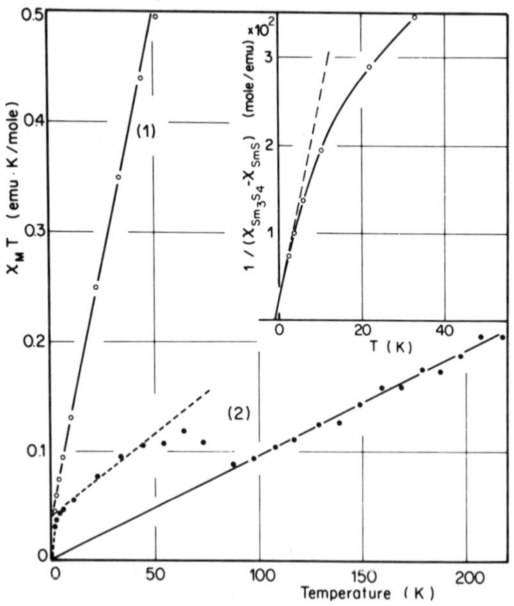

Fig.6 Curve (1): $(\chi_{Sm_3S_4})T$. Curve (2): $(\chi_{Sm_3S_4}-\chi_{Sm^{2+}})T$
Insert: $1/(\chi_{Sm_3S_4}-\chi_{Sm^{2+}})$. Dashed line is theoretical initial slope.

lity and $\alpha_{i,7/2}$ is the magnetic admixture of the J=7/2 level to the crystal field levels of the ground state. $\delta_{i,j}$ are the energies and C_i the Curie constants of the crystal field levels of Sm^{3+}.

$$\chi_{Sm_3S_4} = \frac{L\mu_B^2}{k}\left\{\frac{\frac{8}{E_1}+\left[\frac{4.5}{T}+\left(\frac{15}{E_2-E_1}-\frac{8}{E_1}\right)\right]\exp.(-E_1/T)}{1+3\exp.(-E_1/T)}\right\} + \tag{4}$$

$$+ 2L\mu_B^2\left\{\frac{\sum_{i=1}^{3}\left[\frac{C_i}{k(T-\Theta)}+\sum_{j\neq i}\frac{F_{i,j}}{\delta_j-\delta_i}+\alpha_{i,7/2}\right]\exp.(-\delta_i/kT)}{\sum_{i=1}^{3}\exp.(-\delta_i/kT)}\right\}$$

Below about 80 K the first term of eq.(4) is constant and there will be also a low temperature region where only the lowest crystal field level of the second term of eq.(4) is occupied. In this temperature region the summation extends only over i=1 and the exponentials can be dropped. Thus plotting $(\chi_{Sm_3S_4})\cdot T$ will yield a straight line for a not too large Θ. This is indeed the case and is shown in Fig.6 for curve (1). The intercept with the ordinate permits us to decide quantitatively that the $a|M_J=\pm 5/2\rangle + b|M_J=\mp 3/2\rangle$ level is lowest in energy (25) and with a=0.4 we obtain an intercept of 0.042 emuK/mole in good agreement with the experiment, but in contrast to (15). The contribution of Sm^{2+} to the total $\chi_{Sm_3S_4}$ can be computed rather precisely, because E_1 and E_2 are known from the electronic Raman effect of Sm_3S_4 or the intra-4f luminescence. We find E_1=392 K and E_2=1112 K. Thus we can compute $(\chi_{Sm_3S_4}-\chi_{Sm^{2+}})\cdot T$, shown as curve (2) in Fig.6, which yields the same extrapolated ordinate for $T\rightarrow 0$ as before.

Below about 5 K we observe a significant deviation of the experimental points from the straight line. This is due to a negative paramagnetic Curie temperature Θ and the fit with the second part of eq.(4) yields Θ=-1 K. This indicates antiferromagnetic order which is further supported by a $1/\chi$ plot in the insert of Fig.6.

The second part of eq.(4) predicts a steadily increasing $\chi\cdot T$ with increasing temperature. However, between 50 K and 100 K we measure a departure from the expected behavior:the low frequency part of the susceptibility of Sm^{3+} becomes quenched. At temperatures above 100 K we find again a straight line, now passing through the origin, indicating a J=0 system. This behavior must be related to the valence fluctuations and the fluctuation rate corresponding to say 100 K can be estimated with eq.(1) to be about 10^6cps. This means that for fluctuation frequencies above 10^6 cps the magnetic moment of Sm^{3+} in Sm_3S_4 is quenched. The above experiments thus demonstrate for the first time the transition from a magnetic to a nonmagnetic ground state, which occurs well below the frequencies corresponding to the hybridization width of a metallic mixed valence compound like SmS (about 10^{12} cps). Therefore new models have to be conceived, explaining also the absence of a quenching in χ of Eu_3S_4.

ACKNOWLEDGMENT

It is a great pleasure for the author to acknowledge very valuable discussions with B. Batlogg, A. Schlegel and J. Vitins. The author is also very grateful to Dr. O. Vogt, permitting the use of his susceptibility apparatus. The samples and single crystals have been grown by Dr. E. Kaldis, whom the author is very much obliged.

REFERENCES

1) F. Holtzberg and S. Methfessel, J. appl. Phys. 37, 1433 (1966)
2) B. Batlogg, E. Kaldis and P. Wachter, J. Magnetism and Mag.Mat. 3, 96 (1976)
3) O. Berkooz, M. Malamud and S. Shtrikman, Solid State Commun. 6, 185, (1968)
4) M. Eibschütz, R.L. Cohen, E. Bühler and J. H. Wernick, Phys. Rev. B6, 18 (1972)
5) B. Batlogg, E. Kaldis, A. Schlegel and P. Wachter, Phys. Rev. B12, 3940 (1975)
6) B. Batlogg, E. Kaldis, A. Schlegel, G.v. Schulthess and P. Wachter, Solid State Commun. 19, 673 (1976)
7) B. Batlogg, E. Kaldis, A. Schlegel and P. Wachter, Phys. Rev. B15, Nov. 15.th (1976)
8) P. A. Cox, Y. Baer, C. K. Jorgensen, Chem. Phys. Lett. 22, 433, (1973)
9) J. D. Axe, J. Phys. Chem. Solids 30, 1403 (1969)
10) J. Vitins and P. Wachter, Phys. Rev. B (in print)
11) J. Mulak and K.W.H. Stevens, Z. Physik B20, 21 (1975)
12) J. Vitins, Thesis ETH Zürich, to be published
13) I. Bransky, N. M. Tallan and A.Z. Hed, J. appl. Phys. 41, 1787 (1970)
14) I. A. Smirnov, L.S. Parfen'eva, V.Ya. Khuzmutdinova and V.M. Sergeeva, Sov. Phys. Solid State 14, 2412 (1972)
15) M. Escorne, A. Ghazali, P. Leroux- Hougon and I. A. Smirnov, Phys. Lett. 56A, 475 (1976)
16) J. M. Leger, G. Weill and I. A. Smirnov, phys. stat. sol.(b) 76, K27 (1976)
17) F. L. Carter, J. Solid State Chem. 5, 300 (1972)
18) P. Streit, Phys. kondens. Mat. 15, 284 (1973)
19) J. Vitins and P. Wachter, Phys. Lett. 58A, 275 (1976)
20) F. Hulliger and O. Vogt, Helv.Phys. Acta 39, 199 (1966)
21) E. Görlich, H. U. Hrynkiewicz, R. Kmiec, K. Latka and K. Tomala, phys. stat. sol. (b) 64, K147 (1974)
22) H. H. Caspers, H. E. Rast and J. L. Fry, J. Chem. Phys. 47, 4505 (1967)
23) L. Holmes and M. Schieber, Phys. Rev. 167, 449 (1968)
24) H. H. Davis, J. Bransky and N. M. Tallan, J. Less Common Metals 22, 193 (1970)
25) P. Wachter, Phys. Lett. A (in print 1976)

COMMENTS

JOHANSSON: Can you neglect the magnetic, superexchange interactions in this compound?

WACHTER: The Curie-Weiss behavior suggests that Sm_3S_4 is an antiferromagnet with $T_N \sim 1°K$. But Eu_3S_4 is a ferromagnet because the magnetic level is the divalent one which sits very close to the conduction band and has the same exchange mechanism as in EuS. In Sm_3S_4, the magnetic level is the trivalent Sm one which sits down by the Coulomb correlation energy about 5 eV below the conduction band. Under these conditions you only can have antiferromagnetic exchange.

LUTHI: Structural transitions, cubic to tetragonal, have been found in some praseodymium compounds. One consequence of that is that the rare earth ionic sites are not all equivalent. If your 160°K transition in Eu were structural, you should see more lines in the Raman spectra. Do you?

WACHTER: No: the lattice contraction is minute.

ORBACH: Your susceptibility formula is incorrect. You're using a temperature independent θ. I don't think it will affect these results since θ is so small, but with this level structure you have to account for the changing population of these levels in your calculation of θ.

EIBSCHUTZ: A question put to Dr. Campagna today was, can fluctuations between closed and non-closed shells be seen in 3d metals? For historical purposes I want to report that this type of fluctuation has been observed in the 3d's using the Mossbauer effect of Fe^{57} in Fe^{2+} compounds. This has been observed for many years in many complexes, and recently even in metallic compounds. What happens is the following: Fe^{2+} in the low spin state is in a closed shell, three spins up and three spins down, $^1A_1$ (t_2^6) configuration, a singlet and is not magnetic S=0. As soon as two electrons move from t_2^6 to E_g, Fe^{2+} is in the high spin state an open shell. It's five spins up on one down, $^5T_2$ $(t_2^6e^2)$ configuration, with 15 levels split by spin-orbit coupling and lower symmetry crystal field, and is magnetic S = 2. There is a competition here between cubic crystal field splitting and Hund's rule exchange energy. When these energies are comparable in magnitude you are in an intermediate crystal field and can have hopping between the low spin and high spin states. This hopping can be fast or slow in comparison with the time of measurements. The most dramatic effects as a function of temperature are seen when the singlet is lowest. For this situation the resulting transition may be sharp (actually first order) if the coupling to the lattice is strong enough or "continuous". This type of fluctuation, which still has not been observed using ESCA, may be another answer to Varma's question. This is a different case than Eu_3S_4 where the electrons hop between two different atoms.

EXPERIMENTAL STUDIES OF NARROW BAND EFFECTS IN THE ACTINIDES*

M. B. Brodsky

Materials Science Division
Argonne National Laboratory
Argonne, Illinois 60439

It is the purpose of this paper to present experimental data which emphasize the narrow band behavior of actinides. We will show that phenomena associated with narrow 5*f*-like bands are not rare. The discussion will concentrate on pure actinides and binary cubic intermetallic compounds of the type AnX, AnX_2, and AnX_3.

ORIGIN OF *f*-BANDS

Table 1 summarizes some properties of pure actinide metals.[1] It is seen that their electronic specific heat coefficients,[2] γ, are large, especially for Np and Pu. No magnetic order is found until the heavy actinides, Cm, Bk, etc., where magnetic moments comparable to heavy rare earths are observed.[3] Furthermore, the lighter members of the series, Th, Pa, and U have exhibited superconductivity in at least some of their phases.[4,5] The low temperature electrical resistivities may be approximated by power law dependences of the temperature, T^n. These change from simple transition metal behavior, T^3, to what we believe to be spin fluctuation behavior, T^2, to localized 5*f* electrons (it is believed that Am is trivalent, $5f^6$, with J = 0, and therefore nonmagnetic).[5]

The question then arises: where in the light actinides are the properties we expect from *f*-electrons, viz., magnetism? Part of the answer may be explained by use of Fig. 1. This figure by Freeman and Koelling[6] shows the radial extent of the 5*f* electron wave functions of various actinides scaled by their interatomic separations in the various allotropic phases. For comparison, the corresponding

*Work supported by the U.S. Energy Research and Development Administration.

Table 1. Properties of Pure Actinide Metals

	γ mJ/(mole-K^2)	Magnetic	Supercon.	n ($\rho \propto T^n$)	W (eV)	U (eV)
Th	4.3	No	Yes	3	5	1.5
Pa	-	No	Yes	-	-	1.6
U	9.1	No	Yes	2-3	4	2.3
Np	12.4	No	No	2	3	2.6
Pu	17.0	No	No	2	2	3.5
Am	6.0	No	No	5	0.1	5.0
Cm	-	Yes	No	-	<0.1	-
Bk	-	Yes	No	-	-	-

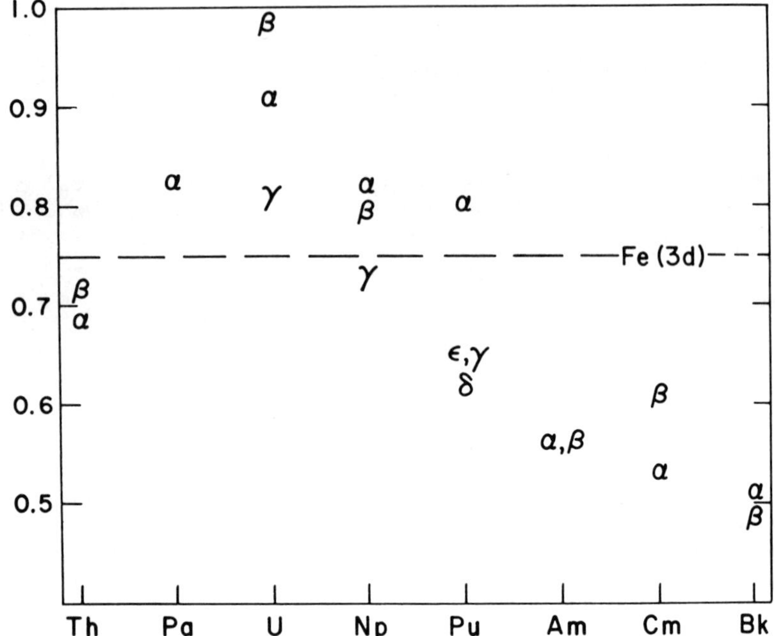

Fig. 1. Values of $[\langle r^4 \rangle]^{1/4}/R$ for the different phases of the actinide metals; R is the half-internuclear separation (from Ref. 6).

value for iron is also shown. From this figure we see that in the lighter actinides, the $5f$ wave functions overlap each other from site to site. Thus, the wave functions are broadened via direct overlap. This result, which strongly contrasts the localized $4f$ electrons in rare earth metals to the actinide $5f$ electrons, has been pointed out be several authors.[6,7] Hill observed, also,[7] that if the actinide-actinide separations were less than 3.5 Å in uranium compounds and 3.4 Å in neptunium compounds, then the compounds would not become magnetic. This was attributed to the formation of a nonmagnetic 5*f* electron band. Hill's rule is nearly always obeyed. There are, however, many intermetallic compounds for which the actinide-actinide distance is larger than the minimum values determined by Hill, but which do not order magnetically.[3]

In the last two columns of Table 1 we see that the polar state formation energy, U,[8] is smaller than the 5*f* band width, W, determined by APW calculations, at least for Th and uranium.[6] Since the U given here has not been corrected for screening, these values for U are all too large. Thus, the band model should be appropriate for the light actinides, as $U/W < 1$.

Figure 2 shows the band structure for UIr_3 (ordered $AuCu_3$ structure) determined by relativistic APW calculations. The U and Ir potentials and the Fermi level have been determined with the aid of deHaas-vanAlphen data.[9] This band structure successfully predicted the μ frequency due to an orbit centered at X, and the n frequency which exists for a short distance about the (111) direction. It would appear from the band structure that there is little *f*-character at the Fermi level. There are nearly three *f*-electrons which may be located at point Γ, and which are strongly hybridized with *d* electrons to form *f*-*d* bands. Yet there is good indication of some *f*-character at the Fermi level since effective electronic masses as large as 2.5 were measured here and $m^*/m_o = 5.1$ was measured in isoelectronic URh_3.[10]

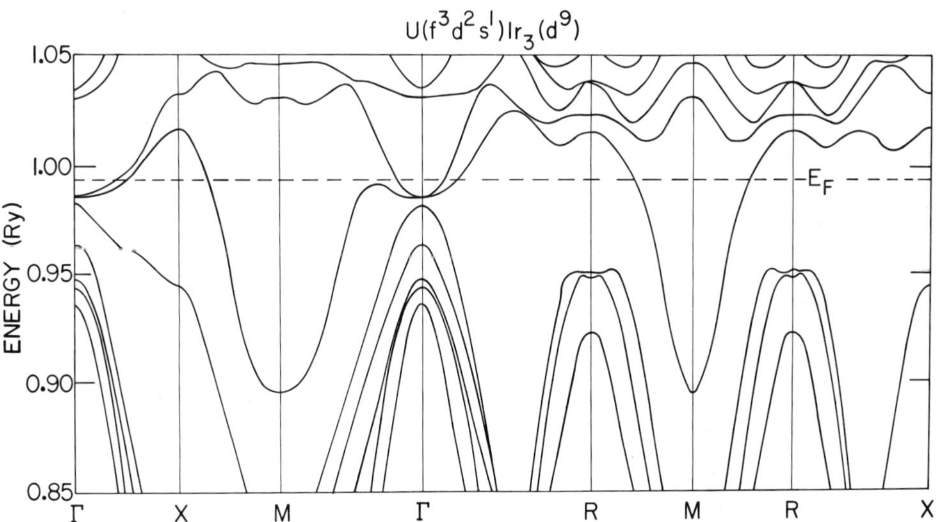

Fig. 2. Energy bands for UIr_3 near the Fermi level (from Ref. 9).

In the case of UIr_3 and URh_3, the existence of hybridized *f*-bands is well established. Their magnetic susceptibilities are nearly temperature independent and the resistivities follow a T^3 law at low temperatures, typical of *d*-band behavior.[11] The measured γ of URh_3 is only 14.5 mJ/(mole-K^2)[12] which is slightly smaller than the value for a fictitious "$RhRh_3$" compound. DeHaas-vanAlphen data and a preliminary band structure are also available for isostructural UGe_3.[9] This compound has a T^2 resistivity dependence and γ = 20.4 mJ/(mole-K^2).[13] Thus, in the case of UGe_3 we have made contact with a microscopic measurement and band calculation for an actinide exhibiting a large density of states due to the actinide (γ of UGe_3-γ of "$GeGe_3$" = 12), and probably a spin fluctuation resistivity due to the narrow *f*-bands. In this compound, the hybridization is presumably of *f*-*s* or *f*-*p* type.

Returning to Table 1, we can now have further confidence in band calculations for pure actinides with $U/W \lesssim 1$. A very dramatic exhibition of the effects of narrow *f*-bands on properties is seen in the resistivity-temperature curves of the pure metals, shown in Fig. 3.[5] Even for uranium there is nonlinearity in the ρ-T curve at room temperature, and the anomalous behavior reaches its peak at plutonium. It has been possible to explain the Np data by a spin fluctuation[14] and narrow band contribution[15] (of course the spin fluctuations rise out of the narrow bands at low temperature). Although similar arguments have been proposed for Pu, they are not totally satisfactory, especially in regards to the resistivity maximum. This is not surprising since $U/W \sim 1$, a very difficult theoretical region.

Apparent spin fluctation behavior, i.e., $\rho \propto T^2$, is observed fairly often in actinides.[16] There are well over 20 system which exhibit this behavior. The χ-T dependence of the compounds fall into two groupings. For some, e.g., $NpIr_2$, etc., there is no temperature dependence of χ. For others, e.g., $PuZn_2$, $NpRh_3$, etc., χ fits a Curie-Weiss plot, with a temperature-independent contribution, down to some $T \gtrsim T_{SF}$, and then nearly becomes temperature independent. This is similar to what is often termed Kondo-type susceptibility.

Relevance of this behavior to narrow *f*-bands is seen in Table 2 where γ's are listed for U and Np $AuCu_3$-type and cubic Laves phase ($MgCu_2$-type) compounds. The γ's are given as "excess", i.e., the contribution for the pure X elements have been subtracted away. UIr_3 is the only compound in Table 2 with $\gamma' \geq 10$ mJ/(mole-K^2) which does not show T^2 resistivity behavior. The same generalization may be made for the pure metals listed in Table 1. Several of the compounds in Table 2 order magnetically and are so indicated. $NpSn_3$ and $NpOs_2$ are discussed further below. If the argument of the present paper is valid, viz., narrow *f*-bands, then in some of the cases listed the bands will contain some very narrow features.

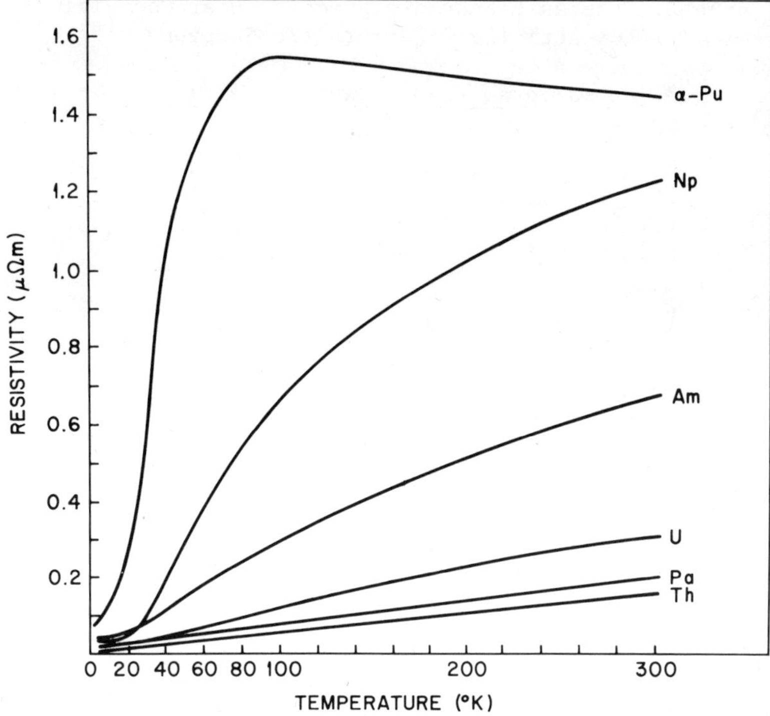

Fig. 3. Electrical resistivity of pure actinide metals (from Ref. 5).

From the discussion given thus far, it could be inferred that the *f*-character (narrow bands leading to localized *f*-electrons) generally develops monotonically in the actinide series. Actually this is not the case. In the band calculations of Koelling, the hybridized *f*-bands change considerably as the *f*-occupation number is changed.[9,10] Failure of simple behavior is also seen in the series UAl_2, $NpAl_2$, and $PuAl_2$. UAl_2 is definitely a spin fluctuation compound[14,17] (see below), $NpAl_2$ behaves as a local-moment ferromagnet[18] and $PuAl_2$ is probably a spin fluctuation compound.[14] However, in the series $AnRh_3$ there is a gradual change over from itinerant to local 5*f* magnetic behavior.[11] The susceptibility data for $ThRh_3$, URh_3, $NpRh_3$, and $PuRh_3$ are shown in Fig. 4. The corresponding exponents for the temperature dependences of resistivity are T^4, T^3, and T^2 for Th, U, and Np, respectively. The systems are described as low density of states transition metal ($ThRh_3$); medium to high density of state transition metal (URh_3); spin fluctuation metal as a consequence of a narrow band [$NpRh_3$, γ = 95 mJ/(mole-K^2)][12]; and a local moment antiferromagnet ($PuRh_3$). The local moment character of $PuRh_3$ is seen in the specific heat where the magnetic

Table 2. Excess Electronic Specific Heat Coefficients γ' of Actinide Intermetallic Compounds [(a)]

mJ/(gram atom of actinide/K^2)			
URh_3	0	$NpRh_3$	85
UIr_3	10	$NpSn_3$	237,83[(b)]
USi_3	10		
UGe_3	16		
UAl_3	38		
UIn_3	45		
UGa_3	50		
USn_3	164		
UIr_2	53	$NpRu_2$	112
UAl_2	138	$NpOs_2$	87,200[(b)]

[a]No corrections have been made for enhancements to γ.

[b]The compounds order magnetically. The first value is for the paramagnetic state and the second value is for the ordered state.

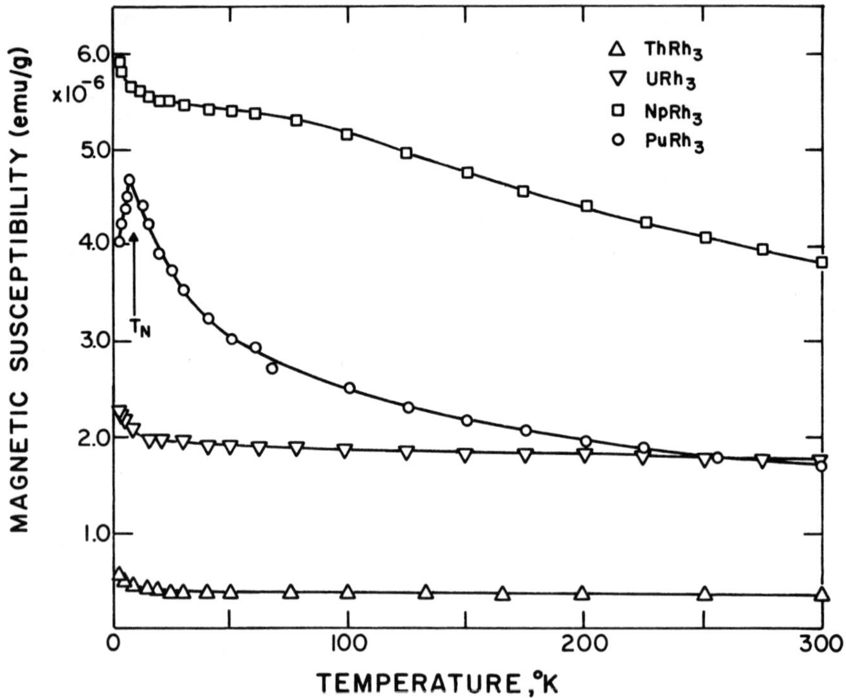

Fig. 4. Magnetic susceptibilities of $ThRh_3$, URh_3, $NpRh_3$, and $PuRh_3$ (from Ref. 11).

entropy is 1.0 Rln2. Of course, local moment magnetism may also be observed when the 5*f* electrons are not truly localized.[15]

f-BAND MAGNETISM IN ACTINIDES

With such high densities of states at the Fermi energy, it is expected that some *f*-band compounds will order magnetically. Three examples of magnetic behavior will be given.

UAl_2, a Spin Fluctuation Compound. In addition to a $\rho \propto T^2$ and a χ-T curve not unlike that of "Kondo compounds"[14,17] the specific heat of UAl_2 clearly shows it to be an *f*-band spin fluctuation compound. The specific heat data are shown in Fig. 5, where it is seen that the spin fluctuation model of Doniach and Engelsberg[19] fits the low temperature data, but they are not fit by magnetic impurity or crystal field models. Most importantly, the $T^2 \ln T/T_{SF}$ spin fluctuation contribution to γ is only affected by $\sim$1% in an applied field of 43 kOe,[17] in agreement with calculations for spin fluctuations.[20] The spin fluctuation contribution to specific heat is seen clearly in this case because $T_{SF} \sim$ 23K is in the right range for this contribution to be observable.

$NpOs_2$, a Weak Ferromagnet. Mössbauer, magnetization, and neutron diffraction measurements show this to be a ferromagnet with $\mu_o \approx 0.25\ \mu_B$ and T_C = 7.5K.[18] Figure 6 shows the specific heat of $NpOs_2$[21] plotted as C/T vs T^2. The peak in C/T occurs at T_C = 7.9K. The magnetic entropy is found to be 0.2 Rln2, which indicates an effective moment about 0.6 μ_B.

The magnetic moment in $NpOs_2$ is probably too small to be due to localized 5*f* electron moments. Yet it seems too large to treat in the Stoner-Wohlfarth-Edwards (SWE) model[22] for moments $\ll 0.1\ \mu_B$, e.g., as in $ZrZn_2$. Yet it has been shown that the magnetization and thermal data are in fair agreement with the SWE model for weak itinerant ferromagnets.[23] The large values of γ in the paramagnetic and ferromagnetic regimes (Table 2) and the large change in γ at T_C shows that *f*-bands are strongly involved in the magnetism of $NpOs_2$.

$NpSn_3$, an Itinerant Antiferromagnet. Figure 7 shows the electronic specific heat of $NpSn_3$.[23,24] The ^{237}Np Mössbauer effect shows the occurence of magnetic order at 9K with a small effective moment, μ=0.29 μ_B/Np.[18] In this case however, the ordering is antiferromagnetic as shown by the two site character of the ^{119}Sn resonance, and a weakly temperature dependent susceptibility. The magnetic entropy is found to be essentially zero, -0.007 Rln2. There undoubtedly are *f*-bands at the Fermi level causing the very large γ's. The gap-like behavior of the transition, as per Cr, is shown by the large discontinuity at T_N and the near-BCS dependence of the data in Fig. 7.

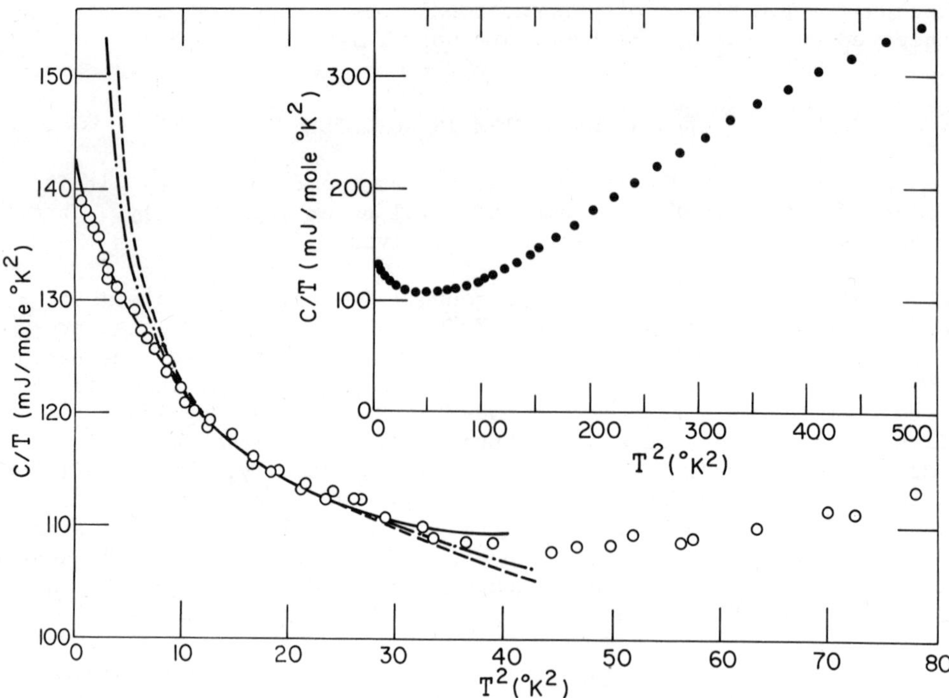

Fig. 5. C/T vs T^2 for UAl_2. Fits using data above 3K were made with $C = At + BT^3 + Df(T)$, where $f(T) = 1/T^2$ (dashed curve), $1/T$ (dash-dotted curve), and $T^3 \ln T$ (solid curve). Inset shows the same data over a larger temperature range (from Ref. 17).

VALENCE INSTABILITIES IN ACTINIDES

Since some pure actinides may have $U/W \sim 1$ (see Table 1), it is possible that this is also true for many of the compounds with high densities of states. Thus, it is reasonable to ask (especially at this conference) if valence instabilities occur in actinides. First we must ask if we know the valence in a metallic actinide. Certainly in the materials where the *f*-electrons are in bands, the concept of valence is misleading. In the rare earth compounds under discussion in most papers at this conference, the hybridization is among ionic states, while in the band picture presented in this paper we are considering one-electron states.[6,9,10]

Yet, van Daal, et al.,[25] have shown that there is a discontinuity in the lattice parameters of $UNi_{5-x}CU_x$ solid solutions at $x \geq 4$. At this composition, anomalies are found in ρ-T and χ-T curves, and there is an abrupt rise in γ over a narrow range of

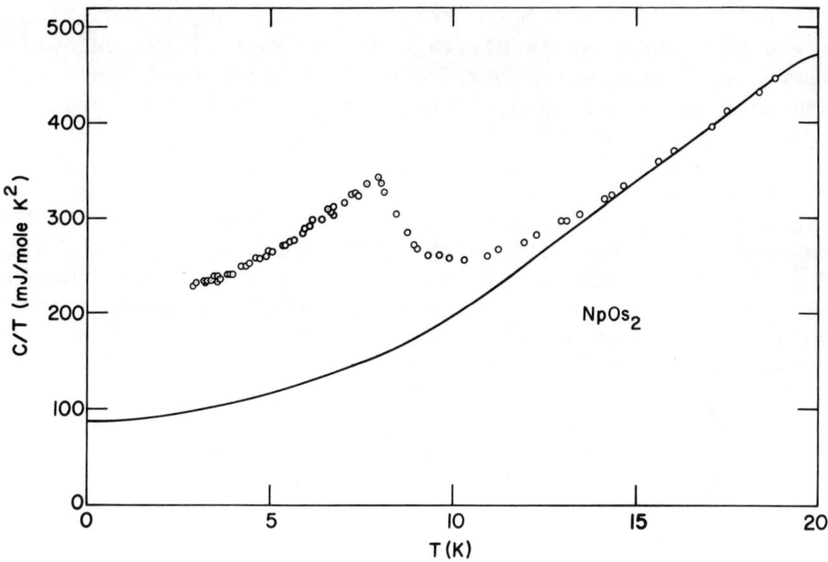

Fig. 6. C/T vs T for $NpOs_2$. The solid curve is the estimated phonon contribution plus the paramagnetic γ (from Ref. 21).

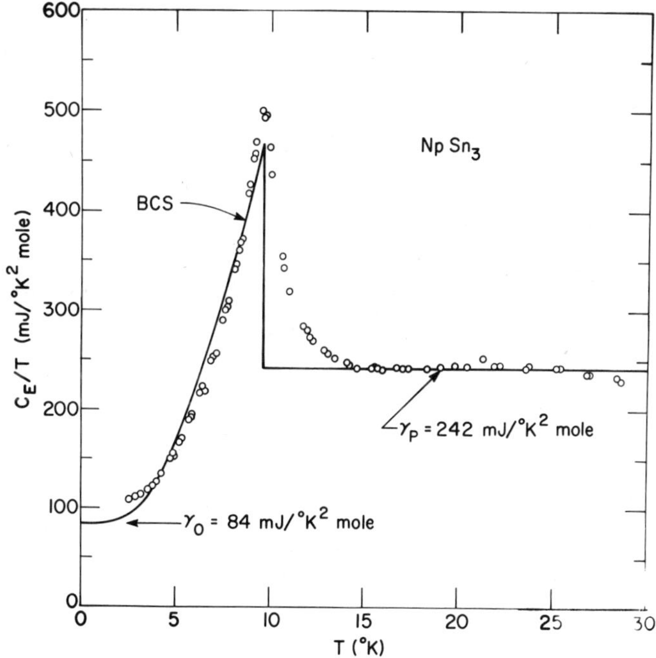

Fig. 7. Electronic specific heat of $NpSn_3$ plotted as C_E/T vs T. The solid curve is the mean-field, BCS prediction (from Refs. 23,24).

composition [$\gamma \sim 45$ mJ/(mole-K^2) for x = 4 and γ = 200 mJ/(mole-K^2) for x = 5]. These properties are reminiscent of the valence instability compounds. As of now, we have no microscopic evidence (XPS, Mössbauer effect, neutron diffraction) to prove the proposal.

Arguments such as those of van Daal, et al., based on effective paramagnetic moments and lattice parameters to establish valency assignments are usually misleading in actinides. Chan and Lam[26] have pointed out the difficulties associated with valence determinations due to nonobeyance of Russell-Saunders coupling. They have used full j-j manifold mixing in their crystal field calculations and do obtain groupings of effective moments for specific $5f^n$ assignments. However, in the case of PuSb, an example of a compound with apparent local moment magnetism, the crystal field parameters used to fit the magnetization below T_C are unable to fit the paramagnetic susceptibility.[26]

Zachariesen[27] and others have attempted to assign valences based on scaling the actinide contraction with the well documented lanthanide contraction. This approach requires a division of the electrons into valence or *f*-electrons. As discussed above, the *f*-band formation (and *f*-electron participation in bonding) makes this division artificial. Carried to its conclusion, this approach assigns a valence of four to Am and cannot explain the lack of magnetic order in Am. Paper 104 in this conference by Johansson discusses this matter further. However, Johansson also shows that in AmS we might expect a valence instability. Perhaps not all Am compounds are trivalent, J = 0, nonmagnetic. It has recently been shown that AmAs is antiferromagnetic[28] with T_N = 13K and $p_{eff} = 1.36\ \mu_B$.

CONCLUSION

In conclusion we have shown that in many actinide metallic systems the *f*-electrons exhibit band behavior. This is a consequence of direct *f-f* wave function overlap or hybridization of *f*-electrons with *s*-, *p*-, and *d*-electrons. The *f*-bands can be responsible for large electronic densities of states at the Fermi level which may lead to band magnetism of various types. Although the concept of valence instabilities must be approached cautiously especially in the light actinides, it would not be surprising to observe them in the future, especially in Am compounds.

ACKNOWLEDGEMENT

The author wishes to thank his colleagues at Argonne and elsewhere for many interesting discussions during the past several years. These include A. T. Aldred, A. J. Arko, S. D. Bader, B. R. Coles,

S. Doniach, B. D. Dunlap, F. Y. Fradin, A. J. Freeman, A. Harvey, D. D. Koelling, D. J. Lam, G. H. Lander, M. H. Mueller, W. J. Nellis, N. J. Rivier, G. K. Shenoy, and R. J. Trainor.

REFERENCES

1. For a general overview see, The Electronic Structure of the Actinides, A. J. Freeman and J. B. Darby, Jr., eds., (Academic Press, NY, 1974).

2. Th: J. E. Gordon, H. Montgomery, R. J. Noer, G. R. Pickett, and R. Tobon, Phys. Rev. 152, 432 (1966); U: S. D. Bader, N. E. Phillips, and E. S. Fisher, Phys. Rev. B12, 4929 (1975); Np and Pu: R. J. Trainor and M. B. Brodsky, to be published; and Am: R. O. A. Hall, M. J. Mortimer, D. L. McElroy, W. Muller, and J. C. Spirlet, in TransPlutonium 1975, W. Muller and R. Lindner, eds., (North Holland/American Elsevier, Amsterdam/New York, 1976), p. 139.

3. D. J. Lam and A. T. Aldred, in Reference I, Volume I, p. 109, and W. J. Nellis and M. B. Brodsky, Reference 1, Vol. II, p. 265.

4. J. A. Lee, in Rare Earths and Actinides, (Inst. of Physics, London and Bristol, 1971), p. 50.

5. M. B. Brodsky, A. J. Arko, A. R. Harvey, and W. J. Nellis, in Reference 1, Vol. II, p. 185.

6. A. J. Freeman and D. D. Koelling, in Reference 1, Vol. I, p. 51.

7. H. H. Hill, in Plutonium 1970, W. N. Miner, ed., (Met. Soc. A.I.M.E., NY, 1970), p. 2.

8. B. Johansson, Phys. Rev. B11, 2740 (1975).

9. D. D. Koelling, 2nd International Conference on Electronic Structure of Actinides, Wroclaw, Poland, 1976, to be published. A. J. Arko, loc. cit.

10. A. J. Arko, M. B. Brodsky, G. W. Crabtree, D. Karim, D. D. Koelling, L. R. Windmiller, and J. B. Ketterson, Phys. Rev. B12, 4102 (1975).

11. W. J. Nellis, A. R. Harvey, and M. B. Brodsky, AIP Conf. Proc. 10, 1076 (1973); M. B. Brodsky, R. J. Trainor, A. J. Arko, and H. V. Culbert, AIP Conf. Proc. 29, 317 (1975).

12. R. J. Trainor and M. B. Brodsky, to be published.

13. M. H. van Maaren, H. J. van Daal, and K. H. J. Buschow, Solid State Commun. 14, 145 (1974).

14. A. J. Arko, M. B. Brodsky, and W. J. Nellis, Phys. Rev. B5, 4564 (1972).

15. F. Y. Fradin, in Plutonium 1975 and Other Actinides, H. Blank and R. Lindner, eds., (North Holland, Amsterdam, 1976), p. 459.

16. M. B. Brodsky, Phys. Rev. B9, 1381 (1974).

17. R. J. Trainor, M. B. Brodsky, and H. V. Culbert, Phys. Rev. Lett. 34, 1019 (1975) and R. J. Trainor, M. B. Brodsky, and G. S. Knapp, in Plutonium 1975 and Other Actinides, H. Blank and R. Lindner, eds., (North Holland, Amsterdam, 1976), p. 475.

18. A. T. Aldred, B. D. Dunlap, D. J. Lam, and I. Nowick, Phys. Rev. B10, 1011 (1974).

19. S. Doniach and S. Engelsberg, Phys. Rev. Lett. 17, 750 (1966).

20. M. T. Beal-Monod, S. K. Ma, and D. R. Fredkin, Phys. Rev. Lett. 20, 929 (1968).

21. M. B. Brodsky and R. J. Trainor, Physica, in press.

22. E. C. Stoner, Proc. Roy. Soc. A165, 372 (1938); E. P. Wohlfarth, J. Appl. Phys. 39, 1061 (1968); and D. M. Edwards and E. P. Wohlfarth, Proc. Roy. Soc. A303, 127 (1968).

23. M. B. Brodsky and R. J. Trainor, Europhysics Conf. on Itinerant Electron Magnetism, Oxford, 1976, Physica, in press.

24. R. J. Trainor, M. B. Brodsky, B. D. Dunlap, and G. K. Shenoy, Phys. Rev. Lett., in press.

25. H. J. van Daal, K. H. J. Buschow, P. B. van Aken, and M. H. van Maaren, Phys. Rev. Lett. 34, 1457 (1975).

26. S. K. Chan and D. J. Lam, in Reference 1, Vol. I, p. 1.

27. W. H. Zachariasen, The Metal Plutonium, A. S. Coffinberry and W. N. Miner, eds., (U. of Chicago Press, Chicago, 1961), p. 99.

28. B. Kanellakopulos, J. P. Charvillat, F. Maino, and W. Muller, TransPlutonium 1975, W. Muller and R. Lindner, eds., (North Holland, Amsterdam, 1976), p. 181.

COMMENTS

WOHLLEBEN: Band structure calculations for those metals which have small values of γ are relatively uninteresting. Are there band structure calculations for systems with $\gamma \simeq 500$?

BRODSKY: No. The reason is because the calculations cannot be made self-consistent. In order to make a self-consistent calculation we need de Haas van Alphen effect data as well as the value of γ. Self-consistent calculations have been made for systems with $\gamma \simeq 25$, e.g., UGe_3.

CAMPAGNA: How difficult is it to do experiments with actinides?

BRODSKY: Working with light actinides is reasonably safe and they are available in moderate amounts. Some of them (e.g., Pu, Np) are sufficiently radioactive that lead box techniques are required. When one goes beyond Am there is the problem of availability of material. This is unfortunate since they have good local moments, i.e., they look like heavy rare earths.

GSCHNEIDNER: I think it is dangerous to obtain γ's from C/T vs. T or C/T vs. T^2 plots and relate them to the density of states, when magnetic ordering or other phenomena may be occuring in addition to the normal lattice and electronic contributions, as found for many of the samples you studied. One should extend the data down to low temperatures ($T<1°K$).

BRODSKY: That may be very difficult. However, for example in $NpSn_3$ there is nothing you can do to that curve that will make γ smaller than 75-80.

THEORETICAL MODELS FOR NARROW BAND EFFECTS IN RARE EARTHS AND ACTINIDES

B. Coqblin, A. K. Bhattacharjee, J.R. Iglesias-Sicardi[+]
and R. Jullien[++]
Laboratoire de Physique des Solides, Université
Paris-Sud, Centre d'Orsay, 91405 Orsay (France)

We review here some theoretical approaches able to describe the properties of anomalous rare-earth (Ce, Yb...) metals, alloys and compounds or of actinide metals and compounds ; in the second case, we emphasize the case of nearly magnetic metals (Np, Pu) or compounds (UAl_2, $NpRh_3$...).

I - INTRODUCTION

It is now well established that the rare-earth metals can be divided in the "normal" ones which have an integer number of 4f electrons and the "anomalous" ones (Ce, Sm, Yb...) which may have a non-integer number of 4f electrons depending on pressure and temperature[1]. Many properties of anomalous rare-earths have been accounted for by assuming that the 4f level is very close to the Fermi level and moves upwards with increasing pressure. Two features of anomalous rare-earth systems have been extensively studied : first, the valence changes or the magnetic-nonmagnetic transitions occuring in Cerium metal[2][3], SmS[4][5] or LaCe[6] under pressure and in particular the phase diagram of Cerium which presents a critical point ; and second, the influence of the crystalline field on the Kondo effect in systems such as $CeAl_3$[7]-[10] or AuYb[11].

On the other hand, in the actinide series, there is a change from transition-like elements to rare-earth-like elements by passing through mixed d-f elements (U, Np, Pu)[12]. Neptunium and Plutonium or also U and Pu compounds are nearly magnetic and their properties can be accounted for within the spin-fluctuation model[13].

The purpose of the present paper is to review some simple theoretical approaches able to describe the properties of anomalous rare-earth or actinide systems. The properties of anomalous rare-earth and actinide systems have been already reviewed elsewhere[14][15].

II - THE ANOMALOUS RARE-EARTH SYSTEMS

2.1. The Phase Diagram of Cerium

Cerium is known to have an anomalous phase diagram, shown on Figure 1, with five solid phases : two f.c.c. α and γ phases, one d.h.c.p. β phase, one high temperature b.c.c. δ-phase and the α'-phase above 50 kbar[1][2][16]. The γ-phase has a Curie-Weiss paramagnetism with a 2.5 μ_B magnetic moment corresponding to the $4f^1$ configuration, while the α-phase has a roughly constant and large Pauli paramagnetism. The α'-phase is superconducting below 1.7 K. The transition between the α and γ phases is a first-order one at normal pressure but the jump in the lattice parameter or in the resistivity decreases with increasing pressure to finally give rise to a continuous transition above the critical point C, shown on figure 1. The valence is changing from 3.1 in the γ-phase, 3.7 in the α-phase at 7kbar, 3.9 in the α-phase at 50kbar and finally 4 in the α'-phase.

The first explanation of the γ-α transition of Cerium is based on the virtual bound state model[2] : the localized 4f level is broadened by mixing with conduction electrons, which yields a lorentzian shape for the 4f density of states with a width of order 0.01 eV. When pressure is applied, the 4f level is shifted upwards compared to the Fermi level and the γ-α-α' sequence of Cerium phases is described by shifting the 4f level from below to far above the Fermi level. When temperature increases, there is an effective increase of the broadening of the 4f level when Δ is of order kT and it results an increase of the effective density of states at the Fermi level, which tends to favour the occurrence of magnetism. So, the virtual bound state model which is based on the Hartree-Fock approximation explains really all the points of the phase diagram of Cerium : the transition between the α and γ phases with a change of valence, the magnetism of the γ phase and the nonmagnetism of the α phase, the existence of first-order and continuous transitions with a critical point in the phase diagram, the existence of a large density of states at the Fermi level in the α-phase. Moreover, reasonable values have been chosen for the parameters, i. e. Δ of order 1/100 ev, the exchange

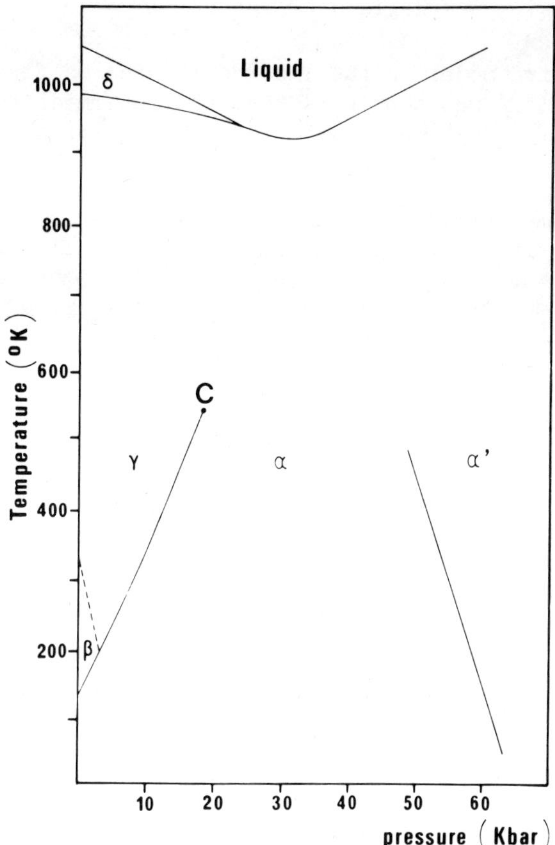

Figure 1 : The phase diagram of Cerium

integrals of order 1 eV and the Coulomb integrals of order some eV.

Ramirez and Falicov(3) have then introduced another model based on an "ionic-like" description of the 4f shell : the 4f electron lie in an infinitely narrow band slightly above the Fermi energy and the f-f Coulomb repulsion is assumed to be infinite. The driving mechanism of the γ-α transition arises here from considering the two-body interaction between 4f and conduction electrons and Ramirez and Falicov have also found a phase diagram with a critical point, separating two magnetic phases with different valences. By including the effect of hybridization of the localized 4f states with the conduction band, Alascio et al.(17) have obtained a roughly constant magnetic susceptibility of α-Cerium, but the method used to obtain a 4f width is in fact not so different from the virtual bound state approach.

A third model is the interconfiguration fluctuation model of Hirst[18] who considers that the intermediate valence is obtained by a fluctuation between two configurations, namely $4f^o$ $4f^1$ in the case of Cerium.

The measurements in the α-phase of Cerium allow us to study the validity of the different models. The electronic specific heat constant and the roughly constant magnetic susceptibility of α-Cerium are large and equal respectively to $\gamma = 12.8$ mJ/mole K^2 and $\chi = 6 \times 10^{-4}$ emu/mole[19]. The high values of χ and γ show clearly the existence of spin fluctuations originating from the large 4f density of states at the Fermi level in nearly magnetic α-Cerium and the exchange-enhancement can be estimated to be of order 10[1][14]. So, the high values of γ and χ of α-Ce can be understood only in a model involving a broadened 4f level such as the virtual bound state model[2] or the model of Alascio et al.[17]. On the contrary, the ionic-type models such as the Ramirez-Falicov one[3] or the Hirst one[18] give magnetic-type situations for α-Cerium and cannot really account for its large χ and γ values.

2.2. The Magnetic-Nonmagnetic Transition in Alloys with Cerium Impurities

A magnetic-nonmagnetic transition has been observed in LaCe[6] or YCe[20] alloys under pressure or in Au-Ag based alloys with Yb impurities[21] and La-Th based alloys with Ce impurities[22] when the relative composition of the host is varied.

In the case of LaCe alloys[6], the decrease $-\Delta T_c/c$ of the superconducting temperature T_c with the concentration c of Cerium impurities increases with pressure up to roughly 15kbar and then decreases up to 125 kbar with a very rapid decrease around 30kbar, as shown on figure 2. Similarly, the Kondo resistivity minimum disappears at high pressures, as observed in YCe alloys[20].

The first interpretation of the behaviour of LaCe under pressure assumes an upwards shift of the 4f level with respect to the Fermi level[23]. In this model, we separate the pressure domain into two regions : the magnetic domain up to 30 kbar where we use an ionic model and treat the resonant scattering in the limit of the Schrieffer-Wolff transformation and the nonmagnetic domain where we use the approach of virtual bound states within the Hartree-Fock approximation ; the theoretical results are shown by the curves (I) and (II) in figure 2. The same method has been used to account for

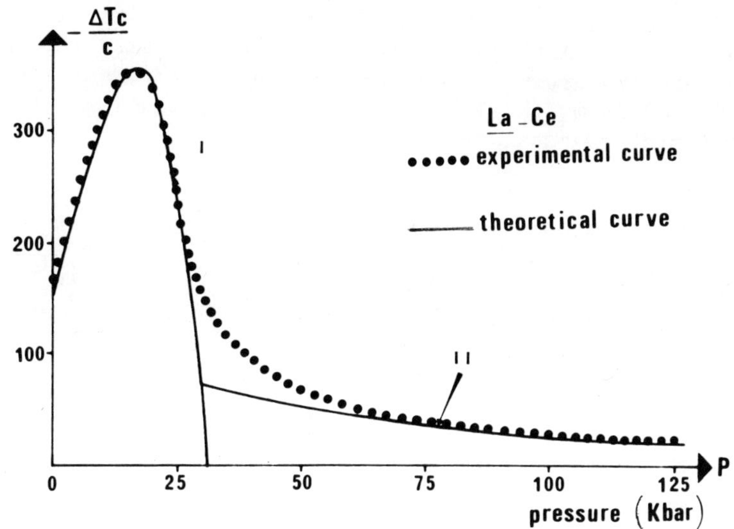

Figure 2 : Plot of $-\Delta T_c/c$ for LaCe alloys under pressure, T_c being the superconducting temperature.

the decrease of the superconducting temperature in LaThCe alloys[22].

Another model to account for the variation of $-dT_c/dc$ in LaCe alloys under pressure[24] or in LaThCe alloys with the Th concentration[25] is to use the Muller-Hartmann-Zittartz formula[26] for the s-f exchange Hamiltonian and to make the Kondo temperature increase with increasing either the pressure or the Thorium concentration. The two methods are clearly equivalent to account for experimental data on superconductivity.

On the other hand, recent magnetization and nuclear orientation experiments[27][28] have shown that the system LaThCe is not equivalent to the system LaCe under pressure, since in the first system there are clearly environment effects, i. e. a Cerium atom is magnetic only if it is surrounded by at least a given number of Lanthanum atoms.

2.3. The Influence of the Crystalline Field on the Kondo Effect

The influence of the crystalline field on the Kondo effect of dilute alloys or compounds with Ce or Yb has been studied within the ionic model in the framework of an effective Hamiltonian which describes the resonant scattering character and takes into account combined spin and orbit exchange scattering[29][8]. For a cerium (or ytterbium) atom, the large spin-orbit coupling leads to a ground state multiplet of total angular momentum $j = 5/2$ (or $j = 7/2$), which is then split by the crystalline field (CF). In cubic cerium compounds, the CF effect gives a Γ_7 doublet and a Γ_8 quartet, while in cubic ytterbium compounds, there is also a Γ_6 doublet in addition to Γ_7 and Γ_8. Let M denote a CF eigenstate of the ground state multiplet, C_M^* the creation operator for a 4f electron in the corresponding M substate, and C_{kM}^* the creation operator for a conduction electron partial wavefunction of wavenumber k, total angular momentum j(5/2 for Ce and 7/2 for Yb) with a magnetic z component $M = \langle j_z \rangle$. Performing the Schrieffer-Wolff transformation on the Anderson Hamiltonian yields an effective Hamiltonian appropriate for the $4f^1$ (or $4f^{13}$) configuration and given by :

$$\mathcal{H} = - \sum_{\substack{k,k' \\ M,M'}} J_{MM'} C_{k'M'}^* C_{kM} (C_M^* C_{M'} - \delta_{MM'} \langle n_M \rangle) + \mathcal{V} \sum_{\substack{k,k' \\ M}} C_{k'M}^* C_{kM} \qquad (1)$$

The exchange integrals $J_{MM'}$ are given by :

$$J_{MM'} = \frac{1}{2} |V_{kf}|^2 \left(\frac{1}{E_M} + \frac{1}{E_{M'}} \right) \qquad (2)$$

V_{kf} is the mixing parameter between conduction and 4f electrons, E_M (<0) designs here the energy of the CF level of M value compared to the Fermi level. $\mathcal{V}$ is chosen here to be independent of M.

The important feature of the Hamiltonian (1) is that it allows matrix elements between all the CF sublevels and that the change in the magnetic quantum numbers $\Delta M = M' - M$ is no longer limited to ± 1 or 0 as for the classical s-f exchange Hamiltonian.

The third-order perturbation theory resistivity[8], thermoelectric power[10], relaxation rate without[11] or with[30] an applied magnetic field have been computed within the Hamiltonian (1). The resistivity behaves as Log T at high and low temperatures compared to the CF splitting Δ and goes through a maximum of order Δ[8]. Figure 3 shows the theoretical fits of the magnetic resistivities of $CeAl_2$ and $CeAl_3$ compounds with parameters given in ref. 8 ; we see that a good agreement is found at temperatures above roughly 5K. Below that temperature, the resistivity is still decreasing regularly with decreasing temperature[31] in contrast to the results of the perturbation theory ; we will discuss this result in the next section. We can finally notice that the absolute value of the slope of the high temperature resistivity curve increases with increasing

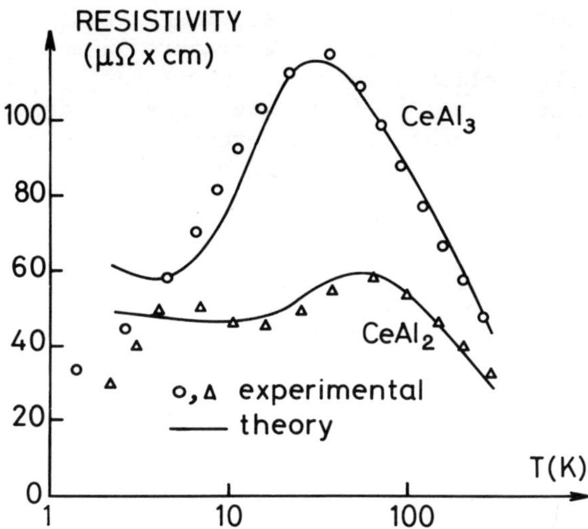

Figure 3 : The experimental and theoretical values of the magnetic resistivities of $CeAl_2$ and $CeAl_3$ versus temperature in Logarithmic scale. The theoretical parameters are given in ref. 8.

pressure, which corresponds to an increase of $|J_{MM'}|$ and to an upward shift of the 4f level compared to the Fermi level(14).

The thermopower has been also computed(10) and a peak is found at a temperature ranging between $\Delta/6$ and $\Delta/3$; the Seebeck coefficient is large and negative for $v = 0$, while it becomes positive for a negative and large v value. A reasonable account of the thermopower results in $Ce_{1-x}La_xAl_3$ alloys(9) is found but the obtained parameters are relatively different from those used to fit the resistivity ; in particular, the value for the CF splitting in $CeAl_3$ is 50 or 255 K according to respectively resistivity or thermopower data.

The third-order perturbation theory relaxation rate W of the doublet ground state Γ_7 in cubic AuYb alloys has been also computed with the Hamiltonian (1) in absence of an applied magnetic field[11]. It is found that W/T behaves as Log T at low temperature, i. e. for kT below roughly $\Delta/15$, and then rises quickly for kT larger than $\Delta/10$; this theoretical calculation agrees quite well with the Mössbauer measurements of W/T in AuYb alloys. More recently, the effect of a magnetic field on the third-order relaxation rate of the ground state of an Ytterbium impurity in gold has been also computed with the Hamiltonian (1)[30]. Figure 4 gives the plot of $R/2\pi T$ versus temperature in Logarithmic scale in the case of AuYb alloys, where R designs the relaxation rate of the Lower Zeeman level split by the magnetic field inside the Γ_7 doublet ; thus, R is clearly different from W. We see on figure 4 that the zero-field T Log T behaviour of the relaxation rate disappears completely for only a 10 kG field in AuYb alloys and

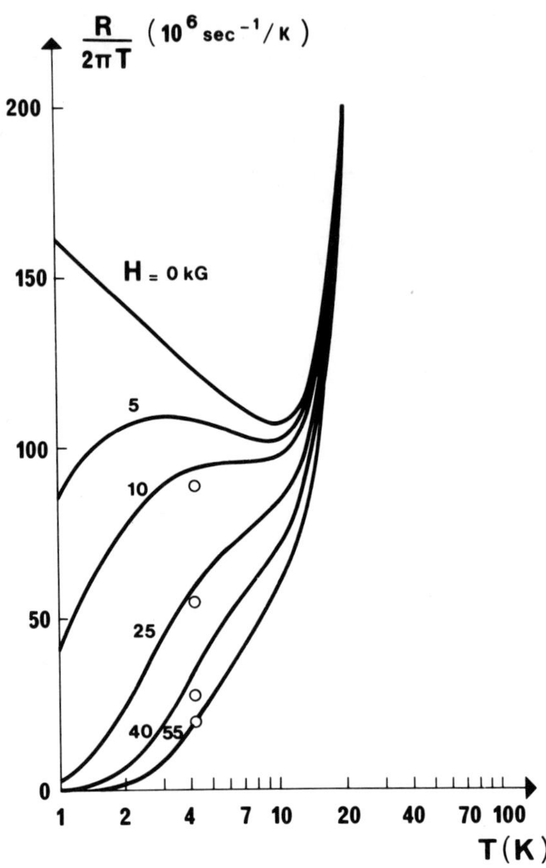

Figure 4 : Plot of the ground state relaxation rate R divided by $2\pi T$ versus LogT in AuYb alloys. The full lines are the theoretical curves as explained in ref. 30 and the circles are the experimental points of Stöhr[32].

the decrease of R/T with increasing field agrees with the results of Stöhr[32] at 4.2 K, as shown on Figure 4.

Thus, the preceding approach based on the ionic model and describing both the CF and Kondo effects is valid at sufficiently high temperatures and fails at low temperatures. For example, in $CeAl_3$, the resistivity decreases continuously with decreasing temperature down to 0.3 K[31] and recent magnetic susceptibility[31] or specific heat[33] measurements indicate that a smooth transition into a non magnetic Kondo state occurs at low temperature with a Kondo temperature of order 5K. At very low temperatures, the magnetic susceptibility χ, of order 3×10^{-2} emu/mole[31] and the electronic specific heat constant γ, of order 1600 mJ/mole K[34], are particularly enormous ; moreover, the ratio $\chi/\gamma = 1.35$ is relatively small and the resistivity behaves as T^2 at very low temperatures[34]. Similar properties have been found in other cerium compounds such as $CeAl_2$ or $CeSn_3$[14]. Pressure effects are also particularly important in such Ce compounds and for example the electronic specific heat constant γ of $CeAl_3$ is decreased by a factor 2 under a 7 kbar pressure[35].

So, in $CeAl_3$, there is apparently at very low temperatures a non magnetic Kondo state which can be described in terms of large spin fluctuations arising from a very narrow 4f band, while at higher temperatures cerium is magnetic and can be described by the high temperature approach of the Kondo effect. The question arises now to know if there is any kind of magnetic ordering at low temperatures in such a "Kondo lattice". This question is presently controversial and is extensively studied both experimentally[36] and theoretically[37]. However, we can think that the transition from the high temperature region to the low temperature non magnetic Kondo state could proceed through an antiferromagnetic ordering at intermediate temperatures in $CeAl_3$ and similar compounds. Any way, the answer is probably not unique, since it depends clearly on the relative strength of the Ruderman-Kittel interaction and of the Kondo effect.

In conclusion, we can thus use at present different models to describe different situations of anomalous rare-earths. On one side, the ionic model, which describes fully a given $4f^n$ configuration and in particular the CF effect, can account for the properties of Cerium compounds at high temperatures or of alloys with Ce (or Yb) impurities even very near the magnetic-nonmagnetic transition. On the other side, a model involving a broadened 4f level, such as the virtual bound state model[2] or the model of Alascio et al.[17], is the only one which can account for the properties of Cerium compounds at very low temperatures or of the α-phase of pure Cerium ; in particular, the virtual bound state model explains fairly well all the properties of the anomalous phase diagram of Cerium.

III - THE ACTINIDE SYSTEMS

3.1. The General Features of Actinides

The magnetic properties of actinide metals are different from those of rare-earth metals, because the 5f electrons are much less localized than the 4f electrons and it results that there is a large d-f hybridization in the beginning of the actinide series and that the localized magnetism occurs only for Curium which has the same position in the f- series as Gadolinium. So, we can separate the actinide metals in three different groups(12) :

(i) Ac, Th and Pa behave as transition metals, Th (and probably Pa) being superconducting.

(ii) U, Np and Pu are non-magnetic metals showing a large mixed d-f character at the Fermi level with almost temperature independent and large magnetic susceptibilities.

(iii) Am, Cm and Bk are rare-earth like metals with a valence close to 3.

In the actinide series, only Cm and Bk are magnetic, with magnetic moments close to those of their ionic $5f^n$ configurations, and they exhibit experimentally a magnetic ordering at low temperatures. The delay in the occurrence of magnetism compared to the rare-earth series has been attributed to a large d-f hybridization which tends to broaden the f-level(12). An extension of this model considering a large spin-orbit coupling yields a reasonable agreement with experiments(38), giving no magnetism for Np, Pu and Am and the order of magnitude for the magnetic moments of Cm and Bk.

On the other hand, Neptunium and Plutonium are nearly magnetic and exhibit some anomalous physical properties which have been attributed to the presence of spin-fluctuations(13)(39)-(41). Many other actinide compounds, such as USn_3, UAl_2, $NpRh_3$, or $PuAl_2$ show similar behaviors which can be accounted for by the spin fluctuation theory(13).

So, we will present here a spin-fluctuation model which takes into account the temperature dependence of the Stoner susceptibility and which is able to explain the main physical properties of these nearly magnetic materials. This model has been already reviewed in the Baden-Baden(15) and Wroclaw(42) Conferences on actinides. So let us recall here briefly the application to transport properties and magnetic susceptibility measurements and at last emphasize more precisely new results concerning the thermoelectric power.

3.2. The Spin Fluctuation Model for Transport Properties

The transport properties, electrical resistivity ρ , thermal resistivity W and thermoelectric power Q, can be derived using the standard variational procedure[43] within the two band model where the conduction electrons of a broad conduction band (index c in the following) are scattered by the spin fluctuations arising from a narrow band (index i) and they are given by :

$$\rho = \frac{\rho_o}{T}\int_o^\infty \frac{\omega \; A_3(\omega)\; d\omega}{(e^{\omega/T} - 1)(1 - e^{-\omega/T})} \tag{3}$$

$$W = \frac{\rho_o}{L_o T^4}\int_o^\infty \frac{\omega T^2 A_3(\omega) + (3/\pi^2)\omega^3 A_1(\omega) - (1/2\pi^2)\omega^3 A_3(\omega)}{(e^{\omega/T} - 1)(1 - e^{-\omega/T})}\, d\omega \tag{4}$$

$$Q = \frac{\pi^2}{3}\,\frac{1}{e}\,\frac{T}{T_{F_c}}\left(\frac{3}{2} + \frac{\alpha}{\beta}\right) \tag{5}$$

with :

$$\alpha = \int_o^\infty d\omega \frac{\left(\frac{\pi^2 T^2}{2} A_3(\omega).\omega - \frac{1}{2}A_1(\omega).\omega^3 - \frac{2}{3}\omega(\omega^2 + 4\pi^2 T^2)\,\mathrm{Im}\chi(2k_{F_c},\omega)\right)}{(e^{\omega/T} - 1)(1 - e^{-\omega/T})} \tag{6}$$

$$\beta = \int_o^\infty d\omega \frac{\left(\frac{\pi^2}{3}\omega T^2 A_3(\omega) + \omega^3 A_1(\omega) - \frac{\omega^3}{6} A_3(\omega)\right)}{(e^{\omega/T} - 1)(1 - e^{-\omega/T})} \tag{7}$$

and :

$$A_n(\omega) = \frac{2}{(k_{F_c})^{n+1}} \int_o^{2k_{F_c}} q^n \,\mathrm{Im}\,\chi(q,\omega)\, dq \tag{8}$$

L_o is the ideal Sommerfeld value for the Lorenz number :

$$L_o = \pi^2 k_B{}^2/(3e^2) \tag{9}$$

and ρ_o is a constant given by :

$$\rho_o = \left(JN_c(\varepsilon_{F_c})/4\right)^2 \frac{m_c}{n_c\, e^2\, \tau_{F_c}}\,\frac{\nu}{n_c} \tag{10}$$

In these formulae, we have made $\hbar = k_B = 1$. J is the coupling constant between electron and spin fluctuation, ν is the number of atom per unit volume, τ_{F_c} is given by :

$$1/\tau_{F_c} = k_{F_c}^2 / 2\, m_c = T_{F_c} \tag{11}$$

$k_F = (3\pi^2 n)^{1/3}$, $N(\varepsilon_F) = mk_F/\pi^2$, m and n (with indices c or i) are respectively the Fermi momentum, the density of states per atom at the Fermi energy, the effective mass and the number of electrons per unit volume.

The spin fluctuation spectrum Im $\chi(q,\omega)$ is the imaginary part of the dynamical paramagnetic susceptibility of the i-electrons given by the RPA :

$$\chi(q,\omega) = \frac{\chi^\circ(q,\omega)}{1 - I\,\chi^\circ(q,\omega)} \tag{12}$$

$\chi^\circ(q,\omega)$ is the i-susceptibility without the interaction I :

$$\chi^\circ(q,\omega) = \sum_{\underset{\sim}{k}} \frac{f(\varepsilon_{\underset{\sim}{k}}) - f(\varepsilon_{\underset{\sim}{k}+\underset{\sim}{q}})}{\varepsilon_{\underset{\sim}{k}+\underset{\sim}{q}} - \varepsilon_{\underset{\sim}{k}} - \omega} \tag{13}$$

The approach to magnetism is quantitatively described by the Stoner enhancement factor S which is the ratio $\chi(o,o)/\chi^\circ(o,o)$ at $T = 0$, i. e. :

$$S = \frac{1}{1 - I\,\chi^\circ(o,o)} = \frac{1}{1 - \frac{I}{2} N_i(\varepsilon_{F_i})} = \frac{1}{1 - \bar{I}} \tag{14}$$

Here, we take into account the temperature dependence of $\chi(q,\omega)$ which is calculated for a narrow parabolic i-band with a small Fermi energy T_{F_i}. The shift of the Fermi level is computed exactly at all temperatures by keeping constant the number of i-electrons :

$$\sum_{\underset{\sim}{k}} f(\varepsilon_{\underset{\sim}{k}}) = \frac{n_i}{2\nu} \tag{15}$$

and then its temperature dependence is used to obtain that of $\chi(q,\omega)$.

The main results for ρ and W can be summarized as follows :
(i) At low temperatures, we recover well-known results : ρ shows a large T^2 term and W shows a large T term, both increasing with S. Also, these effects decrease with increasing values of the ratio $\xi = k_{F_c}/k_{F_i}$ of the two wavevectors for the c- and i- bands.

(ii) At high temperatures, the electrical resistivity ρ saturates to a constant value ρ_∞ while the thermal resistivity W behaves as T^{-1} and the asymptotic formulae are given by :

$$\rho = \rho_\infty\left(1+ \frac{\theta}{T}\left(1 - \frac{4}{3}\frac{\xi^2}{I}\right)\right) \tag{16}$$

$$W = \frac{\rho_\infty}{L_o} T^{-1} \tag{17}$$

with :

$$\rho_\infty = 2\pi\rho_o n_i/\nu$$
$$\theta = \frac{2}{3}\bar{I}\, T_{F_i} \tag{18}$$

The high temperature behavior of ρ is a direct consequence of the behavior of $\chi(0,0)$ which reaches a Curie-Weiss behavior for $T > T_{sf} = T_{F_i}/S$.

The theoretical curves of the electrical resistivity ρ start as T^2 at low temperatures, begin to bend over towards the T axis around the spin-fluctuation temperature $T_{sf} = T_{F_i}/S$ and then tend to a constant value after passing through a maximum or not depending on the values of the parameters S and ξ. On the other hand, the thermal resistivity W starts as T at low temperatures, goes through a maximum around T_{sf} and then decreases as T^{-1} at high temperatures.

In contrast to ρ and W, the thermoelectric power Q, which is negative, does not show an anomalous behavior characteristic of the spin fluctuations. The temperature dependence of Q is almost independent on S and only slightly dependent on ξ. The spin fluctuation thermopower will be discussed in more detail in § 3.3 ; in particular the thermopower due only to normal processes is shown in figure 8.

3.3 Application to the Transport Properties of Actinide Systems

The preceding theory can explain the temperature dependence of both the electrical and thermal resistivities of nearly magnetic systems. When applying the model to actinide systems, we assume that the c-electrons are both the 7s and 6d electrons and that the i-electrons are the 5f electrons and we neglect the 6d-5f hybridization.

We apply now the model to the following actinide systems : Np, Pu, UAl_2, UAl_3, UAl_4, UGa_3, UIn_3, USn_3, $NpRh_3$, $PuRh_2$, $PuAl_2$, $PuZn_2$. All these systems have large enhancement factors as deduced from available susceptibility and specific heat measurements. Their resistivity curves behave as T^2 at low temperatures and saturate at high temperatures with or without a maximum[15].

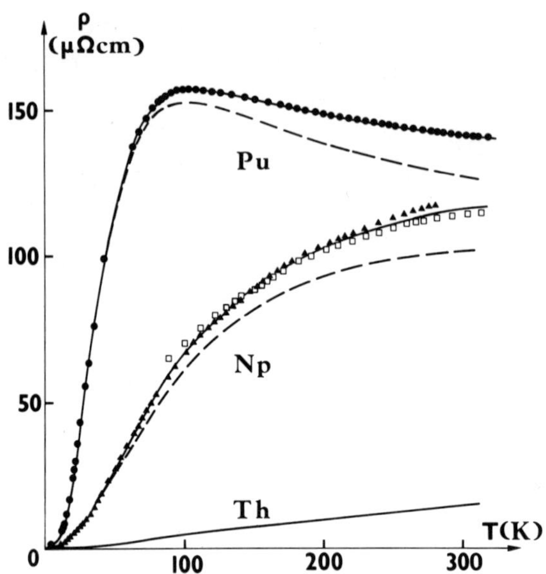

Figure 5 : The electrical resistivities of Np and Pu versus temperature : the full lines are theoretical fits obtained as explained in text.

More precisely, we present in figure 5 a fit of the resistivity curves of Np and Pu[13]. The parameters used are :
$S = 10$, $\xi = 0.4$, $\rho_\infty = 95\mu\Omega \times cm$, $T_{F_i} = 280$ K for Pu and
$S = 9$, $\xi = 0.7$, $\rho_\infty = 105\mu\Omega \times cm$, $T_{F_i} = 620$ K for Np. In order to

obtain a more pronounced maximum for Pu, we have chosen smaller T_{F_i} and ξ values. Also, we have added to the theoretical curves (dashed lines) a small phonon contribution assumed to be equal to the resistivity of Thorium. A similar fit is available in ref. 44 for the electrical resistivity curve of UAl_2. Other quantitative comparisons could have been performed for the other systems listed above. But, anyway, it would be difficult to choose reasonable parameters to account for the rapid change of resistivity in the case of $PuAl_2$ and $PuZn_2$.

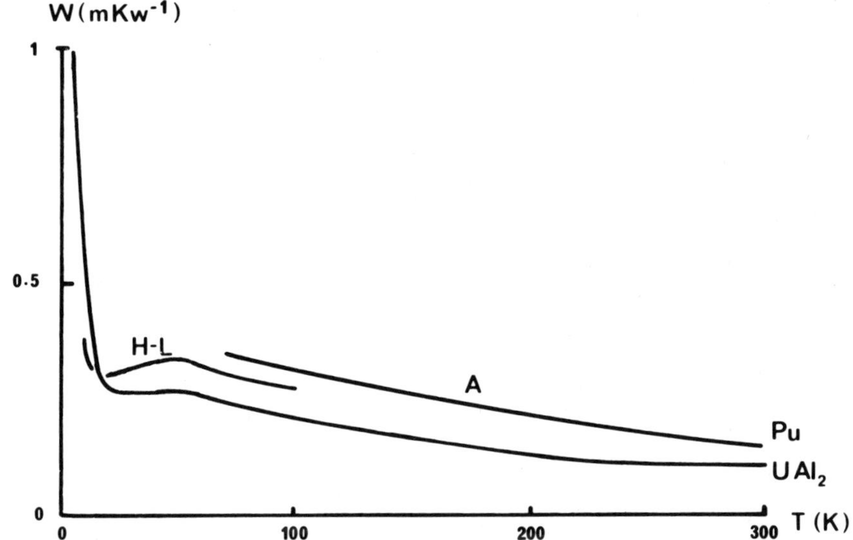

Figure 6 : The experimental thermal resistivity W of Pu and UAl_2 versus temperature. For Plutonium, the experimental curves labelled H-L and A come from ref.45.

The thermal conductivities of Plutonium[45][46] and UAl_2[47] are shown in figure 6. If we disregard the low temperature 1/T-like behavior which is generally attributed to magnetic impurities, both the very large values of W and the general shape of the curves, which present a maximum around 40 K, agree fairly well with the preceding theoretical predictions and thus support the spin-fluctuation model. A more precise analysis of the thermal conductivity data needs to take into account the phonon contribution to the thermal conductivity[46].

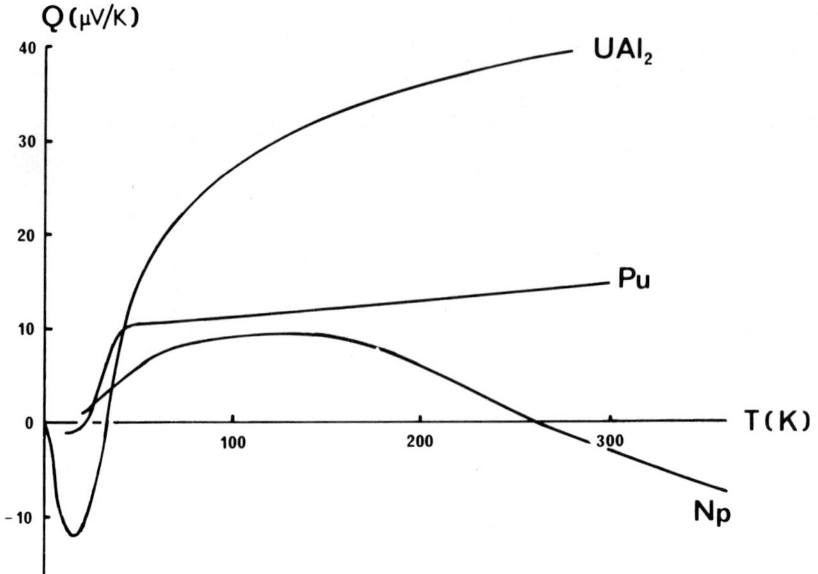

Figure 7 : Experimental plots of the thermoelectric power Q versus temperature for UAl_2, Pu and Np.

The thermoelectric powers of UAl_2[47], Pu[48] and Np[49] are shown in figure 7. Except at very low temperatures, the thermopowers of UAl_2 and Pu are positive and reach relatively large values at high temperatures; on the contrary, the thermopower of Np remains positive only up to 260 K. The theoretical model, which describes only normal processes in the electron-paramagnon scattering (curves labelled $\lambda = 2$ in figure 8), cannot explain at all such behaviours. An improvement of the present model, which takes into account the Umklapp processes, will be presented in the next paragraph.

3.3 Influence of the Umklapp Processes on the Spin-Fluctuation Transport Properties

The preceding calculation is valid only if the radius k_{F_c} of the conduction electron Fermi sphere is much smaller than the dimension of the Brillouin zone. But, in a scattering process, we have in general the relation :

$$\vec{K} = \vec{k}' - \vec{k} = \vec{q} + \vec{G} \tag{19}$$

connecting the incoming and outcoming electron wavevectors $\vec{k}$ and $\vec{k}'$ to the paramagnon wavevector $\vec{q}$ and to the reciprocal lattice vector $\vec{G}$. If $\vec{K}$ lies inside the first Brillouin zone, the process is "normal" ($\vec{G} = 0$), while if $\vec{K}$ lies outside the first Brillouin zone, the process is an "Umklapp" one ($\vec{G} \neq 0$). To perform the calculation of the transport properties due to Umklapp processes, we make several approximations, following previous calculations on the electron-phonon scattering by Ziman(43) and Rössler(50) ; in particular, to perform the $\vec{q}$-integration, we approxime the first Brillouin zone by a sphere of radius G/2 and the allowed $\vec{K}$ values for an Umklapp process are included inside both the sphere of radius $2k_{F_c}$ and the sphere translated by the vector $\vec{G}$ and of radius G/2.

Among the transport properties, the Seebeck coefficient is the most affected by considering the Umklapp processes. We do not reproduce here the calculations which will be reported elsewhere(51) and we just give the final result for the thermoelectric power Q :

$$Q = \frac{\pi^2}{3} \frac{k_B}{e} \frac{T}{T_{F_c}} x \tag{20}$$

where e is the negative electron charge and x is given by :

$$x = x_o + \frac{\alpha}{\beta} \tag{21}$$

with :

$$\alpha = \int_0^\infty \frac{dy}{(e^y-1)(1-e^{-y})} \left[\frac{\pi^2}{3} x_o y A_3(Ty) - \frac{1}{2} y^3 A_1(Ty) - \frac{2}{3} y (y^2+4\pi^2) \mathrm{Im}\chi(2k_{F_c}, Ty)\right] \tag{22}$$

$$\beta = \int_0^\infty \frac{dy}{(e^y-1)(1-e^{-y})} \left[y^3 A_1(Ty) + \frac{\pi^2}{3} y A_3(Ty) - \frac{y^3}{6} A_1(Ty)\right] \tag{23}$$

The functions $A_n(\omega)$ are the sums of a contribution $A_n^N(\omega)$ from normal processes and a contribution $A_n^U(\omega)$ from Umklapp processes, i.e. :

$$A_n(\omega) = A_n^N(\omega) + A_n^U(\omega) \tag{24}$$

with :

$$A_n^N(\omega) = \frac{1}{(k_{F_c})^{n+1}} \int_0^{G/2} q^n \mathrm{Im}\chi(q,\omega)\,dq \tag{25}$$

$$A_n^U(\omega) = \frac{z}{G(k_{F_c})^{n+1}} \int_{G-2k_{F_c}}^{G/2} q\,dq \int_{G-q}^{2k_{F_c}} K^{n-1} \mathrm{Im}\chi(K,\omega)\,dK \tag{26}$$

where z is the coordination number.

In the expressions (21) and (22), x_o is the x value for the case of conduction electrons without any interaction with a narrow band ; so, x_o is equal to 3/2 if the Fermi surface is a sphere of electrons inside the first Brillouin zone, but has a smaller value if the Fermi surface is distorted in the vicinity of the Brillouin zone boundaries, becomes negative if the holes are the most numerous carriers and reaches finally - 3/2 if the Fermi surface is only a sphere of holes inside the first Brillouin zone. Thus, x_o can make a drastic change of sign for the thermopower when the size of the Fermi surface increases but unfortunately, it is difficult to evaluate x_o in our model for actinide systems.

At last, figures 8 and 9 summarize the main theoretical results concerning the thermopower. The two figures are plotted for S = 10 and z = 6 (simple cubic case). Figure 8 gives the plot of $Q\, T_{F_c}/T_{F_i}$ versus temperature for $x_o = 3/2$ and several values of $\xi = k_{F_c}/k_{F_i}$ and $\lambda = G/2k_{F_c}$. The case of normal processes (λ=2) gives negative and almost linear thermopowers. When Umklapp processes are taken into account, the thermopower is larger and a clear curvature is observed in the Q versus T curves. The thermopower increases also when ξ decreases. We have finally to notice that Q remains always negative when x_o is equal to 3/2.

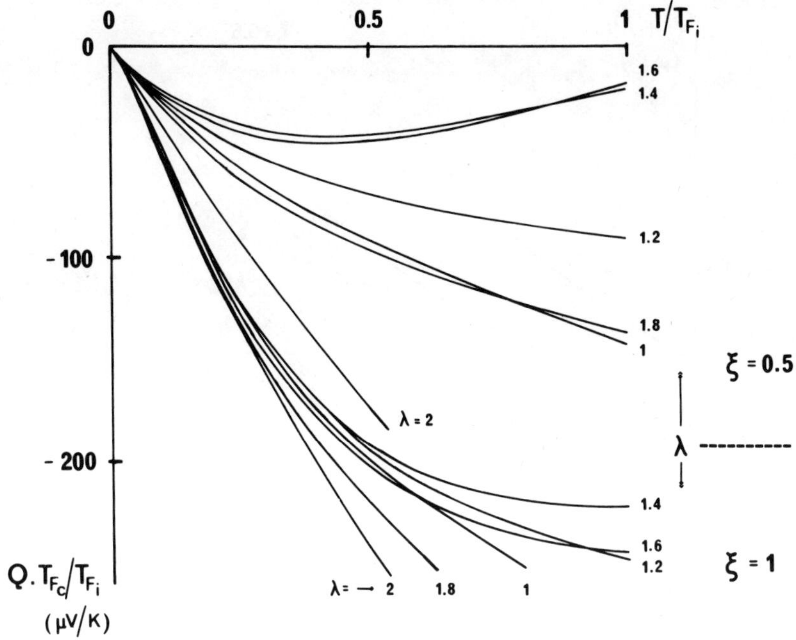

Figure 8 : Theoretical plots of $Q\ T_{F_c}/T_{F_i}$ versus temperature for $S = 10$, $z = 6$, $x_o = 3/2$ and several values of $\xi = k_{F_c}/k_{F_i}$ and $\lambda = G/2k_{F_c}$.

But, when λ is close to 1, the Fermi surface is certainly distorted from a sphere and x_o is smaller than 3/2. To figure out such an effect, we have plotted on figure 9 the values of $Q\ T_{F_c}/T_{F_i}$ versus temperature for $\lambda = 1$ and several ξ and x_o values. So, with x_o values smaller 3/2, we can obtain thermopower curves changing sign with temperature or even remaining always positive.

In contrast to the case of the thermoelectric power, the electrical and thermal resistivities are not really affected by the Umklapp processes which modify only the high temperature limit ρ_∞ of the resistivity.

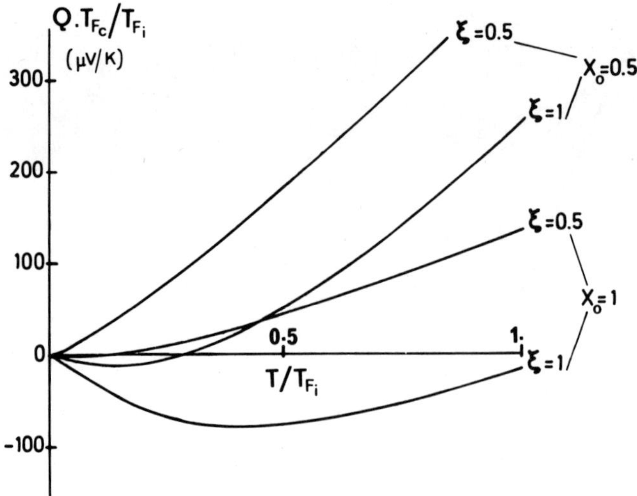

Figure 9 : Theoretical plots of $Q\ T_{F_c}/T_{F_i}$ versus temperature for $S = 10$, $z = 6$, $\lambda = 1$ and several values of $\xi = k_{F_c}/k_{F_i}$ and x_o.

If we compare the theoretical figures 8 and 9 to the experimental figure 7, we see that curves with a change of sign such as those of UAl_2 and Pu can be theoretically obtained, but however we never obtain a saturation of Q when it is positive, in contrast to experiments. Anyway, the introduction of Umklapp processes is apparently important to improve the agreement with experiment for the Seebeck coefficient.

3.4. The Application of the Spin-Fluctuation Model to Other Physical Properties

The spin-fluctuation model can also account for the thermal dependence of $1/T_1T$, if T_1 is the nuclear spin relaxation time measured on the Aluminium nucleus in compounds such as UAl_2 and $PuAl_2$. N.M.R. experiments on these compounds[52] show that $1/T_1T$

decreases by a factor 3 from T = 0 to room temperature. The experimental decrease of $1/T_1T$ can be accounted for by considering the thermal variation of the generalized susceptibility of the f electrons acting through the s-f exchange coupling. A fit of the thermal variation of $1/T_1T$ in the case of UAl_2 has been obtained by use of the same parameters as for the electrical resistivity(44).

Let us discuss now the case of the magnetic susceptibility. The spin fluctuation as it is presented above predicts that the susceptibility must decrease very rapidly around $T_{sf} = T_{F_i}/S$ and follow a Curie-Weiss-like behavior at high temperatures. In fact, experimentally the situation is not unique : some actinide compounds such as USn_3, UAl_2 or $PuAl_2$ have a magnetic susceptibility which presents a Curie-Weiss type behaviour at high temperatures, while on the contrary, Np and Pu metals or compounds such as UAl_3, $PuIr_2$, UIn_3 have a magnetic susceptibility which is almost temperature independent up to room temperature. We have presented an interpretation of these differences by including in the preceding theoretical model the possibility of antiferromagnetic like spin-fluctuations(53). Either we deal with "ferromagnetic" spin fluctuations (as in § 3.2), i.e. with $\chi(\vec{q})$ peaked at $\vec{q} = 0$, and the magnetic susceptibility decreases rapidly with temperature ; or we deal with "antiferromagnetic-like" spin fluctuations, i.e. with $\chi(\vec{q})$ peaked at $\vec{q} = \vec{Q}$ not too close to $\vec{q} = 0$, and the magnetic susceptibility is relatively flat versus temperature. On the other hand, in the two cases, the resistivity behaves as explained before, since there is always a peak in $\chi(\vec{q})$.

The second problem of the magnetic susceptibility concerns its low temperature value : a relatively large enhancement factor of order 10 has been used to explain the transport properties, while, in the examples mentioned above, the ratio of the low temperature experimental values of the magnetic susceptibility and of the electronic specific heat constant is smaller and lies generally between 2 and 4. To explain this discrepancy, we have considered a more realistic spin-fluctuation model which involves two narrow d- and f- bands and which takes into account the two-body d-f coupling parameter J_{df}(54). The calculation has been performed for ferromagnetic spin fluctuations, but it could be in principle extended to the case of antiferromagnetic-like spin fluctuations. We have computed both the exchange enhancement factor S_χ of the magnetic susceptibility and the exchange enhancement factor S_γ of the linear γ-coefficient of the specific heat. These calculations show that, when $S_\chi = 10$, the ratio S_χ/S_γ is of the order 2-2.5 and when S_χ=20, S_χ/S_γ is of order 3.5-4. The theoretical values obtained for S_χ/S_γ are in good agreement with the experimental ones deduced for UAl_2 and USn_3.

IV. CONCLUSION

In conclusion, we have developed simple models which can account for many properties of anomalous rare-earth and actinide systems. For example, it is particularly interesting to notice that the virtual bound state model using the Hartree-Fock approximation can explain the whole phase diagram of Cerium or that the spin fluctuation model based on the R.P.A can account fairly well for the properties of nearly magnetic actinide systems. Indeed, more sophisticated theories[55] would be interesting to improve the models which have been presented here.

REFERENCES

+Address after Jan. 1977 : Instituto de Fisica, U. F. R. G. S. Porto Alegre, Brazil
++Present address : Hansen Laboratories, Dept. of Applied Physics, University of Stanford, Stanford, Ca 94305, USA.

(1) B. Coqblin, J. Physique, Suppl. 32, C1 - 599 (1971) and references therein.
(2) B. Coqblin and A. Blandin, Adv. in Phys. 17, 281 (1968) and references therein.
(3) R. Ramirez and L. Falicov, Phys. Rev. B 3, 2425 (1971)
(4) A. Jayaraman, V. Narayanamurti, E. Bucher and R. G. Maines, Phys. Rev. Lett. 25, 1430 (1970)
(5) M. B. Maple, and D. Wohlleben, Phys. Rev. Lett. 27, 511 (1971)
(6) M. B. Maple, J. Wittig and K. S. Kim, Phys. Rev. Lett. 23, 1375 (1969)
(7) H. J. Van Daal, F. E. Maranzana and K. H. J. Buschow, J. Physique Suppl. 32, C1-424 (1971)
(8) B. Cornut and B. Coqblin, Phys. Rev. B 5, 4541 (1972)
(9) P. B. Van Aken, H. J. Van Daal and K. H. J. Buschow, Proc. 11th Rare-earth Research Conference, Traverse City (USA) october 1974, p. 738.
(10) A. K. Bhattacharjee and B. Coqblin, Phys. Rev. B 13, 3441 (1976)
(11) F. Gonzalez-Jimenez, B. Cornut and B. Coqblin, Phys. Rev. B 11, 4674 (1975)
(12) R. Jullien, E. Galleani d'Agliano and B. Coqblin, Phys. Rev. B 6, 2139 (1972) and reference therein.
(13) R. Jullien, M. T. Béal-Monod and B. Coqblin, Phys. Rev. Lett. 30, 1057 (1973) and Phys. Rev. B 9, 1441 (1974), and references therein.
(14) B. Coqblin, A. K. Bhattacharjee, B. Cornut, F. Gonzalez-Jimenez, J. R. Iglesias-Sicardi and R. Jullien, J. Mag. and Mag. Mat. 3, 67 (1976) and references therein.

(15) R. Jullien and B. Coqblin, "Plutonium and other actinides", ed. by H. Blank and R. Lindner, North-Holland Publishing Company (1976) and references therein.
(16) E. King, J. A. Lee, I. R. Harris and T. F. Smith, Phys. Rev. B 1, 1380 (1970)
(17) B. Alascio, A. Lopez and C. F. E. Olmedo, J. Phys. F : Metal Phys. 3, 1324 (1973)
(18) L. L. Hirst, Phys. Kondens. Materie 11, 255 (1970) and JPCS 35, 1285 (1974)
(19) D. C. Koskimaki and K. A. Gschneidner, Jr., Phys. Rev. B 11, 4463 (1975)
(20) M. B. Maple and J. Wittig, Solid State Comm. 9, 1611 (1971)
(21) V. Allali, P. Donze and A. Treyvaud, Solid State Comm. 9, 1611 (1971)
(22) F. Meunier, S. Ortega, O. Pena, M. Roth and B. Coqblin, Solid State Comm. 14, 1091 (1974).
(23) B. Coqblin, M. B. Maple and G. Toulouse, Int. J. Magnetism 1, 333 (1971).
(24) W. Gey and E. Umlauf, Z. Phys. 242, 241 (1971).
(25) C. A. Luengo, J. G. Huber, M. B. Maple and M. Roth, Phys. Rev. Lett. 32, 54 (1974) ; J. of Low Temp. Phys. 21, 129 (1975).
(26) E. Muller-Hartman and J. Zittartz, Z. Phys. 234, 58 (1970) ; Phys. Rev. Lett. 26, 428 (1971) ; Solid State Comm. 11, 401 (1972).
(27) O. Pena, R. Tournier, A. Benoit and J. Flouquet, to be published.
(28) A. Benoit, Thesis, Orsay (1976).
(29) B. Coqblin and J. R. Schrieffer, Phys. Rev. 185, 847 (1969).
(30) A. K. Bhattacharjee and B. Coqblin, Solid State Comm. 18, 1587 (1976).
(31) A. S. Edelstein, C. J. Tranchita, O. D. Mc Masters and K. A. Gschneidner Jr., Solid State Comm. 15, 81 (1974)
(32) J. Stöhr, Phys. Rev. B 11, 3559 (1975).
(33) J. V. Mahoney, V. U. S. Rao, W. E. Wallace, R. S. Craig and N. G. Nereson, Phys. Rev. B 9, 154 (1974).
(34) K. Andres, J. E. Graebner and H. R. Ott. Phys. Rev. Lett. 35, 1779 (1975)
(35) A. Berton, J. Chaussy, G. Chouteau, B. Cornut, J. Peyrard and R. Tournier, This conference.
(36) R. Tournier and F. Holtzberg, This conference.
(37) S. Doniach, This conference.
(38) R. Jullien, E. Galleani d'Agliano and B. Coqblin, J. of Low Temp. Phys. 10, 685 (1973)
(39) A. J. Arko, M. B. Brodsky and W. J. Nellis, Phys. Rev. B 5, 4564 (1972)
(40) A. R. Harvey, M. B. Brodsky and W. J. Nellis, Phys. Rev. B 7, 4137 (1973)
(41) K. H. J. Buschow and H. J. Van Daal, AIP Conference Proceedings n° 10 "Magnetism and Magnetic Materials" (ed. by D. H. Douglass), p. 1464 (1973)

(42) B. Coqblin, A. A. Gomès, J. R. Iglesias-Sicardi and R. Jullien, To appear in the Proceedings of the "2nd Int. Conf. on the electronic structure of actinides", Wroclaw (Sept. 1976).
(43) J. M. Ziman, "Electrons and phonons", Oxford University Press, England (1960)
(44) R. Jullien and B. Coqblin, J. de Physique, Lettres (France) 35, L 197 (1974)
(45) J. F. Andrew, J.P.C.S. 28, 577 (1967), R. O. A. Hall and J. A. Lee, in "Plutonium 1970" ed. by W. N. Miner (1970) p.35.
(46) R. Jullien and B. Coqblin, J. of Low Temp. Phys. 19, 59 (1975)
(47) F. Steglich, private communication.
(48) R. Lallement, J.P.C.S. 24, 1617 (1963).
(49) G. T. Meaden, Proc. Roy. Soc. A276, 553 (1963).
(50) M. Rösler, Phys. Stat. Sol. 37, 391 (1970)
(51) J. R. Iglesias-Sicardi, R. Jullien and B. Coqblin, To be published.
(52) F. Y. Fradin, M. B. Brodsky and A. J. Arko, AIP Conf. Proceedings n° 10 "Magnetism and Magnetic Materials",(ed. by D. H. Douglass), p. 192 (1973)
(53) R. Jullien and B. Coqblin, J. of Low Temp. Phys. 22, 437 (1976)
(54) J. R. Iglesias-Sicardi, A. A. Gomès, R. Jullien and B. Coqblin J. of Low Temp. Phys. 25, 47 (1976)
(55) See for example H. R. Krishna-Murthy, K. G. Wilson and J. W. Wilkins, This conference.

EPILOGUE

(transcribed from tapes and edited by R. D. Parks and M. C. Croft)

P. W. Anderson

Bell Laboratories, Murray Hill, NJ 07974 and
Physics Department, Princeton University, NJ 08540

When Ron (Parks) asked me to give a summary of the meeting, I said no, because: (A) I couldn't, (B) I would have had to stay awake through all the talks which I am sure everyone has noticed I haven't and (C) I wasn't really sure I wanted to do it anyhow. Nor do I think you need a summary of what everyone has already said. So we decided on the name epilogue which is something you say afterwards, and as far as I know means nothing else and left me quite free. So here are some of the things I wanted to say. First is a private comment. One name on the organizing committee was mispelled. I don't know how this could have slipped by. The name on the organizing committee was actually C. M. Varma; it was spelled P. W. Anderson. Given that reconstituted organizing committee, there is a second comment I wanted to make, which is (now) a message from the organizing committee. If we are to have another good conference on mixed valence compounds, one way we could get it (probably too) well attended is to point out that among other things almost all good catalysts are mixed valence compounds, and perhaps we should begin to think about that; and, even more important, if we want the next conference to be well attended, we should publicize that fact.

Thinking about the conference in general, it seems to me that the experimentalists have been very successful. They have an enormous number of substances which are undoubtedly undergoing valence fluctuations, and there are many phenomenological and experimental ways of looking at these things. The experimentalists have noted that their theoretical colleagues are scratching their heads and not really getting very far with any of this. And naturally when you have got through with your job, and the other guy's standing there and he seems to be all thumbs, the temptation is to look over his shoulder and offer helpful suggestions. It seemed to me that, during the first couple of days at least, there were a good many helpful suggestions offered from the experimentalists, and as a result there were fewer experiments discussed than some of the theorists might have wished. Even the angels have feared to tread in this field, or at least they're treading very lightly, and perhaps we need even more experiments. On the last day the experimental talks were beautifully devoid of any theories except possibly a right one or two.

After that remark, which I probably shouldn't have made, we come to what the theorists are actually puzzled about; and, a typical object which we could take as one of the main subjects of our concerns here is the resistivity curve of $CeAl_3$ or some such compound (See Fig. 1). I've kind of decorated this curve with all the different kinds of things people discussed. I haven't

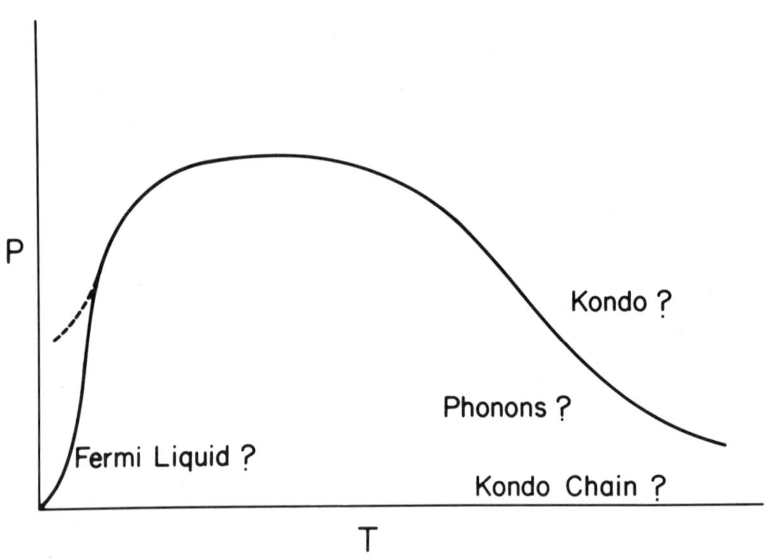

Fig. 1. Resistivity stereotypic of systems such as $CeAl_3$ (see text).

got everything in there but I have got Kondo over here on the high temperature end and the Fermi liquid at the bottom, and in between either phonons or the Kondo chain or both. Thinking about this and the various other theories, reminded me of the parable (Hindu, I think) about the elephant and the seven blind men. Seven blind men come up to an elephant and one of them pulls his tail and says it's a rope, and another one pulls its trunk and says it's a hose, and a third one says it's a tree, and a fourth one says it's a fan and so on, and it's obvious we have that problem here in this field, so I thought I'd make it explicit (Fig. 2) (extended Laughter). It's obvious that at the low temperature end of the elephant you see a Fermi liquid. If you're standing near the mouth, that's where David Sherrington and Peter Riseborough are listening to the phonons. The rest of it is fairly self-evident. (Question from audience: who is the fellow on the trunk? Answer: Well, it's Kondo, obviously. Even though he isn't there, his name has been invoked often enough.) So, very briefly, I am going to run through various more or less serious questions about this elephant and all the different aspects that you can look at.

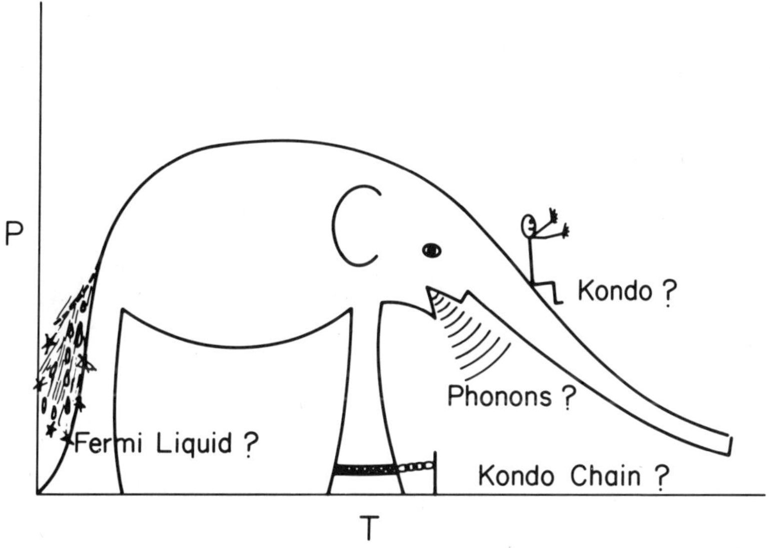

Fig. 2. Elephantine version of Fig. 1 (art work by PWA).

I will now give you one blind theorist's look at the elephant or at least a series of questions I would ask about him. I can't talk very convincingly about what various other people see in this particular elephant. The first question I note is: where is the elephant? And that, we could remark, is a question about phenomenology and quantum chemistry, and more or less under that heading we can include the question of transitions into and out of the mixed valent state. All of these are very interesting and very important questions, particularly for the experimentalist, so that he will know when he is in the presence of this particular elephant and when he isn't. For instance, John Wilson's empirical classification scheme based more or less upon energetics is very useful for understanding to some extent where the different valence states are going to come into rough equality of energy. Basically, one must have as well a quantum chemical picture of how narrow the f bands are and when different configurations are going to come into equilibrium, at which point you will have one or more of the various mixed valence phenomena. It is also interesting to study the transition into and out of the mixed valent state, but from my point of view again, I would put this in the category of approaching the general neighborhood of the elephant as opposed to really studying his intrinsic nature. But, that's one particular blind man and the other people would see that as a much more fundamental thing. So the work on Ce on that kind of thing and the work on the SmS black-to-bronze transitions, which of course are very important, are more or less telling you when you have got into this peculiar state rather than telling you necessarily anything about the state.

Next question: why doesn't he run away. There are a lot of questions about why the elephant doesn't run away, which is a very serious problem for me, at least for the part of the elephant that I see as the core of the problem. The core of the problem is that on a microscopic level you have some atoms at some time which are in one valence and some atoms which are in another valence. They are flipping back and forth at rates that can be either thermally activated, as you heard this morning, or more interestingly, you can have quantum mechanical tunneling back and forth between the different mixed valent states. And yet it is quite clear that when we speak of a mixed valent system we are speaking of a system where the different configurations have in some sense true separate identities. The easy way to see that they have separate idenitities is to look at the kind of XPS data that was presented this morning; that, you understand more or less by the kind of purely theoretical ideas that Hirst has talked about. You get from one configuration into the other by tunneling an electron into or out of the f level, but once you are into it (the mixed valence phase), it is a beast with a lot of internal correlation; and, in particular, it is clear that the two configurations have very different atomic sizes. So, you have this compound, or pure metal, in which objects with different atomic sizes mix at a microscopic level and yet in general do not order. Nor do they segregate. There are two questions in this connection, neither of which has been discussed at this conference. The two questions are: why doesn't the beast run away either (1) by ordering the valences or (2) by segregating the valences. The Friedel, Hume-Rothery theory of alloys would have said that they segregate. The only answer that has been given to that is in a dubious and (I would consider) thoroughly incomplete paper by Chui and myself about phonon effects, which talks about the non-segregation, but it really addresses the question of nonordering to some extent. It's still not obvious why the different valencies don't order. You (almost) obviously gain energy by letting them sit where they want to sit, viz., some ordered way, and if you think hard about the elastic theory, its very tricky that they don't segregate either.

Fundamental model for the elephant. Now we get again to what I consider to be the key aspect of the phenomenon, which is contained in the work of Haldane and similar thoughts by Peter Riseborough (see poster session contributions). Given these ideas you can almost see how the relative position of the two f levels and the fermi level could get pinned in such a way that two stable configurations (stable within conventional Hartree Fock theory) would be pinned to have exactly the same energy within Hartree Fock theory. That I think is an important key to the whole idea of the problem: that within Hartree-Fock theory, looking at things from the point of view of the energetics and time scales of somewhat less than 1 ev, there is a mean field theory which one can think of as Hartree-Fock, which has two separate kinds of states with equal energies on a microscopic level, viz., the non-magnetic state on the one hand and an f level with all the different orbital and spin quantum numbers on the other. At least within Hartree-Fock theory they would be pinned together. Then again you could have various things that you must do with this object in order to understand what really is happening experimentally. This is a fundamental model from which we must all start, and then we must work out a program from this kind of model: various pieces of the program were discussed at this conference. The program's first question is the problem of quenching, which the field theorists might call the idea of confining the f quantum numbers, in some Kondo-like way. One starts doing this of course, with an individual mixed valence impurity, with say one Ce in a background of $LaAl_3$, rather than a whole lot of Ce atoms. But even the one Ce is already a problem, and Wilkins and various other people have thought about this in terms of the asymetrical Anderson model. There is a problem here in even the one impurity problem, for as Duncan Haldane said, "though you do it for one impurity only, its very likely that this is okay in some range of temperatures and in some systems for the concentrated case as well (i.e., for where you have 1 Ce at every site instead of one Ce isolated in a lattice)." The reason being that at high temperatures the fluctuations will be incoherent; and also, because of the admixture affect or some such effect as Tom Kaplan talked about, the actual number of true renormalized f-electrons (magnetic configurations) may be much less than the nominal number of magnetic centers.

Now we come to the heart and core of the elephant, the part which nobody has really done, which was first mentioned at least as a serious problem here in this conference, namely the Kondo lattice, which Seb Doniach made a start on. What you really have is a lattice full of these objects that fluctuate back and forth from one valence to another. And these are all coupled in lots of ways. There are the phonons, there is the fact that the electrons fluctuate by tossing electrons into the d level on the next site which can then go down into the f levels on yet another site. So the things which toss the valences back and forth are definitely coupled between one site and another. The net result of doing this is something that most of the experiments have told us about: that this probably renormalizes to a very heavy Fermi liquid theory with some kind of strong antimagnetic prejudice in that the f-like objects in the Fermi liquid have somehow lost all of their desire to be magnetic and don't very easily order anymore. This is the extremely hard problem, it's a problem in the same category of problems which are failing to be done in field theory these days, and sure enough it is failing to be done properly in our problem too. Then even when you get down to the bottom (of the elephant) no one has even done the fundamental Fermi liquid theory of what you get there. Does it have a d-band and an f-band, what is the nature of the Fermi surface, how might you calculate it and so on? There is this feeling that when you get over the back of the elephant and start-down, at the end you get a Fermi liquid.

In this conference I, at least, began to realize that there is yet one more serious question that we have to consider. Once you get down to this new Fermi liquid, it seems that there is a serious question of then what happens? What does the resulting heavy Fermi liquid do with itself, what further transformation might it undergo? There are several possibilities. One of them which seems to be a very likely one in many cases was mentioned by Kasuya. Namely, this heavy Fermi liquid can localize because the bandwidth is so narrow and because the random potentials can be quite large in these systems. There is no reason at all why it shouldn't localize and maybe there are cases where it localizes. Kasuya gave an argument for one of them. A second possibility a whole series of experiments seem to indicate is that some phase transition takes place in many cases. The question is: what is the nature of these phase transitions? I for one am not ready to accept the idea that they are all simple magnetic phase transitions, e.g., exchange-induced antiferromagnetic phase transitions, as was suggested here by Tournier. He has a theory which shows you in a very concise way how you might get that in a limited range of parameters, but it would (to my mind) be very coincidental that if it does happen, it happens even once, let alone, often. You're getting down to this region where obviously the magnetic character of the f-electrons is dying off; and it's only after you've gotten to a thoroughly non-magnetic region of the f-electrons that all of a sudden the thing decides to order magnetically. They order magnetically in a very nonmagnetic way. If it was going to order magnetically, why didn't it order at some decent temperature with a decent moment? Why would it order at a low temperature with practically no moment? So, maybe it is something else. This heavy Fermi liquid still has some interactions in it other than the ones that were renormalized out in the course of doing the mixed valence job. Maybe there are density wave states in there or maybe there are other kinds of phase transitions. One that nobody mentioned here, fortunately, that I will bring out from under the rug is the possibility of a d to f excitonic phase transition. Maybe there's some kind of d to f excitonic phase transition that either does or does not leave some Fermi surface behind. Maybe there's a density wave. What else?

Another question is brought up by the fact that $CeAl_3$ seems to have in some people's data a phase transition and in other people's data it does not. How many cases are there in which we're quite sure that there is no phase transition and that all we get is the clean renormalization to the state with very low resistance at low temperatures? I mentioned a few possibilities in a slide: $CeAl_3$ perhaps, certainly Yb compounds that weren't much talked about here, etc. An finally, since time is running out, let me mention that last horrid question of what does happen with the phonons: what happens (A) with the optical and (B) with the acoustic phonons? To what extent are the things, which are involved in the screening process that gives the instability in the mixed valence model, phonons and to what extent are they Tomonagons; and how do the phonons and density waves couple? There are lots of fascinating questions here. There are the results of Mook which show that there surely are couplings between phonons and all these phenomena. There's also the questions of whether the Fermi liquid renormalization always take place below the phonon frequency (as it does in many cases) and the Kondo part of it always take place above the phonon frequency. Is that a general rule and is there a reason for it? Those are a number of questions. What is clear is that maybe we have the answer and maybe we're making progress, or maybe the theorists haven't found the elephant at all. So that's my particular view of what is going on (or not going on as the case may be). Clearly there are many more unanswered questions than there are answered ones.

(Note added by PWA: a last slide was omitted in the above which was decorated with a heart and said "Thanks, Ron!")

POSTER SESSION PAPERS

POLARONIC EFFECTS ON THE MAGNETIC SUSCEPTIBILITY OF NONMAGNETIC INTERMEDIATE VALENCE MATERIALS

Peter S. Riseborough

Physics Department, Imperial College, London SW7 2BZ

1. The Introduction

We shall consider some of the possible effects that the crystalline lattice could have on the nonmagnetic properties of intermediate valence materials. These materials are usually described by a multisite, nondegenerate Friedel Anderson Hamiltonian[1,2] $\hat{H}_o$.

$$\hat{H}_o = \sum_{k\sigma} \epsilon_d(k) d^{+}_{k\sigma} d_{k\sigma} + \sum_{i\sigma} \epsilon_f f^{+}_{i\sigma} f_{i\sigma} + \sum_{ki\sigma} \{ V_{ki} f^{+}_{i\sigma} d_{k\sigma} + V^{*}_{k} d^{+}_{k\sigma} f_{i\sigma} \} + \frac{1}{2} \sum_{i\sigma} U f^{+}_{i\sigma} f_{i\sigma} f^{+}_{i-\sigma} f_{i-\sigma} \quad (1)$$

in which $f^{+}_{i\sigma}$ creates an electron of spin σ in the localized level on the i th site; $d^{+}_{k\sigma}$ creates an electron in the itinerant band. We introduce an interaction between the localized 4f electron and the lattice, $\hat{H}_1$

$$\hat{H}_1 = \sum_{i\sigma qs} \lambda_{qs} (a^{+}_{qs} + a_{-qs}) e^{iq \cdot R_i} f^{+}_{i\sigma} f_{i\sigma} + \sum_{qs} \hbar \omega_{qs} a^{+}_{qs} a_{qs} \quad (2)$$

where a^{+}_{qs} represents the creation of a phonon in the s th branch with wave vector $\underline{q}$. This interaction $\hat{H}_1$ represents the coupling of a local deformations of the lattice to the state of ionization of the sites. In section 2 we consider the limit in which the hybridization is zero. In section 3 we shall present a method of interpolation between these limits, and consider the resulting form of the susceptibility. This includes the effect of an exponential reduction on the hybridization matrix elements and the effect of the phonon sidebands.

2. Exact Limits

In the limit where the effective electron-electron interaction is zero we have the problem of hybridizing bands. The Hamiltonian is quadratic in the fermion operators and may be diagonalized by simply changing basis states. The Bloch type of Greens functions are given as

$$\mathcal{G}^{ff}_{\underline{k}\sigma}(\omega_n) = \frac{i\omega_n - \epsilon_{d\sigma}(\underline{k})}{(i\omega_n - \epsilon_{d\sigma}(\underline{k}))(i\omega_n - \epsilon_{f\sigma}) - |V_{\underline{k}}|^2} \qquad 3$$

$$\mathcal{G}^{dd}_{\underline{k}\sigma}(\omega_n) = \frac{i\omega_n - \epsilon_{f\sigma}}{(i\omega_n - \epsilon_{f\sigma})(i\omega_n - \epsilon_{d\sigma}(\underline{k})) - |V_{\underline{k}}|^2}$$

If we assume that the localized and itinerant states have the same Zeeman energy ϵ_σ we may write the susceptibility

$$\chi(\beta) = -\left|\frac{\partial\epsilon_\sigma}{\partial H}\right|\frac{\partial}{\partial H}\left\{\sum_\sigma \sigma \int_c \frac{dz}{2\pi i\beta}\left[\frac{1}{z-\epsilon_{f\sigma}} + \left[1+\frac{|V|^2}{z-\epsilon_{f\sigma}}\right] HT\left[\rho^d_\sigma\left(z - \frac{|V|^2}{z-\epsilon_{f\sigma}}\right)\right]\right] f(z)\right.$$

in which HT $[\rho^d_\sigma(z)]$ is the Hilbert Transform of the itinerant band density of states

$$HT[\rho^d_\sigma(z)] = \int_{-\infty}^{\infty} \frac{\rho^d_\sigma(z')}{z-z'}\, dz'$$

The susceptibilities obtained for an elliptical and Lorentzian density of states are shown in figures 1a and 1b respectively.

In the atomic limit we neglect the effect of hybridization and the long range part of the electroelastic force. We may obtain the real time, thermal Greens function from the definition by direct evaluation of the thermal averages, giving

$$\begin{aligned} g^{ff}_{ii\,\sigma}(\epsilon) = -i\exp\left[-i\left(E_{f\sigma} - \sum_{qs}\frac{\lambda^2_{qs}}{\hbar\omega_{qs}}\right)t\right]\exp\left[-\sum_{qs}\frac{\lambda^2_{qs}}{\hbar\omega_{qs}}(1+2N(\omega_{qs}))\right] Z^{-1}_{\sigma l} \\ \times\Bigg\{\Theta(\epsilon)\left[\exp\left(\sum_{qs}\frac{\lambda^2_{qs}}{\hbar\omega_{qs}}\left\{\epsilon^{-i\hbar\omega_{qs}t} + 2N(\omega_{qs})\cos\hbar\omega_{qs}t\right\}\right)\right] \\ \times\left\{1+\epsilon^{-\beta\left(E_{f\sigma} - \sum_{qs}\frac{\lambda^2_{qs}}{\hbar\omega_{qs}}\right)}\epsilon^{-it\left(U - 2\sum_{qs}\frac{\lambda^2_{qs}}{\hbar\omega_{qs}}\right)}\right\} \\ -(1-\Theta(\epsilon))\left[\exp\left(\sum_{qs}\frac{\lambda^2_{qs}}{\hbar\omega_{qs}}\left\{\epsilon^{i\hbar\omega_{qs}t} + 2N(\omega_{qs})\cos\hbar\omega_{qs}t\right\}\right)\right] \\ \times\left\{1+\epsilon^{-\beta\left(E_{f\sigma} + U - 3\sum_{qs}\frac{\lambda^2_{qs}}{\hbar\omega_{qs}}\right)}\epsilon^{-it\left(U - 2\sum_{qs}\frac{\lambda^2_{qs}}{\hbar\omega_{qs}}\right)}\right\} \\ \times\epsilon^{-\beta\left(E_{f\sigma} - \sum_{qs}\frac{\lambda^2_{qs}}{\hbar\omega_{qs}}\right)}\Bigg\} \end{aligned}$$

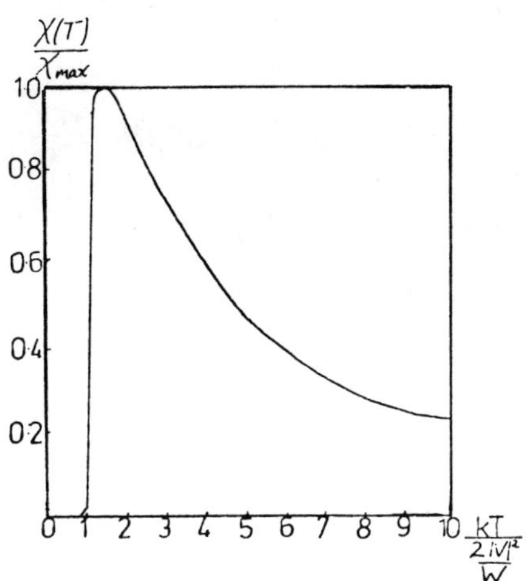

Fig. 1a

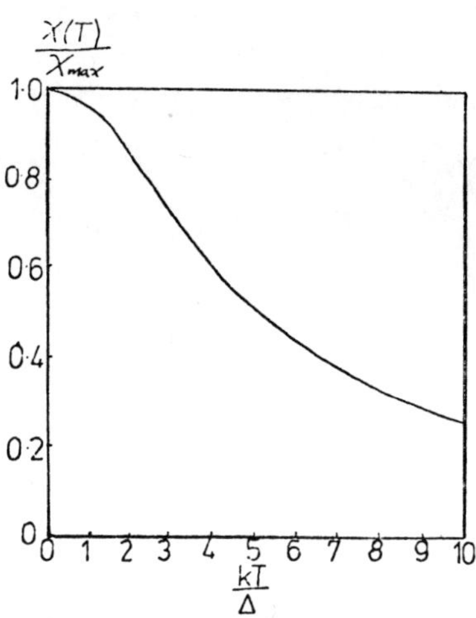

Fig. 1b

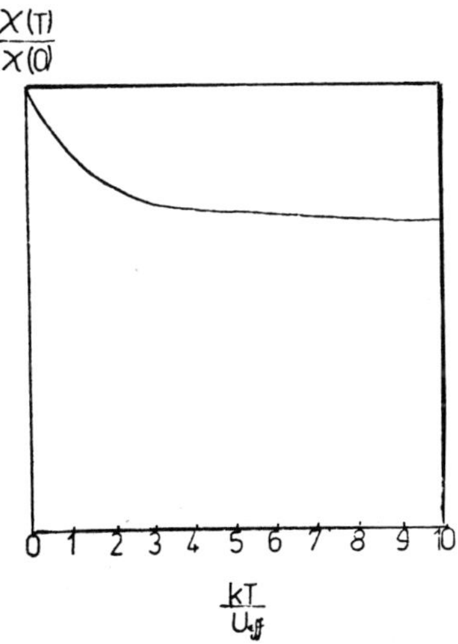

Fig. 2

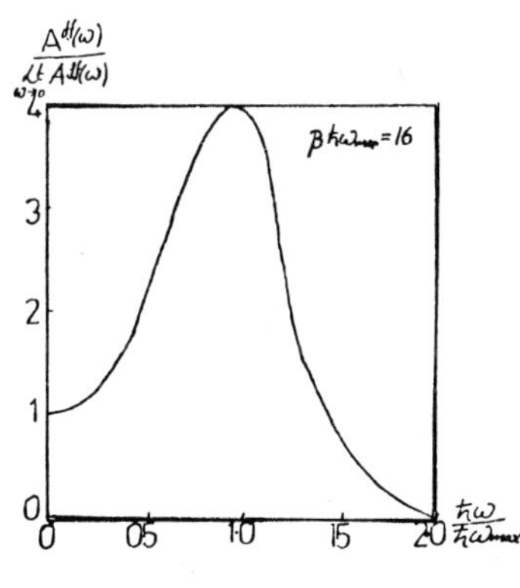

Fig. 3. The 4f spectral density.

where Z_{el} is defined by

$$Z_{el} = \left\{ 1 + 2\epsilon xp\left[-\beta\left(E_f - \sum_{qs} \frac{\lambda^2_{qs}}{\hbar\omega_{qs}}\right)\right] + \epsilon xp\left[-\beta\left(2E_f + U - 4\sum_{qs}\frac{\lambda^2_{qs}}{\hbar\omega_{qs}}\right)\right]\right\}$$

and the magnetic susceptibility

$$\chi(\beta) = Z_{el}^{-2}\left(\frac{\partial\epsilon_0}{\partial H}\right)^2 \beta\, 2\left[\epsilon^{-\beta\left(E_f - \Sigma\frac{\lambda^2}{\hbar\omega}\right)} + 2\,\epsilon^{-\beta 2\left(E_f - \Sigma\frac{\lambda^2}{\hbar\omega}\right)} + \epsilon^{-\beta\left(3E_f + U - 5\Sigma\frac{\lambda^2}{\hbar\omega}\right)}\right] + \chi_{pauli}$$

which is shown in figure 2.

3. A Method of Interpolation

One may argue that the intermediate valence phase corresponds to the case of a large Coulomb Correlation. Physically as U becomes large, only the two configurations corresponding to the unoccupied and singly occupied levels are likely to occur, as in agreement with experiment. On the basis of this we consider a method of interpolation based on the strong correlation atomic limit.

The atomic limit of the Greens Function is Fourier transformed, and an infinite sum of terms is obtained. This is separated into a zero phonon line and phonon side bands which have Poissonian form as shown in figure 3.

The method of interpolation utilizes the separability of the atomic limit Greens Function into electron dynamics and phonon dynamics. The effect of the hybridization width is introduced into the two parts separately, the manner in which they are incorporated are justified by the equations of motion[4], and the shifted harmonic oscillator transformation. The hybridization matrix element is exponentially reduced, the width of the upper and lower Hubbard bands are also narrowed by factors of $\langle f^{\dagger}_{-\sigma} f_{-\sigma}\rangle$ and $(1 - \langle f^{\dagger}_{-\sigma} f_{-\sigma}\rangle)$ respectively. The effect on the phonon evaluation operator may be represented as a reduction in the number of phonons emitted in response to changes in the state of ionization $n_{max} = \frac{\lambda^2}{\hbar^2\omega_0^2} \rightarrow \frac{\lambda^2}{\hbar^2\omega_0^2 + 4\Delta^2}$. In particular the long wavelength phonons are those affected the most. The susceptibility is plotted in figure 4. The large Coulomb repulsion does not give any distinguishable contribution to the magnetic behaviour. The contribution of the zero phonon line is dominant at very low temperatures. The susceptibility is enhanced and narrowed by the polaronic reduction of the hybridization matrix elements. At higher temperatures the susceptibility is dominated by the contribution of the phonon side bands, and eventually falls off in a Curie Weiss manner.

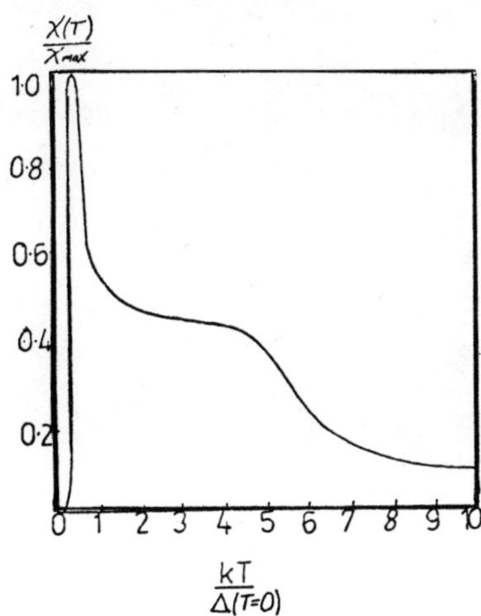

Fig. 4. The susceptibility from the interpolation procedure.

It is interesting to note that the very low temperature structure may be partly masked and partly mistaken as the effect of magnetic impurities.

References

1. Anderson, Phys. Rev. 124, 41 (1961)
2. Varma and Yaffet, Phys. Rev. B 13, 2950 (1976)
3. Sherrington and Riseborough, to be published, Journal de Physique
4. Hewson, Phys. Rev. 144, 420, (1965).

THE FUNCTIONAL INTEGRAL INVESTIGATION OF AN EXTENDED FRIEDEL ANDERSON MODEL

Peter S. Riseborough

Physics Department, Imperial College, London SW7 2BZ

1. The Introduction

In this paper we present an auxiliary field formulation for the partition function for the generalization of the Friedel Anderson Model in which a localized electron phonon interaction is included. In section two we present the model and the auxiliary field formulation of the problem. We make both the time independent auxiliary field approximation, and the external approximation for the thermodynamic potential, these are usually connected to the Hartree Fock Approach[3]. We consider the limits in which the timescale for the relaxation of a local lattice distortion is either much larger or much smaller than the timescale associated with the fluctuations between states of differing valence. In section three we examine the functional form of the thermodynamic potential with respect to the conditions for the existence of local minima, and through this we explore the stability, and the conditions required for spin fluctuations and valence fluctuations to occur.

2. The Formulation

The model is that of a non-degenerate Friedel Anderson system in which we incorporate the effect of the change in ionic size with the degree of ionization of the impurity atom[2]. This results in an interaction which couples the local distortion of the lattice to the occupation number of the localized level. Thus the Hamiltonian may be written as :-

$$H = \sum_{\sigma} \epsilon_f f_{\sigma}^{+} f_{\sigma} + \sum_{k\sigma} \epsilon_d(\underline{k}) d_{\underline{k}\sigma}^{+} d_{\underline{k}\sigma} + \sum_{\sigma} \frac{U}{2} f_{\sigma}^{+} f_{\sigma} f_{-\sigma}^{+} f_{-\sigma}$$

$$+\sum_{\underline{k}\sigma}\{V_{\underline{k}} d^{+}_{\underline{k}\sigma} f_{\sigma} + V^{*}_{\underline{k}} f^{+}_{\sigma} d_{\underline{k}\sigma}\}$$
$$+\sum_{\sigma q s}\lambda_{qs}(a^{+}_{qs} + a_{qs}) f^{+}_{\sigma} f_{\sigma} + \sum_{qs}\hbar\omega_{qs} a^{+}_{qs} a_{qs} \qquad 1$$

where f^{+}_{σ} creates an electron of spin σ in the localized level; $d^{+}_{\underline{k}\sigma}$ creates an electron of spin σ with Bloch number K in the itinerant band; a^{+}_{qs} creates a phonon of wave number q in the s th branch. In this the terms representing the anharmonicity of the lattice have been neglected.

The partition function Z is evaluated in the interaction representation

$$Z = Z_0 \, \mathrm{Trace}\left[T_{\tau'} \exp\left\{ -\int_0^{\beta} H_{int}(\tau')\, d\tau' \right\} \right] \qquad 2$$

where $T_{\tau'}$ is the time ordering operator. The interaction Hamiltonian is written in the form of an effective electron-electron interaction obtained from the Hubbard Coulomb interaction and the electron phonon interaction by using the shifted harmonic oscillator transformation. In the static regime, the lattice couples to the time averaged occupation of the localized level giving the form of the interaction Hamiltonian as :-

$$\frac{1}{4}\left(U - 4\sum_{qs}\frac{\lambda_{qs}^2}{\hbar\omega_{qs}}\right)\left(\sum_{\sigma} f^{+}_{\sigma} f_{\sigma}\right)^2 - \frac{1}{4} U \left(\sum_{\sigma}\sigma f^{+}_{\sigma} f_{\sigma}\right)^2 \qquad 3a$$

In the adiabatic regime the lattice couples to the instantaneous state of occupation of the localized level giving the corresponding interaction as :-

$$-\sum_{qs}\frac{\lambda_{qs}^2}{\hbar\omega_{qs}} f^{+}_{\sigma} f_{\sigma} + \frac{1}{4}\left(U - 2\sum_{qs}\frac{\lambda_{qs}^2}{\hbar\omega_{qs}}\right)\left\{\left(\sum_{\sigma} f^{+}_{\sigma} f_{\sigma}\right)^2 - \left(\sum_{\sigma}\sigma f^{+}_{\sigma} f_{\sigma}\right)^2\right\} \qquad 3b$$

The hybridization matrix element becomes polaronically reduced, but the localized level acquires a width from the phonon sidebands. In either case the width is denoted by $\Delta(\beta)$

The interactions may be expressed in terms of interactions with classical auxiliary fields by using the Hubbard Stratonovich identity

$$\exp\left[a^2\right] = \int_{-\infty}^{\infty} dx \, \exp\left[-\pi x^2 - 2\pi^{\frac{1}{2}} a x\right]$$

On evaluating the trace in equation 2 we obtain the functional form of the thermodynamic potential

$$F_{\{\lambda\}}^{\{s\}} = F_0 + \pi(x^2 + y^2) + \pi^{\frac{1}{2}}\sqrt{U - \left\{\begin{matrix}4\\2\end{matrix}\right\}\sum_{qs}\frac{\lambda_{qs}^2}{\hbar\omega_{qs}}} \; i x$$
$$+\sum_{\sigma}\left\{\frac{1}{\beta}\log_e \Gamma(A^{+}_{\sigma\{\lambda\}^{s}})\Gamma(A^{-}_{\sigma\{\lambda\}^{s}}) - \frac{4}{\beta}\log_e\left|\Gamma\left(\frac{1}{2} + \frac{\beta}{2\pi}(\Delta(\beta) + iE_f)\right)\right|\right\}$$

where Γ is the Eulerian gamma function and

$$A^{\pm}_{\sigma S} = \frac{1}{2} + \frac{\beta\Delta(\beta)}{2\pi} \pm \frac{i\beta}{2\pi}\left\{E_f + \pi^{\frac{1}{2}}\left[\sqrt{U - 4\sum_{qs}\frac{\lambda_{qs}^2}{\hbar\omega_{qs}}}\, ix + \sigma\sqrt{U}\, y\right]\right\}$$

$$A^{\pm}_{\sigma A} = \frac{1}{2} + \frac{\beta\Delta(\beta)}{2\pi} \pm \frac{i\beta}{2\pi}\left\{E_f - \sum_{qs}\frac{\lambda_{qs}^2}{\hbar\omega_{qs}} + \pi^{\frac{1}{2}}\sqrt{U - 2\sum_{qs}\frac{\lambda_{qs}^2}{\hbar\omega_{qs}}}\,(ix + \sigma y)\right\}$$

The auxiliary field x couples to the valence and y to the magnetic moment. Making a Taylor expansion about a stationary point (x_o , y_o) and retaining terms up to quadratic order we obtain the approximation for Z

$$Z\{{}^{S}_{A}\} \simeq \exp\left[-\beta F_{\{{}^{S}_{A}\}}(x_o, y_o)\right]\exp\left[-\frac{1}{2}\log_e \frac{\beta W\{F_{\{{}^{S}_{A}\}}\}}{2\sqrt{\pi}}\Big|_{x_o y_o}\right] \qquad 5$$

where W is the Wronskian of the thermodynamic potential F (x, y).

3. The Stability of the Phases and Fluctuations

The extrema of the functional for the thermodynamic potential F satisfy the equations

$$2\pi x + i\pi^{\frac{1}{2}}\sqrt{U - 4\sum_{qs}\frac{\lambda_{qs}^2}{\hbar\omega_{qs}}}\left\{\sum_{\sigma}\langle f_\sigma^+ f_\sigma\rangle_S\right\} = 0$$
$$2\pi y + \pi^{\frac{1}{2}}\sqrt{U}\left\{\sum_{\sigma}\langle f_\sigma^+ f_\sigma\rangle_A\,\sigma\right\} = 0 \qquad 6(a)$$

in the static approximation. In the adiabatic approximation the corresponding conditions are :-

$$2\pi x + i\pi^{\frac{1}{2}}\sqrt{U - 2\sum_{q\sigma}\frac{\lambda_{qs}^2}{\hbar\omega_{qs}}}\left\{\sum_{\sigma}\langle f_\sigma^+ f_\sigma\rangle_A\right\} = 0$$
$$2\pi y + \pi^{\frac{1}{2}}\sqrt{U - 2\sum_{qs}\frac{\lambda_{qs}^2}{\hbar\omega_{qs}}}\left\{\sum_{\sigma}\langle f_\sigma^+ f_\sigma\rangle_A\,\sigma\right\} = 0 \qquad 6(b)$$

If a particular solution to these equations is a local minima (x_o , y_o) any change in the stability on decreasing the temperature may be found by examining the Wronskian $W\{F_{\{{}^{S}_{A}\}}\}$. When these change sign the lines of equipotential change form, from being approximately elliptical around (x_o , y_o) to being hyperbolic. The only unique solution to the coupled equations is expected from rotational invariance arguments to be a nonmagnetic one $y_o = 0$. This is unique and stable at T = 0 when the inequalities $U - \sum_{qs}\frac{4\lambda_{qs}^2}{\hbar\omega_{qs}} > -\pi\Delta(\beta)$ and $U < \pi\Delta(\beta)$ are fulfilled. When $U - 4\sum_{qs}\frac{\lambda_{qs}^2}{\hbar\omega_{qs}} > -\pi\Delta$ and $U < \pi\Delta$ then transitions between states of differing valence occur. For $U > \pi\Delta(\beta) > \sum_{qs} 2\frac{\lambda_{qs}^2}{\hbar\omega_{qs}}$ both magnetic and nonmagnetic minima exist. The phase diagram is shown in figure 1.

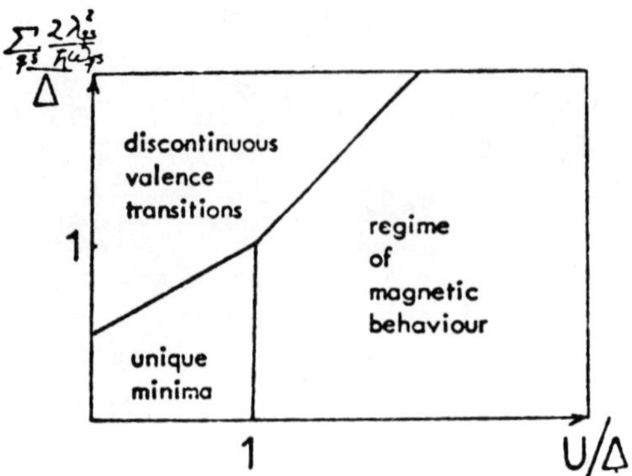

Fig. 1. The static "phase diagram" at T = 0.

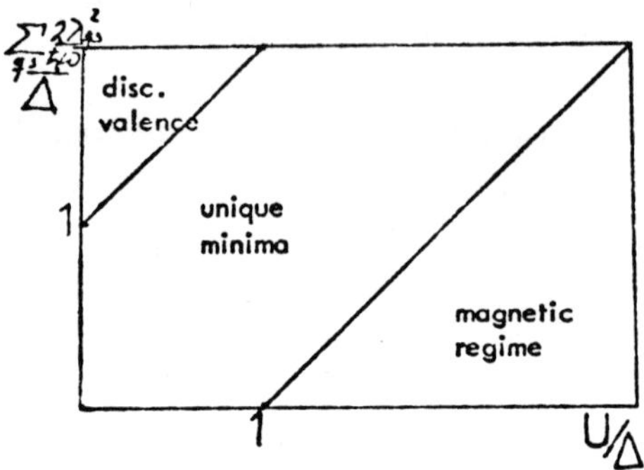

Fig. 2. The adiabatic "phase diagram" at T = 0.

In the adiabatic regime the existence of magnetic minima and of transitions between minima of differing valence are separated by $U-2\sum'_{qs}\frac{\lambda_{qs}^2}{\hbar\omega_{qs}} > \pi\Delta(\beta)$ and $U-2\sum'_{qs}\frac{\lambda_{qs}^2}{\hbar\omega_{qs}} < -\pi\Delta(\beta)$. The phase diagram is shown in figure 2.

In general as $\sum'_{qs}\lambda_{qs}^2/\hbar\omega_{qs}$ is increased the likelihood of magnetic moments forming is decreased and the probability of transitions between minima of differing valence is increased.

Although the single impurity model does exhibit phase transitions in the time independent auxiliary field approximation, they are not expected in the exact solution, as one expects that for large interaction strengths, the itinerant electrons will avoid the region around the single site. This breakdown may be rectified by including a frequency dependence of the auxiliary field. In this one considers the contribution to the partition function of fluctuations between the minima when they are nearly degenerate. In the case of magnetic minima one has zero energy fluctuations which are the approximate Goldstone boson modes which replaces the rotational invariance of the spin system.

For minima of differing valence one has terms representing valence fluctuations.

4. Summary

In conclusion, we have presented a single site model which in the time independent auxiliary field approximation leads to magnetic and or valence transitions. As previously mentioned, these are not features of the exact solution, but may have some applicability to the many site problem, where the argument given breaks down, as, in the manysite case the electron cannot be excluded from all the ionic sites. We may conclude that the inclusion of strong electron-phonon coupling makes differences to the model that are relevant to the intermediate valence problem.

References

1. Anderson P.R. 124; 41; 1961
2. Sherrington and Riseborough J. de Phys., to be published
3. Hubbard P.R.L.

ANOMALOUS THERMAL EXPANSION OF $Sm_{1-x}R_xS$

J. H. JEFFERSON

Department of Physics, University of Nottingham
University Park, Nottingham NG7 2RD, England

In order to simulate the effect of pressure Jayaraman et al. (1973) alloyed SmS with Gd, the latter having a smaller ionic radius than Sm being divalent in the sulphur host. These experiments showed the desired effect and produced crystals which were in a state of 'intermediate valence' at ambient pressure. Upon cooling many of the crystals exhibited anomalous variations of lattice parameter, there being typically a first-order phase transition to another intermediate valence state, closer in character to pure (semiconducting) SmS, and a large volume expansion ($\sim$ 8%). More recent experiments (Campagna et al. 1974, Pollak et al. 1974, Jayaraman et al. 1975, Tao and Hotlzberg 1975) have supported these findings and have also shown similar behaviour for SmS alloyed with a variety of dopants.

The basic physics behind the anomalous thermal expansion can be appreciated by analyzing a simple statistical model (Jefferson 1976). In what follows the results of this analysis will be presented and the mechanisms which give rise to the anomalies pointed out.

In a 'mean field' description of the intermediate valence state there is, on average, a fraction n of the Sm ions in the configuration $4f^5$ which have donated an electron to the Sm-derived conduction band, the remaining ions being in the configuration $4f^6$. The free energy, normalized to a unit cell containing one S ion, is

$$F = E_L + (3W/5)z^{5/3} - z\Delta E + T(1 - x)\left[n\ln n + (1 - n)\ln(1 - n) + n\ln q - \ln \alpha_1\right] + F_p \quad (1)$$

where E_L is the elastic energy which should contain the anharmonic terms that are responsible for stabilizing the intermediate valence.

$(3W/5)z^{5/3} - z\Delta E$ is the energy of the conduction electrons for a parabolic band, ΔE being the energy necessary to remove an electron from the bottom of the conduction band and place it on an ion in the configuration $f^5(\Gamma_7)$ to give the state f^6 (J = 0); $z \equiv (1-x)n$; $-(1-x)\left[n \ln n + (1-n)\ln(1-n) + n \ln q - \ln \alpha_1\right]$ is the entropy due to configuration and spin degeneracy of the ions; $\alpha_1 = 1 + 3e^{-415/T}$, $\alpha_2 = 2(1 + 2e^{-165/T})$. E_p is the free energy of the lattice vibrations (excluding the elastic energy). It may be shown that the entropy of the conduction electrons is negligible.

The coefficient of thermal expansion is

$$K \equiv \frac{1}{v}\left(\frac{\partial v}{\partial T}\right)_p = -\left(\frac{\partial p}{\partial T}\right)_v \Big/ \left(\frac{\partial p}{\partial v}\right)_T ,$$ where v is the volume of a unit cell.

It may be calculated from the free energy using the thermodynamic relations

$$p = -\left(\frac{\partial F}{\partial v}\right)_T = -\left(\frac{\partial F}{\partial v}\right)_{Tz}, \quad \left(\frac{\partial p}{\partial T}\right)_v = \left(\frac{\partial p}{\partial T}\right)_{vz} + \left(\frac{\partial p}{\partial z}\right)_{Tv}\left(\frac{\partial z}{\partial T}\right)_v, \text{ and}$$

$$\left(\frac{\partial p}{\partial v}\right)_T = \left(\frac{\partial p}{\partial v}\right)_{Tz} + \left(\frac{\partial p}{\partial z}\right)_{Tv}\left(\frac{\partial z}{\partial v}\right)_T .$$

For the system to be thermodynamically stable $(\partial p/\partial v)_T$ *must* be negative and so the sign of the thermal expansion coefficient is determined by the sign of $(\partial p/\partial T)_v$. For most materials, of course, K is positive and $(\partial p/\partial T)_v$ comes from the free energy of the phonons, there being a decrease in the phonon frequencies with expansion. With $Sm_{1-x}R_xS$ compounds the contribution to $(\partial p/\partial T)_v$ from the ionic entropy term (particularly the configurational entropy term) swamps the contribution due to the phonons when n is in a certain range, and is negative. *This makes the coefficient of thermal expansion negative at low temperatures.* Increasing the temperature can lead to a number of possibilities depending on the value of n (the intermediate valence) at the low temperature end. Most, if not all, of these possibilities have been realized experimentally. Consider the situation when n is close to zero (black phase like pure SmS). The contribution (to K or $(\partial p/\partial T)_v$) from the ionic entropy is small and the thermal expansion is 'normal' (phonons) or supressed slightly at low temperatures. As n is increased (by increased doping) the thermal expansion curves at low T flatten off and eventually become negative. Hence, as T increases v decreases and n increases causing K to become more negative (a self-generative effect). Eventually a critical lattice parameter is reached at which point there is a thermodynamic instability ($(\partial p/\partial v)_T$ goes to zero). The physics of this phase transition has been discussed by Hirst (1974) for the case of zero temperature and his arguments apply equally well for finite T. The only important difference is that at T = 0 the only way of decreasing the lattice parameter is with the application of pressures whereas for the above compounds it can be done merely by increasing the temperature. (Although the shapes of the p-v isotherms are relatively insensitive to temperature, there will be a critical temperature above which the transition will be rapid but continuous ('quasi-first-order').) We should not be surprised that such a critical lattice parameter can be achieved for the very fact that K is negative at low temperature, which is a consequence of intermediate valence, implies that the critical lattice parameter is very near, provided it can be assumed that the thermodynamic properties of $Sm_{1-x}R_xS$ are not too different from pure SmS at the equivalent pressure. Above the transition K may be negative or positive depending on whether n is less than or greater than some critical value (of order 2/3). If K is negative the lattice continues to contract until a minimum lattice parameter is reached. The experimentally determined temperature at this minimum enables a simple estimation of the corresponding intermediate valence to be made from the equation

$$\ln \frac{1}{q}\left(\frac{1}{n} - 1\right) + \frac{165/T}{1 + \frac{1}{2}e^{165/T}} - \frac{415/T}{1 + \frac{1}{3}e^{415/T}} = 0 ,$$

assuming the phonon contribution is negligible. Explicit calculations give reasonable agreement with independent determinations of n (from lattice constant and XPS measurements). At temperatures above that of the minimum point K is positive but anomalously large. There is no further first-order transition, hence the temperature must be greater than the critical temperature. The entropy tends to retard the high temperature 'transition' for the same reason that it accelerates the low temperature transition.

REFERENCES

Campagna, M., Bucher, E., Wertheim, G. K. and Longinotti, L. D., 1974, Phys. Rev. Lett., 32, 885-9.
Hirst, L. L., 1974, J. Phys. Chem. Solids, 35, 1285-96.
Jayaraman, A., Bucher, E., Dernier, P. D. and Longinotti, L. D., 1973, Phys. Rev. Lett., 31, 700-3.
Jayaraman, A., Dernier, P., Longinotti, L. D., 1975, Phys. Rev. B, 11, 2783-94.
Jefferson, J. H., 1976, J. Phys. C: Solid State Phys., 9, 269-78.
Pollak, R. A., Holtzberg, F., Freeouf, J. L. and Eastman, D. E., 1974, Phys. Rev. Lett., 33, 820-3.
Tao, L. J. and Holtzberg, F., 1975, Phys. Rev. B, 11, 3842-9.

A RENORMALIZATION GROUP APPROACH TO THE MIXED VALENCE PROBLEM

J. H. JEFFERSON

Department of Physics, University of Nottingham
University Park, Nottingham NG7 2RD, England

The Anderson Hamiltonian for an impurity with a non-degenerate localized orbital, embedded in an infinite metallic crystal with a single orbitally non-degenerate conduction band is

$$H = \sum_\sigma \int d^3k \varepsilon_k c^\dagger_{\underline{k}\sigma} c_{\underline{k}\sigma} + \varepsilon_o \sum_\sigma a^\dagger_\sigma a_\sigma + Un\uparrow n\downarrow + \Omega^{-\frac{1}{2}} \sum_\sigma \int d^3k \left[v_o a^\dagger_\sigma c_{\underline{k}\sigma} + \text{h.c.} \right] \quad (1)$$

where Ω is the volume of the Brillouin zone in reciprocal space, ε_o is the energy of the localized orbital and U is the intra-atomic coulomb repulsion between two localized electrons. The energy of the conduction band is, by assumption, only dependent on $|\underline{k}| = k$ and the variation of the hybridization matrix element with $\underline{k}$ is neglected.

There is a special case of the Anderson Hamiltonian which is relevant to the problem of 'mixed valence': when ε_o is close to the Fermi energy (taken as zero) and U is very large. More precisely when $|\varepsilon_o| \lesssim v_o \ll U$. In this region there is a ground manifold of many-electron base states, which are eigenstates of the Hamiltonian

$$H_o = \sum_\sigma \int d^3k \varepsilon_k c^\dagger_{\underline{k}\sigma} c_{\underline{k}\sigma} + \varepsilon_o \sum_\sigma a^\dagger_\sigma a_\sigma + Un\uparrow n\downarrow ,$$

for which the localized level is either unoccupied or is occupied by one electron only, together with small excitations of conduction electrons about the Fermi surface. States for which there are two localized electrons are of order U higher in energy. If U is large (as with f-electrons) it is a good approximation to neglect base-states with two localized electrons as far as the actual low-lying eigenstates are concerned. They will be formally projected out of the problem by setting $U = \infty$ *a priori*. The term $Un\uparrow n\downarrow$ may then be dropped from the Hamiltonian provided that many-electron states with two localized electrons are omitted from the basis set. This is referred to as the infinite correlation limit. The eigenstates of

(1) will be coherent superpositions of the base-states. The relative admixtures of base-states for which the localized level is respectively unoccupied or singly occupied will be comparable since $\varepsilon_o \lesssim v_o$, and a 'mixed valence' or 'mixed configuration' situation results.

As it stands the problem is still intractable. The idea behind the renormalization group approach presented here, initiated by Anderson (1970) in his treatment of the Kondo problem, is to transform the initial Hamiltonian into a sequence of effective Hamiltonians by successively eliminating high energy base-states from the problem with perturbation theory. The hope is that eventually the Hamiltonian will be transformed into one which is soluble. The procedure is as follows. Let $P_{\delta D}$ be the projection operator for all many-electron base-states which have at least one electron and/or at least one hole within energy δD of their respective band edges. The effective Hamiltonian operating in the space of the remaining states (the model sub-space), with projection operator $Q_{\delta D} \equiv 1 - P_{\delta D}$, is given by the Rayleigh –Schrodinger perturbation expansion (Lindgren 1974):

$$\widetilde{H}(\delta D) = Q_{\delta D}HQ_{\delta D} + Q_{\delta D}H_1\sum_{\alpha}G_{\alpha}H_1Q^{\alpha}_{\delta D} + Q_{\delta D}H_1\sum_{\alpha}G_{\alpha}H_1G_{\alpha}H_1Q^{\alpha}_{\delta D}$$
$$- Q_{\delta D}H_1\sum G_{\alpha}(\sum_{\alpha'} G_{\alpha'}H_1Q^{\alpha'}_{\delta D})H_1Q^{\alpha}_{\delta D} + \text{Higher Order Terms}$$

where $Q^{\alpha}_{\delta D} \equiv |\alpha\rangle\langle\alpha|$ is the projection operator for the state $|\alpha\rangle$ in the model sub-space and the sum is over all such states. $G_{\alpha} \equiv P_{\delta D}/(E_{\alpha} - H_o)$ is the 'resolvent' for the state $|\alpha\rangle$. Substituting

$$H_1 = \Omega^{-\frac{1}{2}}\sum_{\sigma}\int d^3k\left[v_o a^{\dagger}_{\sigma}c_{\underline{k}\sigma} + \text{h.c.}\right]$$

it may be shown, after some manipulation, that to third-order in v_o $\widetilde{H}(\delta D)$ may be written

$$\widetilde{H}(\delta D) = \sum_{\sigma}\int d^3k\varepsilon_k c^{\dagger}_{\underline{k}\sigma}c_{\underline{k}\sigma} + \varepsilon\sum a^{\dagger}_{\sigma}a_{\sigma} + \Omega^{-\frac{1}{2}}\sum_{\sigma}\int d^3k\left[v_1 a^{\dagger}_{\sigma}c_{\underline{k}\sigma} + v_2 c^{\dagger}_{\underline{k}\sigma}a_{\sigma}\right] \quad (2)$$

where ε, v_1 and v_2 are constants proportional to δD (see Equations 3 below). A systematic investigation of fourth and higher order terms shows that further contributions of the type $a^{\dagger}_{\sigma}a_{\sigma}$ or $a^{\dagger}_{\sigma}c_{\underline{k}\sigma}$ are proportional to δD^n, $n > 1$, and thus negligible. Potential and spin-flip scattering first occur in fourth order and more complicated multiple scattering in higher orders. For $D_o >> v_o$ these higher-order terms are negligible.

Since the effective Hamiltonian to third order has the same form as the original Hamiltonian it follows that repeating the whole operation for a further reduction $\delta D'$ in the band edges will again result in an effective Hamiltonian of the same form to third order. This applies, of course, to all successive iterations. These transformations of the Hamiltonian are said to belong to the renormalization group. Comparing effective Hamiltonians with bandwidths 2D and 2(D + dD) leads directly to the following differential scaling equations for ε, v_1 and v_2

$$\frac{d\varepsilon}{dD} = -\frac{v_1 v_2}{2D_o}\left(\frac{2}{D+\varepsilon} - \frac{1}{D-\varepsilon}\right), \frac{dv_1}{dD} = \frac{v_1^2 v_2}{2D_o D(D-\varepsilon)}, \frac{dv_2}{dD} = \frac{v_2^2 v_1}{D_o D(D+\varepsilon)} . \qquad (3)$$

In deriving these equations the variation of v_1 and v_2 with k has been neglected by setting $\varepsilon_k = 0$ in the energy denominators. The restriction to third order is expected to be valid when $D \pm \varepsilon \gg v_1$ or v_2. In this region $D \gg \varepsilon$ when $\varepsilon_o \lesssim v_o$ and so the denominators $D \pm \varepsilon$ in (3) may be replaced by D. It follows that $v_2 = v_1 |v_1/v_o| \leq v_1$ and $d\varepsilon/dv_1 = -(D/v_1)$. Hence, as the conduction band cut-off is decreased the localized level scales to higher energies and the hybridization interaction becomes weaker, with ε scaling much faster than v_1 (or v_2). Unfortunately the absolute changes in ε and v_1 are small compared with v_o as can be seen by explicit integration of the scaling equations. For example, making the choice $v_o/D_o = 0.1$ and setting $D = v_o$, which is already outside the expected region of validity, one gets $v_1 = 0.95v_o$ and $\varepsilon - \varepsilon_o = 0.12v_o$. Extending the range of integration of the scaling equations seems pointless since higher-order terms in the effective Hamiltonian are no longer negligible.

Within the limits set by perturbation theory the original problem scales to equivalent problems of the same type in which the conduction band cut-off is reduced at the expense of increasing the energy of the localized state and decreasing the magnitude of the hybridization. If this trend is extrapolated one has the possibility (which is unlikely to be proved within the framework of perturbation theory) that with ε_o small the problem may eventually scale to one in which $\varepsilon \gg v_1$ or v_2. The solution to *this* problem is well known. Since ε is now above the Fermi surface the localized level is unoccupied and the impurity is non-magnetic giving rise to ordinary potential (non-spin-flip) scattering only.

The special case of the Anderson Hamiltonian for which $U + \varepsilon_o \equiv U - |\varepsilon_o| \ll U$ is of the same form as (2) in the limit of large U. It corresponds to a mixed valence situation with eigenstates comprised of base-states in which the localized level is either singly or doubly occupied (one or zero localized holes). Scaling again leads to a curtailment in the probability of occupation of the magnetic state.

REFERENCES

ANDERSON, P. W., 1970, J. Phys. C: Solid State Phys., 3, 2346.

LINDGREN, I., 1974, J. Phys. B: Atom. Molec. Phys., 7, 2441.

STATISTICAL MECHANICS OF INTERMEDIATE VALENCE SmS

J. H. JEFFERSON and K. W. H. STEVENS

Department of Physics, University of Nottingham
University Park, Nottingham NG7 2RD, England

At a pressure of 6.7 kbar the semiconductor SmS undergoes a first-order phase transition to a metallic state of 'intermediate valence'. The transition occurs because the application of pressure causes the interconfiguration excitation energy (the energy necessary to promote an f-electron from an ion in the state $4f^6$ $(^7F_0)$ to the bottom of the conduction band leaving the ion in the state $f^5(^6H_{5/2})$) to approach zero, leading to a thermodynamic instability (Hirst 1974). The transition is finally arrested when the lowering of energy due to population of the conduction band is balanced by the increase in elastic energy. In a model which neglects the hybridization between f and d electrons, correlations between conduction electrons, interactions between f-electrons on different ions and electron-lattice interactions, the description of the intermediate valence state is quite simple. In thermodynamic equilibrium there are n (say) conduction electrons per molecule and hence a fraction n Sm ions in the configuration $4f^5$, the remaining fraction (1 - n) being in the configuration $4f^6$. Since the system arranges itself so as to minimize the free energy the position of the conduction band relative to the ionic levels is such that the interconfiguration excitation energy is of order kT (see below). Apart from this constraint the system is equivalent to three non-interacting sub-systems of two types of Sm ions and conduction electrons (the role of the S ions is not important for the present discussion). The thermodynamic behaviour of the system is now trivially deduced in this approximation. For example the susceptibility has Curie contributions from the ions and a Pauli contribution from the conduction electrons.

Now whilst the above description is reasonable at high temperatures it must break down at low temperatures ($\lesssim$ 100 K) since experimentally there is no divergence of the susceptibility (except at very low temperatures due to impurities) nor any sign of large configuration and spin entropy. The interaction which is most likely responsible for this new behaviour is the hybridization interaction between the localized f-electrons and the conduction electrons. To investigate this question further we have analyzed the statistical mechanics by obtaining a thermodynamic perturbation expansion of the Gibb's free energy (G) to second-order in the hybridization interaction (H_1) (Jefferson and Stevens 1976). H_1 is obtained by extending the Anderson (1961) Hamiltonian of a magnetic impurity in a metal to one which allows configurational fluctuations of a concentrated system of ions. For convenience the temperature is assumed to be sufficiently low that thermal population of the J = 1 level of $4f^6$ and the J = 7/2 level of $4f^5$ is negligible. The crystal field splitting of the J = 5/2 level is also neglected.

From G all thermodynamic quantities may be determined. For example, the entropy and susceptibility are

$$S = -\left(\frac{\partial G}{\partial T}\right)_{\mu v} \qquad \text{and} \qquad \chi = -\left(\frac{\partial^2 G}{\partial \mathcal{H}^2}\right)_{\mathcal{H}=0}$$

respectively. Explicit calculation confirms that S_o and χ_o (H_1 = 0) have the expected form referred to earlier and also yields the relation

$$\varepsilon^{(5)} + \mu(T) - \varepsilon^{(6)} \simeq kT$$

for the interconfiguration excitation energy: this expression is exact when T = 0.

The second-order contributions to G, S and χ are

$$G_2 = -N\rho_o|v|^2\left(1 - \frac{5}{6}n\right)\ln\frac{D}{kT}$$

$$S_2 = \left(-N\rho_o|v|^2\left(1 - \frac{5}{6}n\right)\right)/T$$

and

$$\chi_2 = \frac{-T_F(T)}{T}\chi_o^{(i)}$$

where:

N is the number of Sm ions, ρ_o is the (assumed constant) density of states for the conduction band of width 2D, $|v|^2$ is an 'averaged' square of the hybridization matrix elements,

$$T_F(T) = \frac{5\rho_o|v|^2}{6k}(1 - n)\ln\frac{D}{kT}$$

and

$$\chi_o^{(i)} = \frac{NnC}{T} \qquad \text{with} \qquad C = \frac{35g_{5/2}^2\mu_B^2}{12k},$$

the Curie constant for the J = 5/2 level.

When T is large compared with T_F the susceptibility may be written approximately (omitting the Pauli contribution from the conduction electrons)

$$\chi = \chi_o^{(i)}\left(1 - \frac{T_F}{T}\right) \simeq \frac{NnC}{(T + T_F)} .$$

Hence, since $T_F(T)$ varies only slowly (logarithmically) with T, the susceptibility of the Nn $f^5(J = 5/2)$ ions is approximately antiferromagnetic Curie-Weiss-like over a large high temperature range. Further, since ln D/kT is of order unity in this range we have

$$kT_F \sim \Delta = \pi\rho_o |v|^2$$

where Δ is the 'linewidth' associated with the f-conduction electron mixing interaction (Anderson 1961). These results add theoretical support to the speculations of Wohlleben and co-workers who predicted a Curie-Weiss-law with $kT_F \sim \Delta$ on the basis of semi-quantitative uncertainty principle arguments.

Within its range of mathematical validity the second-order expansion is in reasonable accord with experiment. For very high temperatures the theory may easily be modified to include higher-lying ionic levels and should be quite adequate for all high temperatures ($T \gg T_F$) of interest. The breakdown of second-order perturbation theory at low temperatures is hardly surprising for if H_1 removes the entropy and the divergence in the susceptibility meaningful results are only expected if the perturbation expansion is summed to infinite order.

REFERENCES

ANDERSON, P. W., 1961, Phys. Rev., 124, 41-53.

HIRST, L. L., 1974, J. Phys. Chem. Solids, 35, 1285-96.

JEFFERSON, J. H. and STEVENS, K. W. H., 1976, J. Phys. C: Solid State Phys., 9, 2151-2165

USE OF STANDARD-BASIS OPERATORS IN THE CONFIGURATION-BASED MODEL OF NARROW BAND SOLIDS

John M. Robinson

Indiana University-Purdue University at Fort Wayne

Fort Wayne, Indiana 46805

We show that the standard-basis (or atomic) operators are well suited for determining the Fermion excitation spectra which arise when interatomic f-d hybridization and f-f overlap are considered within the configuration-based model. The temperature dependences of the magnetic susceptibility and electrical resistivity of the model show qualitative agreement with experimental data on many narrow band rare earth and actinide materials.

Conventional band theory is inadequate for describing the states of 4f and 5f electrons in rare earth and actinide materials such as YbC_2 and $UA\ell_4$, because the intraatomic interactions in the f-shell are probably larger than the energy widths of the f states. This means that the appropriate theoretical starting point is the localized or atomic[1] limit, in which the energy levels of the f electrons are discrete, atomic, several-electron energies obtained by the methods of atomic physics. In the configuration-based model (CBM) of Hirst[2], the f electrons make transitions to an energetically overlapping band of itinerant d-electron states, but the quantum-mechanical admixture of the f and d states is not explicitly considered. In contrast, we consider the latter admixture by writing the Hamiltonian H of the solid as follows[1,3-5],

$$H = \sum_{pi} (\varepsilon_p - \zeta n_p) L^i_{pp} + \sum_{pqrs} \sum_{ij} B_{ij}(pqrs) L^i_{pq} L^j_{rs} \quad . \qquad (1)$$

In Eq. (1), the sums over i and j run over the N lattice sites of the rare earth or actinide atoms, each having partly filled outer s, d, and f shells. Each of the indices p, q, r, and s represents a set of quantum numbers (e.g., LSJM) of an atomic state. The standard-basis[4] (or atomic[1]) operator L^i_{pq} causes a transition of the atom at site i from the atomic state q to the atomic state p of energy ε_p. Note that the numbers n_p and n_q of electrons in the states p and q, respectively, are allowed to differ. The first term in H describes the localized limit, i.e., an ensemble of non-interacting atoms which weakly exchange both electrons and energy in attaining thermal equilibrium. The second term in H describes one-electron intersite hopping which leads to energy bands. The effect of the intraatomic correlations on the itinerant motion is contained in the hopping energies $B_{ij}(pqrs)$, which depend on the quantum numbers of the initial and final atomic states on the two sites i and j between which the electron is transferred. The equations of motion for the Green's functions $\langle\langle L^i_{pq};L^j_{rs}\rangle\rangle$ of the standard-basis operators in Eq. (1) are now written down and solved in the RPA approximation as described elsewhere.[3-5] The poles of the Fourier-transformed Green's functions yield the excitation spectra (energy bands), which are used to calculate the physical observables of interest.

As a simple example of the present theory, we have considered[3] a model of a simple cubic lattice of rare earth or actinide ions, each having two exterior s (or d) orbitals and two f orbitals. The hopping matrix $B_{ij}(pqrs)$ in Eq. (1) is assumed to equal a constant parameter t for electronic hopping between nearest neighbor s orbitals and to equal a parameter V for s-f hopping. Otherwise, $B_{ij}(pqrs)$ is assumed to be zero. The energy t leads to a conduction band of width W, and V yields the s-f hybridization. The intraatomic repulsion U between two f-electrons and the repulsion G between an s and an f electron on the same site are also considered. The calculations in Fig. 1 are for the case in which $U \to \infty$ and n=1, where

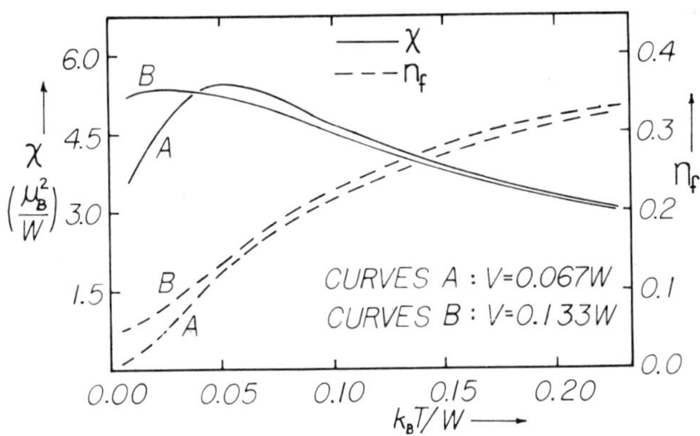

Fig. 1. The theoretical paramagnetic susceptibility χ (solid curves) and the number n_f of f electrons per atom (dashed curves) vs. the temperature T for two values of the s-f hybridization V. The quantity W is the width of the s-band. The energy of the f orbital is 0.067W above the center of the s-band.

n is the number of valence electrons per atom. The paramagnetic susceptibility χ of the model is shown in Fig. 1(a) as a function of the temperature T for two values of the hybridization V. Note the transition from a non-divergent (band-type) susceptibility at low T to a Curie-Weiss behavior as T increases. The χ(T) curve in Fig. 1 for the larger value of V is typical of narrow-band actinide intermetallics such as $UA\ell_4$,[6] while the curve for the smaller value of V resembles the experimental χ(T) data on intermediate valence rare earth compounds such as YbC_2.[7] The physical reason for the transition in the model is that at low T the f states are coherently hybridized with the s (or d) band, yielding narrow bands intersecting the Fermi level ζ. As T increases, the model predicts a decrease in the effect of the hybridization and an increase in the localization of the f electrons, resulting in the Curie-Weiss behavior. Similar χ(T) results were recently obtained by Varma and Yafet.[8] The number n_f of f electrons per atom is also shown in Fig. 1 as a function of T. The n_f(T) curve has been correlated with the electrical resistivity ρ(T) under the assumption that the localized f electrons scatter the conduction electrons.[3] The results show qualitative agreement with experimental ρ(T) data[6] on many rare earth and actinide intermetallics. A second application of the theory improved the above results by taking into account interatomic f-f hopping as well as spin-orbit and crystal field interactions.[3]

The standard-basis operator Green's function method provides a systematic method of calculating the excitation spectrum and thermodynamic properties of the CBM. In this way, the magnetic susceptibilities of mixed valence systems may be explained without resorting to the customary assumption of real-time fluctuations. The method also interpolates between the Boltzmann statistics of the localized model and the Fermi-Dirac statistics of the band model. The fundamental difference between the latter two statistics has hampered the development of the CBM.

1. J. Hubbard, Proc. Phys. Soc. A, 277, 237(1964); 285, 542(1965).
2. L. L. Hirst, A.I.P. Conf. Proc. 24, 11 (1975).
3. J. M. Robinson, A.I.P. Conf. Proc. 29, 319 (1976); Proc. Joint MMM-Intermag Conference (Pittsburgh, 1976), to appear.
4. S. Haley and P. Erdos, Phys. Rev. B5, 1106 (1972).
5. Y. F. Irkhin, Sov. Phys. - JETP 39, 490 (1974).
6. K. Buschow and H. van Daal, A.I.P. Conf. Proc. 5, 1464 (1972).
7. B. C. Sales and D. K. Wohlleben, Phys. Rev. Lett 35, 1240(1975).
8. C. M. Varma and Y. Yafet, Phys. Rev. B13, 2950 (1976).

SOME RARE-EARTH COMPOUND BAND MODELS RELATING TO CONFIGURATION CROSSOVER AND THE I.C.F. REGIME

J. A. Wilson

Bell Laboratories

Murray Hill, New Jersey 07974

Best use of the space here available seems to be as an illustrative addendum to a long paper by the present author, shortly to appear in the Springer-Verlag series 'Structure and Bonding'. Numbers in square brackets below refer to subsections in that text. It deals principally with understanding which R.E. compounds suffer configuration crossover, and modelling the circumstances in which interconfiguration fluctuation behaviour occurs.

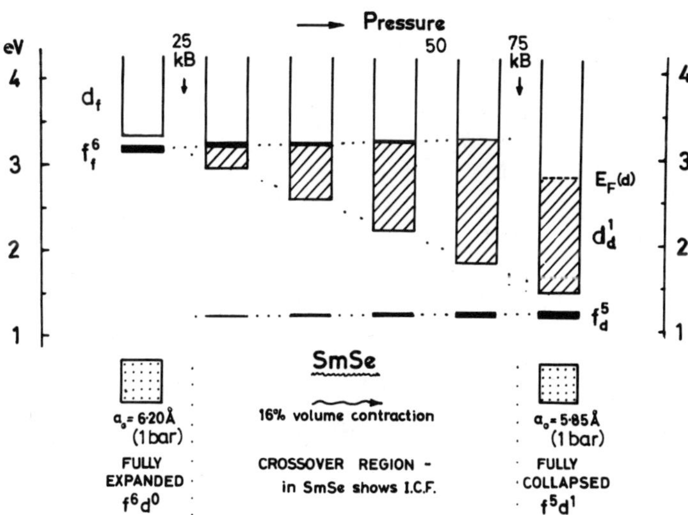

FIG. 1. Course of crossover $f^6d^0 \rightarrow f^5d^1$ in SmSe under pressure. [§1.4] Over second order transition, from 25 to 75 kB, t_{2g} band falls 2 eV. While f^6 level is pinned to $E_F(d)$, I.C.F. phenomena are supported.

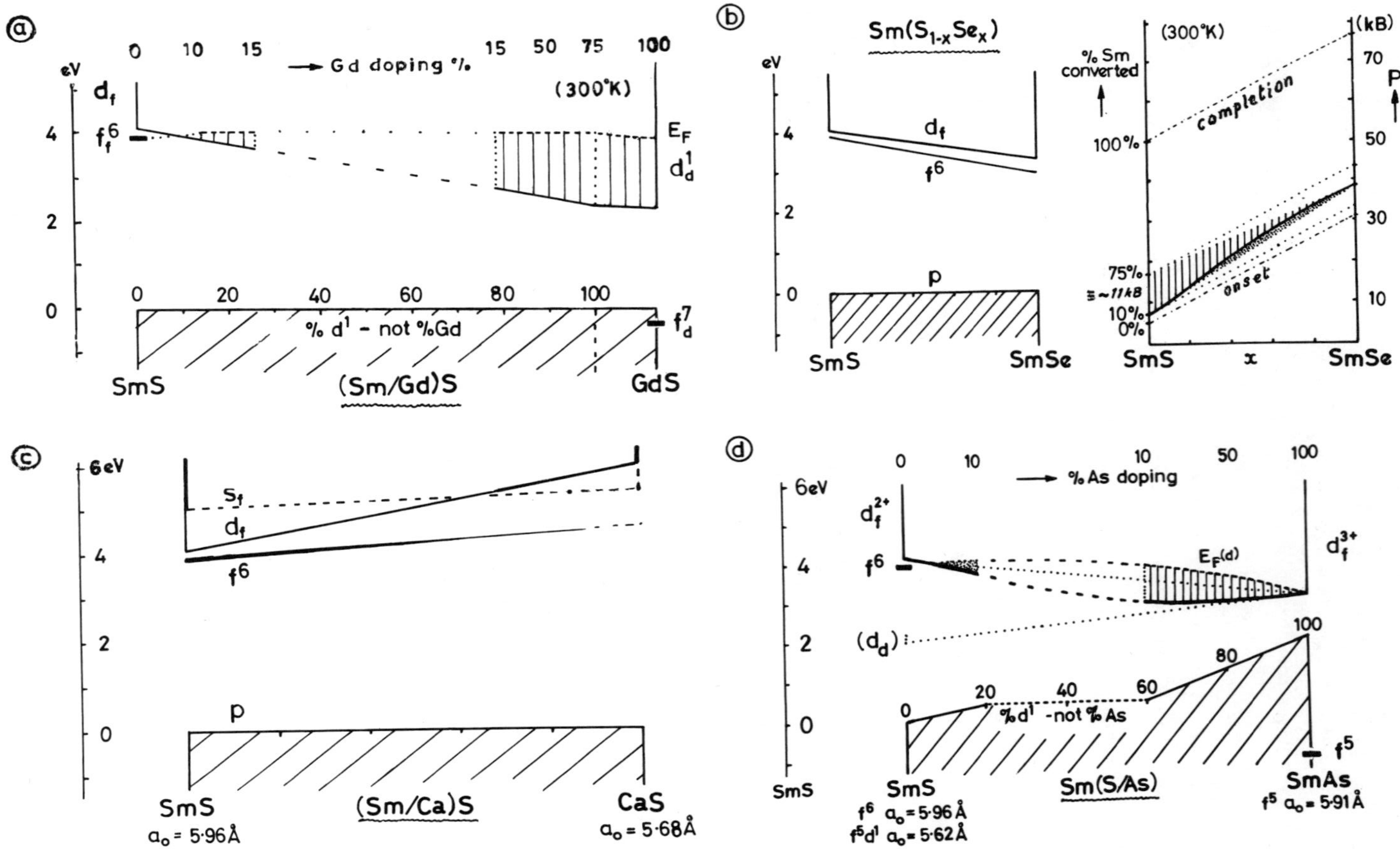

FIG. 2. Models concerning configuration crossover in SmS-based materials. (a) (Sm/Gd)S [§2.1,2.2]; (b) Sm(S/Se) [§1.2,2.4]; (c) (Sm/Ca)S [§2.4]; (d) Sm(S/As) [§2.4].

Figure 2a models $f^6 \rightarrow f^5d^1$ crossover in $Sm^{2+}S$ under Gd substitution (f^7d^1). Figure 2d shows the related circumstance of joint $Sm^{3+}(f^5)/As^{3-}(p^6)$ substitution. Unlike SmSe, these SmS systems shortly after d-band delocalization skip over much of the crossover region. Figure 2b shows this jump (vertical shading) in sulpho-selenides to shrink to zero just before SmSe. Figure 2c demonstrates why CaS does not collapse SmS despite the reduction in a_o.

Figure 3a illustrates with the cerium pnictides, (as the samarium chalcogenides), the covalent origin of the reversal in $f \rightarrow d$ gap energy. Covalent p/d mixing both increases the d-band width and reduces the $f \rightarrow d$ centroid separation [App. 2].

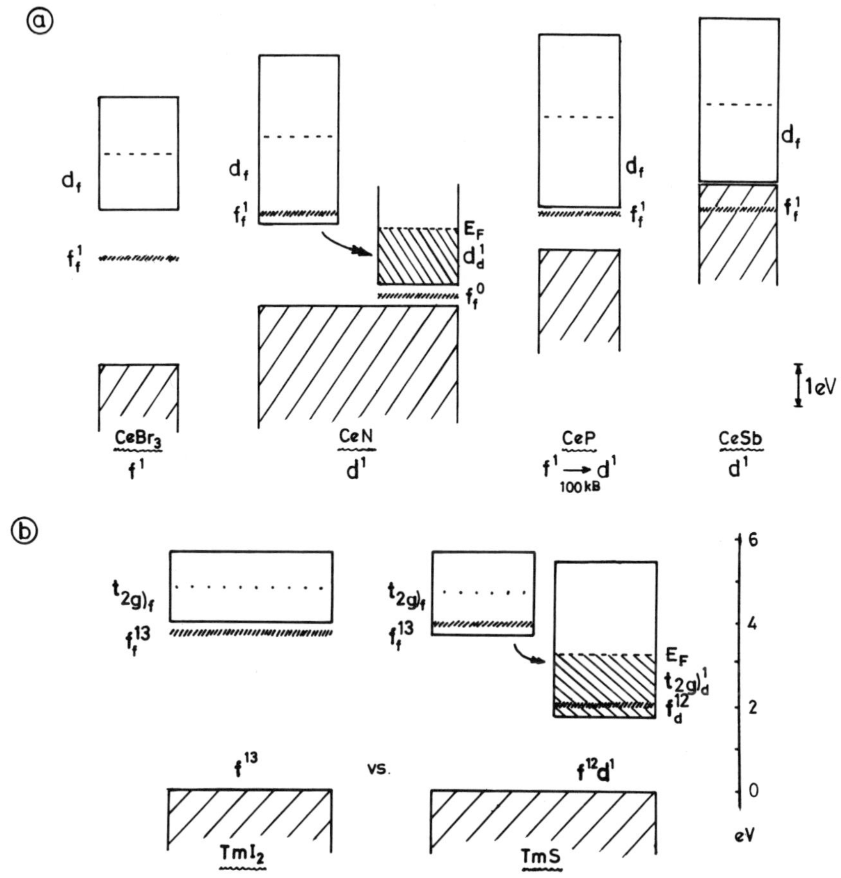

FIG. 3. Crossover as governed by covalency. The $f \rightarrow t_{2g}(d)$ gap depends on both the $f \rightarrow t_{2g}$ (centroid) separation, and the width of the t_{2g} band. (a) shows differential nephelauxetic effect [§1.2] in Ce^{3+} compounds yielding choice sequence $f^1 \rightarrow d^1 \rightarrow f^1$ [§1.6]. (b) shows f^{13} in TmI_2 being replaced by $f^{12}d^1$ in TmS [§1.5].

VOLUME COLLAPSE IN SOLIDS*

Mustansir Barma, S. D. Mahanti and T. A. Kaplan

Physics Department, Michigan State University
East Lansing, Michigan

Isostructural phase transitions have been observed in several rare earth compounds, and we recently studied a model[1] exhibiting such a transition as the force λ per surface atom was varied. The Hamiltonian (appropriate to a constant λ ensemble) was

$$H = \sum_i \frac{P_i^2}{2m} + \sum_{<ij>} \{\frac{1}{2} k x_{ij}^2 + (\Delta+\gamma x_{ij})(n_i+n_j) + \lambda x_{ij}\} \qquad (1)$$

The second sum is over nearest-neighbor bonds on a simple cubic lattice. n_i is either 0 or 1 (corresponding, e.g. to f^6 and f^5d states of a Sm^{++} ion in SmS). x_{ij} is the projection of the difference of the displacements of atoms i and j on the unit vector from the equilibrium positions j to i at $\lambda = T = 0$. As discussed in Ref. 1, H incorporates the physical effect of the change in ionic size when the electronic state is changed.

H was related to an Ising Hamiltonian with a lattice-induced exchange constant and a λ-dependent magnetic field. Consequently there was a first order transition at $\lambda = \lambda_0$, a certain value, and the lattice spacing a jumped abruptly. However the behavior of the model under pressure p remained unknown, as the relation $p = \lambda/(a)^{d-1}$ broke down in the 2-phase region.

The purely elastic part of H resembles the elastic part of the Baker-Essam model[2] for an Ising magnet. If one replaces the external force term in (1) by a hydrostatic pressure term, then the model would give strongly p-dependent elastic constants[3].

Below we consider a related but different model (one with p-independent elastic modulii) and analyze its behavior as a function of p. In a constant p ensemble, the Landau-Ginzburg form [4] of the Hamiltonian is taken to be

$$\mathcal{H} = \int d^dx \; \{ \frac{1}{2} c_{11} \sum_\alpha e^2_{\alpha\alpha} + \sum_{\alpha<\beta} (c_{12} e_{\alpha\alpha} e_{\beta\beta} + \frac{1}{2} c_{44} e^2_{\alpha\beta})$$

$$+ p \sum_\alpha e_{\alpha\alpha} + T[-h\Psi - g\psi\nabla.u + \frac{1}{2} r \psi^2 + \tilde{u} \psi^4]\} \quad (2)$$

Here the strain $e_{\alpha\beta}$ is $1/2\ (\partial u_\alpha/\partial x_\beta + \partial u_\beta/\partial x_\alpha)$ where u is the displacement field at x and α,β are cartesian labels. T is the temperature. The c's are elastic constants. The p term arises as we are in a constant p ensemble, and $\int d^dx \sum_\alpha e_{\alpha\alpha}$ is the volume change. ψ is a continuous Ising (1-component) field. An Ising variable arises because we associate a spin with the electronic variable $n_i (s_i = \frac{1}{2} - n_i)$; Eq. (2) is the Landau-Ginzburg form [4] of a Hamiltonian involving an Ising variable. There is no $\nabla\psi$ term, as there is no direct coupling between spins. The magnetic field term is the analog of the single-site term involving Δ in Eq. (1), and favors $\psi>0$, i.e. the spin up (f^6) state. The g term is a spin-phonon interaction arising from the γ term in (1).

$\mathcal{H}$ is somewhat similar to compressible Ising Hamiltonians. [3,6-8] The difference is that here $\nabla\cdot u$ couples to the order parameter ψ rather than to ψ^2 thus providing a lattice-dependent magnetic field.

We decompose $e_{\alpha\beta}$ into a uniform and non-uniform part [6,7]

$$e_{\alpha\beta} = e^o_{\alpha\beta} + V^{-1/2} \sum_q ie^{iqx} [q_\beta u_\alpha(q) + q_\alpha u_\beta(q)] \quad (3)$$

and further treat u(q) as normal mode coordinates. Such an approximation has proved not unreliable in the past--it was made in connection with the study of critical behavior of compressible Ising models on isotropic lattices [6,7] and gave results agreeing with the more complete treatment of Ref. 8.

Using Eq. (3) in Eq. (2) and tracing over the lattice variables [3], we obtain an effective spin Hamiltonian $\mathcal{H}_s$ where

$$\frac{\mathcal{H}_s}{T} = \sum_q \frac{1}{2} r \; \psi_q \psi_{-q} + \frac{\tilde{u}}{V} \sum_{q_1 q_2 q_3} \psi_{q_1} \psi_{q_2} \psi_{q_3} \psi_{-q_1-q_2-q_3}$$

$$- T \sum_{q\neq 0} K(q) \psi_q \psi_{-q} - \frac{Tg^2}{2B} (\psi_{q=0})^2 - \sqrt{V}\ (h - \frac{g}{B} p)\ \psi_{q=0} \quad (4)$$

Here $\psi_q = V^{-1/2} \int d^dx \; e^{iqx} \; \psi(x)$, and $K(q) = \frac{1}{2} g^2 (q\cdot D^{-1} \cdot q)$ where D

is, the dynamical matrix and B is the bulk modulus. As noted in Ref. 3, K(q) is, in general, singular, and this complicates the study of $\mathcal{H}_s$. However, for isotropic lattices ($C_{11} = C_{12} + 1/2 C_{44}$) and Baker-Essam-like lattices ($C_{44} = 0$), $K(q)$ is independent of q ($K(q) = K_o$) and equals[3] $(g^2/2C_{11})$ and $(g^2/2B)$ respectively. Writing

$\sum_{q\neq 0} K(q)\psi_q\psi_{-q}$ as $\sum_q K(q)\psi_q\psi_{-q} - K_o\psi^2_{q=o}$, we find that the $K_o\sum_q\psi_q\psi_{-q}$ term changes the effective value of r in Eq. (4) whereas the term $-K_o\psi^2_{q=o}$ leads to an effective long range interaction $\sim(K_o-g^2/2B)\psi^2_{q=o}$. For an isotropic lattice $K_o-g^2/2B < 0$, leading to a long range ferromagnetic interaction. In this case when the effective magnetic field h-gp/B passes through zero, there is a first order phase transition. Thus for $T<T_c$, there is a line p=Bh/g of first order transitions in the P-T plane.

For the Baker-Essam-like lattices ($C_{44}=0$, $K_o-g^2/2B = 0$), the effective long range interaction between the spins vanishes. In this case one has to incorporate in Eq. (2) the contributions of the type $\sum_\alpha \frac{\partial u_\alpha}{\partial x_\alpha}\frac{\partial\psi}{\partial x_\alpha}$ (referred to in footnote 5). These contributions will lead to couplings between derivatives of ψ in $\mathcal{H}_s$ on tracing over the elastic variables, so that spins at different points in space are coupled. We are investigating this coupling.

Finally we comment on the difference in the behavior of the models of Eqs. (2) and (1). It is similar to the difference between the Baker-Essam-like model (BE2) discussed in Ref. 3 and the original Baker-Essam model (BE1). The difference is that while the elastic part of BE2 and the model of Eq. (2) have p-independent elastic constants, BE1 and the model of Eq. (1) have elastic constants which depend strongly on the pressure.[3]

We thank Professor B. I. Halperin for pointing out the similarity of the lattice part of the models of ref. 1 and ref. 2.

* Supported by the National Science Foundation.

References

1. M. Barma, T. A. Kaplan, S. D. Mahanti, Phys. Lett. 57A, 168 (1976).
2. G. A. Baker and J. W. Essam, Phys. Rev. Lett. 24, 447 (1970).
3. D. J. Bergman and B. I. Halperin, Phys. Rev. B13, 2145 (1976).
4. K. G. Wilson and J. Kogut, Phys. Rep. 12C, 77 (1974).
5. In addition to the g term in Eq. (2) there are contributions of the type $\sum_\alpha \frac{\partial u_\alpha}{\partial x_\alpha}\frac{\partial\psi}{\partial x_\alpha}$. The importance of these terms is discussed later in the text.
6. A. I. Larkin and S. A. Pikin, Zh Eksp. Teors. Fiz. 56, 1664 (1969) [Sov. Phys.-JETP 29, 891 (1969)].
7. J. Sak, Phys. Rev. B10, 3957 (1974).
8. F. J. Wegner, J. Phys. C7, 2109 (1974).

VALENCIES IN LANTHANIDE AND ACTINIDE SYSTEMS

Börje Johansson

FOA, 104 50 Stockholm

Most lanthanide (Ln) and actinide (An) atoms have a divalent configuration, $f^{n+1}s^2$. In the condensed phase, however, most of the lanthanides are trivalent metals. Divalent metals like Ca, Sr and Ba have a cohesive energy of about 40 kcal/mol, while trivalent metals like Y, La and Lu have a cohesive energy ≈ 100 kcal/mol. For Gd and Cm (atomic configuration f^7ds^2) the cohesive energy is about 95 and 90 kcal/mol, respectively. To study the question about possible divalent f-metals we compare the following condensation processes (units = kcal/mol):

$$\boxed{f^{n+1}s^2(\text{atom})} \begin{cases} \xrightarrow{\sim 40} \boxed{f^{n+1}\text{-metal}} \\ \xrightarrow{E(f\to d)} \boxed{f^n ds^2(\text{atom})} \xrightarrow[\sim 90 (An)]{\sim 95 (Ln)} \boxed{f^n\text{-metal}} \end{cases} \quad (1)$$

By making the metal trivalent (assuming non-bonding f-electrons) we gain about 50 (An) or 55 kcal/mol (Ln) in binding, but lose the excitation energy $E(f \to d)$. These energies are compared in Fig. 1. Thereby we find that among the lanthanides only Eu and Yb will be +2-metals. For the actinides we notice that Am must be a +3-metal. For the heavier actinides not only No will be divalent but also Md, Fm and Es. Cf is found on the borderline between a +2 and +3 behaviour. Recent experiments[1] confirm this, and a mixed valence state is observed.

For compounds, the energy balance in eq. (1) will be somewhat modified. From trends in heats of formation of monosulphides one might conclude that the number of about 50 - 55 kcal/mol for the metals should be somewhat reduced, say to about 40 kcal/mol. From Fig. 1 we then find that for the monosulphides

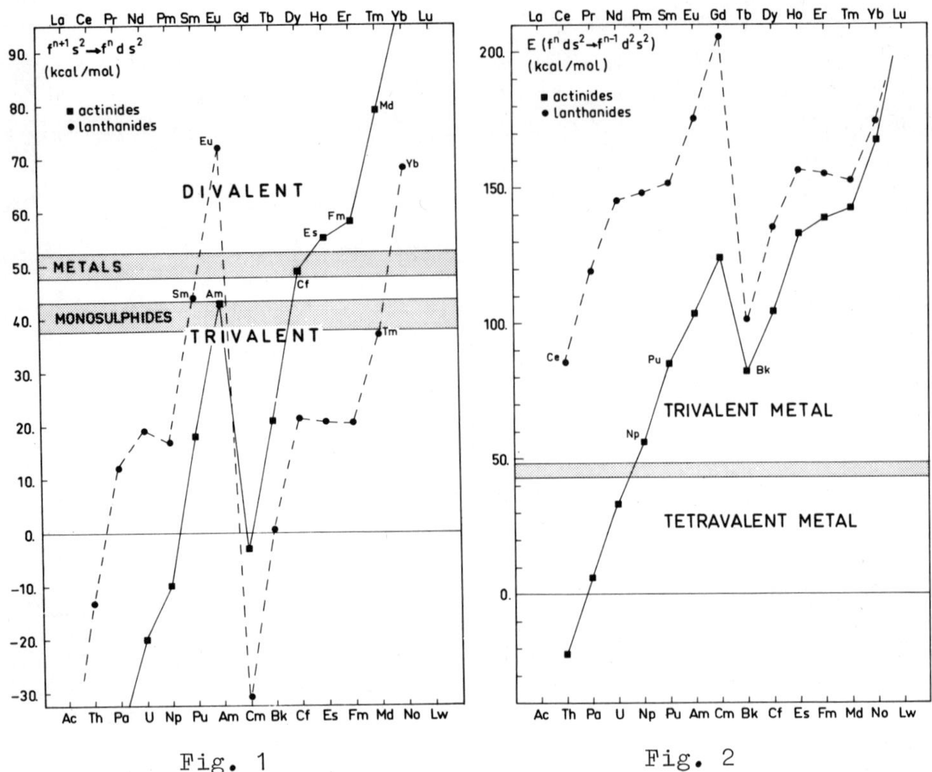

Fig. 1 Fig. 2

SmS will be divalent. TmS, on the other hand, falls on the trivalent side. Of interest is also to notice that AmS is on the borderline for a valence transition.

Trivalent contra tetravalent f-metals can be studied in the same way. The +4-metals Zr, Hf and Th all have a cohesive energy close to 145 kcal/mol. The corresponding value for +3-metals is close to 100 kcal/mol. Therefore, by studying the following two competing reactions;

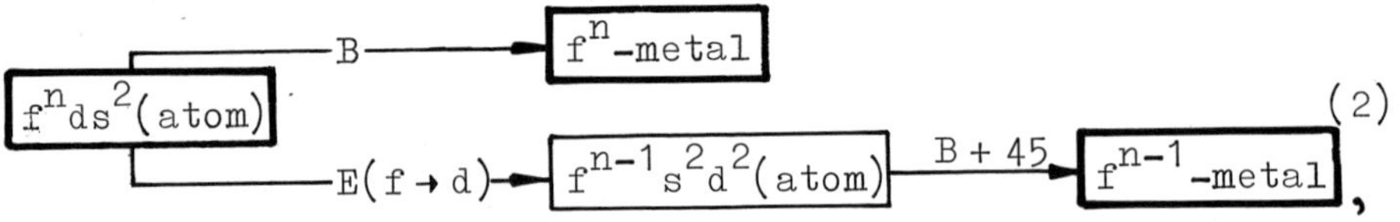

(2)

the question about +3 versus +4 can be investigated. In Fig. 2 the appropriate comparison is made. Thereby we find that neither Np nor Pu can be tetravalent metals. However, their atomic volumes deviate strongly from a trivalent behaviour. These conflicting evidences point to that the assumption about non-bonding f-electrons is no longer valid[2,3]. Thus, in Np and Pu (as well as in U and Pa) the 5f-electrons are itinerant and contribute with a metallic binding in the solid. Consider the formation of Pu-metal as taking place in three steps;

$$Pu(f^6s^2) \rightarrow Pu(f^5ds^2) \rightarrow Pu\ (+3\text{-metal}) \rightarrow Pu\ (\text{true metal}). \quad (3)$$

In the formation of the +3-metal (non-bonding f-electrons) there would be an energy gain of about 72 kcal/mol[2]. However, the experimental cohesive energy is 84 kcal/mol. Thus the metallic f-electrons give an extra binding of about 12 kcal/mol (last step in eq. (3)). Similarly, one finds that in Np the metallic 5f-electrons contribute to the binding by about 23 kcal/mol. In U the corresponding number is even larger. Thus, between Pu and Am there is a drastic change in the behaviour of the 5f-electrons. In Pu they are itinerant and in Am they are localized.

From Fig. 2 we notice that none of the lanthanides can be tetravalent metals. This holds also for cerium, which is in sharp contrast to what is commonly believed. Therefore, the occurrence of the dense α-phase in cerium must have the same origin as the dense phases of the earlier actinides, namely that the 4f-shell drastically changes its localized nature to become an itinerant 4f-state[4].

The energy difference between trivalent cerium and hypothetical tetravalent cerium has been derived to be about 40 - 45 kcal/mol. In $CeRu_2$, Ce appears to be IV-valent (lattice constant, superconductivity). In $CePt_2$, on the other hand, Ce appears to be +3-valent (lattice constant, magnetism). Miedema[5] has derived an empirical equation for calculation of heat of formation of intermetallic systems. Using his method we compare the following two possibilities:

$$\begin{aligned} Ce^{III}\text{-metal} + 2\,Ru\text{-metal} &\rightarrow Ce^{III}Ru_2\ (\text{trivalent}) \\ Ce^{IV}\text{-metal} + 2\,Ru\text{-metal} &\rightarrow Ce^{IV}Ru_2\ (\text{tetravalent}). \end{aligned} \quad (4)$$

For the first reaction, we calculate $\Delta H_f(III) = -10$ kcal/mol, and for the second $\Delta H_f(IV) = -53$ kcal/mol. The difference 43 kcal/mol, is about enough to overcome the excitation of Ce^{III}-metal → → Ce^{IV}-metal (∼40 kcal/mol). Replacing Ru in eq. (4) with Pt we obtain the following numbers; $\Delta H_f(III) = -62$ kcal/mol and $\Delta H_f(IV) = -93$ kcal/mol. The difference of 31 kcal/mol is here too small to bring Ce into the tetravalent state.

1. M. Noé and J.R. Peterson, Proceedings of the 5th Int. Conf. on Plutonium and Other Actinides, H. Blank and R. Lindner, eds. (North Holland, Amsterdam, 1976).
2. B. Johansson and A. Rosengren, Phys. Rev. B11, 1367 (1975).
3. B. Johansson, Phys. Rev. B11, 2740 (1975).
4. B. Johansson, Philos. Mag. 30, 469 (1974).
5. A.R. Miedema, R. Boom and F.R. de Boer, J. Less-Common Met., 41, 283 (1975).

CONFIGURATION SHIFTS IN NARROW BANDS DUE TO COUPLING WITH A BROAD CONDUCTION BAND

Samuel P. Bowen

Department of Physics, VPI & SU

Blacksburg, VA 24061

As an application of a thermodynamic perturbation theory[1], I have studied the effects of couplings between many electron configurations on atoms in metals and broad conduction electron bands. Of particular interest has been those situations in which many electron states with different numbers of electrons participate in relatively strong fluctuations. One of the main characteristics of this study is the attempt to avoid the one-electron picture implicit in the usual HF approach and deal with the many-electron configurations in a fashion that allows, at least in principle, the treatment of atomic correlation effects in zeroth order. The simplest conceptualization of our results is obtained in the S-shell Anderson model[2] of a single impurity atom in contact with a host conduction electron band. Extensions of these considerations to compounds like $CeAl_3$ with very narrow f-shell bands will be discussed at the end of this paper.

The first question of interest involves the determination of what criteria must be applied to the atomic configuration energies E_N and E_{N+1} (for N and N+1 electrons respectively) in order that the two configurations would participate in significant fluctuations. This question can be easily answered in two equivalent perspectives: either the canonical ensemble for composite states of the atom and conduction electrons, or the grand canonical ensemble (GCE) for the atom in equilibrium with the conduction electron resevoir at chemical potential μ. For brevity, we adopt the latter perspective, and examine the condition in the GCE that configuration energies E_N and E_{N+1} represent almost equally probable states. Evidently, this requires $E_N-N\mu\simeq E_{N+1}-(N+1)\mu$ or that the energy difference

$$\varepsilon(N,N+1)=E_{N+1}-E_N-\mu\simeq 0 \qquad (1)$$

The condition for strong fluctuations is that the atomic configurations (extant in the metal) have energies which differ by the chemical potential μ of the conduction electrons. A second relationship suggested by (1) is that $\varepsilon(N,N+1)$ is the interconfigurational excitation energy (ICEE) which appears in the one electron Green's function for the atom, or equivalently, in the conduction electron t-matrix reflecting their fluctuation induced scattering.

An examination of the effect of a many-electron mixing matrix element $V_{\vec{k}}(N\rightarrow N+1)$ which gives the strength of a conduction electron jumping onto the atomic shell and changing the atomic configuration has been examined in the context of the Anderson model of a single atom shell. The first effect of such a coupling is to introduce a lifetime broadening $\Delta=\pi N(0)|V_{\vec{k}}(N\rightarrow N+1)|^2$ (2) into the resonances of the t-matrix. Here $N(0)$ is the conduction electron density of states at the Fermi energy. The second and more important effect of the coupling is to introduce a temperature dependent shift $\delta\varepsilon$ of the ICEE resonance, which can cause a relative "motion" of the t-matrix resonance through the conduction electron Fermi energy. If the atomic configurations satisfy (1) and the $N\rightarrow N+1$ configuration resonance dominates the t-matrix, then

$$\mathrm{Im}t(0)\propto\Delta^2/((\varepsilon(N,N+1)+\delta\varepsilon)^2+\Delta^2) \tag{3}$$

If the ICEE is small then the behavior of the resistivity as well as other properties of the system will reflect the temperature dependence of $\delta\varepsilon$. The temperature dependence of $\delta\varepsilon$ in the simplest model can be shown by temperature dependent second order perturbation theory, or by more sophisticated techniques[3] to behave in a "high" temperature range like log (T) and in a low temperature range like $(1-T^2)$ for the simple S-shell Anderson model the ICEE shift depends on two temperatures which are characteristic of the energies of the ICEE's in the atom

$$T^{(\pm)} = \frac{1}{\pi}((\varepsilon_{ds}+U/2 \pm U/2+\delta\varepsilon)^2+\Delta^2)^{\frac{1}{2}} \tag{4}$$

For the situations of interest here one of the $T^{(\pm)}$ is of order Δ/π and the other is possibly large. For definiteness, let us assume $T^{(-)}<T^{(+)}$, then the ICEE shift is $\delta\varepsilon=\Delta\Lambda(T)/\pi$ where (5)

$$\Lambda(T) = \begin{cases} 0 & T>T^{(+)},T^{(-)} \\ +\ell n(1.13\pi T^{(+)}/T), & T^{(-)}<T<T^{(+)} \\ +\ell n(T^{(+)}/T^{(-)})+O(T^2), & T<T^{(+)},T^{(-)} \end{cases} \tag{6}$$

Of particular note in (6) is the result that in the log (T) range for Λ the numerator of the argument of the log is not the band width D as the standard perturbative result obtains[4], but is $T^{(+)}$, which can be significantly smaller than D and reflects properties of the atomic configurations rather than the host band width. This result is clearly more in accord with many experimental results for alloy systems[4].

The application of these results to the resistivity of alloys and finally of $CeAl_3$ rests on the temperature dependent shifting of the t-matrix Lorentzian (3) through the Fermi energy of the conduction electrons. For the high temperature range $T^{(-)}<T<T^{(+)}$ where the shift varies like $\ell n(T)$ behavior while the wing of the Lorentzian passes through the Fermi energy. When the peak of the ICEE resonance reaches the Fermi energy the resistivity will saturate. If this peak temperature is higher than $T^{(-)}$, the resistivity will be seen to further decrease until Λ achieves its low temperature limit. This passing of the ICEE peak through the Fermi energy reflects the changeover in the relative thermodynamic weighting of the two atomic configurations whose GCE energy difference is the ICEE. As long as the ICEE energy (1) remains the same order of Λ, the relative probability of occupation of the two configurations will be comparable and the system will exhibit a statistical mixture of both configurations. In the dilute alloy problem the host lattice seems to oppose a complete interchange of impurity configurations. However in intermetallic compounds such as $CeAl_3$ and $CeAl_2$, the Ce atom would appear in this picture to quite completely interchange ionic configurations ($f^1 \rightarrow f^0$) as the temperature goes to zero.

The zero pressure magnetic resistivity[5] of $CeAl_3$ is quite well reproduced by this picture if $T^{(+)} \simeq 800K$, $T^{(-)} \simeq 1.6K$ and $\Delta \simeq 2K$ and the resistivity is assumed to be a simple factor times the Ce atom t-matrix. This procedure which is carried over from the dilute alloy problem is not really appropriate for the concentrated system, but because of the narrowness of the f-shell bands, it is not perhaps too far wrong[4].

The application of pressure to the $CeAl_3$ system yields a very simple result in this picture which illustrates the necessity of taking into more careful account the crystal field splitting and the Ce atom state degeneracies. The pressure changes the magnitude of the ICEE energy and thus changes the temperature at which the ICEE energy is shifted to zero. This is easily rationalized by the simple Anderson model. However, the lower temperature behavior reflects thermal shifts which are due to the other crystal field split states of the Ce configurations, and these are outside of the context of the S-shell model. Work on a more realistic model Hamiltonian is proceeding and will be reported elsewhere.

Footnotes

1. Samuel P. Bowen, J. Math. Phys 16, 620 (1975).
2. P. W. Anderson, Phys. Rev. 124, 41 (1961).
3. Samuel P. Bowen, submitted to Physical Review.
4. B. Cornut and B. Cogblin, Phys. Rev. B5, 4541 (1972).
5. A. Percheron, et al, Solid State Comm. 12, 1289 (1973).

SPIN SUSCEPTIBILITY OF CHEMICALLY COLLAPSED CERIUM*

J. M. Lawrence† and R. D. Parks

Dept. of Physics and Astronomy, University of Rochester

Rochester, N. Y. 14627 U.S.A.

Due to the connection between the valence instability problem and that of magnetic moment formation in metals, the magnetic susceptibility χ of intermediate valence materials has been studied extensively. Near the critical point $\{x_o, T_o\}$ in $Ce_{1-x}Th_x$ it is relatively easy to vary the fundamental energy parameters which determine the behavior of χ. In Fig. 1 we show $\chi(x;T)$ for several concentrations bracketing x_o[1]. The influence of the critical point is readily apparent; to discuss the energetics, however, we concentrate first on the regions of x-T space well away from the phase transition.

The intermediate valence state has been described as a narrow-band fermi liquid[2]; one expects a temperature independent susceptibility inversely proportional to the bandwidth, $\chi \propto 1/\Gamma$. In Ref. 1 we analyze the results for the low temperature intermediate valence α-state; we show that for $0.10 < x < 0.27$ and $4.2 < T < 50K$ the raw data fits the expression

$$\chi(T) = \chi^{corr}(0) + \eta_\gamma(A + (C/T))$$

where $\chi^{corr}(0) = 3.75\times10^{-6}$emu/g-Ce, $A = 288\times10^{-6}$emu/g-Ce, and $C = 5760\times10^{-6}$emu-K/g-Ce (the J = 5/2 free ion value). We argue that there is an "impurity" contribution arising from the presence of a fraction η_γ(typically ~ 0.005) of untransformed magnetic cerium ions. (The fraction is too large to arise from foreign impurities and does not reproduce if the sample is recycled through the phase transition - i. e. a different fraction remains untransformed during any cycle.) We take, then, the true value of $\chi(T)$ to be $\chi^{corr}(0)$ independent of x or T in the stated interval. Naively applying $\chi = C/\Gamma$ we obtain an estimate of the bandwidth, $\Gamma \sim 0.1$eV. For x = 0.41 we have

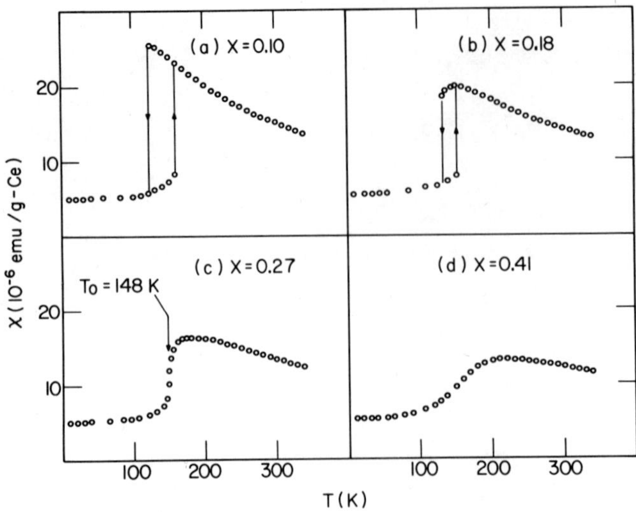

Fig 1. Magnetic Susceptibility $\chi(x;T)$ of $Ce_{1-x}Th_x$ for four concentrations x in the vicinity of the critical concentration x_o.

Fig 2. $1/\chi$ in the γ-state of $Ce_{1-x}Th_x$. The solid lines represent the temperature dependence $\chi = C'/(T+\theta(x))$ where $C' = 6000$emu-K/g-Ce and the values utilized for $\theta(x)$ are shown in b). [From Ref. 1.]

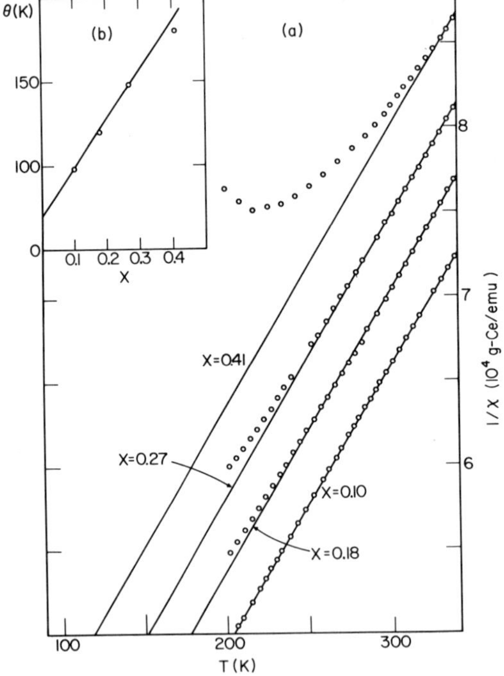

$\chi^{corr}(x=0.41, T=0) \cong 4.3\times10^{-6}$emu/g-Ce, indicating a slowing down of the valence fluctuations in the increasingly disordered alloy environment.

The integral valence state of cerium presumably involves a localized magnetic level E_f degenerate with the 5d6s band, and residing near the fermi level $\mathcal{E}_F$; hence spin fluctuations and/or valence fluctuations may be important. In Fig. 2 we see that the data seem to fit the formula $\chi = C'/(T + \theta(x))$; C' is 4% enhanced over the free ion value and $\theta(x)$ roughly tracks the limit of stability of the γ-state, defined as the first-order transition temperature on cooling $T_c^-(x)$. Certain reservations are here in order: the region of fit needs to be extended to far higher temperatures; the data has not been corrected for a background Pauli term, which could easily lead to 5% shifts in C' and θ; and the hypothesis that $\theta(x) \sim T_c^-(x)$ needs to be tested on other cerium-based alloys. Nevertheless it is reasonable to speculate that $\theta(x)$ reflects a fundamental energy parameter. This might be a valence fluctuation width, as in the phenomenology[3] where $\chi \sim C\eta_f/(T + T_{VF})$; or it might represent spin fluctuations or enhanced RKKY exchange, arising in the limit of small $\mathcal{E}_F-E_f$. We note that if the parameters Γ and θ represent level widths, then an important feature of the valence transition is an order of magnitude increase in this parameter.

The behavior near the critical point can give a clue as to how the susceptibility is related to the occupation number η_f, which is presumably the microscopic order parameter. Plots of $T\chi-T_o\chi(T_o)$ follow $(T-T_o)^{1/\delta}$ with $\delta\simeq3$ for the mean-field transition. On the basis of the phenomenological formula given above one might expect $(T + \Delta)\chi/C$ to exhibit this power law for some Δ with $\theta\leq\Delta\leq\Gamma$; however one obtains acceptable fits only for $\Delta \to 0$. The fit is good over a much larger region than that for the resistivity. We feel that the origin of this linear coupling of $T\chi$ to η_f needs to be established.

* Work supported by the Army Research Office.

† Present address: Physics Department, University of California, Irvine, CA 92717.

1. J. M. Lawrence and R. D. Parks, Journal d'Physique, Suppl., to be published.
2. C. M. Varma and Y. Yafet, Phys. Rev. B 13 (1976) 2950.
3. B. C. Sales and D. K. Wohlleben, Phys. Rev. Lett. 35 (1975) 1240.

EFFECT OF ADDITIVES ON THE VALENCE TRANSITION OF CERIUM *

M. A. Manheimer and R. D. Parks

Physics and Astronomy Department, University of Rochester, Rochester, New York 14627

In previous papers,[1] we reported on the mean field character of the Cerium γ-α valence transition in the $Ce_{1-x}Th_x$ system. It was suggested that long range forces arising from anharmonic strain effects dominate the critical behavior. In order to elucidate more information on the mechanisms which drive the transition, we have undertaken a study of the effects of the other rare earth additives on it. Specifically, we wanted to determine if other solute systems show the same critical behavior as the $Ce_{1-x}Th_x$ system, and also to distinguish between effects due to solute size and valence.

The Cerium transition has been studied extensively both as a function of pressure and rare earth additive.[2,3] It is believed that the transition from the high temperature-low pressure γ phase to the α phase is associated with the promotion of the cerium 4f electron into the hybridized 5d6s conduction band. The valence of the γ phase has been estimated at 3.1 and changes to 3.6 on transformation to the α phase, along with jumps in volume, resistivity, and magnetic susceptibility characteristic of a first order transition. As one moves further along the coexistence line by increasing pressure (or by alloying), the discontinuities in these quantities decrease and reach zero at a critical pressure (or concentration) beyond which they vary continuously as the coexistence line extension is crossed. For the $Ce_{1-x}Th_x$ system, the derivatives of these properties ($V,\rho,\chi T$) diverge at the critical point with a mean field exponent. Interpretations of results of other previous studies have been clouded by the presence of hexagonal β phase Cerium. The proportion of β phase present is strongly dependent both on sample history and solute, additions of those heavier rare earths favoring more strongly the formation of a hexagonal phase.

Since Thorium concentrations of greater than 10% are known to eliminate β phase Cerium,[2] it was decided to use $Ce_{0.9}Th_{0.1}$ as a starting point for this study. Pseudobinary systems were formed by addition of $R_{0.9}Th_{0.1}$, (R=Rare earth). Thus, any changes in transformation properties could be imputed to the additional solute, and not to the formation of β Cerium.

Samples were prepared by arc melting constituents of highest available purity and treated as in Ref. (1). Nominal compositions were checked qualitatively against measured room/helium temperature resistance ratios. Sample quality was also verified by X-ray analysis, only γ phase Cerium being found.

Transitions for all the samples were determined from resistivity ρ(T) measurements. Transition temperatures T_c for first order samples were defined as the temperature midpoint of the hysteresis, and for continuously transforming samples as the temperature where the maximum in dρ/dT occurred. Plotted in Fig. 1 are T_c for the various systems versus substituant concentration in $Ce_{0.9-x}R_xTh_{0.1}$. In all cases, the samples either transformed to the α phase, or became too mixed valent before a magnetic ordering temperature was reached, precluding the existence of the magnetically ordered γ phase postulated by Alascio, Lopez, and Olmedo.[4]

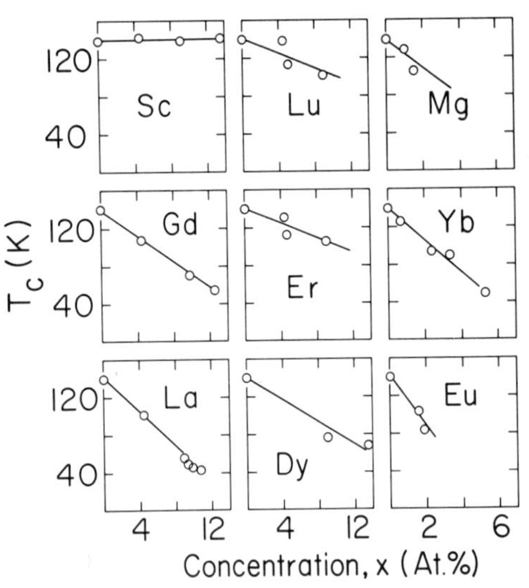

Fig. 1. Phase diagram for $Ce_{0.9-x}R_xTh_{0.1}$. The coexistence lines separate the high temperature γ phase from the lower temperature α phase. See Ref. (1) for the $Ce_{1-x}Th_x$ phase diagram.

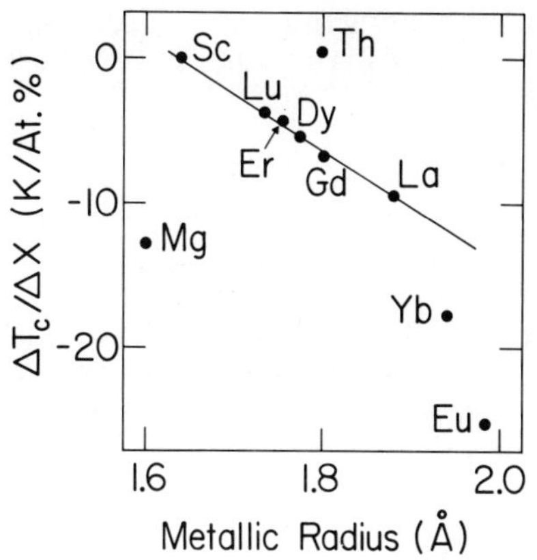

Fig. 2(a). Slope of phase diagram, $\Delta T_C/\Delta x$, for each system as a function of the metallic radius of additive.

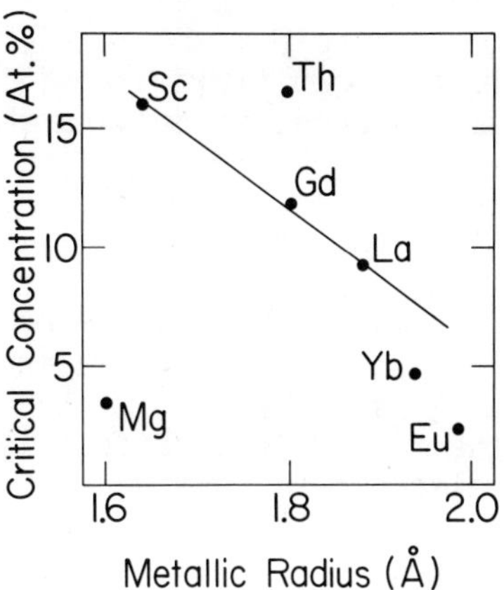

Fig. 2(b). Critical concentration, x_C of those systems for which it has been determined.

A summary of the phase diagram slopes appears in Fig. 2(a) where $\Delta T_C/\Delta x$ versus metallic radius of additive (R_M) is plotted. The most striking feature of this plot is that the data for 3-valent additives fall on a straight line. The point for 4-valent Thorium falls above this line while those for the (nominally) 2-valent substituents fall below. Ytterbium appears mixed valent in this system, an interpolation of the data giving its valence as about 2.3. This result is not totally unexpected since Yb occurs as the mixed valent form in many of its compounds. However, room temperature lattice constant measurements of the $Ce_{0.9-x}Yb_xTh_{0.1}$ samples proved inconclusive, and it is also not known how the valence of Yb in this system changes with temperature.

Figure 2(b) shows the critical concentrations x_C for several of the systems studied plotted against R_m. Except for the Magnesium point, there is a strong qualitative similiarity to Fig. 2(a). It might be argued that since Magnesium is not a rare earth or rare earth like, its $3s^2$ electrons locally disturb the 5d6s conduction band in a significant way; in a way that differs when one considers the effects of Magnesium on either $\Delta T_C/\Delta x$ or x_c.

The exponent δ was measured for the La and Yb additive systems. The δ exponent measures the behavior of the order parameter on a path not parallel to the coexistence line. As expected, the mean field character of the transition was unchanged. The values obtained were $\delta_{La} = 3.1 \pm .2$ and $\delta_{Yb} = 2.9 \pm .4$ which compare favorably with δ_{Th} =3.0±.2 obtained previously (1) and δ = 3.0 for mean field transitions. (This confirms the role of long range forces in driving the transition.)

The data of Fig. 2 indicates that size effects and valence effects of the additive are, to a first approximation, independent of one another. Additives with a valence larger than that of the host ions makes delocalization of the Ce ions easier whereas additives with a smaller valence make delocalization more difficult (viz., T_C is depressed). This is consistent with the explanation that the main valency effect is to alter the screening of the Ce 4f electrons. Thorium, e.g., adds an extra electron to the s-d band of the system and this leads to an increase in the screening between the Ce 4f electron and the Ce nucleus, thereby increasing the size of the 4f orbital and, consequently increasing the s-f mixing parameter. The effect of adding different size solutes is to change the average lattice spacing. The net effect of decreasing solute size is to act as positive pressure, broadening and deforming the bands in such a way as to move the f level closer to the Fermi level, here again increasing s-f mixing. The line formed by the data for the 3-valent solutes in Fig. 2(a) shows this effect: the larger solutes favor the higher volume higher temperature γ phase, so that T_C is depressed. Quantification of this picture must await detailed knowledge of the band structure.

REFERENCES

*Work supported in part by the U. S. Army Research Office and the U. S. Office of Naval Research.

1. J. M. Lawrence, M. C. Croft and R. D. Parks, Phys. Rev. Letters 35, 289 (1975), and these proceedings.
2. K. A. Gschneidner, Jr., R. O. Elliott and R. R. McDonald, J. Phys. Chem. Solids 23, 555, 1191, and 1201 (1962).
3. W. N. Miner, R. O. Elliott and F. W. Clinard, J. Less. Com. Met. 11, 301 (1966).
4. B. Alascio, A. Lopez, and C. F. E. Olmedo, Phys. Rev. B 7, 1274 (1973).

CALORIMETRIC STUDIES OF THE γ-α TRANSITION IN $Ce_{1-x}Th_x$*

J. M. Markovics and R. D. Parks

Department of Physics and Astronomy, University of Rochester, Rochester, New York 14627

One of the salient features of the valence transition in Cerium is the disappearance of magnetism when the material transforms from the α to the γ state. The importance of the magnetic entropy in determining the nature of the γ-α transition may be easily inferred from the phase diagram and is dramatically evident in the specific heat at the critical point.

The phase boundary separating the integral valent γ from the intermediate valent α form of Ce is almost perpendicular to the temperature axis in the T-x plane (where x is the concentration of Th, which plays the role of pressure). Since the transition into the α state is heralded by a 15% volume reduction, the Clausius-Claperyon relation, $\Delta S/\Delta V = dP/dT$ at the phase boundary, tells us that the entropy loss is large in the Ce γ-α transition. (In the SmS system, where the transition is also marked by a large volume reduction upon transformation to the intermediate valence phase, the phase boundary is very nearly parallel to the temperature axis, implying only a minor entropy change.) Our measurements on samples of $Ce_{1-x}Th_x$ at the critical point show (Fig. 1) a strong symmetrical divergence not unlike that seen in the liquid gas transition, an analogous isomorphic transition in which density is the macroscopic order parameter. However, we find that the critical behavior is essentially describable in a mean field framework.

Lawrence, et. al.,[1] has shown that the reduced resistivity difference $\Delta R/R_0 = (R-R_0)/R_0 \sim \Delta T^{1/3}$, couples linearly to the order parameter, the reduced volume $v = (V-V_0)/V_0$; so in the light of a Landau free energy expansion, the entropy term, TS, can be

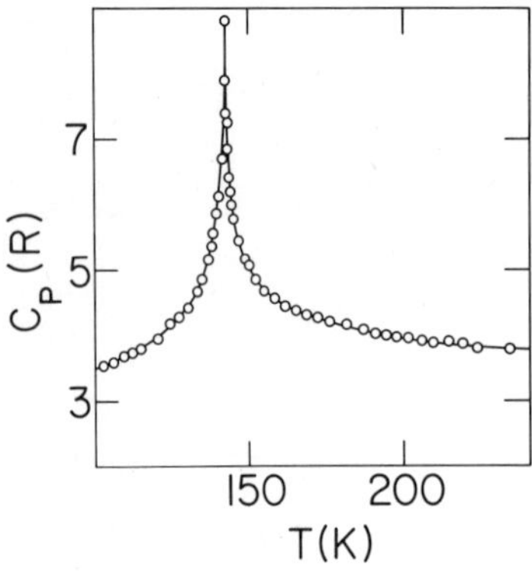

Fig. 1. Heat capacity, C_p, of $Ce_{1-x}Th_x$ near the critical concentration. Note the symmetric divergence and the large high temperature value of C_p.

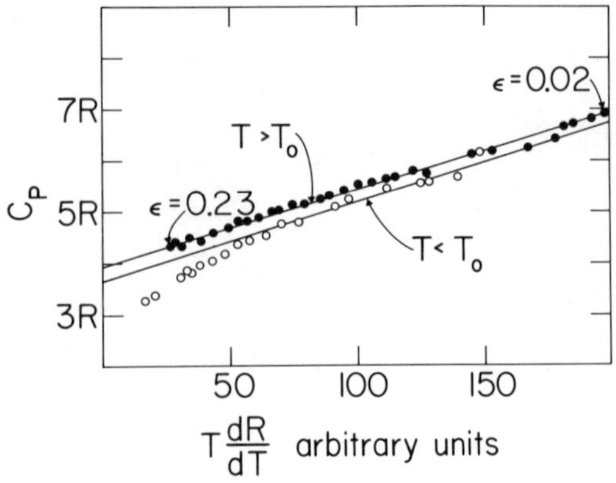

Fig. 2. Heat capacity, C_p, plotted against T dR/dT near the critical concentration, demonstrating the linear coupling between entropy and the order parameter.

written as $\Delta T \Delta R$ (apart from a normalization factor) implying that the critical part of the heat capacity should be proportional to T dR/dT. In Fig. 2 we see that C vs. T dR/dT fits a straight line both above and below T_0 with the same slope but with different intercepts, reflecting the changing background as the material transforms from γ, with θ_D = 135 K, to α, with θ_D = 179 K. The poorer fit below T_0 is attributed to the higher Debye temperature of the low temperature phase. The direct correspondence between C and T dR/dT demonstrates at once the linear coupling between entropy and the order parameter, and, solving the equation of state for the temperature dependence of dR/dT, that the heat capacity diverges with the mean field exponent $\varepsilon = (1-1/\delta) = 2/3$. The strong as opposed to weak exponent is to be expected since C_p represents the thermodynamic derivative with respect to the field orthogonal to the coexistence curve. The ε exponent results since C_p is measured along a pathway essentially normal to the coexistence curve.[2]

The entropy under the peak, less a smoothed lattice background, enables us to determine the slope of the phase boundary in T-P space, (a_{1p} in the notation of ref. 4), at the critical point without doing a pressure measurement. Recalling that $C_p - C_v = \alpha^2 T/K$, where α and K are the isobaric thermal expansion and isothermal compressibility, and from the free energy written in terms of the reduced volume change,[3] one can show that the divergent part of C_p is $(3a_4{}^{1/3}\ a_{1p}\ (T-T_c)^{2/3})^{-1}$.[4] Subtracting the lattice background and integrating C_p/T yields 24.3 deg/kbar for a_{1p}. This slope has been measured independently by Jayaraman[5] and Kutsar[6] who obtain the values 26.5 and 23.7 deg/kbar respectively.

We note that the entropy released at the transition, approximately 0.5R, is a large fraction of the free spin entropy $R\ell n(2J+1)$, ($J = 5/2$ for Ce^{3+}). Some of the remaining entropy appears in the enhanced electronic specific heat (γ=12.8 mJ/mol-deg for α-Ce)[7] reflecting the pinning of the 4-f level at the Fermi energy in the α state. The rest shows up as an excess heat capacity at high temperature. Figure 1 shows that even well above the transition, the heat capacity is 0.75 R in excess of the Debye lattice and only very slowly decreasing. Apparently, the delocalization of the 4-f electron occurs over a very broad temperature range, albeit with varying rates. If we set the remaining entropy equal to the integral of the extrapolated non-divergent heat capacity in excess of the Debye 3R, we find a saturation temperature on the order of 1000 K. Alternatively, this excess heat capacity may be indicative of a compensation mechanism at work which compensates the magnetic moment even at very high temperatures, as in the Kondo effect. At low enough temperatures the reduced spin entropy loses out to the entropy to be gained by delocalization, and the transition to an intermediate valence state occurs more or less precipitously.

REFERENCES

*This work was supported by the Office of Army Research and the Office of Naval Research.

1. J. M. Lawrence, M. C. Croft, J. M. Markovics, and R. D. Parks, A.I.P. Conf. Proc. 29, 429 (1976).
2. R. B. Griffiths and J. C. Wheeler, Phys. Rev. A 2:3 1047 (1970).
3. J. M. Lawrence, M. C. Croft and R. D. Parks, these proceedings.
4. M. C. Croft, private communication.
5. A. Jayaraman, Phys. Rev. 137A, 179 (1965).
6. A. R. Kutsar, Technical Preprint, Institute of Metal Physics of the Bazdyn Central Research Institute of Ferrous Mettalurgy, Moscow, U.S.S.R. (1974).
7. D. C. Koskimaki and K. A. Gschneidner, Jr., Phys. Rev. B 10:5, 2055 (1974).

ELASTIC MEASUREMENTS NEAR A VALENCE TRANSITION*

M. C. Croft and R. D. Parks

Physics and Astronomy Department, University of Rochester, Rochester, New York 14627

The valence transitions in the Ce and SmS based systems involve gigantic discontinuous volume changes with no change in lattice symmetry. An appropriate order parameter for such a transition is the reduced volume $(V-V_c)/V_c$ (V_c = critical volume) and the appropriate instability in the elastic constant matrix is the vanishing of the bulk modulus[1] (divergence of the isothermal compressibility k_T). Velocity of sound measurements across first order valence transitions in $Sm_{1-x}Y_xS$ by Penney, Melcher, Holtzberg and Guntherodt[2] and in Ce by Vornov, et al.[3] and Rosen[4] reflect the insipient critical divergence in k_T. The large slope of volume versus pressure curves for the high temperature continuous transition in Ce[5] is directly related to the k_T anomaly. The purpose of this work is to examine the form of the divergence of k_T at a valence transition.

We have measured the temperature dependece of Young's modulus (Y) near the critical concentration for the γ-α transition in the $Ce_{1-x}Th_x$ alloy system. The Young's modulus must soften to zero at the critical point along with the bulk modulus and is related to the compressibility via $Y^{-1} = k_T/9 + 1/3\mu$ where μ is the sphere modulus. A sample of dimensions $\ell \simeq 0.24$ cm, $t = 0.089$ cm, $b = 0.508$ cm mounted as a cantelever spring has the spring constant $Yb/4(t/\ell)^3 \simeq Y(4.23)(10)^8$/cm. A portion of the free end of the sample formed one plate of a capacitor. The sample was flexed by a dc voltage of up to 500 V across the capacitance gap and the response was monitored with the off balance signal from a capacitance bridge. The apparatus was calibrated against a silver sample mounted in the same geometry. The advantages of this method lie in the large geometrical amplification factor multiplying Y

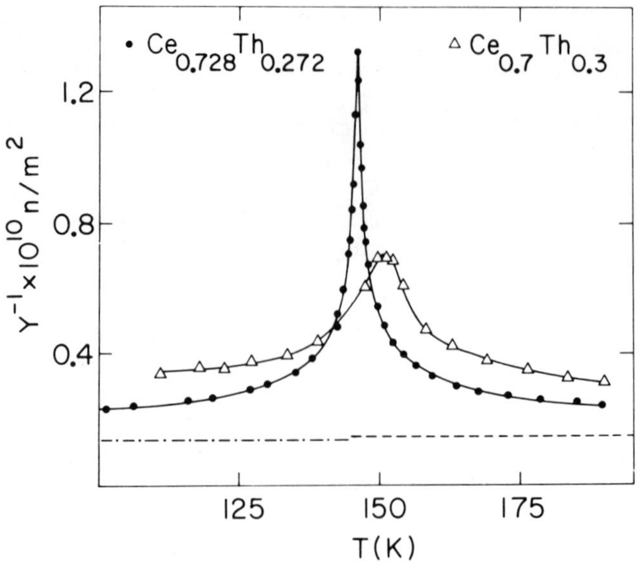

Fig. 1. The Young's modulus Y(T) for two $Ce_{1-x}Th_x$ samples with continuous transitions.

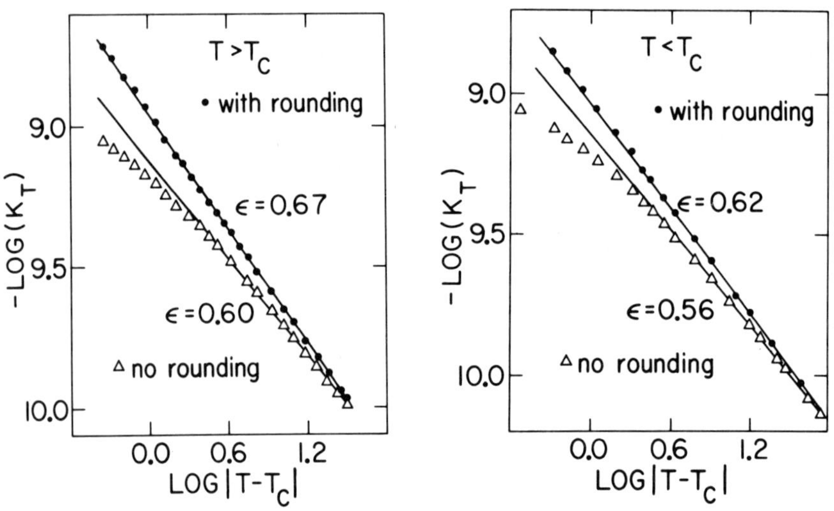

Fig.2(a) Power law dependence of K_T, $T>T_c$ for $Ce_{.728}Th_{.272}$.

Fig.2(b) Power law dependence of K_T, $T<T_c$ for $Ce_{.728}Th_{.272}$.

in the spring constant combined with extreme precision available from a capacitance bridge measurement. The principal disadvantages of this technique are: inter-sample comparisons of the absolute value of Y are difficult due to the crude geometrical estimates and a small capacitance gap is manditory both to exert force on and to measure the response of the sample.

The temperature dependence of Y^{-1} for two transcritical samples of $Ce_{1-x}Th_x$ are shown in Fig. 1. The dramatic sharpening of the response, and narrowing of the temperature region of the activity, typical of the $Ce_{1-x}Th_x$ system (or of any critical system along this path of approach), are evident. Using the value of the ratio of the ratio of k_T to μ at room temperature and pressure for pure Ce[3,4] one finds the sphere modulus contribution to Y^{-1} for $T>T_c$ represented by the dashed line in Fig. 1. The resulting power law dependence of $k_T \sim |T-T_c|^{-\varepsilon}$, after subtraction of this effective background, is shown in Fig. 2(a), for the x = 0.272 sample. The triangles illustrate the effect of the intrinsic rounding in a sample beyond the critical concentration. The closed circles represent the fit obtained by optimizing the rounding parameter. Both with and without rounding the exponents are in good agreement with the mean field exponent of $1-1/\delta$ with $\delta=3$, when compared with the analogous non-mean field exponent for the liquid gas transition of $1-1/\delta$ with $\delta=4.2$. Since the sphere modulus in the α state will differ from that in the γ state, the data below T_c was fit using the rounding determined from above T_c and optimizing the sphere modulus. The resulting sphere modulus is represented by the dot dashed line in Fig.1(a). The exponents below T_c with and without rounding, represented in Fig. 2(b), are again consistent with mean field theory. The compressibility obtained using these sheer moduli increases by a factor of 13 between room temperature and the peak.

Close to the transition a dramatic analystic (time dependent and recoverable) response was observed. The amplitude of this response became larger and its time constant seemed to become smaller as the transition temperature was approached. The latter of these tendencies is inconsistent with critical slowing down. Two possible explanations of this behavior have occurred to the authors. The growth or dimunition of one phase about inhomogeneities in the bulk phase of the order could be involved. A more interesting possibility is that of a diffusive motion whose activation energy becomes dramatically smaller in a softening lattice. Two candidates for the diffusing entity are interstitials, principally hydrogen, or lattice vacancies.

REFERENCES

*This work was supported by the U. S. Army Research Office and the U. S. Office of Naval Research.

1. R. A. Cowley, Phys. Rev. B 14, 1758 (1976).
2. T. Penney, R. L. Melcher, F. Holtzberg and G. Guntherodt, AIP, Conf. Proc. 29, 392 (1976).
3. F. F. Voronov, L. F. Vereshchagin, and V. A. Goncharova, Sov. Phys. 135, 1280 (1960).
4. M. Rosen, Phys. Rev. 181, 932 (1969).
5. R. I. Beecroft and C. A. Swenson, J. Phys. Chem. Solids 15, 234 (1960).

THERMOELECTRIC POWER OF CERIUM-BASED SYSTEMS*

I. Zorić and R. D. Parks

Physics and Astronomy Department, University of Rochester, Rochester, New York 14627

A study of the valence instability in the $Ce_{1-x}Th_x$ system was carried out recently via resistance and volume measurements.[1] In this work we report measurements of the thermoelectric power (TEP) in the $Ce_{1-x}Th_x$ and $Ce_{0.9-x}Th_{0.1}La_x$ systems. Figure 1 shows the characteristic shape of TEP curves in the region of a continuous transition (x=0.290), at the second order instability (x=0.269) and across a first order instability (x=0.175). In all cases the TEP is large and positive with a cut-off occurring at the valence instability temperature. One can see the signature of critical rounding on the 0.269 curve, but it is not possible to separate the critical contribution from the normal background which is strongly temperature dependent. In addition, below the valence instability temperature, i.e., in the mixed valence phase, the TEP is small $\simeq$7-8 μV/deg and weakly temperature dependent. To get more insight into the TEP of the γ-phase we have tried to extend the γ-phase region by alloying CeTh with La. It is known that the addition of La lowers the instability temperature ∿9°K/at % La and that above 9 at % La the transition becomes continuous.[2] Figure 2 shows the TEP for a CeThLa sample. One can see that the TEP keeps increasing with the sharp-cut off occurring again at the instability temperature.

DISCUSSION

There is no satisfactory model for transport properties of mixed valence systems. Recent neutron scattering studies[3] on the $Ce_{1-x}Th_x$ system have shown that the γ as well as the α phase is of mixed valent nature. The f-electron spectrum is reminiscent of the paramagnon type spectra. One of the differences between

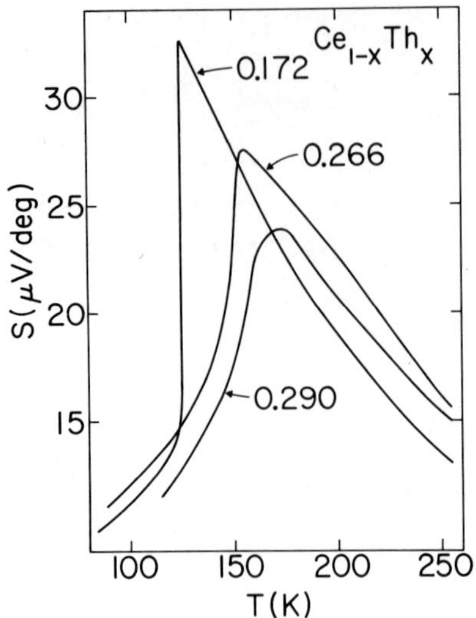

Fig. 1. Temperature dependence of the TEP of $Ce_{1-x}Th_x$.

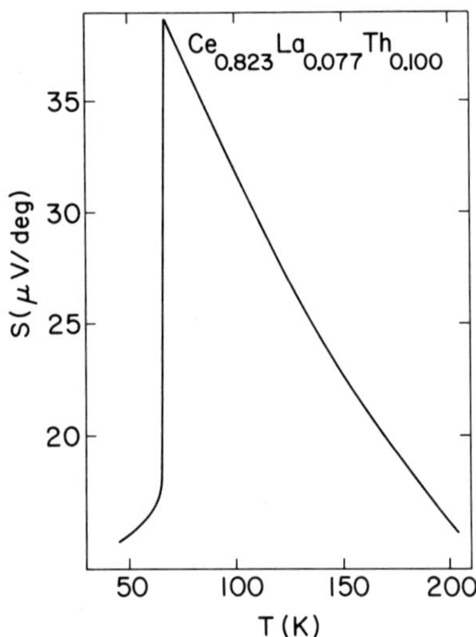

Fig. 2. TEP of $Ce_{0.823}L_{0.077}Th_{0.100}$ as a function of temperature.

the γ and α phase seems to be the characteristic spin fluctuation energy which is 20 meV in the γ phase and 70 meV in the α phase. In addition, in the γ phase the f-electron spectrum is temperature independent and the value of Im $\chi(k\omega)$ at the spin fluctuation frequency exceeds the corresponding value in the α phase by $\simeq$ 3 times. The spectrum in the α phase seems to show a weak temperature dependence. At the present time there is no information about the conduction electron spectrum. A standard expression for TEP, based on the assumption that transport electrons are scattered dominantly by external scatterers (e.g., localized spins, spin fluctuations etc.) and that the scattering is elastic, is given by

$$Q \propto \frac{\partial \ln(n\tau)}{\partial \varepsilon}\Big|_{E_f}$$

i.e., the energy dependence of the density of states $n(\varepsilon)$ and the relaxation time $\tau(\varepsilon)$ at the Fermi surface is reflected in TEP. In the case of $Ce_{1-x}Th_x$ at room temperature, the 4f level resides below the Fermi energy and as one approaches the valence instability, the 4f level moves closer to E_f and finally it becomes pinned to E_f at the instability temperature, yielding a strong energy dependence of the renormalized density of states and the relaxation time. Another possible speculation concerns the f electron spectrum. If the f electron spectrum is indeed of the paramagnon type then a spin fluctuation drag contribution[4] to the TEP will cause a strong temperature dependence before the valence instability takes over. This will remain a very speculative idea until all the details of the conduction electron spectrum in the presence of the $f^{n-1}d$ configuration are determined.

REFERENCES

*Work supported in part by the U. S. Office of Naval Research and the U. S. Army Research Office.

1. J. M. Lawrence, M. C. Croft and R. D. Parks, Phys. Rev. Lett. 35, 289 (1975).
2. M. Manheimer and R. D. Parks (these proceedings).
3. R. Birgeneau and S. Shapiro (these proceedings).
4. A. B. Kaiser, A.I.P. Conf. Proc. 29, 364 (1976).

X-RAY PHOTOEMISSION FROM γ- AND α-CERIUM AND FROM γ- AND α-$Ce_{0.731}Th_{0.269}$

R. A. Pollak, S. P. Kowalczyk, and R. W. Johnson

IBM Thomas J. Watson Research Center

Yorktown Heights, New York 10598

We report X-ray photoemission valence band and Ce 4d core level spectra for the α- and γ-phases of cerium metal and $Ce_{0.731}Th_{.269}$ alloy. X-ray photoemission spectroscopy (XPS or ESCA) is being used to study the electronic structure of cerium metal in an effort to better understand the complex nature of the Ce 4f electronic states and how it changes between the magnetic, fcc γ-Ce phase at room temperature and the 14% contracted, nonmagnetic, fcc α-Ce phase at low temperatures. Many quite different and incompatible models exist which attempt to explain the differences in physical properties of α- and γ-Ce.[1] From the variety of theories that presently exist it is seen that the basic question of whether an f electron is promoted to an (sd) level in the α-phase is as yet controversial.

Ultraviolet[2] ($h\nu$ = 5.7 and 11.1 eV) and X-ray[3,4] ($h\nu$ = 1486.6 eV) photoemission valence band spectra of Ce at room temperature have been previously reported. At low photon energies the Ce 4f photoelectric cross section is negligible with respect to that of the Ce $5d^16s^2$ electrons. At X-ray energies the Ce $4f^1$ cross section is approximately equal to the sum of cross sections for the three other valence electrons and it is difficult to assign the spectral features of valence bands originating specifically from the 4f level. Baer and Busch tentatively assign structure at 0.9 eV below the Fermi level to the Ce 4f level.[3]

Lawrence, Croft, and Parks[5] using resistivity and density measurements studied the γ- to α-phase transition in $Ce_{1-x}Th_x$ which is first order up to x=0.264 where it becomes second order. Addition of Th to Ce stabilizes the fcc γ-phase with respect to the magnetic double-hcp β-phase which is formed in pure cerium after cold working and temperature cycling.[6] The spectra presented here were obtained with a HP5950A ESCA photoelecton spectrometer which has an Al Kα ($h\nu$ = 1486.6 eV) monochromated X-ray source. The base pressure of the spectrometer and sample preparation chamber is 10^{-11} torr. The cerium metal (99.9%) sample was cut from an as cast ingot. The $Ce_{0.731}Th_{.269}$ sample was cut from an ingot obtained from Ron Parks[5].

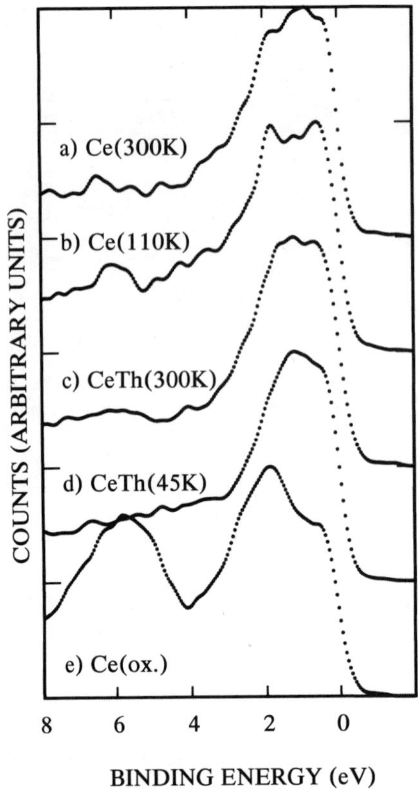

Fig. 1. Valence bands of (a) Ce at 300K, (b) Ce at 110K after spending several hours at 45K, (c) Ce.731Th.269 at 300K, (d) Ce.731Th.269 at 45K, (e) partially oxidized Ce at 300K.

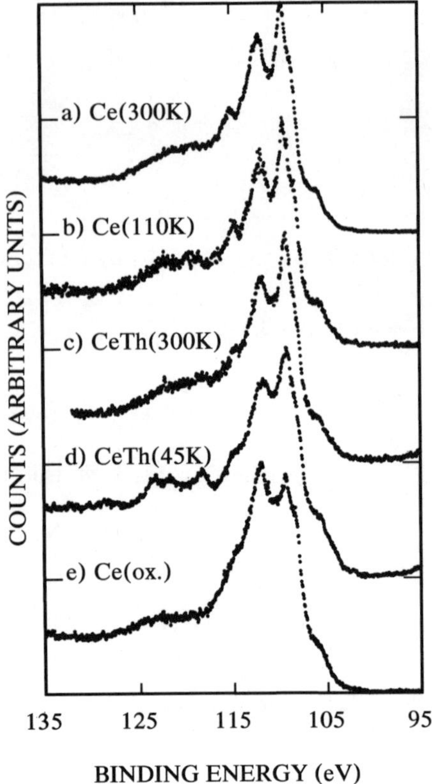

Fig. 2. Ce 4d core levels of samples in Fig. 1.

The samples were scraped with a tungsten carbide milling tool *in situ* periodically to maintain a clean surface. This sample preparation undoubltedly produced some β-phase material.

The valence bands (2.3 eV FWHM) shown in Fig. 1 which have been three point smoothed ten times show only small differences between the γ-phases (Fig. 1a and 1c) and α-phases (Fig. 1b and 1d). A small shoulder at 4 eV binding energy appears in the α-phase for both Ce and $Ce_{0.731}Th_{.269}$. The partially oxidized Ce metal spectrum (Fig. 1e) agrees with the UPS spectra in Ref. 2. Notice that no peak appears at 4 eV for the oxidized sample.

The Ce 4d core level spectra are shown in Fig. 2. The structure between 105 and 115 eV can be assigned as multiplet structure resulting from the electrostatic exchange coupling for a Ce $(4d^9 4f^1)$ final hole state.[7] The origin of the shoulder in the Ce 4d spectra between 115 and 125 eV is not presently well understood. It may originate from the same (4d;nf) ^1P excited final states observed in Ba and La photo-absorption experiments.[8] Wendin points out that maximum complications exist for

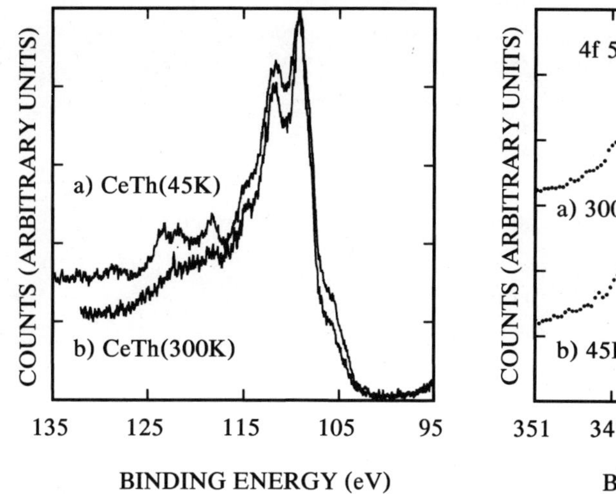

Fig. 3. Comparison of Ce 4d level in Ce.731Th.269 at (a) 45K and (b)300K.

Fig. 4. Th 4f core levels in Ce.731Th.269 at (a) 300K and (b) 45K.

the $4d^{10}$ shell where collective effects, relaxation, relativistic effects, and strong configuration mixing may all be important.[8] On cooling to the α-phase new discrete structure appears in this shoulder region. Partial oxidation of Ce at room temperature decreases the intensity of this shoulder (Fig. 2e). A detailed comparison of the γ- and α-$Ce_{0.731}Th_{.269}$ Ce 4d spectra is shown in Fig. 3.

The Th 4f core levels taken at room temperature and 45°K are shown in Fig. 4. An unexpected splitting of 2.1 eV is observed which may signal the presence of inequivalent thorium sites, collective excitations in the final state or possibly thorium mixed valence.

We thank Ron Parks for giving us the $Ce_{0.731}Th_{0.269}$ sample.

References

1. Börje Johansson, Phil. Mag. **30**, 469 (1974) and refs. therein.
2. C. R. Helms and W. E. Spicer, Appl. Phys. Lett. **21**, 237 (1972).
3. Y. Baer and G. Busch, J. Elec. Spect. **5**, 611 (1974).
4. S. P. Kowalczyk, Ph. D. Thesis, unpublished.
5. J. M. Lawrence, M. C. Croft, and R. D. Parks, Phys. Rev. Lett. **35**, 289 (1976).
6. K. A. Gschneider, Jr., R. O. Elliott, and R. R. McDonald, J. Phys. Chem. Solids **23**, 555, 1191, 1201 (1962).
7. A. J. Signorelli and R. G. Hayes, Phys. Rev. **B8**, 81 (1973).
8. G. Wendin, in *Vacuum Ultraviolet Radiation Physics*, (Pergamon Vieweg, Braunschweig, 1974) p.225.

MAGNETIC BEHAVIOR OF Ce IN $(Ce_x(La_{1-z}Y_z)_{1-x})Al_2$: DYNAMICAL SUSCEPTIBILITY AND ITS RELATIONSHIP TO TRANSPORT PROPERTIES

F. Steglich[+], M. Loewenhaupt[x]

[+]II. Physikalisches Institut der Universität zu Köln and SFB 125, D 5000 Köln 41
[x]Institut für Festkörperforschung der KFA Jülich and SFB 125, D 5170 Jülich

ABSTRACT

The dynamical susceptibility of single crystals of the isostructural metallic systems $(Ce_xLa_{1-x})Al_2$ and $(Ce_xY_{1-x})Al_2$, $0.07 \leq x \leq 1$, has been measured by neutron scattering between 2 and 300 K. The energy spectra are analyzed in terms of quasielastic and inelastic lines of finite widths. The results are compared with resistivity and thermopower data on dilute $(Ce_x(La_{1-z}Y_z)_{1-x})Al_2$ alloys ($x \leq 0.15$, $z \leq 1$).

By neutron scattering experiments on the dilute Kondo alloy $(C\underline{e}La)Al_2$ [1] and the "Kondo-compound" $CeAl_2$ [2] the authors recently demonstrated that the $j = 5/2$ state of the Ce^{3+} ions is crystal field (CF) split with $\Delta = 9$ meV and that the spin relaxation rates of these ions exhibit an unusual temperature dependence. According to the thermal population of the excited Γ_8 quartet and the Γ_7 doublet ground state the relaxation rate which is $\sim 10^{-13}$ cps at 300 K shows a distinct drop below 100 K. At low temperatures a Korringa behavior for the quasielastic linewidth has been observed in the dilute $(C\underline{e}La)Al_2$ alloy - in accordance with the low Kondo temperature $T_K \simeq 0.4$K determined from a variety of physical quantities. However in the $CeAl_2$ compound no diffuse quasielastic scattering could be detected below 4 K. This is consistent with the occurence of a phase transition in this compound around 4 K in several of its macroscopic properties. Above 4 K however quasielastic scattering is observed exhibiting a linewidth which is larger by an order of magnitude when compared to the dilute $(C\underline{e}La)Al_2$. This increase of linewidth which is observed when going from the dilute $(C\underline{e}La)Al_2$ to the $CeAl_2$

compound could be caused by interactions between Ce ions and/ or by an increasing interaction between the Ce ions and the conduction electrons. The latter effect is expected since the lattice parameter (at room temperature) of $CeAl_2$ is smaller by ~1% than that of dilute (CeLa)Al_2. An appropriate chance to discriminate between the two effects is provided by the dilution of $CeAl_2$ with YAl_2, which is accompanied by a further decrease of the mean lattice parameter. In this paper we present a comparison between the dynamical susceptibilities of some $(Ce_xLa_{1-x})Al_2$, $CeAl_2$, and $(Ce_xY_{1-x})Al_2$ single crystals. We also relate these results to those of recent transport measurements.

Figs. 1 to 3 show typical TOF-spectra (incident neutron energy 3.53 meV) of the magnetic scattering from the Ce ions

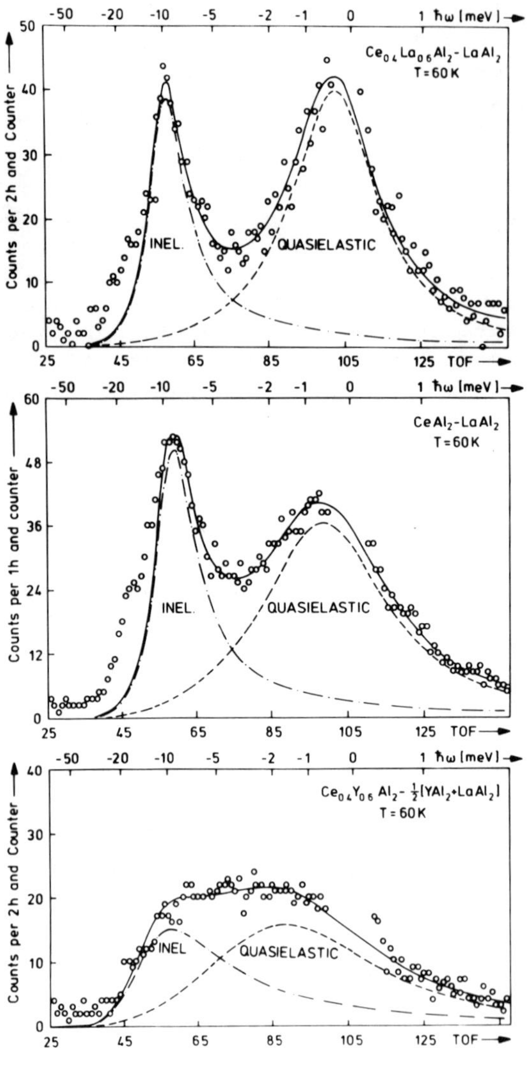

Figs. 1 to 3
Magnetic TOF-Spectra at T = 60 K (average over 12 counters, $q_{el} \simeq 0.7$ Å^{-1})

Fig. 1: $Ce_{0.4}La_{0.6}Al_2$

Fig. 2: $CeAl_2$

Fig. 3: $Ce_{0.4}Y_{0.6}Al_2$

at T = 60 K. Besides a Bose factor these spectra immediately display the properties of the dynamical susceptibility of the magnetic moments. For $Ce_{0.4}La_{0.6}Al_2$ and $CeAl_2$ the inelastic CF transition $\Gamma^8 \rightarrow \Gamma^7$ at $\hbar\omega = -9$ meV is well separated from the quasielastic scattering which has contributions of nearly equal intensity from the Γ^7 and Γ^8 level at 60 K. For $Ce_{0.4}Y_{0.6}Al_2$ however the linewidths are considerably larger and no sharp CF transition can be observed. We analyse the spectra as previously explained[1] in terms of quasielastic and inelastic lines and assume the same CF scheme for all three systems. We then obtain for the halfwidths $\Gamma/2$ the values 3, 3.5, 8 meV (inelastic) and 1, 1.6, 3 meV (quasielastic) for $Ce_{0.4}La_{0.6}Al_2$, $CeAl_2$ and $Ce_{0.4}Y_{0.6}Al_2$, respectively.

Comparing the spectra of all the single crystals we have investigated a systematic increase of the linewidths is observed when going from dilute $(Ce_xLa_{1-x})Al_2$ to $CeAl_2$ and then to dilute $(Ce_xY_{1-x})Al_2$ (x = 0.07, 0.15, 0.40 and 0.70). This result indicates that possible contributions of Ce-Ce correlations to the quasielastic scattering are of minor importance. The observed increase of the linewidths is therefore caused by an increase of the Ce ion-conduction electron interaction, which is due to a systematic decrease of the mean lattice parameter. We note that these results of the neutron scattering experiments are consistent with conclusions from recent resistivity measurements on dilute $(Ce_x(La_{1-z}Y_z)_{1-x})Al_2$ samples ($x \leq 0.15$)[3]. From the low temperature T^2-law an increase of the Kondo temperature T_K by two orders of magnitude was inferred for $z \rightarrow 1$. In particular from the resistivity data, $T_K \simeq 50$ K was inferred for $(Ce_{0.15}Y_{0.85})Al_2$ and recent thermopower measurements exhibit a large peak at this same temperature[4]. Further evidence for such high T_K's (large residual linewidths) is also found in a strong suppression of the magnetic scattering intensity for the two most dilute $(Y,Ce)Al_2$ crystals at low temperatures.

We gratefully acknowledge critical discussions with Prof. W. Schmatz and Dr. K. Fischer. Thanks should also be given to Dr. W. Just (ILL/Grenoble) for his valuable help during the experiments.

REFERENCES

1. M. Loewenhaupt, F. Steglich: Proc. 2nd Int. Conf. on Crystal Field Effects in Metals and Alloys, Zürich 1976, in press
2. M. Loewenhaupt, F. Steglich: Proc. Int. Conf. on Magnetism, Amsterdam 1976, in press
3. F. Steglich, W. Franz, W. Seuken, M. Loewenhaupt: Proc. Int. Conf. on Magnetism, Amsterdam 1976, in press
4. W. Franz: private communication

SPECIFIC HEAT AND MAGNETIZATION MEASUREMENTS UNDER PRESSURE ON $CeAl_2$ AND $CeAl_3$ COMPOUNDS

A. Berton, J. Chaussy, G. Chouteau[§], B. Cornut, J. Peyrard[§*], and R. Tournier

Centre de Recherches sur les Très Basses Températures, CNRS, BP 166 Centre de Tri, 38042 Grenoble-Cedex, France

Both $CeAl_2$ and $CeAl_3$ present Kondo behavior at high temperatures, as seen by resistivity measurements.[1] Their properties are also modified by applying pressure, primarily because the exchange between conduction and 4f-electrons is increased. This exchange is responsible for the Kondo behavior. In addition, pressure increases (for $CeAl_3$) or decreases (for $CeAl_2$) the overall crystalline field splitting.

Up to now, the low temperature state of both compounds is not understood. Namely no peak has been observed in the susceptibility and resistivity of $CeAl_3$[2] whereas a large specific heat anomaly occurs around 5 K.[3] Concerning $CeAl_2$ no magnetic order has been observed by neutron experiments. But in both cases, the entropy at 300 K is greater than the maximum crystalline field entropy. Thus one can say that the twofold degeneracy of the ground state of Cerium is lowered at low temperature. Consequently we may ask the following questions. What is the origin of this extra entropy, magnetic order, Kondo effect or a competition between them ? What is the nature of magnetic ordering if it exists ?

Figure 1 shows the magnetization of $CeAl_3$ at 4.2 K. We have also plotted the specific heat between 1.3 K and 12 K of $CeAl_3$ (Figure 2) and $CeAl_2$ (Figure 3) at 0 and 7 kbars. The quasi-hydrostatic pressure is obtained by putting the samples in small clamps. We use a new vibrating coil system for magnetization measurements. The accuracy of all these experiments is about 0.5 %. More experimental details will be published elsewhere.

For $CeAl_3$ we observe a remanent magnetization of about 0.05 μ_B in one of our samples (nbr. 2), whereas no such behavior has been observed on sample 1. The specific heat of sample 2 is plotted in figure 2. We observe two peaks at 2.6 K and 6.0 K whereas Mahoney et al. have obtained a third peak at about 4 K, which can be attri-

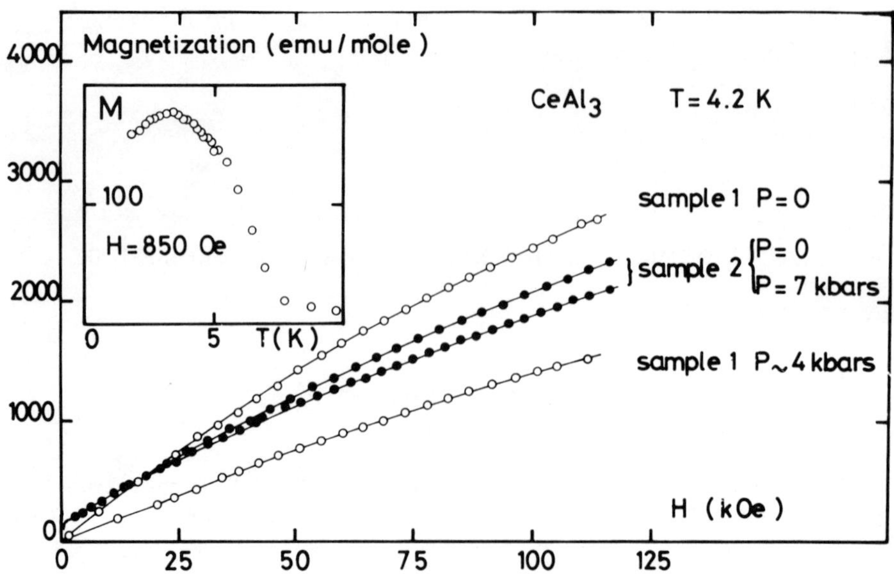

Fig.1 : Magnetization of two samples of Ce Al_3 versus field. The inset shows the magnetization in emu/mole of sample 2 at 7 kbar ,in 850 Oe.

buted to a small fraction of $CeAl_2$ present in their sample. The different behavior of our $CeAl_3$ samples which do not contain $CeAl_2$ shows the sensitivity of the magnetic properties to the stoechiometry or defects as in a great number of rare earth compounds and alloys.

The remanent magnetization is a proof of the existence of magnetic order. This order disappear above 7 K (inset of figure 1). Thus we attribute the 6 K peak in the specific heat to the occurence of this order.

However, in the ordered region, both the susceptibility, χ, and the linear specific heat coefficient γ remain very large. The existence of a large linear specific heat, the value of the χ/γ ratio and the T^2 law of the resistivity[4] are a proof of Kondo behavior. Thus both magnetic order and Kondo effect are simultaneously present in the $CeAl_3$ compound. But the entropy associated with magnetic ordering is small compared to that associated with the Kondo effect. This entropy represents a weak magnetism coexisting with the Kondo condensation. The remanent magnetization could be the uncompensated moments of antiferromagnetic domains, so the type of ordering could be in fact a weak antiferromagnet.

The pressure of 7 kbar which decreases the γ value and the χ value indicates an increase of the Kondo temperature. The same feature has been observed on the high temperature Kondo temperature by resistivity measurements.

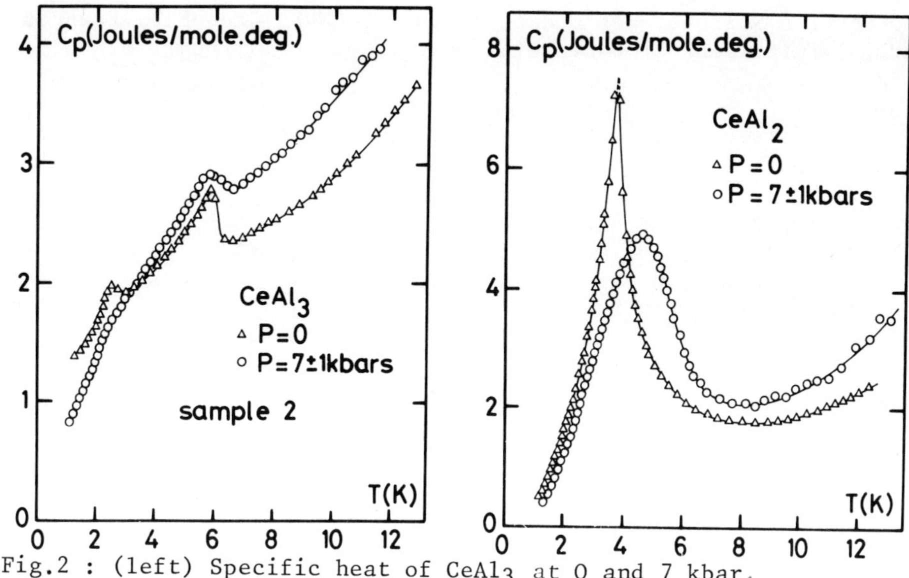

Fig.2 : (left) Specific heat of $CeAl_3$ at 0 and 7 kbar.

Fig.3 : (right) Specific heat of $CeAl_2$ at 0 and 7 kbar.

Concerning $CeAl_2$ we observe the same fondamental disparity as in $CeAl_3$: We find (Fig. 3) a specific heat anomaly, the peak of which is at 3.8 K whereas other authors find it at 3.36 K or 3.4 K. Moreover the amplitude of the peak we observe is much greater than that found by Walker et al.[5] and Hill et al.[6]

From linear extrapolation of the Curie temperature of the ferromagnetic $Pr_xCe_{1-x}Al_2$ alloys (which crystallize all in the same $MgCu_2$ structure) we expect a ferromagnetic 5 K temperature for $CeAl_2$. But no remanent magnetization occurs even at 0.05 K. The rapid increase of the magnetization at 40 kOe indicates, however, a field induced ferromagnetism. The effect of pressure on $CeAl_2$ is very drastic : The temperature of the peak is shifted from 3.8 K to 4.6 K between 0 and 7 kbar. Moreover the width of the transition increases and the amplitude of the peak decreases (figure 3). At the same time the critical magnetic field increases, the width of the transition increases and the saturation magnetization seems to decrease.[7]

As a preliminary interpretation we suggest that the specific heat anomaly of $CeAl_2$ is related to the Kondo effect. Pressure reinforces the Kondo character of Cerium thus the Kondo transition appears at higher temperatures. The magnetization behavior characterises an induced ferromagnetic transition like the Wohlfarth's one (due to a minimum of the density of states at the Fermi level). When the ferromagnetism occurs, the moment of Cerium takes a value between 0 and the full ionic value depending on the competition between the Kondo temperature and the exchange interactions.

§and Service National des Champs Intenses, CNRS, Grenoble, France.
*Partial fulfilment of Doctorat d'Etat thesis of J. Peyrard.

References

1 - M. Nicolas-Francillon, A. Percheron, J.C. Achard, O. Gorochov, B. Cornut, D. Jerome and B. Coqblin, Solid State Com. 11, 845 (1972)
A. Percheron, J.C. Achard, O. Gorochov, B. Cornut, D. Jerome and B. Coqblin, Solid State Com. 12, 1289 (1973).
2 - A.S. Edelstein, C.J. Tranchita, O.D. Mc Masters and K.A. Gschneidner, Solid State Com. 15, 81 (1974).
3 - J.V. Mahoney, V.U.S. Rao, W.E. Wallace, R.S. Craig, and N.G. Nereson, Phys. Rev. B9, 154 (1974).
4 - K. Andres, J.E. Graebner, and H.R. Ott, Phys. Rev. Lett. 35, 1779 (1975).
5 - E. Walker, H.G. Purwins, M. Landolt, and F. Hullinger, J. Less Comm. Metals 33, 203 (1973).
6 - R.W. Hill, J.M. Machado da Silva, Phys. Lett. 30A, 13 (1969).
7 - B. Barbara, H. Bartholin, D. Florence, M.F. Rossignol, and E. Walker, to be published in the ICM 76 Proceedings.

WITHER GOEST THOU $CeAl_2$?

M. Croft, I. Zorić, J. Markovics and R. Parks*

Physics and Astronomy Department, University of Rochester, Rochester, New York 14627

Germane to this conference is the theoretical and experimental question of how a periodic array of rare earth atoms can pass from a local moment (magnetically ordered), to a "Kondo" nonmagnetic, to a mixed valent nonmagnetic ground state. At present it is unclear whether the ground state of the intermetallic compound $CeAl_2$ falls in the ordered magnetic or the "Kondo" nonmagnetic regime. At about 3.9° K there are a set of anomalies in thermal expansion, specific heat and susceptibility[1] which appear to be indicative a second order phase transition. Cornut,[2] Tournier and Holtzberg,[3] and Loewenhaupt and Steglich[4] have discussed Kondo effects in this compound. In this paper we will discuss the measurements of thermal expansion, magnetostriction and specific heat near the transition into the low temperature phase (LTP) of $CeAl_2$.

Figure 1 shows the results of polycrystalline length measurements both longitudinal and transverse to magnetic field near the transition into the LTP. The phase diagram as determined from the peaks in the derivatives of these curves has been presented elsewhere[5] and appears as an inset in the paper of Zorić et al. in these proceedings. In fields above 3.5 k gauss the sample contracts longitudinally and expands transverse to the field direction as the transition temperature is crossed. Below 3.5 k gauss this longitudinal contraction turns into an expansion. The details in this crossover in the sign of the anomaly in longitudinal thermal expansion coefficient (α_ℓ) are shown in Fig. 2a. Magnetostriction measurements above the LTP (4.2°K) and in the LTP (2°K) are shown in Fig. 3. The basic points we wish to make regarding these three figures are (1) in the presence of field the sample distorts, (2) the distortion is strongly temperature dependent,

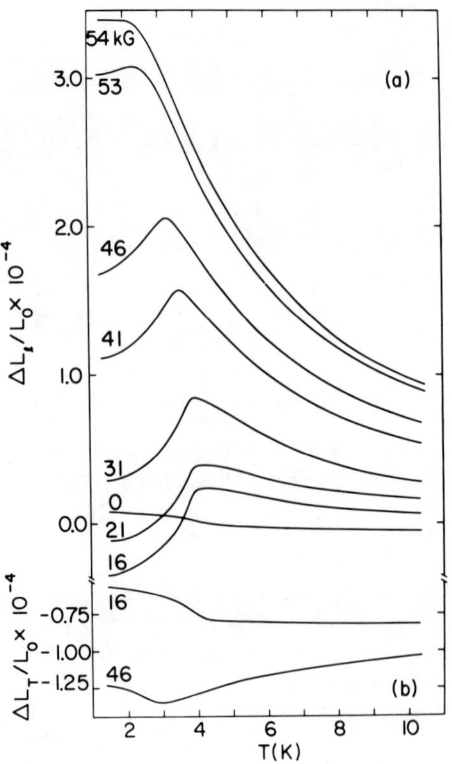

Fig. 1. Sample length for $CeAl_2$ (a) longitudinal to and (b) transverse to magnetic field. All lengths (L) are expressed as $\Delta L/L_0 = (L-L_0)/L_0$ where L_0 is the length in zero field at 4.2°K.

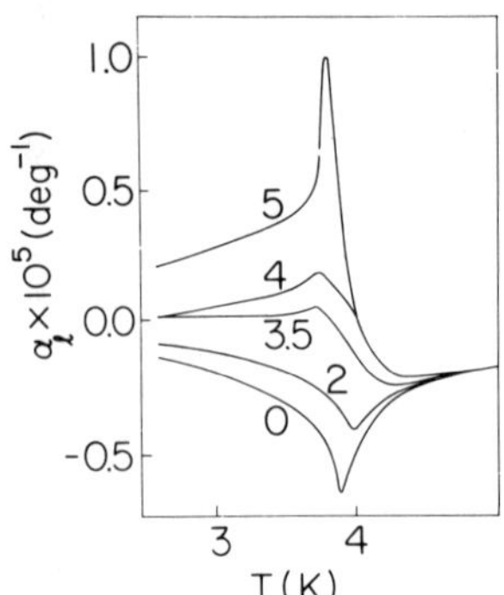

Fig. 2a. Longitudinal thermal expansion coefficient α_ℓ showing low field crossover in the sign of the anomaly.

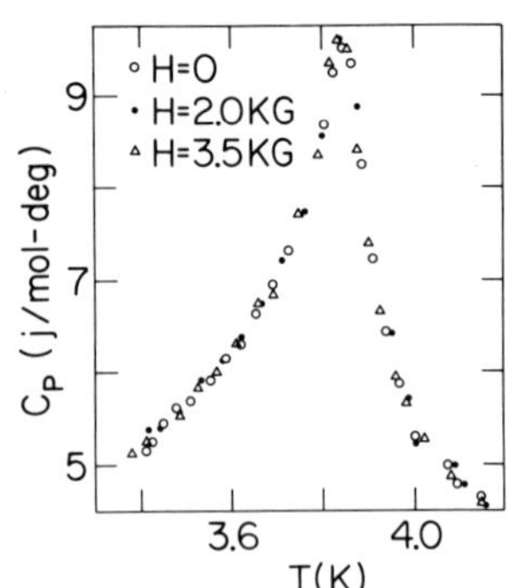

Fig. 2b. Specific heat in the low field crossover region.

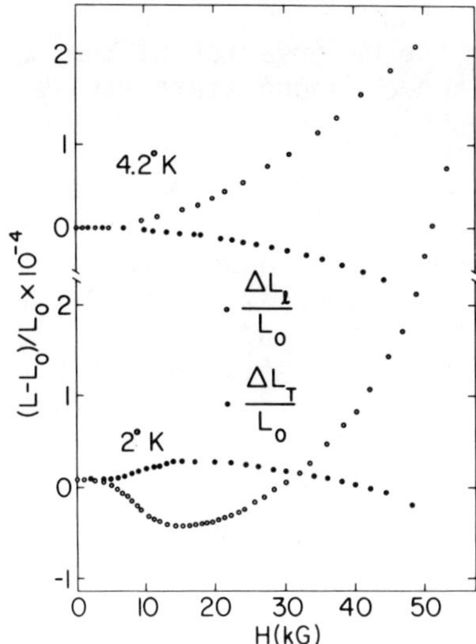

Fig. 3. Magnetostriction of $CeAl_2$ above the LTP (4.2° K) and in the LTP (2° K).

reversing sign at the transition into the LTP in fields above 3.5 k gauss and (3) the field where the crossover in the sign of the α_ℓ anomaly occurs (3.5 k gauss) corresponds roughly to the onset of the nonlinear distortion in the magnetostriction in the LTP.

Assuming that the low temperature length anomalies are indeed indicative of a phase transition, one must consider the possibility that the crossover region of point (3) is caused by the presence of two distinct low temperature phases. If the low and high field phases had different physical origins one would expect their phase boundaries to meet in some non-smooth manner. Accordingly, we have performed careful specific heat measurements (see Fig. 2b) in the crossover region and have found the phase diagram smooth and the specific heats identical in this region. We believe therefore, that only a single phase is present and that the crossover is caused either by domain effects or by continuous modifications in the LTP in field. The increase in zero field transition temperature under pressure as seen by Berton et al.[6] would seem to support the ascribing of the zero field length expansion to poly-domain averaging in a distorting system.

Our results, viz., the sharp phase boundary, high field distortion and apparent domain effects seem to argue strongly for ordering in the LTP. To what extent Kondo compensation is present in and above the LTP and how this compensation participates in the LTP are two pressing questions. Moreover, with the apparent increase of both the Kondo temperature and the ordering temperature[6] under pressure, the observation of the modifications in this participation as the Kondo ground state regime is approached present an exciting possibility.

For a system with long range magnetic order one can speculate that the tendency toward singlet formation might produce either a homogeneous moment reduction or a periodic moment amplitude modulation. On the other hand grossly inhomogeneous moment reduction might destroy long range order and create a mictomagnetic type of short range order.

REFERENCES

*Work supported in part by the U. S. Office of Naval Research and the U. S. Army Research Office.

1. I. Zorić, J. Markovics, L. Kupferberg, M. Croft and R. D. Parks, these proceedings.
2. B. Cornut, Ph.D. Thesis, Grenoble, 1976 (unpublished).
3. R. Tournier and F. Holtzberg, these proceedings.
4. M. Loewenhaupt and F. Steglich, Proc. Int. Conf. on Magnetism, Amsterdam 1976, in press.
5. M. C. Croft, I. Zoric and R. D. Parks, to be published.
6. A. Berton, T. Chaussy, G. Chouteau, B. Cornut, T. Peyard and R. Tournier, these proceedings.

THERMODYNAMIC AND TRANSPORT PROPERTIES OF $CeAl_2$ IN MAGNETIC FIELD*

I. Zorić, J. Markovics, L. Kupferberg, M. Croft and R. D. Parks

Physics and Astronomy Department, University of Rochester, Rochester, New York 14627

Despite the Malthusian growth of interest in $CeAl_2$, its ground state remains enigmatic. We have studied $CeAl_2$ at low temperatures in magnetic field in order to shed some light on the transition from the high temperature crystal electric field (CEF) modified Kondo behavior[1] into the unknown ground state. Our measurements of thermal expansion and heat capacity have determined a phase diagram in the H-T plane consisting of a single low temperature, low field region bounded by a smooth line of second order transitions which approaches the T axis perpendicularly (see inset in fig. 4). We have studied some thermodynamic, magnetic and transport properties above and across the phase boundary in order to get more information about the ground state properties of $CeAl_2$.

As can be seen in Fig. 1, the heat capacity in zero field is characterized by a huge anomaly which appears to diverge at 3.85°K with a small positive exponent characteristic of a weak divergence in magnetic systems.[2] The shape of the divergence and the transition temperature do not appear to be affected by fields up to 15 kG. The entropy under the C_p curve amounts to 0.95 ln 2 by 10°K and is still increasing.

Magnetocaloric effect measurements were made on both sides of the phase boundary. Above the transition temperature a small positive magnetocaloric response is observed, characteristic of paramagnetic systems. Below the transition we observe an initial warming of the sample which by 2 kG becomes negative (cooling) and continues to increase in magnitude as H^2.

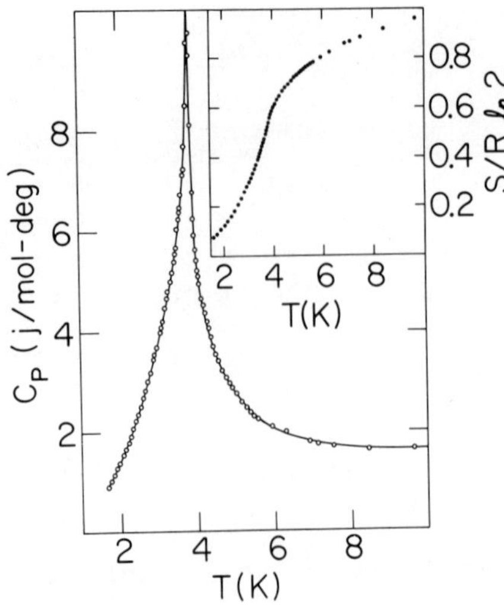

Fig. 1. Zero field specific heat of $CeAl_2$. Inset shows the entropy vs. temperature.

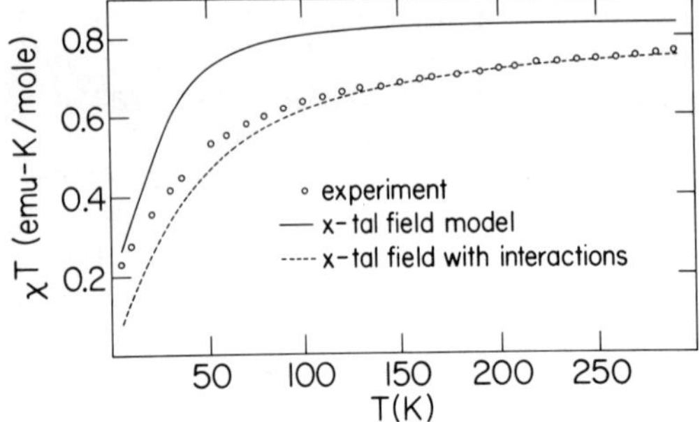

Fig. 2. Temperature dependence of magnetic susceptibility of $CeAl_2$ together with predictions of different CEF models.

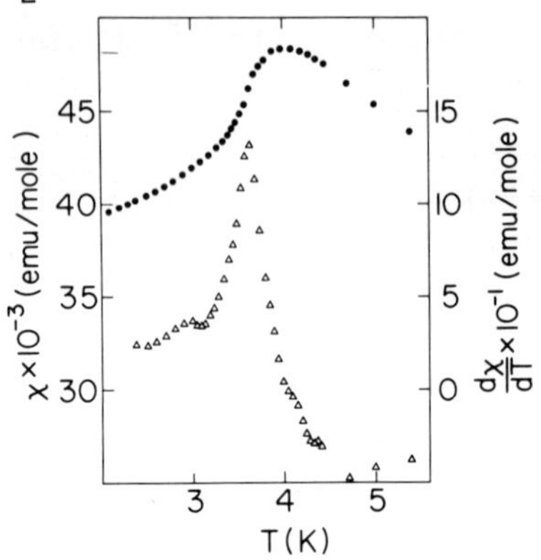

Fig. 3. Magnetic susceptibility of $CeAl_2$ in the transition region and its derivative.

Magnetic susceptibility measurements for $CeAl_2$ are shown in Fig. 2. At high T, χ is of Curie-Weiss form with θ = -32°K and $\mu \cong 2.52\mu_B$. By assuming a ground state doublet and excited state quartet as a CEF configuration scheme with splitting $\approx$ 90°K[3] one cannot fit the experimental curve. Much better fits for T>100°K are obtained if one takes into account the tendency towards magnetic ordering in a phenomenological way, i.e., $\chi \propto [F(T)/(T+\theta F(T))]$,[4] where F(T) represents the reduction of $\langle\mu^2\rangle$ due to the CEF effect and θ is the Curie-Weiss temperature. Below 100°K this fit breaks down since it seems as if θ is decreasing. The behavior of χ in the transition region is shown in Fig. 3. An inflection point (divergence in $d\chi/dT$) is present at the same temperature at which C_p and α show divergent behavior.

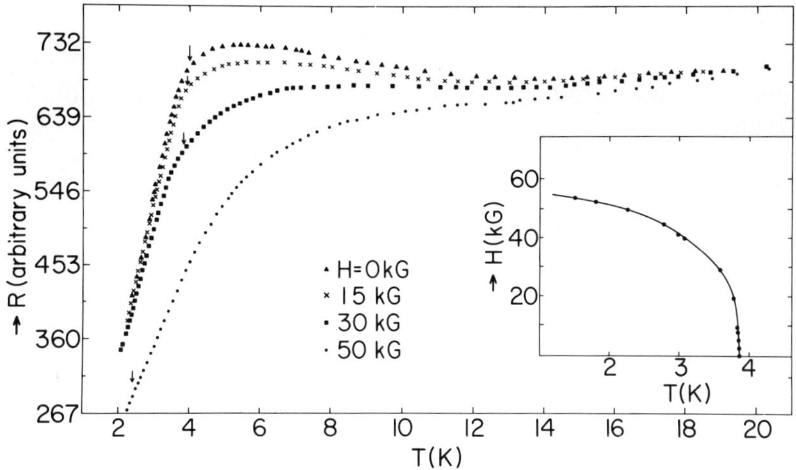

Fig. 4. Resistance versus temperature for $CeAl_2$. Inset shows the phase diagram in magnetic field. Arrows indicate transition temperatures as determined from specific heat measurements.

The resistivity of $CeAl_2$ as a function of temperature for different applied magnetic fields is shown in Fig. 4. The following features are apparent in our data: a well defined linear region below 4°K, a large negative magnetoresistance and washing out of the resistance minimum in increasing magnetic field. Perhaps the most surprising result is the absence of a divergence of dR/dT at the transition temperature.

DISCUSSION

The weight of the calorimetric evidence suggests that the nature of the ground state can be best characterized by some sort of magnetic order. The entropy removal in the low temperature region corresponds roughly to the removal of the Γ_7 doublet degeneracy, which would be consistent with magnetic ordering. It was shown recently[5,6] that upon the transition into the ground state in field, $CeAl_2$ exhibits a lattice distortion which is often characteristic of ordered magnetic systems. Perhaps the most puzzling aspects are the failure to see any magnetic structure in neutron scattering experiments[7] and the absence of a divergence in dR/dT at the transition. Both facts can be attributed perhaps to a complicated form of magnetic order (e.g. involving spatial variations in both moment amplitude and direction as observed in some actinide monopnictides).[8]

REFERENCES

*This work was supported by the Army Research Office and the Office of Naval Research.

1. B. Cornut, Ph.D. Thesis, Grenoble, 1976, and references therein.
2. T. S. Chang, A. Hankey and H. E. Stanley, Phys. Rev. B 8, 346 (1973).
3. F. Steglich and M. Loewenhaupt, these proceedings.
4. T. Murao and T. Matsubara, Prog. Theor. Phys. 18, 215 (1957).
5. M. C. Croft, I. Zorić and R. D. Parks, to be published.
6. M. C. Croft, I. Zorić, J. Markovics, and R. D. Parks, these proceedings.
7. F. Steglich (private communication).
8. A. T. Alfred, B. D. Dunlap, A. R. Harvey, D. T. Laru, G. H. Lander, and M. H. Mueller, Phys. Rev. B 9, 3766 (1974).

MAGNETIZATION OF TmS AND TmSe AND THE FIELD INDUCED TRANSITION[1]

G. Chouteau[§], F. Holtzberg[*], J.M. Mignot, O. Peña[§] and R. Tournier

Centre de Recherches sur les Très Basses Températures, CNRS, BP 166 Centre de Tri, 38042 Grenoble-Cedex, France

Various physical properties of TmS and TmSe have been studied by Bucher et al[2] and by others using similar samples. In particular magnetization[2,3] and magnetostriction[4] measurements of TmSe exhibit a jump in their amplitudes, indicative of a magnetic phase transition at a small critical field. In Fig. 1, we show some of the earlier results as well as our results on various samples. The amplitude of the jump is not the same for all samples. We believe that this jump is due to the presence of Tm^{2+} in a Tm^{3+} host, as indicated by the lattice constant. The two TmSe samples with a large jump in the magnetization have a lattice constant, a_o, of 5.70 Å indicating a larger amount of Tm^{2+} than the samples with no jump which have a_o = 5.68 Å. The amplitude of the jump is larger when the Kondo residual resistivity is large.[5] An abrupt decrease in the resistivity[6] at the same critical field can be correlated with the jump.

For TmS the residual resistivity is an order of magnitude lower[5] and only a very small jump of magnetization is observed (Fig. 1).

The low field susceptibility is shown in Fig. 2. The susceptibility peak indicates antiferromagnetic ordering at different temperatures depending on the sample. These results confirm the observations of Bucher et al. The magnetic ordering is also characterized by the appearance of a remanent magnetization which increases when the temperature decreases. A reverse field of 0.4 kOe is necessary to eliminate the remanent magnetization of sample H 75.I.32 at the lowest temperature. A saturation magnetization is associated with this remanence, both disappear at the ordering temperature. They are related to the antiferromagnetic domain structure and measure the uncompensated moments of the domains due to defects in the ideal antiferromagnetic lattice of Tm^{3+} ions. The low field satura-

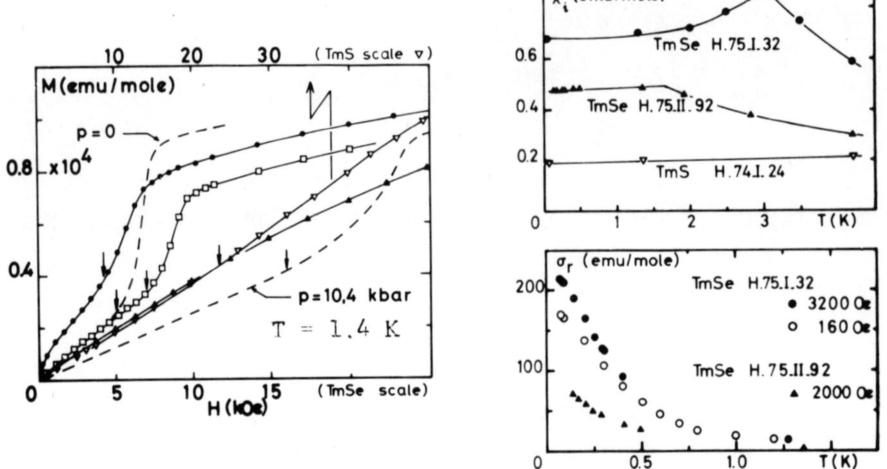

Fig. 1. Magnetization measurements on various TmSe and TmS samples. Lower scale : TmSe 75.I.32 (●); TmSe 75.II.92 (▲); TmSe B 999 (□). Upper scale : TmS 74.I.24 (▽). Broken lines are TmSe measurements under pressure (ref. 3). Arrows show the beginning of the jump (measurements made at T = 1.4 K).

Fig. 2. a) Slope of the magnetization in a 1 kOe field versus T, after cooling under a field of 2 kOe.
b) Remanent magnetization versus temperature for two TmSe samples cooled under various fields.

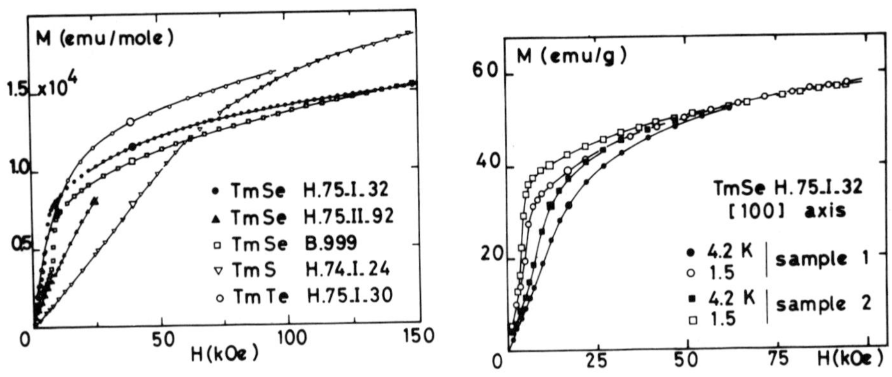

Fig. 3. Magnetization curves up to 150 kOe of several Tm chalcogenides at T = 1.4 K.

Fig. 4. Influence of the demagnetizing factor of 2 specimens cleaved from the same sample.

tion magnetization is caused by the replacement of Tm^{3+} by vacancies or by Tm^{2+} which carry no magnetic moment at T = 0 and H = 0. The saturation obtained in low field at T = 0 is 0.09 μ_B in the sample (75.I.32) which has many Tm^{2+} impurities and is 0.01 μ_B in the sample (75.II.92) which shows no jump in the magnetization curve and which therefore contains a much smaller concentration of Tm^{2+}

The crystal field ground state of Tm^{3+} is generally a singlet. However a moment can be induced by exchange interactions. In such induced moment systems, the Van Vleck susceptibility, the local magnetic moment and the ordering temperature are strongly dependent on the crystal field splitting. For example, in Fig. 3 the slope of M vs. H at 125 kOe is different for the two TmSe and for TmS even though we attribute it to the Tm^{3+}. Since the slope and the saturation moment are determined by the crystal field splitting and the strength of the magnetic interactions, this observation is consistant with the fact that the ordering temperature and the local moment[7] of Tm^{3+} depend on the sample. The TmSe and TmS sample have approximately the same saturation moment, 2 μ_B, in high field. The magnetization of divalent TmTe above its ordering temperature (0.21 K) follows a Brillouin function with a saturation moment of a similar value and a Van Vleck contribution of the same order of magnitude.

We attribute the jump in the magnetization curve observed for some TmSe samples (Figs. 1, 3) to a transition from an exchange induced antiferromagnet[3] to an exchange induced ferromagnet. The width of the jump is determined by the demagnetizing field and probably some inhomogeneity in the Tm^{2+} concentration. For the sphere, the slope dM/dH, of the critical region is five times the inverse of the demagnetizing factor ($3/4\pi$) at 3 K and only three times larger at 1.4 K. The effect of the different demagnetizing fields of samples of different shape is to change the width of the transition and the low field behavior but not the high field behavior (Fig. 4)

In conclusion, since the jump appears only on samples containing Tm^{2+} impurities we believe that in a field there is a magnetic moment induced on the Tm^{2+} which generates a ferromagnetic contribution to the interaction between the Tm^{3+}. When this field dependent ferromagnetic interaction exceeds the antiferromagnetic one, the transition occurs. Since the amplitude of the magnetization jump is an order of magnitude larger than the possible contribution of Tm^{2+} by itself, it must in some such way include a change in Tm^{3+} moment.

This phenomenon is possible because Tm^{2+} is non magnetic in zero field in the Tm^{3+} ordered magnetic state as shown by the large residual resistivity.[5] If the Tm^{2+} were ordered below T_N, along with the Tm^{3+} host, the Kondo resistivity would go to zero. This interpretation suggests that the antiferromagnetic order is similar to that found in MnO where the molecular field of the nearest neighbors sums to zero and the antiferromagnetic order is caused by next nearest neighbors. The fact that the transition in TmS occurs at higher field than in TmSe may reflect the fact that the Kondo temperature of Tm^{2+} in TmS is higher than in TmSe.[5]

We thank P.W. Lockwood for technical assistance.

§and Service National des Champs Intenses, CNRS, Grenoble, France.
*Permanent address : IBM Research Center, Yorktown Hts, N.Y. 10598.

1 - Part of the thesis of O. Peña
2 - E. Bucher, K. Andres, F.J. Di Salvo, J.P. Maita, A.C. Gossard, A.S. Cooper, G.W. Hall, Phys. Rev. B11, 500 (1975).
3 - R.P. Guertin, S. Foner, F.P. Missell, Phys. Rev. Lett. 37, 529 (1976).
4 - H.R. Ott, K. Andres, E. Bucher, AIP Conf. Proc. 24, 40 (1974).
5 - A. Berger, E. Bucher, P. Haen, F. Holtzberg, F. Lapierre, T. Penney, and R. Tournier, This Conference.
6 - P. Haen, F. Holtzberg, F. Lapierre, T. Penney, and R. Tournier, This Conference.
7 - F. Holtzberg, J.C. Lasjaunias, J.M. Mignot, A. Ravex, M. Vandorpe, This Conference.

HYPERFINE SPECIFIC HEATS OF TmS AND TmSe

F. Holtzberg*, J.C. Lasjaunias, J.M. Mignot, A. Ravex,
and M. Vandorpe

Centre de Recherches sur les Très Basses Températures,
CNRS, BP 166 Centre de Tri, 38042 Grenoble-Cedex, France

Since the compounds TmS and TmSe have been reported to exhibit unstable valence states[1,2], a lot of experimental work has been devoted to some unusual features of their low temperature properties. It seems now well established that Tm^{3+} induced-moment ordering occurs in both of these singlet ground state systems in the liquid helium temperature range. However, the local moment induced on the thulium ions cannot be obtained simply from a magnetization measurement since an external field changes not only its orientation but also its absolute value.

In such a case, the nuclear moment is a very appropriate probe to determine the induced moment in the ordered state without applying an external field. The splitting of the nuclear energy levels can be measured either by a spectroscopic technique or, as in this work, by a static thermodynamical method.

We have studied the specific heat of TmS and TmSe between 50 mK and 2 K. The specimens used are those labeled 74.I.24 and 75.I.32 respectively in reference 3. The results are shown on figures 1 and 2 with the higher temperature data of Bucher et al.[1] First of all, we emphasize the striking similarity between the two curves : Both of them exhibit a rather weak peak centered at the Néel temperature, a decrease faster than T on cooling down to about 0.3 K, then a hyperfine anomaly, the maximum of which lies below our experimental range.

This similarity can be understood in a model proposed recently[3] to account for the magnetization jump, the Kondo-like resistivity and the negative magneto-resistivity of these compounds : Both systems are essentially described as a matrix of singlet ground state Tm^{3+} ions but also containing some Tm^{2+} located in positions where they are only weakly coupled to the Tm^{3+} host. Moreover, the Tm^{2+} ions are assumed to be in a Kondo state at low temperature. We think that the general features of the specific heat curves reflect the induced-moment ordering and nuclear properties of the Tm^{3+} ions which are qualitatively the same in both cases. As for the Tm^{2+}, they are expected to give a T contribution, larger in TmSe than in TmS but this term is difficult to extract from the curve.

The hyperfine fields associated with the induced moments in TmS and TmSe can be deduced from the specific heat data : We have fitted our results below the temperature of the minimum to the Schottky anomaly for I = 1/2 and μ_N = -0,231 nuclear magnetons taking H_{hf} as an adjustable parameter (fig. 1 and 2). We obtain 3,55 and 2,63 MOe for TmS and TmSe respectively. This analysis requires that the magnetic hyperfine term is the only contribution to the a/T^2 slope. This assumption is reasonable in this case since the quadrupolar interaction is expected to vanish in cubic symmetry and with a nuclear spin 1/2. Another question is the possibility of a non-uniform value of H_{hf}, which is likely to occur if some thulium ions are divalent and behave as Kondo impurities. However, their amount is certainly small and we can neglect them as a first approximation.

The larger hyperfine field found in TmS is consistent with the higher ordering temperature of this compound[1]. As for TmSe the only determination of H_{hf} available up to now was that deduced by Triplett et al[2] from a Mössbauer experiment : These authors obtained 1,93 MOe in the low temperature limit instead of 2,6 MOe from our measurement. However, the ordering temperature of their sample is 2,8 K whereas ours orders a little above 3 K (susceptibility[4] and resistivity[3] results). These observations are consistent and confirm that the conditions of induced-moment ordering can vary appreciably from one sample to another. This effect is probably related to the sensitivity of the crystal field to valence instabilities which also appears in the position of the resistivity bump associated with the population of the excited crystal field levels[3]. The rather high value of T_N in our TmSe sample is also responsible for the shift of our data compared to Bucher's in the 2 K region.

If we assume the hyperfine constant A_J to have its ionic value (-18.7 mK in Tm^{3+})[5], we can calculate the local moment of the rare earth by the formula :

$$\mu/\mu_B = - H_{hf} \cdot g_J \cdot g_N \mu_N / A_J$$

Here, we obtain μ = 3,75 μ_B in TmS and 2,80 μ_B in TmSe. The fact that these values are much larger than those deduced from the "saturation magnetizations" in high fields[4] may indicate that the hyperfine constant does not retain its ionic value in such unstable valence systems. Alternatively the induced moment measured in a high external field could be reduced by negative interactions.

We thank T. Penney for his critical reading of the manuscript and R. Tournier, and D. Thoulouze for many stimulating discussions.

* Permanent address : IBM Research Center, Yorktown Hts, N.Y. 10598.

1 - E. Bucher, K. Andres, F.J. Di Salvo, J.P. Maita, A.C. Gossard, A.S. Cooper, and G.W. Hull, Jr., Phys. Rev. B11, 500 (1975).
2 - B.B. Triplett, N.S. Dixon, P. Boolchand, and S.S. Hanna, J. Physique Coll. 35, C6-653 (1974).
3 - A. Berger, E. Bucher, P. Haen, F. Holtzberg, F. Lapierre, T. Penney, and R. Tournier, This Conference.
4 - G. Chouteau, F. Holtzberg, J.M. Mignot, O. Pena, and R. Tournier, This Conference.
5 - B. Bleaney, in Magnetic Properties of Rare Earth Metals (Plenum Press, London and New York, 1972), p. 383.

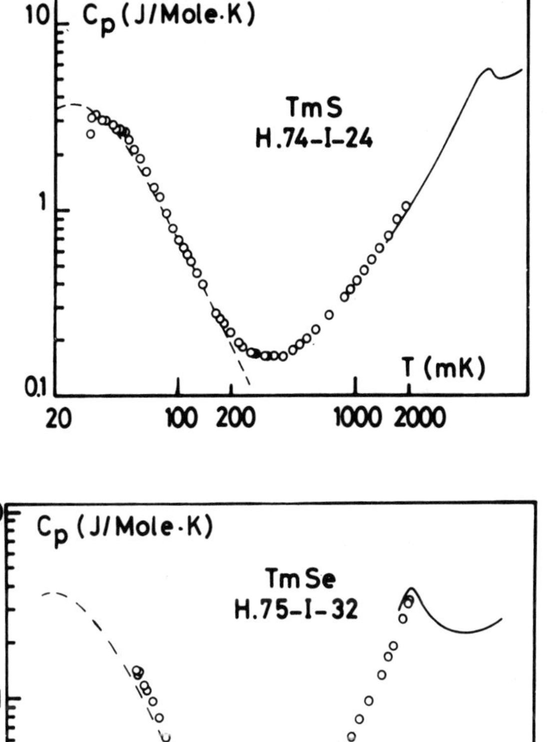

Figs. 1 & 2. Specific heats of TmS and TmSe. The solid lines represent the results of ref. 1. The dashed lines are the best fits by a Schottky anomaly. The circles represent the results of this work.

KONDO RESISTIVITY OF TmSe AND TmS DOWN TO 20 mK

A. Berger[†], E. Bucher[§], P. Haen, F. Holtzberg[*],
F. Lapierre[†], T. Penney[*], and R. Tournier

Centre de Recherches sur les Très Basses Températures,
C.N.R.S., BP 166 X, 38042 Grenoble-Cedex, France

The experiments presented here will be interpreted within a model which assumes that TmSe and TmS are essentially the same and the Tm ion predominantly trivalent. Because the Tm valence is unstable, there is always Tm^{2+} present which acts as a Kondo impurity and dominates the resistivity behaviour. From measurements of the lattice constant at room temperature, the concentration of Tm^{2+} in TmS is very small but in TmSe it can vary from near 0 to 20 %.

Both TmSe and TmS have similar metallic values for their room temperature resistivities, 200-700 μΩ.cm for TmSe and about 100 μΩcm for TmS. These values are not far from those of $CeAl_3$ which is well known as a Kondo compound. With decreasing temperature the resistivities (fig. 1) increase gradually in general agreement with previous results.[1] The resistivities would show approximately log T behaviour if we were able to subtract the lattice contribution as in $CeAl_3$.[2]

A maximum is observed in the low temperature resistivity of TmS (fig. 2) which can be attributed to the Kondo resistivity of Tm^{3+}. Normally we would expect that the resistivity would go to zero because the system is magnetically ordered at $T \sim 6$ K. Furthermore, without ordering, the magnetic resistivity of Tm^{3+} would go to zero for temperatures smaller than the splitting between the singlet ground state and the excited levels. In fact we observe a large residual resistivity ρ_o in both compounds (fig. 1). Since ρ_o is much larger in TmSe than in TmS it is natural to attribute it to the Kondo residual resistivity of Tm^{2+} impurities in the Tm^{3+} compound with different Tm^{2+} concentrations depending on the sample.

In such an interpretation the Tm^{2+} moment must be independent of the antiferromagnetic order of the Tm^{3+}. Such a situation would be reasonable if the antiferromagnetic order were the same as found in MnO where there are 12 nearest neighbors, 6 of each spin orienta-

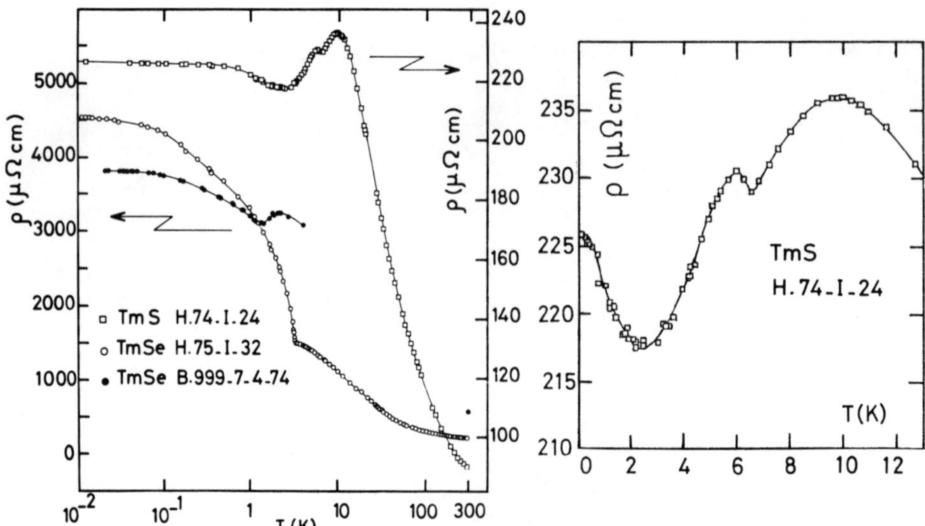

Fig. 1. (on the left). Resistivity vs. Log T of TmS (right scale) and TmSe (left scale). The numbers of the sample batch are indicated, B. Bucher, H. Holtzberg.

Fig. 2. (on the right). TmS, Resistivity vs. T, data of Fig. 1.

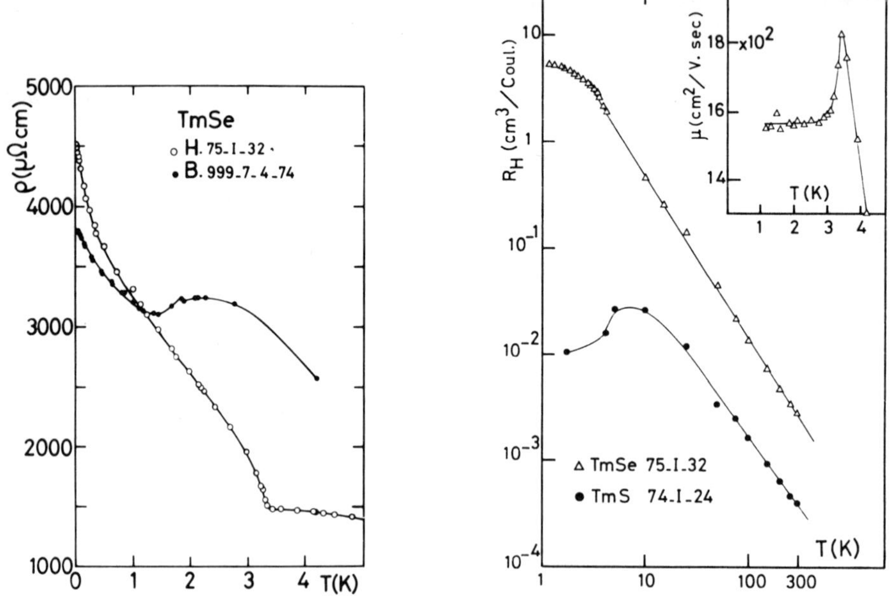

Fig. 3. (on the left). TmSe, Resistivity vs. T, data of Fig. 1

Fig. 4. (on the right). Hall coefficient vs. T for TmS and TmSe. Insert : Low Temp. Hall mobility (R_H/ρ) of TmSe.

tion, and hence no molecular field due to the nearest neighbors. The effect of magnetic ordering on the Tm^{3+} lattice may be to change the Kondo temperature of Tm^{2+} causing the change in the resistivity of TmSe below the ordering temperature (fig. 1, 3). The fact that the Tm^{2+} resistivity saturates at very low temperature indicates that the Tm^{2+} moment has disappeared below its Kondo temperature and that the Kondo temperature of Tm^{2+} is an order of magnitude larger in TmS than in TmSe. Since the Kondo temperature depends exponentially on the density of states at the Fermi level and on the exchange coupling of the Tm^{2+} moment with the conduction electrons we deduce that TmS and TmSe are similar hosts for the Tm^{2+}. In this case, the residual resistivity ρ_o, is a measure of the number of Tm^{2+} impurities present in each sample of both compounds. In fact ρ_o is largest in those samples which show the largest lattice constant. It is also possible to correlate ρ_o with the magnetoresistance[3] and with the jump in the low temperature magnetization which are also caused by Tm^{2+} impurities.

The Hall effect (fig. 4) has been measured but is difficult to interpret. It is independent of field for small fields and always negative. Between room temperature and 5 K its magnitude increases much more strongly than the resistivity. The simplest one band interpretation gives a carrier concentration at room temperature $N_H \equiv 1/R_H e$ of 2.2×10^{21} for the TmSe and 1.5×10^{22} for the TmS. For comparison the concentration of Tm calculated from the lattice constants 5.70 and 5.42 Å are 2.2×10^{22} and 2.5×10^{22} for TmSe and TmS respectively. The agreement in the case of TmS is undoubtedly accidental since it is unlikely that the large temperature change in Hall constant actually reflects a large change in carrier concentration. It is interesting to note that in the case of TmSe below 3 K the change in ρ and R_H are the same and so their ratio, the Hall mobility, $\mu_H = R_H/\rho$, is constant (fig. 4).

Instead of the simple metal approach the Hall effect should be interpreted with the framework of the model : a highly concentrated Tm^{3+} Kondo system with Tm^{2+} impurities at high temperature and at low temperature a dilute alloy of Tm^{2+} Kondo impurities coexisting and uncoupled with ordered Tm^{3+} moments.

We thank O. Gorochov for making ultrasonic contacts and P.W. Lockwood, D. Point, and J.M. Rigotty for technical assistance.

†Partial fulfillment of the thesis of A. Berger and F. Lapierre.
*Permanent address : IBM Research Center, Yorktwon Hts N.Y. 10598.
§Address : University of Constance, D-775 Constance, Germany

1 - E. Bucher, K. Andres, F.J. Di Salvo, J.P. Maita, A.C. Gossard, A.S. Cooper, G.W. Hull, Phys. Rev. B11, 500 (1975).
2 - B. Cornut and B. Coqblin, Phys. Rev. B5 , 4541(1972).
3 - P. Haen, F. Holtzberg, F. Lapierre, T. Penney, R. Tournier, this conference.
4 - G. Chouteau, F. Holtzberg, J.M. Mignot, O. Peña and R. Tournier, This Conference.

NEGATIVE MAGNETORESISTANCE, HALL EFFECT AND FIELD INDUCED TRANSITION IN THE KONDO SYSTEMS TmSe AND TmS

P. Haen, F. Holtzberg*, F. Lapierre†, T. Penney*, and R. Tournier

Centre de Recherches sur les Très Basses Températures, CNRS, BP 166 Centre de Tri, 38042 Grenoble-Cedex, France

In a companion paper, it is argued that the low temperature resistivity of TmSe and TmS is dominated by the Kondo resistivities of both Tm^{2+} impurities and the Tm^{3+} host. Below the magnetic ordering temperature (T_N) of the Tm^{3+} moments only the Tm^{2+} impurities contribute to the resistivity. Fig. 1 shows that the resistivity of a TmSe sample both above and below T_N can be quenched by an external field to below the room temperature value. In fact, below the ordering temperature the resistivity decreases abruptly at a critical field and is nearly constant above it. For TmSe (H 75.I.32) this critical field is about 4 kOe when the demagnetizing factor is small (Fig. 2, $H_{\parallel}$). The critical field is higher for a different orientation with a larger demagnetizing factor (Fig. 2, $H_{\perp}$). The magnetization measured on the same material shows a jump in the magnetization curve at H_c which has the same dependence on the demagnetizing factor.[2] Approximately the same resistivity limit is obtained in high field for the different orientations. Various TmSe samples show rather different behavior which we attribute to differing amounts of Tm^{2+} impurity. The TmSe sample (H 25.II.92) with the smallest residual resistivity, ρ_o, the smallest negative magnetoresistance, $\Delta\rho$, and no jump in the magnetization curve has the smallest lattice constant (a_o = 5.68) indicating the smallest concentration of Tm^{2+}. The other sample of TmSe (H 75.I.32) has large ρ_o and $\Delta\rho$, exhibits a jump in the magnetization curve[3] at H_c and a larger lattice constant (a_o = 5.70). On the other hand TmS, which from its lattice constant has only a small concentration of Tm^{2+}, shows little or no jump in the magnetization, a smaller residual resistivity and smaller magnitude negative magnetoresistance. Since the magnetoresistance of TmSe saturates in a small field while TmS is unsaturated at 70 kOe, Tm^{2+} in TmS has a higher Kondo temperature than in TmSe, in agreement with the temperature variation of the low

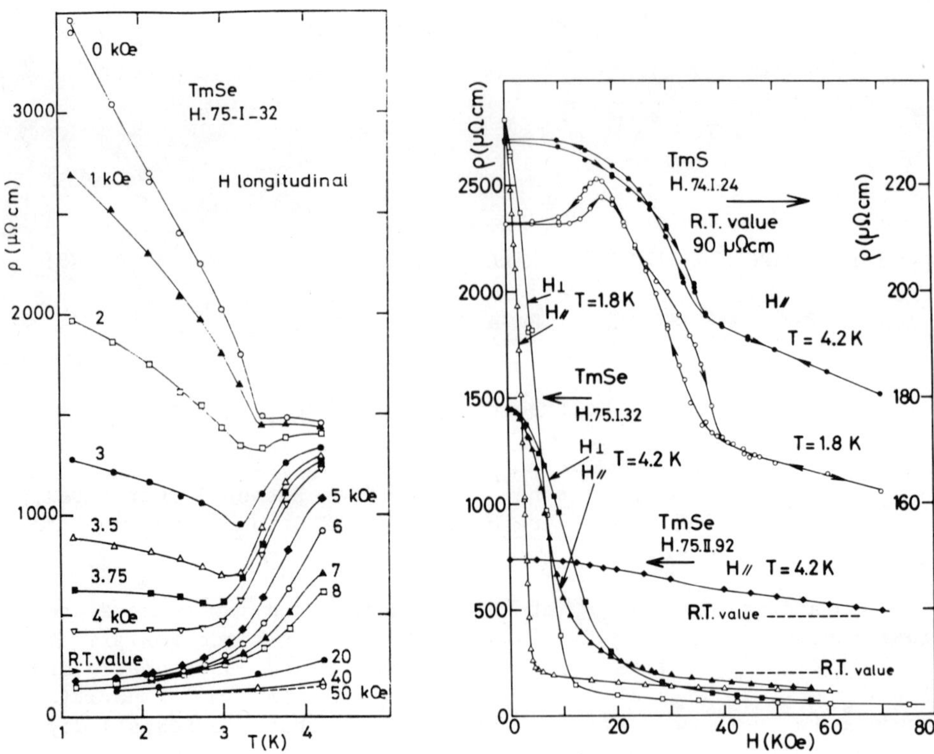

Fig. 1. Resistivity vs. temperature for TmSe in a magnetic field.

Fig. 2. (on the right). Resistivity vs. field for TmS and two types of TmSe sample identification as in ref. 1.

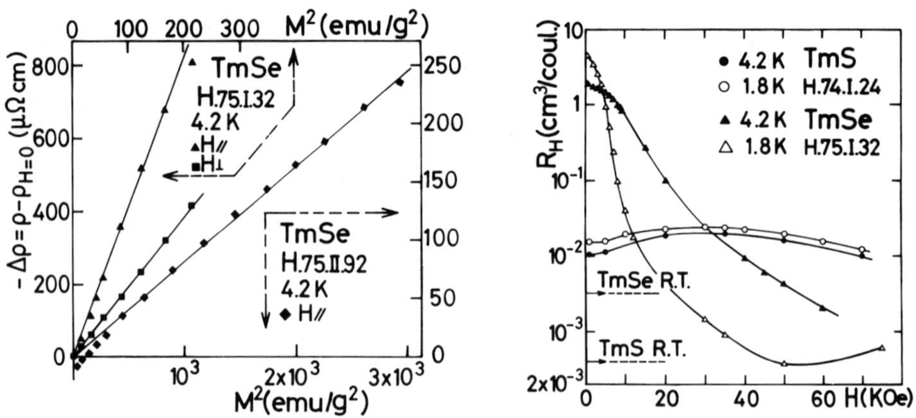

Fig. 3. (left). Negative magnetoresistance vs. magnetization squared.

Fig. 4. (right). Hall constant vs. magnetic field for TmS and TmSe.

temperature resistivity.[1]

In order to test a specific aspect of the Kondo effect, we have plotted (Fig. 3) the negative magnetoresistance of TmSe versus the magnetization[3] squared in the paramagnetic region for two different samples. The square law is observed in each case even though the magnitude of the effect is very different due to the different Tm^{2+} concentration, since both Tm^{2+} and Tm^{3+} contribute their own Kondo resistivity.

The critical field, H_c, is also observed in the Hall effect of TmSe below T_N (Fig. 4). The Hall constant, R_H, decreases gradually for small fields, but for $H>H_c$ it drops by three orders of magnitude to the room temperature value. Although the magnitude of the change is larger than for the resistivity, the general behavior, i.e. high field low temperature saturation at the room temperature zero field value, is the same. The variation of R_H with field is much weaker and unsaturated in the case of TmS just as the magnetoresistance was weaker and unsaturated.

In conclusion, the saturation of the magnetoresistance of TmSe H 75.I.32 in high field indicates that ferromagnetic order has been attained. We therefore attribute the sharp decreases in resistivity and Hall effect and the jump in magnetization to a transition from antiferromagnetism to ferromagnetism at H_c. A possible explanation for such a transition follows. The antiferromagnetic order observed is due to interactions between the Tm^{3+} ions. If the magnetic structure is of the MnO type then there is no nearest neighbor molecular field at each Tm site, only the second neighbor field. Therefore, the coupling between Tm^{3+} and Tm^{2+} is very weak, consistent with the observation of Tm^{2+} Kondo resistivity below T_N. However, in an external field alignements of the Tm^{3+} moments produce a large nearest neighbor molecular field which induces a moment on the Tm^{2+} Kondo impurity. Assuming that the Tm^{2+}-Tm^{3+} interaction is ferromagnetic, there can be a critical field at which the induced field dependent ferromagnetic interaction exceeds the original antiferromagnetic interaction and a transition will take place.

We thank P.W. Lockwood and D. Point for technical assistance.

† Partial fulfilment of a thesis of F. Lapierre.

* Permanent address : IBM Research Center, Yorktown Hts N.Y. 10598.

1 - A. Berger, E. Bucher, P. Haen, F. Holtzberg, F. Lapierre, T. Penney and R. Tournier, This Conference.
2 - G. Chouteau, F. Holtzberg, J.M. Mignot, O. Peña, and R. Tournier, This Conference.
3 - The magnetization measurements were made using samples from the same batch but not the identical crystals.

SAMPLE DEPENDENT MAGNETORESISTANCE AND MAGNETIZATION ANOMALIES IN TmSe

W. M. Walsh, Jr., K. Andres, L. W. Rupp, Jr. and
L. D. Longinotti
Bell Laboratories
Murray Hill, New Jersey 07974

Previous measurements on TmSe, which appears to be an intermediate valence compound, have indicated an ordered state occurs below 3 K characterized by magnetostriction and magnetization anomalies (1,2) at "critical" magnetic fields of $\sim$2 and $\sim$5 kOe. Even above 3 K weak anomalies in magnetostriction (2) and magnetization (3) are observed at higher magnetic fields. The magnetic anomalies have recently been shown to be very pressure sensitive, the upper critical field increasing rapidly versus pressure. (3) We had performed microwave and radiofrequency surface resistance experiments which indicated a large decrease in electrical resistance at the upper critical field. Subsequent D.C. resistance measurements in this laboratory and elsewhere have confirmed that the resistivity decreases strongly ($\times 10^{-1}$-10^{-2}) at the upper critical field. (4)

Here we summarize extensive investigations of these resistivity and magnetization anomalies in several single-crystal samples. The critical fields prove to be quite sample dependent, even among samples taken from a given boule. They are also markedly anisotropic with respect to the cubic crystal axes. ( ) The upper critical field anomaly exhibits weak hysteresis ($\sim$200 Oe) and appreciable shape dependence due to demagnetization effects. Finally the widths of the anomalies, as measured by the peak in the field derivative of the radiofrequency loss, are markedly sensitive to strains due to sample mounting. These widths also broaden significantly in the microwave range (10-35 GHz).

The sample dependence of both resistivity and differential magnetization is illustrated in Fig. 1 for pieces cleaved from different regions of a slowly crystallized boule. The correlation of the two physical properties is clear. The variations between

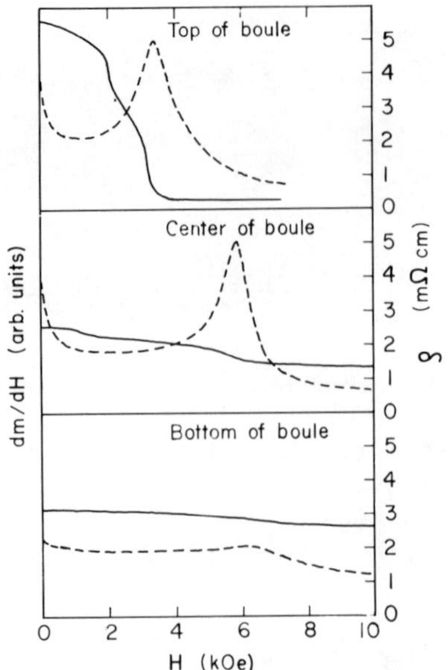

Fig. 1 Magnetic field dependence of the resistivity (solid lines righthand scale) and the differential magnetization (broken lines) at 1.3 K of three samples of TmSe cut from the same boule.

samples may result from slight changes in stoichiometry though no systematic lattice parameter variation has been detected thusfar. The resistivity anomaly, as monitored via rf loss, is very sensitive to strains induced by sample mounting but this proves reversible so it is unlikely that the sample-to-sample variation in the critical fields are simply due to strain.

Evidence of anisotropy is shown in Fig. 2 and 3 for the differential magnetization and the rf loss peak (35 MHz resistivity) respectively. While exhibiting symmetry relative to the principal cubic axes the curves of Fig. 3, particularly the 110 runs, are necessarily somewhat distorted due to demagnetizing fields ($4\pi M \approx 2$ kOe).

The most striking result we have observed is summarized in the plot of log ρ versus log T in various magnetic fields (Fig. 4). The monotonic increase in resistivity as T decreases in zero magnetic field is marked by a plateau and sudden rise at the ordering temperature. In successively higher applied fields the low temperature resistance drops very rapidly - decreases of 200x have been observed in some samples. Qualitatively one might ascribe the effect to spin fluctuation scattering suppressed by a field but the detailed behavior is not understood. Since valence fluctuations might also

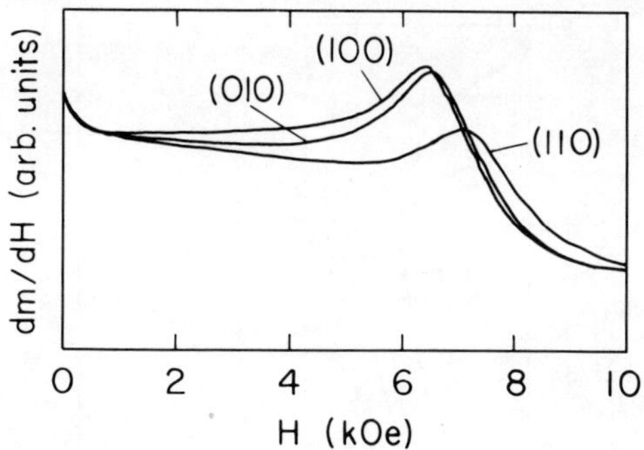

Fig. 2 Anisotropy in the field dependence of the differential susceptibility for fields in the [100] plane at 1.3 K.

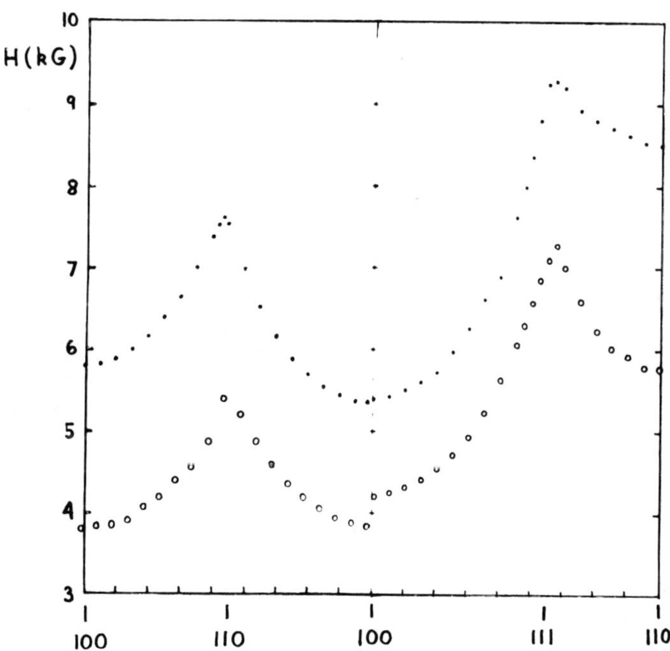

Fig. 3 Anisotropy of the radiofrequency loss peaks of two samples (solid & open circles) from different boules. These plots are uncorrected for demagnetizing effects.

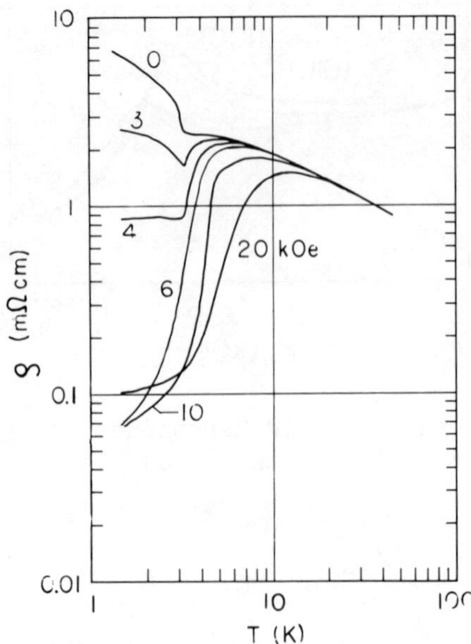

Fig. 4 Temperature dependence of the resistivity of a TmSe sample cut from the top of the boule, in different magnetic fields. The field is oriented along the (010) axis, the current along the (100) axis.

be thought to play a role the optical reflectivity of a sample was monitored at 1.6 K in fields up to 10 kOe. The absence of any visible change suggests no field-induced valence change to the fully trivalent limit occurs.

1. E. Bucher, K. Andres, F. J. DiSalvo, J. P. Maita, A. C. Gossard, A. S. Cooper and G. W. Hull, Phys. Rev. B 11, 500 (1975).
2. H. R. Ott, K. Andres and E. Bucher, AIP Conf. Proc. 24, 40 (1975).
3. R. P. Guertin, S. Foner and F. P. Missell, Phys. Rev. Letters 37, 529 (1976).
4. H. R. Ott (private communication).
5. The importance of anisotropy was brought to our attention by S. Foner (private communication and this Conference).

Pressure Studies of Rare Earth Singlet Ground State Systems†

J. E. Schirber and H. T. Weaver

Sandia Laboratories

Albuquerque, New Mexico 87115

Our studies of the pressure dependence of the rare earth NMR in PrAs, PrP, and TmP have shown that the point-charge model gives the wrong sign for the volume dependence of the crystal field splitting in these materials. Similar conclusions were reached from pressure studies of the susceptibilities of closely related singlet ground state materials by Guertin et al. In a recent paper in which pressure derivatives were obtained as a function of temperature in order to vary the contributions of crystal field and exchange effects, we showed that this unexpected volume dependence could not be attributed to exchange. We believe that these effects have their origin in pressure-induced changes in conduction electron character. As a test for this hypothesis, we have initiated studies on the isomorphous cogener PrN which is, however, an insulator. We would predict that the point-charge model would be qualitatively correct for this material. We have not been able to detect a shift in the Pr Knight shift (which is +290%) with 4 kbar at 4 K in PrN. It will be necessary to measure the compressibility of PrN before we can assess the significance of this result.

The rare earth group VA compounds have provided an exceedingly fertile field for study of the effects of the presence of 4f electrons on the magnetic and electronic properties. These compounds are generally metallic and crystallize in the NaCl structure. Of particular interest are those situations in which the lowest lying crystalline electric field sublevel is nonmagnetic. In this discussion we will be concerned with situations in which this lowest lying state is a singlet. The nature of a singlet ground state system depends both upon the crystalline field and on the rare earth-rare earth exchange.

The Pr and Tm pnictides were first shown by Jones[1] to exhibit extremely large Knight shifts directly proportional to the susceptibility.

$$K_{orb} = A_{orb}\ C_J f_J(\Delta,T)/[T - \theta\ f_J(\Delta,T)]\ , \qquad (1)$$

where A_{orb} is the coupling constant proportional to $\langle r^{-3}\rangle$ with r_f the f-electron radius, C_J is the Curie constant, Δ is the crystal field splitting of the lowest spin orbit state defined by the total angular momentum J, and θ is proportional to the Curie temperature for $\Delta = 0$. The quantity f_J is defined in Ref. 1 and contains the crystal field effects.

At low temperatures $K \propto \Delta^{-1}$, so our measurements[2] of the Knight shifts of PrP and TmP as a function of pressure gave strong indications that the crystal field splitting decreases with pressure. This is in contradiction to the often successful point-charge model (PCM) where Δ is proportional to $1/R^N$ (R is the interatomic spacing and $N \sim 5$) and, consequently, Δ should increase with pressure. Subsequent independent studies[3] of the susceptibility versus pressure for other rare earth pnictides were consistent with our result.

A possible salvation for the point-charge model would be to postulate changes in x, the ratio of the fourth-to-sixth order crystal field interactions. Within the framework of the PCM, $d\ \ell n\ x/dP = 2/3\ K_T$ where K_T is the compressibility. This leads to $d\ \ell n\ \Delta/dP \sim \pm 0.1\ K_T$ for Tm and Pr, respectively. This is far too small (we observe $\sim -(2$ to $4)\ K_T$ for $d\ \ell n\ \Delta/dP$) and inconsistent with the fact that Δ decreases for both compounds. We feel that it is highly unlikely that x would change at a very high rate in opposite directions in these materials.

As pointed out by a number of investigators,[4,5] exchange effects are important in these systems.

We can write from Eq. (1):

$$\frac{d\ \ell n\ K_{orb}}{dP} = \frac{d\ \ell n\ A_{orb}}{dP} + \frac{T\Delta}{T - \theta f_J}\ \frac{d\ \ell n\ f_J}{d\Delta}\ \frac{d\ \ell n\ \Delta}{dP} + \frac{\theta f_J}{T - \theta f_J}\ \frac{d\ \ell n\ \theta}{dP} \qquad (2)$$

Measurement of the pressure derivatives of the rare earth Knight shifts as a function of temperature can in principle separate the crystal field and exchange contributions, as described by the second and third terms, respectively. We have carried out such measurements on PrP, PrAs, and TmP and the results indicate that, at least with the simple treatment of exchange outlined above, the pressure derivative of Δ is negative.

Our conclusion is that the effects we are observing stem from pressure-induced variations in the conduction electron character. To test this hypothesis, we have attempted to repeat our Knight shift measurements as a function of pressure on PrN, which is isomorphous to PrP, but is an insulator at low temperatures. Our picture would then predict that the PCM would be valid in this case.

We have observed a large positive Knight shift for Pr in PrN of 290%, which agrees with the known[7] crystal field splitting of 91 MeV. However, we are unable to detect any change in the magnitude of this shift with 4 kbar. While it is tempting to ascribe this very small effect to cancellation between a positive contribution from exchange and the predicted negative contribution of the PCM, it is inappropriate to do so without a determination of the compressibility of the material. The lattice constant of PrN is ~ 5.2 Å compared to ~ 5.9 Å for PrP so it is possible that the compressibility is small and our sensitivity is just insufficient to detect the changes. We plan to extend our measurements to higher pressures and to measure the compressibility in order to make a meaningful comparison of the PrN results with those for its metallic cogeners.

We acknowledge the technical assistance of B. R. Hansen and R. L. White and helpful discussions with A. Narath. We are indebted to D. E. Lavalle, Analytical Chemistry Division, Oak Ridge National Laboratory, for kindly providing us with the PrN sample.

---

† Research prepared for the U. S. Energy Research and Development Administration under Contract AT(29-1)-789.

1. E. D. Jones, Phys. Rev. 180, 445 (1969).
2. H. T. Weaver and J. E. Schirber, AIP Conf. Proc. 24, 49 (1974).
3. R. P. Guertin, J. E. Crow, L. D. Longinotti, E. Bucher, L. Kupferberg, and S. Foner, Phys. Rev. B12, 1005 (1975).
4. C. Rettori, D. Davidov, A. Grayevsky, and W. M. Walsh, Phys. Rev. B11, 4450 (1975).
5. K. Sugawara, C. Y. Huang, and B. R. Cooper, Phys. Rev. B11, 4455 (1975).
6. H. T. Weaver and J. E. Schirber, Phys. Rev. B14, 951 (1976).
7. H. L. Davis and H. A. Mook, Magnetism and Magnetic Materials, ed. C. D. Graham Jr. and J. J. Rhyne (AIP, New York, 1972), 1548.

LOCALIZED Sm ION COLLAPSE IN SmS As

F. Holtzberg*, O. Peña§[1], T. Penney*, R. Tournier

Centre de Recherches sur les Très Basses Températures,

CNRS, BP 166 Centre de Tri, 38042 Grenoble-Cedex, France

In this paper we will discuss the results of lattice constant, magnetization and transport measurements for small (1-5 %) concentrations of As in SmS. We find evidence for a local collapse from Sm^{2+} to Sm^{3+} of the 6 Sm which are the nearest neighbors of each As. These Sm^{3+} are coupled to form a superparamagnetic cluster and the clusters are coupled forming a spin glass. Many of the electrons which are removed from the f states in this Sm^{2+}-Sm^{3+} transition remain localized and non conducting at low temperature.

The SmS lattice collapses upon substitution of 5-10 % arsenic for sulfer with a volume change of over 10 % but with no change in symmetry[2]. For As concentrations of 10 to 50 % the lattice constant is somewhat larger than that expected for Sm^{3+}, indicating an intermediate valence of 2.8 to 3.0. Photoemission experiments[3] have detected both Sm^{2+} and Sm^{3+} but do not distinguish between a mixture and intermediate valence. For As concentrations between 50 and 100 % the lattice constant indicates that the samarium is integral valent Sm^{3+}.

The variation of the lattice constant for small As concentrations is shown in Fig. 1. The shaded region indicates the phase transition which separates the initial region of local collapse from the approximately homogeneous intermediate valent region. If each As^{3-} produced one Sm^{3+}, in the otherwise divalent Sm^{2+} S^{2-} material, then the lattice constant would be close to the line between SmS and SmAs, marked $n(Sm^{3+}) = n(As^{3-}) = X$ on the figure. Since the slope of the data is much larger, Kasuya suggested that each As induced a transition to the Sm^{3+} configuration on each of the 6 nearest neighbor sites. To test this idea, the line marked $n(Sm^{3+}) = 6X$ was constructed to pass through the data but also to intersect the "all Sm^{3+}" line at $X = 1/6$. The good agreement supports this idea. The possibility that a larger number of Sm are converted to an intermediate valence or that the system is essentially homogeneous at these small concentrations will be discussed later.

The magnetization at 4.2 K up to 150 kOe is shown in Fig. 2. The slope at high field is the Van Vleck susceptibility of the Sm^{2+}. Surprisingly there is, in addition, a saturation magnetization which is too large to be explained by the presence of other rare earth impurities. If we plot this saturation moment versus arsenic concentration (inset Fig. 1) we find a slope of 1.4 μ_B per arsenic ion. Such a large moment can not be explained by the existence of one Sm^{3+} for each As^{3-} because the maximum possible magnetic moment for Sm^{3+} is the free ion value, gJ = 0.71. The ground state will probably have a lower value, of order 0.2 μ_B as is often observed in samarium intermetallics.[4] In fact our value of 1.4 μ_B per As is just 6 times the value gS = 0.24 expected for a Γ_7 doublet in a cubic crystalline field. Even though the exact Sm^{3+} moment is not known, it seems clear that a minimum of two and probably about 6 Sm^{3+} are induced by each As. Because these moments saturate easily (about 50 kOe at 4.2 K) it is probable that the individual Sm^{3+} moments are coupled forming a superparamagnetic cluster.

In order to learn more about these magnetic moments we have studied the magnetization down to 50 mK in fields to 3 kOe (Fig.3). If the clusters were uncoupled and the susceptibility obeyed a Curie law, a comparison of the Curie constant with the saturation moment would determine the size of the cluster moment and confirm the superparamagnetic behavior.

We find, however, that the clusters are coupled, for we observe spin glass ordering. A remanent magnetization appears at 4 K for the 4.5 % As concentration (inset fig. 3) and at 0.5 K for the 0.9 % concentration. This remanent magnetization represents about 10 to 12 % of the total saturation moment in agreement with observations in spin glasses composed of rare earth dilute alloys.[6] A susceptibility peak is observed at the onset of the remanent magnetization. Its relative amplitude is small because of the large Van Vleck contribution which hides the effect. However, with the onset of the remanent magnetization a small decrease of the initial slope can be seen in Fig. 3; In a 3 kOe field the magnetization is far from saturation as is generally observed for a spin glass.

Because of the similarity in ordering temperatures, it appears that the long range interaction which couples the clusters to each other is as large as the RKKY interaction observed for dilute rare earth impurities in noble metals. The rapid increase in the ordering temperature with concentration implies a strong dependence of the interaction on distance. Spin glass behavior is observed well before the As concentration reaches the percolation threshold for electronic conduction and well before the lattice collapse.

The results of the lattice constant and magnetization indicate that each As^{3-} induces its 6 Sm nearest neighbors to convert to Sm^{3+}. In such a process, 6 electrons are removed from the narrow Sm f band but only one is taken up by the As^{3-}. Assuming a mobili-

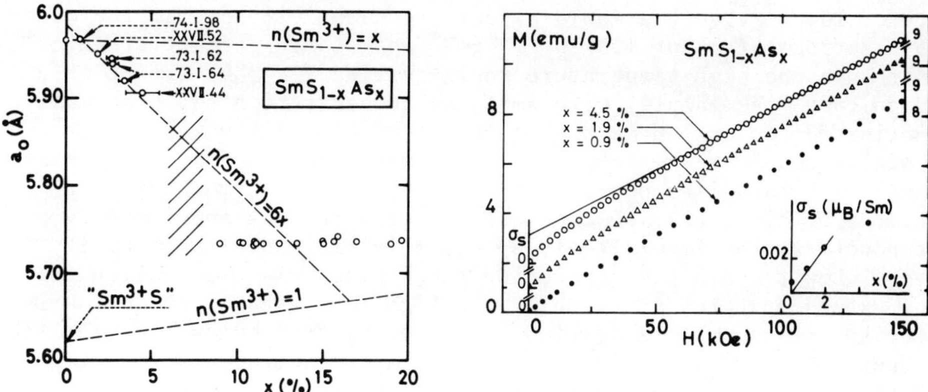

Fig. 1. (Left) Lattice constant vs. composition.

Fig. 2. (Right) Magnetization vs field at 4.2 K. Inset : saturation moment per mole samarium vs. atomic per cent arsenic.

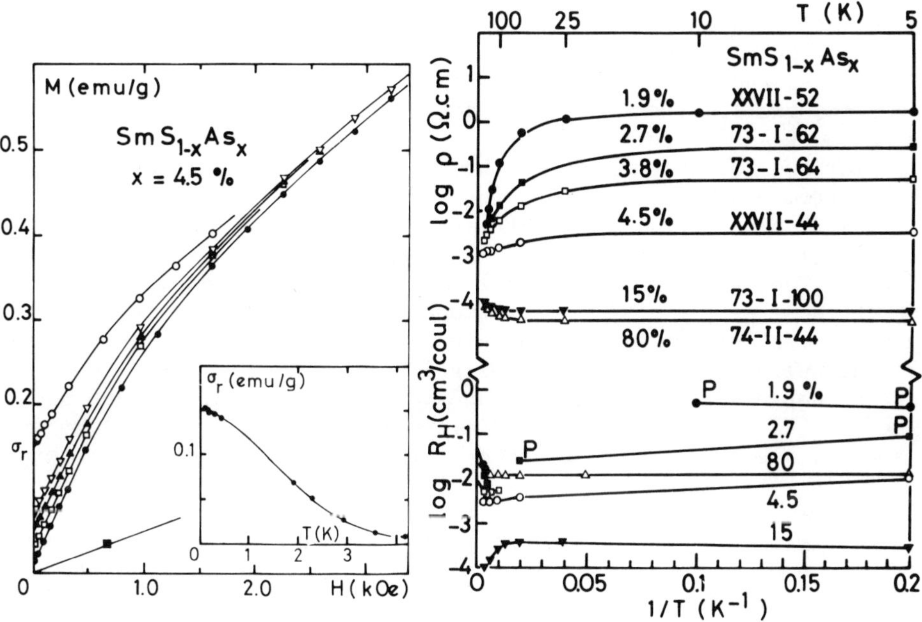

Fig. 3 (Left) Magnetization vs. magnetic field at 70 mK (O), 1.90 K (▽), 2.27 K (▲), 2.92 K (◻), 4.2 K (●). The susceptibility of SmS at 100 kOe is also shown ■ . Inset : Remanent magnetization (after cooling in a 3 kOe field) vs. temperature.

Fig. 4. (right) Resistivity and Hall constant vs. inverse temperature for various compositions.

ty of 10 $cm^2/v.sec$ we could expect a conductivity due to the other 5 electrons of about $1.5x10^3$ $\Omega^{-1}cm^{-1}$ per atomic percent arsenic. Although the high temperature resistivities (Fig. 4) are of this magnitude, the samples with small As concentration show thermally activated conduction at high temperature, indicative of electronic localization. At low temperature the metallic behavior is consistent with band like non-activated conduction of a much smaller number of electrons. In the language of amorphous and impurity band conductors, the Fermi level, E_F, is slightly above the mobility edge. Low temperature conduction takes place via the small number of extended states at E_F while the high temperature conduction is due to the thermal excitation of electrons from below the mobility edge. Electronic localization in such a concentrated system implies that the electrons are very tightly bound spatially. A measure of their extent can be inferred from Fig. 5 which shows the resistivities at high and low temperatures as a function of composition. The difference between these resistivities vanishes at about 5 % just before the lattice collapse (which occurs somewhere in the shaded region, Fig. 5). At this point, if there are 6 Sm^{3+} for each As, then the Sm^{3+} concentration is 30 % which is about the percolation threshold for site percolation.[6] Beyond this concentration the clusters of 6 Sm^{3+} are contiguous. If the electrons were to spread beyond a cluster of 6 Sm^{3+} then, because of their overlap, the limit of thermally activated conduction would occur at a lower As concentration. The 5 % concentration appears to be the maximum for which most of the clusters of six Sm^{3+} are insulated from each other by Sm^{2+} ions.

For comparison the resistivity of an intermediate valence sample (15 %) and a purely Sm^{3+} sample (80 %) are also given in Fig. 4. They are metallic at all temperatures. The Hall constant of the 80 % sample (Fig. 4) is temperature independent and gives a Hall electron concentration $n_H = 1/R_He = 5x10^{22}$ cm^{-3} very close to the sulfer concentration, which is the value expected from charge conservation if the valences are purely Sm^{3+}, S^{2-} and As^{3-}. The Hall effect of the intermediate valent case (15 %) is complicated by the overlapping f and d states and is not readily interpretable.

A similar problem exists for the Hall effect at small arsenic concentrations. As the temperature is reduced from room temperature, the Hall constant of the small concentration samples decreases (Fig. 4). In a simple one band picture, this decrease would imply an increase in carrier concentration, contrary to the observed increase in resistivity. In addition, for the 1.9 and 2.7 % samples, the sign of the Hall effect changes from n-type to p-type at low temperature (marked p on the figure). Such behavior could result from an impurity band with a density of states which decreases with energy. Another possibility is hole conduction in the nearly full f band which is broadened by hybridization with the Sm d-band and the As p-band.

Finally, we discuss the possibility of intermediate valence

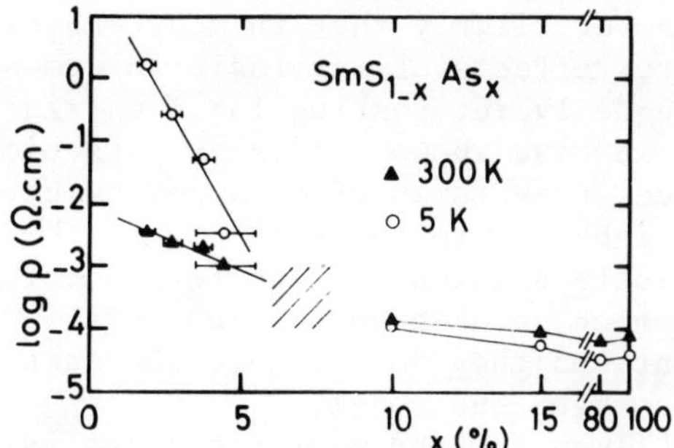

Fig. 5. Resistivity vs. composition at 5 K and 300 K.

as compared to complete conversion to integral valent Sm^{3+}. Homogeneous theories for SmS[7] and its alloys[8] show that the valence transition depends on a balance between electronic and lattice energies. After the phase transition, the intermediate valence state is stabilized by an anharmonic bulk modulus which increases with volume decrease, thus preventing a complete collapse to the Sm^{3+} state. Similar considerations will apply to the dilute case of the 6 Sm ions around an As impurity. Whether the local collapse is to a purely Sm^{3+} state or to an intermediate valence state must depend on a balance of these effects. The evidence suggests that the purely Sm^{3+} case is more likely. Since the change in lattice constant is equivalent to six fully collapsed Sm^{3+} per As, if the collapse were to an intermediate valence, it must spread out to more than 6 Sm. In this case it would be difficult to have electronic localization up to 5 % As. The susceptibility χ of intermediate valent SmS, produced by pressure or alloying is considerably less than χ of Sm^{2+} or Sm^{3+} at low temperature[10]. In general the intermediate valence corresponds to an f level strongly mixed with the Fermi sea. The consequence is that the local magnetism disappears by Kondo effect so the moment of intermediate valent ion is harder to saturate than ionic Sm^{3+}. If this is also true for clusters of 5 electrons, then our observation of the large easily saturable moment (Fig 2) is evidence for ionic behavior of Sm^{3+} and against intermediate valence.

Local environmental effects in SmS and its alloys have been discussed previously both theoretically[11] and in relation to EPR[12], photoemission[13], and a combined study[14] of resistivity, magnetization and specific heat. In the case of the EPR study of Gd impurities in SmS, it was argued that for concentrations up to 1 % the extra electron from each Gd^{3+} remained localized. It seems to be generally true that systems with an unstable valence will be particularly sensitive to the presence of impurities, deviations from stoechiometry and defects of all sorts. Whether the defects will cause local or homogeneous effects will depend on their nature and concentration. It is very likely that in many experimental situations the saught after effects of intrinsic intermediate valence are accompanied by equally interesting local extrinsic effects.

In conclusion, we have shown that small arsenic concentrations in SmS cause the local conversion of a larger number of Sm (probably the 6 nearest neighbors) to the Sm^{3+} state, the 6 Sm^{3+} form superparamagnetic clusters which in turn form a spin glass. The localized behavior which we observe implies that in this region all Sm are not equivalent and they do not have the same valence as one would suppose in a homogeneous model.

We thank S.O. Ellman for the microprobe analysis and J.M. Rigotty for the transport measurements.

§and Service National des Champs Intenses, CNRS, BP 166 X, 38042 Grenoble-Cedex, France
*Permanent address : IBM Research Center, Yorktown Hts, N.Y. 10598.

1 - Part of the thesis of O. Peña.
2 - S. Von Molnar, T. Penney, F. Holtzberg, J. Physique Coll. 37 C4- (1976).
3 - R.A. Pollak, F. Holtzberg, J.L. Freeouf, D.E. Eastman, Phys. Rev. Lett. 33, 820 (1974).
4 - K.H.J. Buschow, A.M. Van Diepen, H.W. de Wijn, Phys. Rev. B11, 5134 (1974).
5 - F. Holtzberg, J.L. Tholence, R. Tournier, Proc. 2nd Amorphous Magnetism Conference, Troy N.Y. (1976) Plenum.
6 - S.Kirkpatrick, Rev. Mod. Phys. 45, 574 (1973).
7 - L.L. Hirst, J. Phys. Chem. Solids 35, 1285 (1974).
8 - T. Penney, R.L. Melcher, J. Physique Coll. 37, C4- (1976).
9 - M.B. Maple, D. Wohlleben, Phys. Rev. Lett. 27, 511 (1971).
10 - L.J. Tao, F. Holtzberg, Phys. Rev. B11, 3842 (1975).
11 - T. Kasuya, J. Physique Coll. 37, C4- (1976) and this conf.
12 - W.M. Walsh Jr, L.W. Rupp Jr, E. Bucher, L.D. Longinotti, AIP Conf. Proc. 18, 535 (1974).
13 - M. Campagna, S.T. Chui, G.K. Wertheim, E. Tossati, Phys. Rev. B14, 653 (1976).
14 - G. Chouteau, O. Peña, F. Holtzberg, R. Tournier, T. Penney, S. Von Molnar, J. Physique Coll. 37, C4- (1976).

KONDO EFFECT IN METALLIC SmS FILMS

A. Chenevas-Paule[§], P. Haen, F. Lapierre[†], R. Tournier, and L. Vieux-Rochas[§]

Centre de Recherches sur les Très Basses Températures, CNRS, BP 166 Centre de Tri, 38042 Grenoble-Cedex, France

The existence of an increase of the resistivity at low temperature in the metallic intermediate valence phase of pure SmS under pressure has been shown previously[1]. We suggest that this effect is not intrinsic but rather due to magnetic Sm^{3+} impurities. The presence of such Sm^{3+} impurities has been observed in the semiconducting phase of SmS by EPR[2] and in $Sm_{1-x}La_xS$ alloys by resistivity and magnetization measurements[3]. At helium temperatures the specific heats of metallic SmS[1] and of $Sm_{1-x}Y_xS$ alloys[4] exhibit, in addition to the "electronic" linear term, anomalies which we believe are caused by the presence of a Kondo temperature of the Sm^{3+} impurities or to a magnetic ordering of such localized moments.

An alternative explanation is that the collapsed phase of SmS is an intrinsic Kondo compound. The linear term of 145 mJ/mole.K^2 in its low temperature specific heat[1] could be attributed to the Kondo condensation of all Sm ions present in the compound, as it is observed in many cerium compounds[5]. The magnitude of this term implies a Kondo temperature of about 50 K. In such an interpretation, however, one expects that the resistivity would reach a low residual value at zero K, because the Sm ions form a regular lattice, contrary to the results presented here.

The resistivity of the gold-collapsed phase of SmS can be studied on stable films, converted by means of polishing[6] or rubbing[7] the surface of a semiconducting film. We have prepared SmS films of various thickness (1000 to 5000 Å) by coevaporation and deposition on heated glass substracts (350°C). The as grown films were black, single phase and had the lattice constant of bulk semiconducting SmS. Their room temperature resistivities $\rho(300)$ varied from the order of 10^{-3} to 10^{-2} Ω.cm. The gold phase was obtained by rubbing with a 2 mm dia. glass sphere. The $\rho(300)$ values were reduced by

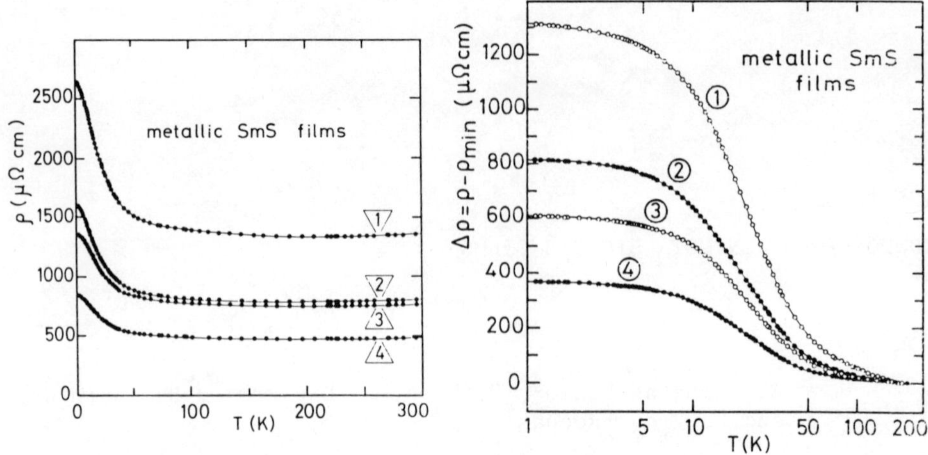

Fig. 1. (left). Resistivity vs. T of metallic SmS films of thickness 1) = 4700 Å, 2) = 5000 Å, 3) = 2000 Å, 4) = 1600 Å.

Fig. 2. (right). Variation of $\Delta\rho = \rho(T) - \rho(T_{min})$ vs. Log T.

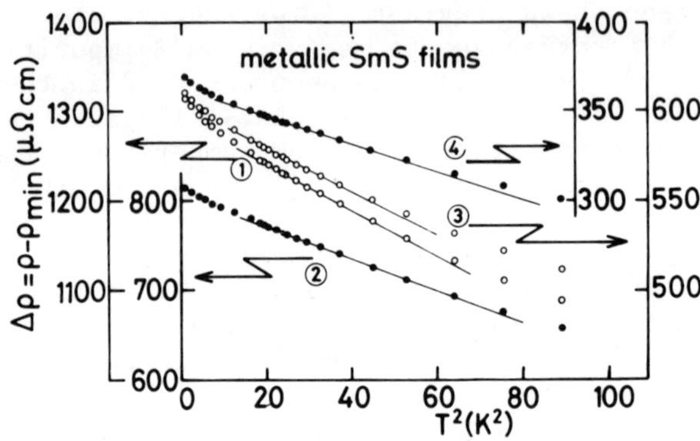

Fig. 3. Variation of $\Delta\rho$ vs. T^2 in the range 1 K - 10 K.

factors up to 5 which depending on both the initial value and the thickness of the film. These results are in good agreement with previous ones on bulk[1,8] or film specimens[6,7]. We have chosen four of such metallic films and measured their resistivity from 1 K to 300 K. Their room temperature resistivities extend from 5×10^{-4} to 1.4×10^{-3} Ω.cm for thickness extending from 1600 to 5000 Å. They exhibit at 1.2 K a value $\rho(1.2) \simeq 2 \times \rho(300)$ and a slight minimum near 200 K (fig. 1). A plot of $\Delta\rho = \rho(T) - \rho(T_{min})$ against Log T shows that $\Delta\rho$ varies as Log T in the range 15 K - 35 K, tends to a saturation below about 3 K (fig. 2) but does not show the onset of a decrease to a lower residual resistivity. From this data the Kondo temperature can be estimated to lie between 20 K and 50 K. A plot of $\Delta\rho$ between 1 and 10 K as a function of T^2 shows a good T^2 behavior (fig. 3) which is characteristic of a Kondo behavior below T_K. The deviation from the T^2 variation observed at the lowest temperatures may be due to the contribution of some Sm^{3+} impurities having a lower Kondo temperature than others. Such deviation arises in many dilute Kondo alloys, such as Cu-Fe[9] and are correlated with interactions between impurities[10]. The existence of a small negative magnetoresistivity at 1.5 K (in 50 kOe we found a relative variation of ρ of $\simeq 10^{-3}$) confirms the presence of a high T_K.

This T_K is in apparent agreement with that deduced from the "electronic" specific heat. However, the constant low temperature resistivity eliminates the possibility of an intrinsic effect : metallic SmS is not a Kondo compound similar, for example, to $CeAl_3$. The linear term in the low temperature specific heat suggests strongly that it is a nearly magnetic metal such as α cerium or $CeSn_3$. In this case one expects that in the absence of impurities, its resistivity would be an increasing function of temperature. The opposite is observed for the collapsed phase at low temperature. We conclude that this anomaly is due to Sm^{3+} which are present in all Sm compounds with or without pressure.

[†]Partial fulfilment of a thesis of F. Lapierre.

[§]Laboratoire d'Electronique et de Technologie de l'Informatique, CENG, BP 85 X, 38041 Grenoble, France.

1 - S.D. Bader, N.E. Phillips, D.B. McWhan, Phys. Rev. B7, 4686 (1973).
2 - W.M. Walsh, Jr., E. Bucher, L.W. Rupp, Jr., L.D. Longinotti, AIP Conf. Proc. 24, 34 (1975).
3 - G. Chouteau, F. Holtzberg, O. Peña, T. Penney, R. Tournier, S. Von Molnar, J. Phys. Coll. 37, C4-001 (1976).
4 - S. Von Molnar, F. Holtzberg, AIP Conf. Proc. 29, 394 (1976).
5 - B. Coqblin et al. Proceedings of the 1975 discussion meeting on magnetic semiconductors, Editor W. Zinn, North-Holland Publishing Company 1976.
6 - S.G. Shul'man, T.L. Bzhalava, T.B. Zhukova, T.T. Dedegkaev, I.A. Smirnov, Soviet Phys.-Solid State 17, 1984 (1975).
7 - D.W. Pohl, R. Jaggi, K. Gisler, H. Weibel, Sol. State Comm. 17 705 (1975).
8 - A. Jayaraman, N. Narayanamurti, E. Bucher, R.G. Maynes, Phys. Rev. Lett. 25, 1430 (1970).
9 - W.M. Star, F.B. Basters, G.M. Nap, E. de Vroede, C. Van Barle, Physica 58, 585 (1972).
10 - J.L. Tholence, R. Tournier, Physica 84B, 3 (1976).

EFFECT OF PRESSURE ON THE RESISTIVITY AND MAGNETIZATION OF SmS, $(Sm_{0.75}Y_{0.20}Eu_{0.05})S$ AND $(Sm_{0.95}Eu_{0.05})S$

B. Cornut, F. Holtzberg*, O. Peña, J. Voiron§

Centre de Recherches sur les Très Basses Températures,

CNRS, BP 166 Centre de Tri, 38042 Grenoble-Cedex, France

With increasing Y concentration $Sm_{1-x}Y_xS$ shows a collapse from a black phase to a gold intermediate valence phase. We have added Eu with the hope of observing a transition between Eu^{2+} and Eu^{3+} using both chemical pressure and external pressure. We did not find a change in Eu valence but did find that the Eu^{2+} induces a moment of order 1 μ_B on the Sm which decreases substantially when the SmYS host collapses to the intermediate valence phase under pressure. Quasi hydrostatic pressure is obtained using a small CuBe clamp. Figure 1 shows the magnetization versus field curves of, at 7 kbars, $(Sm_{.75}Y_{.20}Eu_{.05})S$. This magnetization is dominated by the Eu contribution. At 0 pressure and high field we measure a susceptibility $\chi_f = 9.8\ 10^{-3}$ emu/mole Sm. This value corresponds to the Van Vleck susceptibility of Sm^{2+} in the SmYS alloy which is in the non collapsed phase[1]. Pressure decreases χ_f to 6.4 10^{-3} emu/mole Sm, which shows that we have obtained the collapsed phase. From the saturation magnetization and the nominal (5 %) concentration of Eu we deduce a moment σ of 8.7 μ_B/Eu ion at p = 0 and 7.6 μ_B/Eu ion at p = 7 kbar. Since the Eu^{2+} moment is 7 μ_B/Eu ion, the Eu must induces a weak moment on its Sm neighbors, which decreases by 1.1 μ_B/Eu ion when the lattice collapses under pressure. We have also measured the magnetization in the semiconductor $(Sm_{.95}Eu_{.05})S$ with the same Eu concentration and find the normal value $\sigma_{Eu} = 7\ \mu_B$/Eu ion. We conclude that the magnetic interaction between the Sm and the Eu is due to the conduction electrons and that this coupling is more effective in the simpler black phase than in the high pressure intermediate valence phase. Previously a coupling has been observed on the Van Vleck susceptibility of SmS which increases when conduction electrons are introduced in the black phase by Y substitution.[1] The mean internal field H_i we deduce (from $\sigma - \sigma_{Eu} = \chi_f H_i$) is about 60 kOe at p = 0 and 35 kOe at p = 7 kbar.

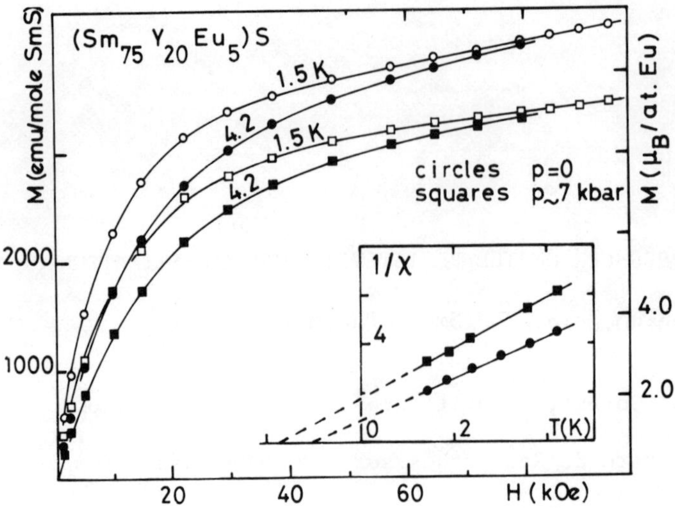

Fig. 1. Magnetization curve of $(Sm_{.75}Y_{.20}Eu_{.05})S$. The inset shows the initial reciprocal susceptibility at p = 0 (circles) and p = 7 kbar (squares).

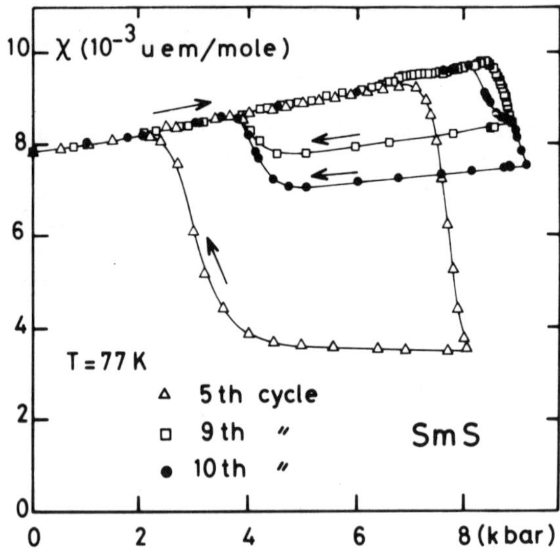

Fig. 2. Susceptibility of SmS versus hydrostatic pressure at 77 K.

The inset (fig. 1) shows the inverse initial susceptibility of the sample at 0 and 7 kbar. We observe a Curie Weiss like behavior which indicates a coupling between the magnetic clusters and prevents an evaluation of the exact nature of these clusters. However the large Curie Constant we deduce is consistent with the idea of an Sm induced moment.

Finally we indicate here that we have also observed the structural collapse of $Sm_{.95}Eu_{.05}S$ at 8 kbar by room temperature resistivity measurements.

SmS under hydrostatic pressure

With pressure, SmS collapses to an intermediate valence state with a large decrease in susceptibility[2]. Initially we applied pressure up to 8 kbar at about 70 K, and no transition was observed. However after applying 8 kbar at room temperature we observed the collapse of SmS. We conclude, therefore, that the slope, $\frac{dP}{dT}$, of the low temperature phase boundary is negative at 8 kbar as indicated previously[3].

After this first transition the single crystal exploded (even with hydrostatic pressure) indicating a spatial variation of critical pressure. Another pressure cycles have been performed. The non infinite slopes of the transition seen on figure 2 confirm the existence of this inhomogeneity which has also been observed on the collapsed phases of $Sm(S_{1-x}As_x)$ which explode on cooling[4].

Subsequent cycles at temperatures around 150 K show a small decrease in the width of the transition. At the same time the pressure where the transition begins, increases with the number of cycles. In fact, during the 8th cycle at 300 K, the transition was not observed up to 8 kbar. This indicates that defects affect greatly the critical pressure which depends of the history of the sample.

The last two cycles (9th and 10th) performed at 77 K without annealing at 300 K between the cycles show definitively that each piece of the sample has a particular critical pressure. Namely during the 9th cycle only the part of the sample which has the lowest critical pressures is collapsed. Their mean critical pressure is reduced whereas the other part of the sample has not changed as seen on 10th cycle.

In conclusion, we have found that the critical pressure depends on the temperature and the defects in the sample. The critical pressure is changed by the collapse itself and by temperature cycling which presumably change the defects.

We have used the high pressure equipment of the Service National des Champs Intenses. We wish to thank the high pressure group of CNRS for helpfull discussions and technical assistance.

*Permanent address : IBM Research Center, Yorktown Hts, N.Y. 10598.
§Service National des Champs Intenses, CNRS, BP 166 Grenoble,France

References

1 - L.J. Tao and F. Holtzberg, Phys. Rev. B11, 3842 (1975).
2 - M.B. Maple, D. Wohlleben, Phys. Rev. Letters 27, 511 (1971).
3 - A. Jayaraman , V. Nayaranamurti, E. Bucher, R.G. Maines, Phys. Rev. Letters 25, 1430 (1976).
4 - F. Holtzberg, AIP Conference n° 18, 478 (1974).

PRESSURE AND TEMPERATURE EFFECTS IN SOME MIXED VALENCE SYSTEMS

T.G. Ramesh and V. Shubha

Materials Science Division

National Aeronautical Laboratory, Bangalore, India

We recently reported the 4f-5d electronic phase transition in SmS as investigated by Thermoelectric power[1] (Q). Some new data on the low temperature behaviour of chemically collapsed $Sm_{0.8}Tb_{0.2}S$ and further evidence for the existence of B' phase after the explosive phase transition is reported here. Figure 1, presents the variation of Q with temperature ranging from -200°C to +200°C. The most notable feature is the large negative values of Q at lower temperatures. The explosive phase transformation[2,3] is seen by the large discontinuity in Q occuring near -170°C. The sign of Q in B and B' phases are opposite lending support to the fact that their electronic structures are different. The decrease of Q at higher temperatures leading to a change of sign around +200°C is associated with the continuous phase transformation from metallic to B-phase.

Figure 2 gives the pressure dependence of Q in $Sm_{0.8}Tb_{0.2}S$ and in the metallic phase of SmS. The important feature is the anomalous increase in Q_{SmS} and $Q_{Sm_{0.8}Tb_{0.2}S}$ over a narrow range of pressures.

The anomalous temperature and pressure dependence can be qualitatively understood on the basis of the model where in the 4f band with high density of states lies at the Fermi level. The negative value of Q and its rapid variation between -170°C to -50°C can be understood on the basis of the hypothesis that with increase of temperature the centre of 4f resonance gets shifted downwards with respect to the Fermi level. This would make $(d\tau/dE)_{E_F}$ strongly positive there by making a large and negative contribution to Q. The decrease in Q at higher temperatures can be correlated with the

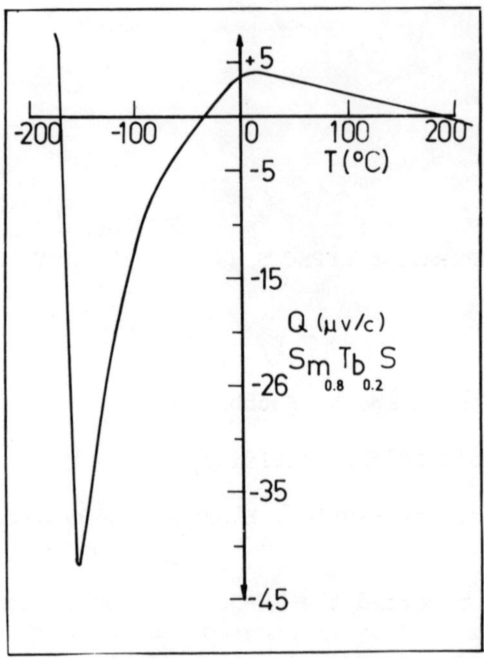

Figure 1. Explosive phase transition in $Sm_{0.8}Tb_{0.2}S$.

continuous change of valence of Samarium ions towards the divalent state.

The anomalous pressure behaviour (Fig.2) can be explained if one assumes that with the increase of pressure the Fermi level probes the region lying below the 4f resonance. This is just opposite to the temperature effect discussed earlier.. This gives rise to $(d\tau/dE)_{E_F}$ becoming negative thus making a positve contribution to Q. It is of interest to study the low temperature behaviour of the high pressure phase of SmS and experiments in this direction are in progress in our laboratory.

We are grateful to Dr.A.Jayaraman for providing us with the samples and to Dr.S.Ramaseshan for his keen interest in this work. One of us (V.S) would like to acknowledge C.S.I.R for the grant of

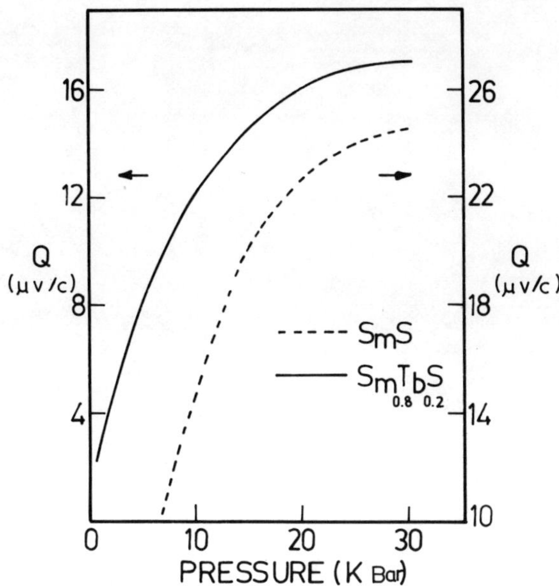

Figure 2. Q Vs P in SmS and $Sm_{0.8}Tb_{0.2}S$.

a fellowship.

References:

1. Ramesh,T.G and Shubha,V., Solid State Comm.,19, 591,(1976).
2. Jayaraman,A, Dernier,P.D, and Longinoth,L.D, Phys.Rev. B11, 2783 (1975).
3. Tao,L.J. and Holtzberg,F., Phys.Rev., B11, 3842 (1975).

ELECTRONIC CONFIGURATIONS OF Sm^{2+} IN SmS AND SmB_6, AND ELECTRICAL PROPERTIES

X. Oudet

Laboratoire de Magnétisme, C.N.R.S.

1, Place Aristide Briand - 92190 Meudon (France)

In Dirac's theory the 4f shell is divided into two subshells : $4f_{5/2}$ with six quantum states and $4f_{7/2}$ with eight quantum states. Assuming that these two subshelles can be filled up successively, we had successfully obtained the saturation magnetic moment of the garnets $Ln_3Fe_5O_{12}$ (1,2). This study is devoted to establish the role of the subshells in the case of SmS phase transformation under pressure P (3). Particularly for Sm^{2+} the low spin configuration $4f^6_{5/2}\ 4f^0_{7/2}$ which with six 4f electrons has a filled subshell $4f_{5/2}$ leading to a zero magnetic moment.

But as the insulator $\rightleftarrows$ metal I $\rightleftarrows$ M transition exhibited with this phase transition, is explained through a valence change, we first consider the localized electron model proposed in the case of Fe_3O_4 (4) to show that large variations of electrical resistivity of SmS and SmB_6 also can be understood through this model. A result which allows us to consider the ion Sm^{2+} in both phases of SmS.

ELECTRICAL PROPERTIES

The I $\rightleftarrows$ M transition is not a specific property of SmS. Large variations in the electrical conductivity associated with a phase transformation P.T. are also to be found in other compounds such V_2O_3 (5,6) and Fe_3O_4 (7). Even if no P.T. occurs a compound can pass progressively from the I type into the M type as in the case for SmB_6 (8,9). This leads us to think that the large variations in the electrical conductivity of a compound are rather related to a single mechanism than to the particular compound itself.

We have already proposed this mechanism in the case of Fe_3O_4 (4). The electron responsible for the conduction is to be thought

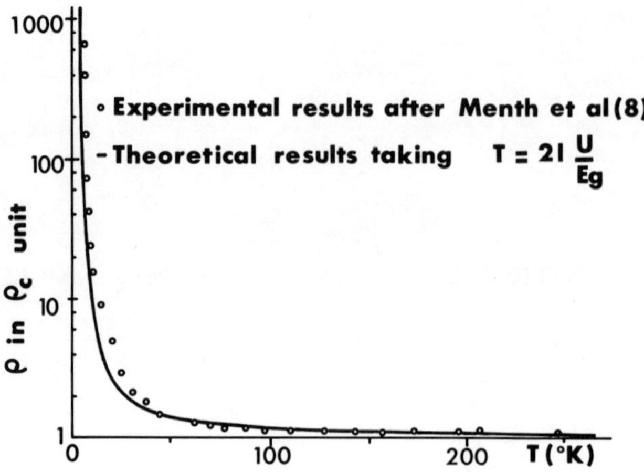

Fig.1 Theoretical and experimental resistivity of Sm B_6

as localized on its ions. The absence of magnetic ordering in the two phases of SmS (10,11) as well as in SmB_6 (9) corroborate this hypothesis, which allows us to consider the ion Sm^{2+} with the $4f^6_{5/2}\ 4f^0_{7/2}$ configuration, the probability of which tends towards one when the temperature tends towards zero (§ II).

We write $\rho = \rho_c/P(Eg,U)$ (4). P(Eg,U) is the probability for an electron with energy higher than the potential barrier Eg to leave its atom, U the mean thermal energy per electron an ρ_c the resistivity which the material would have if all the electrons of the level Eg were to contribute to the current flow. The statistical analysis gives

$$P(Eg,U) = B \ln [1+\exp.-\alpha(Eg/U -1)] \quad B=0.5864, \quad \alpha=1.5049.$$

The comparison of $\rho = \rho_c/P(Eg,U)$ with the experimental results, more easy without phase transformation, is given in figure I for SmB_6. As the variations of ρ are determined experimentally as a function of T, the knowledge of U(T) is necessary for calculating ρ (T). We are faced here with a new problem for which there are no experimental data available. Since U(T) is not known, we take U(T) to be linear with T. From the Dulong and Petit law, this approximation is reasonable in the higher temperature region. But in the low temperature region from the curve of specific heat (12) the linear approximation above could make the increase of ρ with the decrease of T more slow than it would be for the exact function U(T). Taking U/Eg = T/21 we obtain a good fit. From the measurements of the specific heat (9) we find $Eg < 10^{-3}$ eV. Thus metallic type is due to semiconductor state with a feeble value of Eg. When U/Eg = 2, then almost all the electrons contribute to the electrical conductivity and the compound is rather like a metal. When T tends zero, U also tends to zero and ρ increases. This is verified not only with SmB_6 (8,9) but also for SmS in the M phase (12).

THE PHASE TRANSFORMATION OF SmS

To explain the nature of the phenomenon of the lattice constant shrinkage we suppose that the third valence electron keeping its 4f character promotes a particular binding leading to this shrinkage. Then let us consider two $4f^6$ configurations :

$$4f^6_{5/2}\ 4f^0_{7/2} \text{ low spin} \quad 4f^5_{5/2}\ 4f^1_{7/2} \text{ high spin.}$$

In the first configuration the subshell $4f_{5/2}$ is filled making Sm^{2+} similar to the one having a filled shell like Zn^{2+} for instance. Such configuration without unpaired electron does not favor the type of binding responsible for the shrinkage of the lattice constant, but favors the phase I. On the contrary the high spin configuration with two unpaired electrons one of them being the third valence electron favors this binding, and consequently the phase M. Now let E_p be the energy necessary for the electronic transition

$$4f^6_{5/2}\ 4f^0_{7/2} \longrightarrow 4f^5_{5/2}\ 4f^1_{7/2}$$

E_p is known to diminish when the atoms are drawn together (10), the SmS phase transition can be explained as follows : let us consider SmS in the phase I at 0 K, the amount of heat received is zero. All the Sm^{2+} ions are in the low spin configuration. But when the temperature increases, the Sm^{2+} ions receive an amount of heat a part of this, U(T) in mean value, induces the high spin configuration. The concentration C_f of the Sm^{2+} ions having this configuration increases with T, as the high spin configuration favors the phase M the pressure for the P.T. decreases when T increases. This aspect is borne out by the study of the ρ(P,T), (12). In the same way the increase of P decreases Ep and hence induces P.T. (3).

REFERENCES

1. X. Oudet Phys. Let 48A, 252, (1974)
2. X. Oudet 11th rare Earth Research conf. U.S.A.E.C., technical infor. center, Traverse City - Michighan (U.S.A.), 1, 97,(1974)
3. A. Jayaraman, et al., Phys. Rev. Let. 25, 1430, (1970)
4. X. Oudet I.C.F. II to be published in J. de Phys. (France)
5. M. Foëx, CR. Hebd. Séances Acad. Sci. 223, 1126, (1946)
6. E.P. Warekois, J. App. phys. 31, 346S, (1960)
7. B.A. Calhoun, phys. Rev. 94, 1577, (1954)
8. A. Menth, et al. Phys. Rev. Let. 22, 295, (1969)
9. J.C. Nickerson, et al. Phys. Rev. B 3, 2030, (1971)
10. E. Bucher, et al. J. Appl. phys. 42, 1741, (1971)
11. M.B. Maple, et al. Phys. Rev. Let. 27, 511, (1971)
12. S.D. Bader, et al. Phys. Rev. B 7, 4686, (1973)

INTERMEDIATE VALENCE PROPERTIES OF A SERIES OF Ln_4X_3 COMPOUNDS WITH ANTI-Th_3P_4 STRUCTURE

E. Bucher & A. S. Cooper
Bell Laboratories, Murray Hill, N.J. o7974/USA
and
D. Jaccard & J. Sierro
University of Geneva, Geneva/Switzerland

1. INTRODUCTION

Recently, Gambino[1] has explored crystallographically a number of Ln_4X_3 compounds. In particular, he has discovered considerable deviations in $Sm_4(Bi_3, Sb_3)$ and $Yb_4(Bi_3, Sb_3)$ from the trivalent systematics. From the lattice constant of Yb_4Sb_3 e. g. it was concluded that this compound could be written as $Yb^{3+}_{4/3}Yb^{2+}_{8/3}Sb_3$. This is very close to $Yb^{3+}Yb^{2+}_3Sb^{3-}_3$ in an ionic binding model. The question thus arises whether these compounds belong to the inhomogeneous or homogeneous mixed valence compounds. We have tried to give an answer to this question by measuring magnetic and transport properties. Furthermore, one of the underlying motivations was to achieve giant Seebeck-coefficients for possible application as thermoelectric generator materials or thermoelectric cooling substances.

2. EXPERIMENTAL PROCEDURE

All 6 compounds listed in table I were obtained by direct fusion in sealed tantalum tubes (1/4"-3/8" o. d.) by resistance heating. The Sm-, Eu- and Yb-compounds appear to melt congruently. They do not wet the Tantalum. Therefore they were cooled down relatively fast within about 1 hr from the melting point to room temperature.

Susceptibilities were taken in a Faraday balance with constant gradient pole pieces in fields of up to 9.6kOe. Seebeck-coefficients were taken on spark-cut square samples of 7.5 - 13 mm length in pulsed temperature gradients.

3. RESULTS

Experimental results are summarized in table I. The Sb-compounds show a trend to be closer to the trivalent state than the Bi-compounds, e. g. Sm_4Sb_3 exhibits a relatively high ordering temperature of 14oK as is typical for the 3+ state of Sm, whereas Sm_4Bi_3 exhibits a delocalization of the 4f-electrons and Eu_4Bi_3 a magnetic transition at low temperature with an effective moment close to the 2+ state. Resistivities range typically in the region between 1 - 5 mΩ cm at room temperature, however their values drop at 77K by about a factor of 6 - 7. Seebeck-coefficients below room temperature reach values up to 45 μV/K. Except for Sm_4Bi_3 all samples exhibit negative values at room temperature. Sm_4Sb_3, La_4Sb_3 and Yb_4Bi_3 undergo a change in sign at lower temperature.

4. DISCUSSION

For space reasons, we will confine here our discussion to the magnetic properties, which are the most conclusive ones.

<u>Sm_4Sb_3, Sm_4Bi_3:</u> In agreement with lattice constants and atomic size considerations between Sb and Bi, compounds with Bi tend more towards the 2+ state of ambivalent rare earth ions, whereas Sb compounds are closer to the 3+ state. Sm_4Sb_3 exhibits a relatively high ordering temperature of 14oK, which is typical for the 3+ state of the Lanthanide. In contrast, Sm_4Bi_3 shows a temperature dependence of the susceptibility closely resembling $Sm_4^{2+}Bi_3$, however with a value about 3o % lower over the whole temperature range of 1 - 3ooK and a weak Curie-like anomaly superposed at low temperature. From the ferromagnetic exchange interaction in Gd_4Bi_3 (2) and the nearly divalent Eu_4Bi_3 we must conclude that the bare 4f-susceptibility is even lower than the experimental values.

We therefore are led to the conclusion that the 4f-levels in Sm_4Sb_3-Sm_4Bi_3 offer the possibility to study physical properties at any degree of delocalization between the two end members. In particular, recently studied lifetime effects of valence states related to magnetic properties could be of considerable interest. (3)

<u>Eu_4Bi_3</u>: Although this composition is reported to melt congruently, we encountered difficulties in obtaining single phase material. This is reflected in two transition temperatures at 12 and 36.7 K, and a number of extra lines in the Debye-Scherrer pattern. The effective moment was slightly reduced, compared to the Eu^{2+} state. Nevertheless the much lower transition temperature of Eu_4Bi_3 (either 12 or 36.7 K) is striking if compared to a Curie temperature of 346 K of the isomagnetic counterpart Gd_4Bi_3. It is possible, that Eu_4Bi_3 has similar properties like TmSe[4] which undergoes a magnetic transition in an ICF state.

<u>Yb_4Sb_3, Yb_4Bi_3</u>: Both compounds exhibit complex magnetic behavior, in the temperature range 1.3 - 300K. Yb_4Sb_3 shows a weakly paramagnetic susceptibility, however with a sharply increasing tail below about 20K but without magnetic ordering above 1.3K. The de Gennes function leads to an expected transition at 7K. Yb_4Bi_3 exhibits a susceptibility about an order of magnitude lower.

1) R. J. Gambino, J. Less-Common Metals <u>12</u>, 344 (1967)
2) F.Holtzberg, T. R. McGuire, S. Methfessel and J. C. Suits, J. Appl. Phys. <u>35</u>, 1033 (1964)
3) M. Campagna, G. K. Wertheim and E. Bucher, Structure and Bonding <u>30</u>, 125 (1976)
4) E. Bucher et al., Phys. Rev. B <u>11</u>, 500 (1975)
5) K. Andres, unpublished

TABLE I (Properties of Ln_4Sb_3, Ln_4Bi_3)

Compound	Lattice const.	Magn.prop.	Seebeck-coeff.(r.t.)
La_4Sb_3	9.63 Å[1]		-5.2μV/K
Sm_4Sb_3	9.317 Å	T_N=140K	-4.4μV/K
Sm_4Bi_3	9.815 Å	χ(77)= 5.4x10⁻³cm³/mole	+40μV/K
Eu_4Bi_3	9.944 Å	T_N=12K, 36.7K (two phase)[5]	
Yb_4Sb_3	9.320 Å	χ(300)=0.89x10⁻³cm³/mo.	-23μV/K
Yb_4Bi_3	9.572 Å	χ(300)=0.054x10⁻³cm³/mo.	-16μV/K

OPTICAL PROPERTIES OF SmB_6

J. W. ALLEN

XEROX PALO ALTO RESEARCH CENTER

PALO ALTO, CALIFORNIA 94304

SmB_6 is of interest as one of a group of mixed configuration compounds in which there is noninteger occupation of the 4f-shell of the rare earth ion. The reflectivity of SmB_6 has been measured at room temperature in the energy range of .02 to 5 eV, and at 15K in the energy range .35 to 5 eV. Samples used consist of small flux-grown, single-crystal flakes obtained from Z. Fisk, and large polycrystalline melt pieces obtained from J.C. Mikkelsen. The polycrystalline samples were measured in the entire energy range, but the single crystals were too small for use between .02 and .35 eV. Between .35 and 5 eV data from both types of samples agreed well.

Fig. 1 shows the reflectivity data for the polycrystalline sample, the dielectric constant obtained from a Kramers-Kronig analysis of the room temperature data, and high and low energy extrapolations of the data used in the Kramers-Kronig analysis. The most prominent reflectivity feature is the edge at 1.5 eV, which has the general appearance of a metallic plasma edge associated with intraband absorption. But the behavior of ε_1 shows large departures at low energy from simple metallic intraband behavior. In the vicinity of the reflectivity edge minimum ε_1 passes through zero, as would be expected, but it does not continue to increasingly large negative values as would usually be the case. Instead it turns back through zero and acquires a large positive value down to the lowest energy for which data exists. The details of the shape of ε_1 near .02 eV and at lower energies obviously depend on the low energy extrapolation used in the reflectivity, and hence are presently unknown. But it is clear that there is intense low energy absorption causing the positive value of ε_1. Fig. 1 also shows that the shape and position of the edge are largely unchanged at T=15K. The reflectivity data are generally very similar to those reported for the

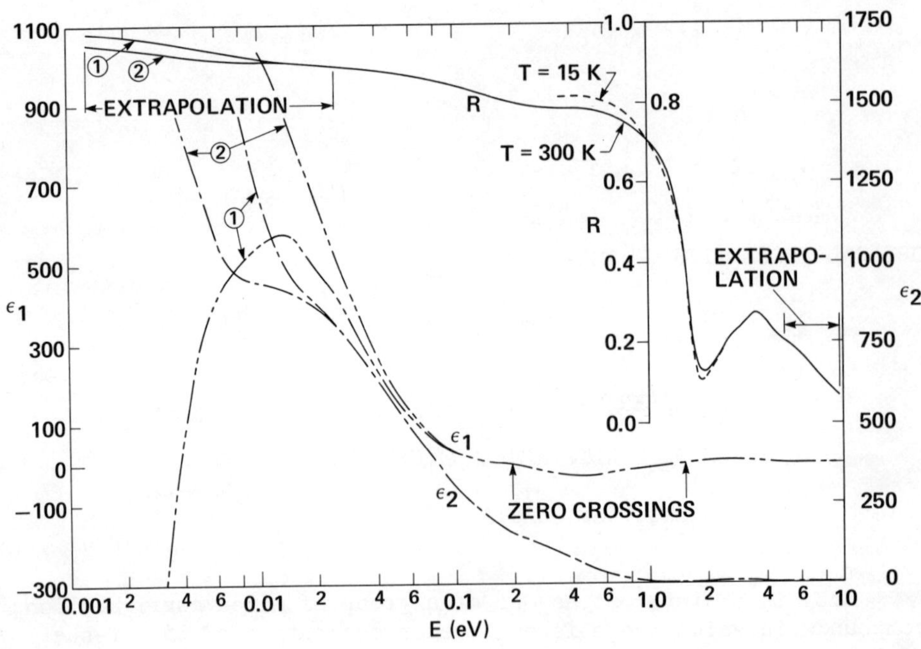

Fig. 1. Reflectivity data for SmB_6, dielectric constant obtained by Kramers-Kronig analysis of room temperature data, and high and low energy extrapolations used for Kramers-Kronig analysis.

mixed configuration phase of SmS and for TmS and TmSe.[1]

Discussing the optical results should be carried out in the context of the other properties of SmB_6, especially the d.c. resistivity. Above about 60K the resistivity is temperature independent or slightly increasing but below 60K it has a minimum and then increases rapidly as the temperature decreases. Although Nickerson et. al.[2] have reported a saturation of this increase at about 10K, data by Menth et. al.[3] and data from the samples used in this reflectivity study show no saturation down to 4K. The lack of a magnetic moment at low temperatures[3] is an added constraint on models for SmB_6.

The interpretation to be given to the optical spectrum is uncertain at present. There is a paucity of theoretical predictions and it is not clear what parts of experience with other solids are relevant. The assignment of the reflectivity edge is not straightforward. The possibility of employing a semiconductor model, as opposed to the metallic one[1] suggested by the shape of the edge, will be explored briefly here. The fast low temperature increase of the d.c. resistivity is compatible with a very small gap semiconductor model.[2] If the increase were due to a lifetime effect not involving impurities, as in a metallic model, then it should saturate at some low temperature. But the preponderance of the present data is that there is no saturation down to 4K. The lack of a strong temperature dependence of the position or shape of the reflectivity edge also complicates an assignment of the edge to intraband absorption as in a metallic model. But in a semiconducting type of picture the edge could be the high energy portion of interband absorption across a very small gap between a nonmagnetic ground state and higher lying states that carry current and have a magnetic moment. Then the edge would have no direct relation to the d.c. conductivity. In this case there should be somewhere in the room temperature spectrum a Drude type absorption due to carriers thermally excited. This is either obscured by the strong low energy absorption that renders ε_1 positive or it lies at lower energies than have been measured. Explaining the details of the shape of the low energy absorption entails having a theory for the ground and low lying electronic states of the coupled 4f-5d-6s system. Strong electron-phonon coupling may also play an important role.

Thus there is a need for more sophisticated models of the ground state and the associated transverse excitation spectrum. Experimentally it is important to know the temperature dependence of the infrared optical properties and this is currently being pursued.

It is a pleasure to acknowledge G. Lucovsky for encouragement, helpful discussions, and cooperation in obtaining the infrared data. L.M. Falicov, Y. Liang, and R.M. Martin directed my attention to the possibility of a semiconducting type of picture.

REFERENCES

1. B. Batlogg, E. Kaldis, A. Schlegel and P. Wachter, Phys. Rev., to be published, and Physics Letters 56A, 122 (1976).
2. J. C. Nickerson, R.M. White, K.N. Lee, R. Bachman, T.H. Geballe, and G.W. Hull, Jr., Phys. Rev. B 3, 2030 (1971).
3. A. Menth, E. Buehler, and T.H. Geballe, Phys. Rev. Letters 22, 295 (1969).

LOW ENERGY ANOMALIES IN THE OPTICAL SPECTRUM OF SmS WITH INTERMEDIATE VALENCE

B. Batlogg and P. Wachter

Laboratorium für Festkörperphysik, ETHZ Hönggerberg

8093 Zürich, Switzerland

The short history of homogeneously mixed valent compounds is characterized by two classes of experiments (1). The first one, to which belong the determination of the lattice constant, the magnetic susceptibility and the isomer shift, shows the system in a state corresponding to a non-integer occupation of the 4f-shell of the Rare Earth ions. Therefore the term "mixed valence" has been introduced. However, the same compounds respond in the second class of experiments with a coexistence of two types of RE ions with different, but integer number of electrons in the 4f shell (X-ray photoelectron spectroscopy, optical spectroscopy in the visible and the UV region).

At least in solid state physics, this behavior is novel and is now explained with a hybridization of the originally localized 4f states with 5d and 6s band states. The electronic ground state becomes a combination of (s,d) and f wave functions, where the number of 4f electrons plays no longer the part of a good quantum number. The energy of the 4f electrons is broadened and is given by the interaction energy with the band electrons (typically ~0.1 to 0.01 eV).

It has been shown recently in a very simple conceptual model that the two groups of experiments mentioned above are only the limiting cases of a more complicated physical system (2). Its response to an external probe depends in a nontrivial manner on the difference of three energies: the energy difference of the quantum states of the non hybridized system, the hybridization energy and the measuring energy. Evidently, the region where all these energies are comparable promises both the most interesting and the most complex behavior. Therefore, we can expect some novel observations using low energy-long wavelength photon spectroscopy.

The metallic high pressure phase (p > 6.5 kbar) of SmS is a subject of many investigations, because the Sm ions have an intermediate valence between 2 and 3. The electronic ground state of each Sm is a hybridization between $4f^6$ and $4f^5(5d6s)^1$ states. We have induced the high pressure phase by mechanical polishing well defined semiconducting single crystals. The surface transforms into the frequently quoted "golden" phase and remains single crystalline (3). The lattice constant yields an intermediate valence of ~2.9.

We have measured the optical reflectivity in the photon energy-range between 0.03 eV and 12 eV and have calculated the dielectric function $\varepsilon_1(\omega) - i\varepsilon_2(\omega)$ by means of a Kramer-Kronig analysis. For detailed interpretation of the results above ~1 eV, we refer to (3).

In this contribution, we want to concentrate on the low energy part $\hbar\omega < 1$ eV. Due to the coupled plasmon mode, being a longitudinal excitation, the real part ε_1 of the dielectric function crosses the photon energy axis at 2.45 eV with a positive slope. On going towards lower energy, ε_1 becomes more and more negative. This steady decrease is normal for free electron oscillations (plasmons) and has been observed in all the RE chalcogenides with one free electron. Therefore, we have drawn in fig. 1a the "normal" behavior, as known for example from GdS and TmS (dashed line). The real part ε_1 behaves quite strange below ~.8 eV and the deviation from the dashed line is not influenced by the extrapolation of the reflectivity towards $\omega \to o$ used in the KK procedure, at least above ~.15 eV. The crossing point T of ε_1 with the photon axis at ~.2 eV indicates a transversal mode, because the slope is negative. ε_1 cannot increase indefinitely, but must turn back again due to the free electron contribution, which gives a negative value at zero energy. Though this low energy part is not accessible to our measurement, we have represented it in fig. 1 at an arbitrary energy. Anyway, ε_1 has a positive slope at the crossing point L and correspond to a longitudinal mode, similar to a plasmon. In fact, we may ascribe it to the (screened) plasma oscillation of the heavy 4f electrons. A rough estimate of the effective mass yields a lower limit in the order of magnitude of $\sim 10^3\ m_e$.

Now we are confronted with a new and unexpected phenomenon in a homogeneously mixed valence system. For the moment, we don't propose a decicive explanation, but we show three approaches. Using a general model of the system (2), the new mode in SmS could be interpreted with a resonance produced by the fact, that all typical energies, mentioned in the introduction, are in the same order of magnitude. The two other approaches are more concrete ones. Because elec-

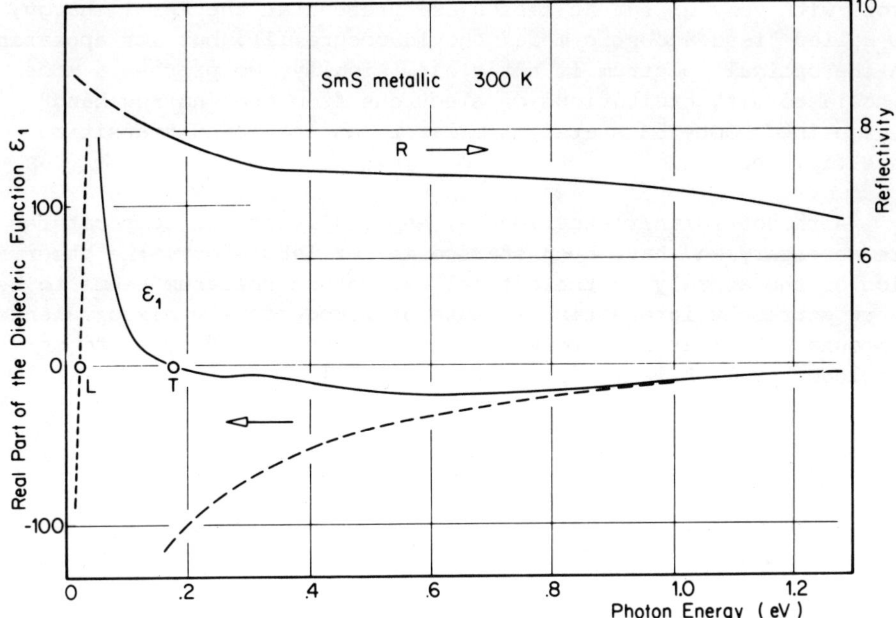

Fig. 1a

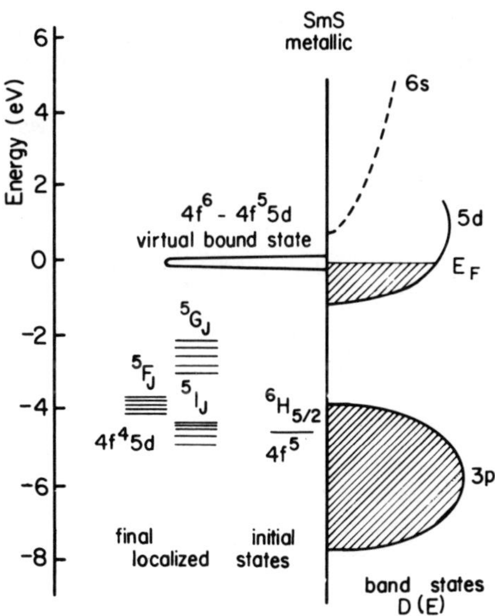

Fig. 1b

Metallic SmS at 300 K: a) Low energy part of the reflectivity and of ε_1. The dashed line respresents the normal behavior like in TmS and GdS. b) Energy level diagram as deduced from optical spectroscopy.

trons with very different masses are present at the Fermi energy, a so called "f-sd acoustic mode" should occure (1), but its appearance in the optical spectrum is not clear. Finally, we propose a mode associated with excitations of electrons from the "narrow band" regime into empty 5d states, similar to an interband transition (see fig. 1b).

Although further experiments, especially at low temperatures are necessary and have been started in our laboratory, the observation of the anomaly in the optical excitation spectrum seems to be a new extremely interesting feature of homogeneously mixed valence compounds. Similar measurements on SmB_6 have been brought to our knowledge by J. Allen, Xerox Corp., Palo Alto.

ACKNOWLEDGEMENT

The authors are grateful to PD Dr.E. Kaldis for supplying the crystals and to Mr.A. Schlegel for help with the Kramers-Kronig program.

REFERENCES

1) for references see e.g. C.M. Varma, Rev. Mod. Phys. 48, 219 (1976)
2) C.E.T. Gonçalves da Silva and L.M. Falicov, Phys. Rev. B 13, 3948 (1976)
3) B. Batlogg, E. Kaldis, A. Schlegel and P. Wachter, Phys. Rev. B 15, Nov. 1976

VALENCE RELATED OPTICAL STUDY OF (Sm,Tm) COMPOUNDS

R.Suryanarayanan

Laboratoire de Physique des Solides, C.N.R.S.

92190 - BELLEVUE , FRANCE

The existence of a mixed valence(m.v.) of Sm and Tm has been invoked to explain certain unusual properties exhibited by $Sm_{1-x}Ln_xS$ (1,2) and TmSe(3).These present challenging problems to the current theoretical models(4).We present here some optical absorption data in polycrystalline films indicating the effect of m.v. of Sm andTm.

The samples were obtained by co-evaporation of the elements on glass(5).The optical density(o.d.) of semiconducting SmS reveals the well known transitions of Sm^{2+} $4f^6 \rightarrow 4f^5 5d(t_{2g},e_g)$(fig1a,$E_1$ toE_5). For Ln=Tm, for x∼0.3, a_o = 5.86Å, the film turns golden indicative of a semiconductor-metal transition(SC-MT).The o.d. does not show maxima E_3 to E_5(fig1b) and the definition of E_1 and E_2 seems to vanish resulting in a broad maximum.The o.d. decreases from 1to 0.5eV indicating that it is not a free carrier absorption.The broad maximum is reminiscent of an absorption due to Sm^{2+} and resembles the o.d. of a metallic SmS film(a_o=5.72Å,fig1c)(6).Thus a metallic state is reached before all Sm^{2+} are converted to Sm^{3+}. The observed maximum could indicate the crystal field effects become screened by carriers in d-band and or a broadening of 4f levels of Sm^{2+} into 5d band.For Ln=Yb, a_o passes through the critical value of 5.72Å, when a SC-MT takes place for other Ln (1,2).However, the o.d. show well defined maxima (fig1e) E_1 and E_2 identifying the stability of Sm^{2+} for Ln=Yb;hence one does not observe a SC-MT.

Sm_2S_3 has a Th_3P_4 type structure(a_o=8.44Å).It is an insulator with a valence-conduction band gap of∼2.2eV. Sm^{3+} f→f transitions can be clearly identified in visible and infrared(in spite of the interference fringes).One such transition identified as $^6H_{5/2} \rightarrow {}^6F_{9/2}$ is shown in fig2d.Substituting As into Sm_2S_3 results in a non-linear increase in a_o. The gap shifts to lower energies(fig2b,c). The f→f transitions in visible cannot be seen. However, those in infrared are seen though with reduced intensity(after taking into account the thickness) and displaced to lower energies($\sim$260cm^{-1}) wrt Sm_2S_3(fig2). Further the band broadens;the half width being 80,160 and 200 Å for Sm_2S_3 with As content of 0,0.2 and 0.4 respectively. Though the shift of f→f transitions in Sm_2S_3 wrt free ion can be explained due to crystal field effecfts, the observed effects with As substitution cannot be. Our data would rather suggest the presence of Sm^{2+} ions. The electron microprobe analysis for As rich sample gave the composition as $Sm_{2.2}S_{2.6}As_{.4}$ confirming the simultaneous presence of Sm^{2+} and Sm^{3+}.

Tm is divalent in TmTe, a semiconductor(7). The o.d. of TmTe reveals absorption bands(fig3a) at 0.7,1.4 and 1.8eV identified as Tm^{2+} $4f^{13}$ - $4f^{12}(^3H_6)5dt_{2g}$, $4f^{12}(^3H_4)5dt_{2g}$ and $4f^{12}(^3H_6)5de_g$ respectively(7). For $TmTe_{1-x}Se_x$, the o.d. does not change considerably upto $x \sim 0.3$ with a_o=6.15 Å showing Tm is still stable as divalent(fig3b). However, for $5.8 < a_o < 5.9$Å, with $0.6 < x < 0.8$, the o.d. is considerably different. For the sample with a_o=5.88Å the first maximum seems to merge with the second and third forming a very broad band (fig3c). This is very similar to $Sm_{.7}Tm_{.3}S$ for which a SC-MT and a m.v. of Sm sets in. Thus the broadening of the band in the present case can be interpreted in a similar way and one could speak of a m.v. of Tm setting in with a SC-MT. Note further pure TmTe shows a SC-MT at 30kbar with a_o reaching 5.94Å(8), a value very close to that observed here with Se substitution. The o.d. of TmSe shows a

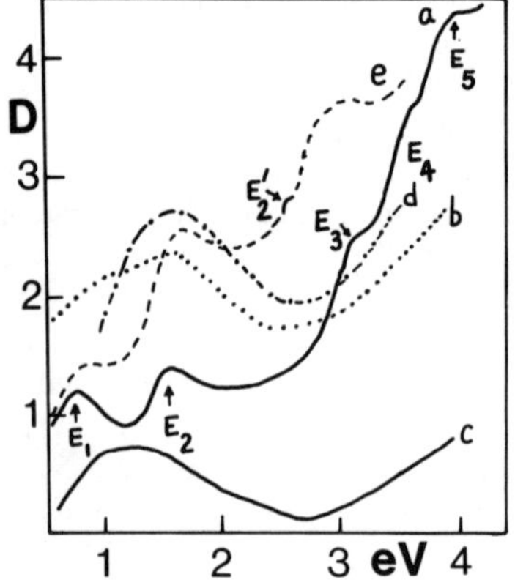

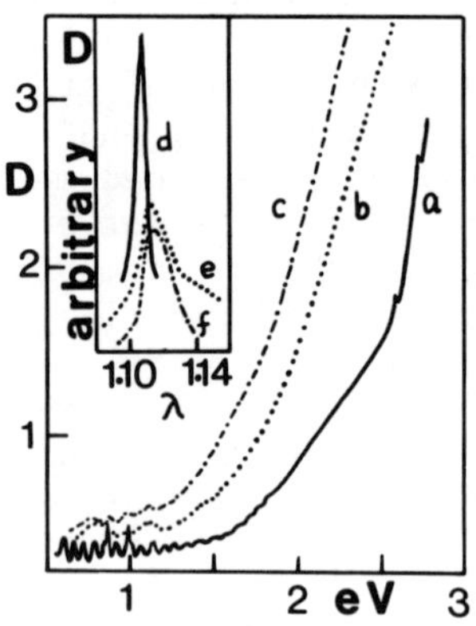

Fig 1,2. Optical density as a function of photon energy.
Fig1(a) SmS(t=0.6μ, a_o=5.96Å,8K) (b) $Sm_{.7}Tm_{.3}S$(0.4μ,5.86Å,300K) (c) metallic SmS(0.1μ,5.72Å,300K) (d) $Sm_{.4}Tm_{.6}S$(0.6μ,5.64Å,300K) (e) $Sm_{.4}Yb_{.6}S$(0.6μ,5.74Å,300K)E'_2 is due to Yb^{2+} f → d transition
Fig2(a,d)Sm_2S_3(3.2μ,8.44Å) (b,e) with $As_{.2}$(0.7μ,8.47Å) (c,f)$Sm_{2.2}S_{2.6}As_{.4}$(0.8μ,8.51Å) all at 10K; λ in microns.

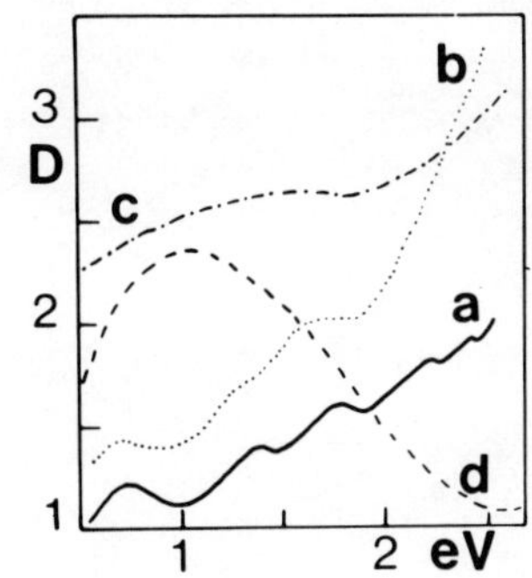

Fig.3. Optical density as a function of photon energy.
(a)TmTe(arbitrary units of D,300K) (b)$TmTe_{.7}Se_{.3}$(0.8μ,6.15Å,10K)
(c)$TmTe_{1-x}Se_x$(0.4μ, 5.88Å,10K) (d)TmSe(0.2μ,5.68Å,300K)

broad maximum around 1.1eV with a minimum at 2.7eV(fig3d) and increases in u.v.(not shown in fig).These feautres resemble those of the metallic SmS(fig1c) where Sm is considered to be in a m.v. state. In a similar way, our data indicate Tm to be in a m.v. state in TmSe.Our TmSe samples had a_o ranging from 5.68 to 5.695Å and all of them showed this maximum at 1.1eV though the minimum was between 2.7 and 2.3eV.The reflectivity of a single crystal of TmSe shows a minimum around 2.7eV(9).However,this cannot rule out a m.v. state of Tm in TmSe as was pointed out from XPS measurements(3). It would be certainly interesting to study XPS of TmSe for different values of a_o and of $TmTe_{1-x}Se_x$ for different values of x.

I would like to thank G.Brun for help in films preparation, Mrs.M.Rommeleure for analysis and R.Krishnan for some comments.

References

1.A.Jayaraman et al Phys.Rev. 11B,2783(1975) and references therein.
2.F.Holtzberg A.I.P.Conf.Proc.18,478(1974)
3.M.Campagna et al Phys.Rev.Lett.32,885(1974)
4.C.M.Varma Rev.Mod.Phys.48,219(1976) and references therein.
5.R.Suryanarayanan et al Thin Solid Films 35,263(1976)
6.I.A.Smirnov et al Phys.Stat.Solidi(b)75,K153(1976)
7.R.Suryanarayanan et al Phys.Rev.12B,4215(1975)
8.A.Jayaraman et al 8th Rare Earth Conf.(U.S.,G.P.O.,Washington D.C.) vol.1,333(1970)
9.B.Batlogg et al Phys.Lett.56A,122(1976)

METAL-SEMICONDUCTOR TRANSITION IN $Tl_2Ru_2O_7$

HS Jarrett, AW Sleight, JF Weiher, JL Gillson
CG Frederick, GA Jones, RS Swingle,
D Swartzfager, JE Gulley, PC Hoell

E. I. du Pont de Nemours & Company
Central Research & Development Department
Wilmington, Delaware 19898

$Tl_2Ru_2O_7$ possesses two valence states for Tl and since the seventh oxygen atom is not necessary to maintain crystal structure ($A_2B_2O_6$ pyrochlors are known), $Tl_2Ru_2O_7$ tends to be off stoichiometry viz. $Tl^{+3}_{2-\epsilon}Tl^{+1}_{\epsilon}Ru_2O_{7-\epsilon}$, where $0<\epsilon<0.5$. However, ϵ cannot be determined chemically because of low accuracy for oxygen analysis. Instead, the cell parameter $\underline{a}$ at room temperature provides the independent variable to differentiate samples. Approximately 40 samples spanning this range of composition have been synthesized. A representative set is presented.

The resistivity for several compositions is shown in Fig. 1. Mott and others have suggested that when the semiconducting state is comprised of localized states $\ln\rho\sim T^{-1/4}$. The inset in Fig. 1 shows the data fit better $T^{-1/4}$ (solid line) than T^{-1} (dashed line) from 25K-125K. Below 25K, resistivity varies from $T^{-1/4}$ toward higher conductivity from sample to sample. Thermal hysteresis is also observed indicating the transition is first order.

The magnetic susceptibility is shown in Fig. 2. Far from stoichiometry ($\epsilon\sim0.5$), susceptibility increases monotonically with decreasing temperature. As $\underline{a}$ decreases, the susceptibility at the higher temperatures is larger, and at low temperatures, an enhancement appears, although metallic-like conductivity remains. At still smaller $\underline{a}$, the enhancement disappears and electrical conductivity deviates from metallic behavior. The onset of the M/S transition is

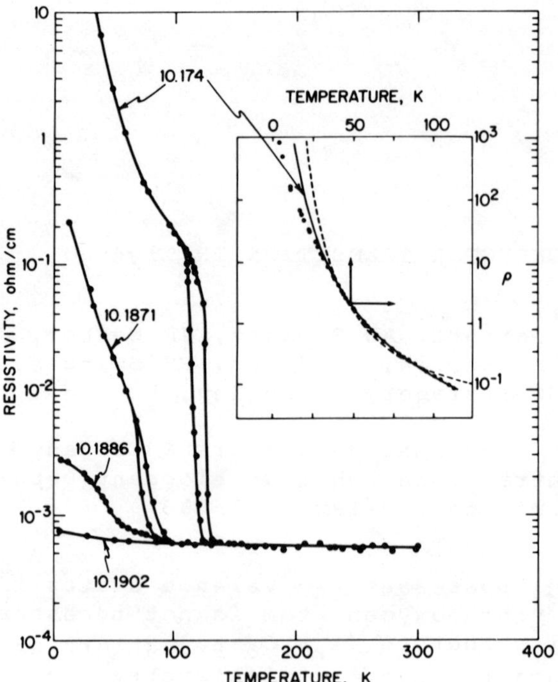

Fig. 1. Resistivity vs. temperature compositions are designated by cell dimension.

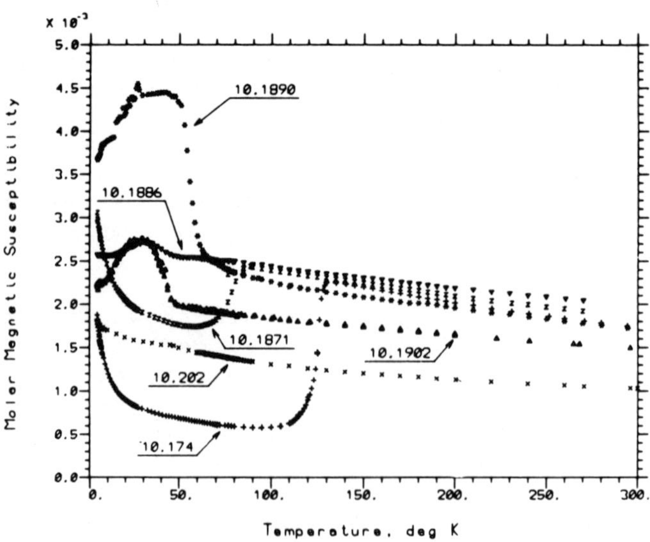

Fig. 2. Magnetic susceptibility vs, temperature,

accompanied by an abrupt decrease in susceptibility. Thermal hysteresis, not shown, is also observed in the susceptibility at the transition.

Antiferromagnetic order at the transition seems unlikely since no shift in the Tl nuclear magnetic resonance attributable to an internal molecular field was found. The Knight shift, which is 0.65% at room temperature in the metallic samples, decreases to 0.45% at 120K in the semiconducting region, and the nmr line width narrows from 23G to 14G. This decrease is consistent with the decrease in the susceptibility between the metallic and semiconducting regions.

Tl 4f x-ray photoelectron spectra (XPS) are shown in Fig. 3. The two major peaks with integrated intensity ratio of 4/3 separated by the spin-orbit energy of 4.35eV are doublets separated by 1.4eV. By convoluting an instrument broadening function of Gaussian shape whose width is 1.2eV with a Lorentzian spectral line shape of 0.4eV width for the lower energy component of the doublet and 0.3eV for the higher energy component, the solid curves are obtained with the relative amplitudes of the two components being the only adjustable parameter. Agreement is excellent, the only deviation occurring for sample 10.202, which is far from stoichiometry (ϵ=0.5). Inelastic scattering is removed from each spectrum, and all spectra are normalized to unit area. The relative peak intensity vs. lattice dimension is given in Fig. 4. The minimum is at the stoichiometry for which the M/S transition first appears.

These doublets are not due to impurities, since such large impurity content would have been apparent in the x-ray diffraction pattern. Second, the Tl 5d doublet for sample 10.1902, whose Tl 4f doublet intensity ratio is 0.77, becomes 0.12, suggesting that the intensities depend on the angular momentum of the hole remaining after photoemission. By subtracting valence band spectra a slight redistribution of the optical density of states becomes apparent away from the bottom of the valence band and into the Ru 4d band. Such shifts may represent the final filling of the Ru 4d band that leads to semiconductivity. Note that stoichiometric rare earth pyrochlors $(RE)_2Ru_2O_7$, which contain Ru^{+4} cations and similar valence band spectra, semiconduct at all temperatures.

We assume a peak in the differential thermal analysis represents a latent heat, of 0.55 $\pm$ 0.04 cal/g. The pressure dependence of the transition in sample 10.174, measured by the shift of the transition in the

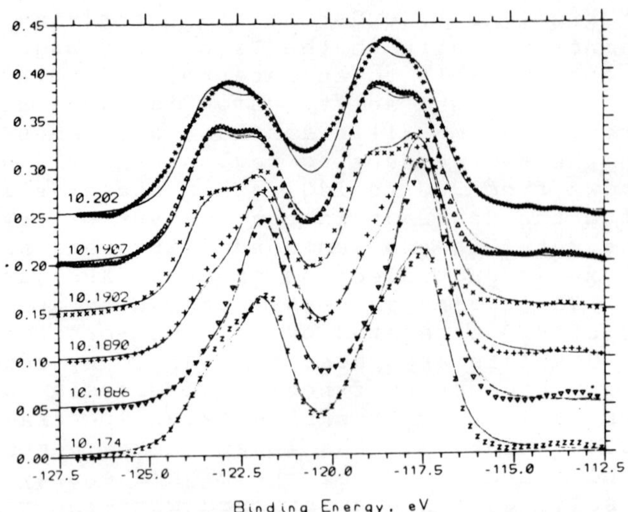

Fig. 3. Tl 4f x-ray photoelectron spectra, Spin-orbit separation 4.35 eV; satellite separation 1.4 eV.

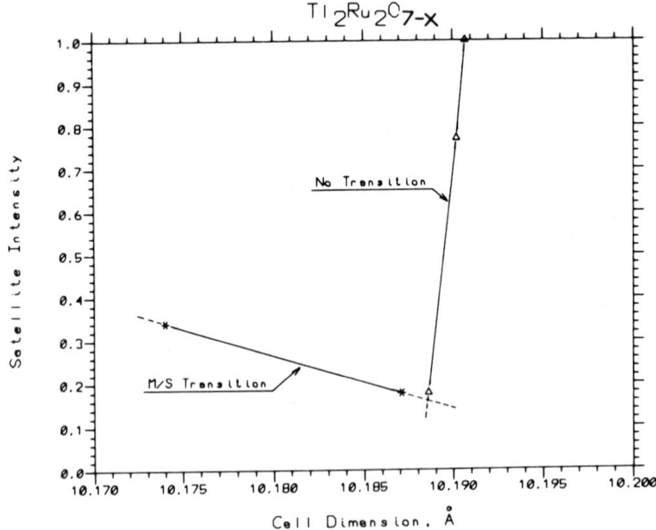

Fig. 4. Ratio of XPS satellite to parent peak intensity vs. cell dimension.

resistivity and extrapolated to P=0, is 2600 kb/cm^2 deg. The discontinuous change in cell dimension expected from the Clapeyron relation is 0.0017A. X-ray diffraction measurements were taken in effort to investigate this anomaly. However, scatter in our cell dimension data was only about a factor of two smaller than the expected dimensional change, and although an abrupt change can be reconciled with the data, it is not sufficiently quantitative to draw conclusions. Slight broadening of some of the diffraction lines was also seen, but no change from pyrochlor symmetry was found. Since both oxygen deficient and Ru^{+5} pyrochlors are known, it might be expected that a symmetry change need not occur even if there is charge redistribution at the transition.

DISCUSSION

Three regions are identifiable as *a* decreases: 1) metallic, 2) enhanced susceptibility, and 3) M/S. We suggest that this behavior is associated with filling of the narrow Ru 4d band, upon approaching stoichiometry. If this analysis is correct, Tl^{+3} must also oxidize Ru^{+4} so that formally the stoichiometry becomes $Tl^{+3}_{2-(\varepsilon+\delta)}Tl^{+1}_{\varepsilon+\delta}Ru^{+4}_{2-2\delta}Ru^{+5}_{2\delta}O_{7-\varepsilon}$, which is equivalent to removal of 2δ electrons from the Ru 4d band. Ru^{+5} is known in $Cd_2Ru_2O_7$, so Ru^{+5} in $Tl_2Ru_2O_7$ is not unrealistic. In fact, for $\delta\sim1$, $Tl^{+1}Tl^{+3}Ru_2^{+5}O_7$ would be isoelectronic with $Cd_2Ru_2O_7$. The doublet for the 4f XPS may indicate the coexistence of the two Tl valence states. Such a coexistence is known to occur in thulium chalcogenides.

Several observations fail to be answered satisfactorily. For example, why does the Tl 5d doublet show a different intensity ratio than Tl 4f? Second, why does the high binding energy member of the Tl 4f doublet become smallest at the onset of the M/S transition, (Fig. 4) and then increase again for those compositions exhibiting the transition.

Finally, the sequence of events in the susceptibility is not clear. Brinkman and Rice suggest that a susceptibility enhancement should arise from increased effective mass of the carriers near an M/S transition. The temperature dependence of the semiconductivity suggests localization of the carriers and hence increased mass. If the density of states is small compared to the reciprocal of the average band energy, magnetic ordering should not occur, and the M/S transition could arise between paramagnetic states.

CHARGE TRANSPORT IN NiO: AN ELECTRO-ELASTIC AND SPIN-DEPENDENT MODEL FOR THE EXPERIMENTS[†]

J. E. Keem, J. M. Honig, and L. L. Van Zandt

Purdue University, West Lafayette, IN 47907 U.S.A.

ABSTRACT

Results of Seebeck effect and Resistivity measurements in carefully grown, pure single crystals of NiO indicate the presence of localized carriers in this material. A new model for hopping conduction is proposed which includes aspects of models for valence transitions in rare earth compounds (differences in ionic volume and electro-elastic interactions) and also introduces a new dependence of the hopping activation energy on spin ordering. The calculation of the magnitude and temperature dependence of the resistivity for NiO based on this model is in good agreement with experiment.

TEXT

Because of the very narrow d-bands in many transition metal oxides, charge transport in these materials is assumed to take place by diffusion of localized charge carriers, i.e., by valence fluctuations of the participating cations. As was pointed out by Sherrington and von Molnàr (1), the rate at which these fluctuations take place depends on the difference of ionic radii in the two configurations. (For NiO the Ni^{+3} ion has a 10% smaller radius than the Ni^{+2} ion.) We provide an elementary derivation of the conductivity to be expected from the valence fluctuation mechanism, based on the above concept and on the uncertainty principle. We then extend the derivation to include the effects of antiferromagnetic order by considering the magnetic contributions to carrier site energies. The case of NiO will be used to illustrate these effects. Certain aspects of this work are similar to earlier work by various authors (2,3); the present derivation differs in that fewer formal assumptions and manipulations are required, and that magnetic interaction effects are included.

We assume harmonic lattice vibrations $E_\ell = \frac{1}{2}m\,\omega^2 x^2$ and a linear coupling of the carrier to the lattice distortion $E_{c\ell} = -Ax$, where A is the carrier-lattice coupling constant, x the distortion from the equilibrium position, ω the single phonon frequency associated with the distortion, and m the effective mode mass associated with ω. On minimizing the total energy of the coupled systems with respect to x one obtains a configuration stabilization energy:

$$E_b = -A^2/2m\,\omega^2, \tag{1}$$

where E_b is the net energy gained when the lattice relaxes around the smaller ion.

In order for a valence fluctuation, i.e., charge carrier transfer, to take place the energy of the carrier on the initial and the final site, 1 and 2, must be momentarily identical; this situation, termed a "coincidence event" (2), lifts the orthogonality between the two distinct ionic configurations. In order for a valence fluctuation, i.e. charge carrier transfer, to take place the orthogonal ionic configurations (in our case Ni^{+3} and Ni^{+2}) must be made degenerate. This occurs when a thermally excited lattice distortion momentarily gives rise to the equality $x_1 = x_2 \equiv x_f$, where x_1 and x_2 respectively are the initial and final site distortion parameters. The minimum (therefore, most probable) energy required to produce such a coincidence event is calculated by minimizing the combined energy on sites one or two, $E_t(x_f) = E_1(x_f) + E_2(x_f) = (\frac{1}{2}m\omega^2 x_f^2 - Ax_f) + (\frac{1}{2}m\omega^2 x_f^2)$, with respect to the coincidence event distortion parameter x_f. The magnitude of the minimum energy is $E_t = E_b/2$ where E_b is defined by Eq. (1). The period of fluctuation is $2\pi/\omega$ and the probability that a coincidence event of energy E_f occurs is given by a Boltzmann factor. Correspondingly, the mean time between coincidence events is

$$t = (2\pi/\omega)e^{+E_f/kT} \tag{2}$$

The rate at which a configuration fluctuation actually occurs is given by

$$R = \frac{\text{number of coincidence events}}{\text{sec}} \times \frac{\text{probability of carrier transfer}}{\text{coincidence event}} \tag{3}$$

For small time intervals, Δt, the duration of the coincidence event, time-dependent perturbation theory predicts the probability of transfer as

$$|a_{12}|^2 = (J/\hbar)^2\,(\Delta t)^2 \tag{4}$$

where J is the site overlap integral appropriate to the coincidence event.

To obtain a measure of Δt we use the uncertainty principle, $\Delta E \Delta t = h$, where ΔE is uncertainty in site energy during a coincidence event of duration Δt. ΔE is related to the small change in distortion by $\Delta E = A\Delta x = Av\,\Delta t$, where v is the average relative velocity of the vibrating masses, approximately specified according to the equipartition law $mv^2 = kT$. On solving for v, substituting to obtain ΔE, and rearranging, the uncertainty principle leads to the result

$$(\Delta t)^2 = h(A^2kT/m)^{-\frac{1}{2}} \tag{5}$$

Eq. (4) may now be rewritten as

$$|a_{12}|^2 = \frac{2\pi J^2}{\hbar A} \frac{1}{(kT/m)^{\frac{1}{2}}} = \frac{2\pi J^2}{\hbar\omega} \frac{1}{(2E_b\ kT)^{\frac{1}{2}}} \tag{6}$$

where (1) has been used to eliminate A.

Substituting (2) and (6) into (3) yields

$$R = (J^2/\hbar)(4E_f\ kT)^{-\frac{1}{2}} e^{-E_f/kT} \tag{7}$$

and when this is substituted into the standard expression for the mobility, μ, associated with the diffusion process, $\mu = ea^2\ R/kT$, (where a is the jump distance and e the electronic charge), one obtains

$$\mu = \frac{ea^2J^2}{\hbar\ kT(4E_f kT)^{\frac{1}{2}}} e^{-E_f/kT} \tag{8}$$

The conductivity is given by $\sigma = ne\mu$, and the temperature dependence of the charge carrier density reads $n = n_o e^{-E_d/kT}$, where E_d is the number activation energy. Eq. (8) thus leads to

$$\sigma = ne\mu = [n_o e^2a^2J^2/\hbar\ kT(4E_f\ kT)^{\frac{1}{2}}]e^{-(E_d + E_f)/kT} \tag{9}$$

which differs from the result cited by Holstein (2) for non-adiabatic hopping processes by a factor $\sqrt{\pi}$.

From the forgoing discussion, it should be clear that any mechanism which alters the carrier energy on a site also alters the transfer rate. For example, there is a magnetic contribution to the site energy of a carrier with a spin in a magnetically ordered system. As the magnetic disorder increases, there is an increasing probability that a high energy spin configuration is encountered, which results in an increase in the site energy of the carrier. This reduces the lattice distortion required to produce a coincidence event and thereby decreases the valence fluctuation activation energy. Through such a mechanism E_f then becomes a function of temperature:

$$E_f(T) = E_f - E_m(T) \tag{10}$$

where $E_m(T)$ represents the magnetic contribution. In an antiferromagnetic material such as NiO, E_m is given by the standard expression

$$\begin{aligned} E_m &= 2 \sum_j J_{ij}\{1 - (1 - T/T_N)^{2/3}\} \qquad T \leq T_N \\ &= 2 \sum_j J_{ij} \qquad\qquad\qquad\qquad\qquad\; T > T_N \end{aligned} \tag{11}$$

where T_N is the Néel point and J_{ij} is the superexchange integral between the ion on site i and its antiferromagnetically coupled nearest neighbor on site j. Use of Eq. (11) in (10) and (9) then leads to a temperature-dependent activation energy for conductivity below the Néel point.

Eqs. (9)-(11) have been tested against our resistivity measurements on NiO (4) by use of the input parameters required in Eq. (9) which are available in the literature. Quantitative agreement was attained between the theoretically calculated and experimentally observed conductivities as a function of temperature between 250 and 700 K which brackets the Néel point T_N = 523 K. The theory thus successfully explains the constant conductivity activation energy encountered above the Néel point, the gradual change in this quantity near $T = T_N$, and the variable activation energy at lower temperatures.

REFERENCES

†This work was supported by the National Science Foundation on Grant No. DMR 76 - 12583.

1. D. Sherrington and S. von Molnàr, Solid State Comm. 16, 1347 (1975).

2. T. Holstein, Ann. Phys. (N. Y.) 8, 343 (1959).

3. I. G. Austin and N. F. Mott, Adv. Phys. 19, 41 (1969).

4. J. E. Keem. Thesis, Purdue University, May 1976, unpublished.

Index

A

B

C

T

U

X